Concanavalin A
as a Tool

Concanavalin A
as a Tool

Edited by

H. BITTIGER

Research Department of the
Pharmaceuticals Division
CIBA-GEIGY Limited,
Basel, Switzerland

and

H. P. SCHNEBLI

Friedrich Miescher-Institut
Basel, Switzerland

A Wiley – Interscience Publication

JOHN WILEY & SONS

London · New York · Sydney · Toronto

Library of Congress Cataloging in Publication Data:
Main entry under title:

Concanavalin A as a tool.

  'A Wiley—Interscience publication.'
  Includes index.
  1. Concanavalin A. I. Bittiger, H. II. Schnebli,
H. P.
QP935.C6C66   599'.08'7028   75—37841

ISBN 0 471 01350 1

Typeset by Preface Ltd, Salisbury, Wiltshire and
printed in Great Britain by The Pitman Press, Bath

Contributing Authors

Allan, David (Chapter 44)

Dept. of Biochemistry, University of Birmingham, P.O. Box 363, Birmingham B 15 2TT, England.

Andersson, Jan (51)

Dept. of Immunology, Wallenberg-laboratory, P.O. Box 562, 75122 Uppsala 1, Sweden.

Aunis, Dominique (39)

Centre de Neurochimie de CNRS, 11 rue Humann F-67085 Strasbourg, France

Avrameas, Stratis (7)

Unité d'Immunocytochimie, Dépt. de Biologie Moléculaire, Institut Pasteur F-75015 Paris, France.

Bächi, Thomas (13, 14)

Institut für Med. Mikrobiologie, Universität Zürich, Postfach, CH-8028 Zürich, Switzerland.

Becker, Joseph W. (3)

The Rockefeller University, 1230 York Avenue, New York, N.Y. 10021, USA

Bevan, Michael (52)

The Salk Institute, P.O. Box 1809, San Diego, Calif, 92112, USA

Bittiger, Helmut (10, 15, 47)

Research Dept. of the Pharmaceuticals Division, CIBA-GEIGY Ltd., CH-4002 Basel, Switzerland.

Bornens, M. (7)

Unité d'Immunocytochimie, Dépt. de Biologie Moléculaire, Institut Pasteur, F-75015 Paris, France.

Burger, Max M. (28)

Dept. of Biochemistry, Biocenter of the University of Basel, Klingelbergstrasse 70, CH-4056 Basel, Switzerland.

Catt, K. J. (38)

Reproduction Research Branch, National Institute of Child Health, NIH, Bethesda, Md. 20014, USA.

Chang, Kwen-Jen (18, 20, 23, 45, 58)

Borroughs Wellcome Co., 3030 Cornwallis Road, Research Triangle Park, NC 27709, USA.

v

Cohen, Gary H. (40)   *Dept. of Microbiology, School of Dental Medicine, University of Pennsylvania, Philadelphia, Penn. 19104, USA.*

Crumpton, Michael J. (44)   *National Institute for Medical Research, Mill Hill, London NW7 1AA, England*

Cuatrecasas, Pedro (18, 20, 23, 45, 58)   *Burroughs Wellcome Co., 3030 Cornwallis Road, Research Triangle Park, N.C. 27709, USA.*

Culp, Lloyd A. (61)   *Dept. of Microbiology, Case Western Reserve University, School of Medicine, Cleveland Ohio 44106 USA.*

Cunningham, Bruce A. (3)   *The Rockefeller University, 1230 York Avenue, New York, N.Y. 10021, USA*

Dufau, Maria L. (38)   *Reproduction Research Branch, National Institute of Child Health and Human Development, NIH, Bethesda, Md. 20014, USA.*

Edelman, Gerald M. (3, 48, 57)   *The Rockefeller University, 1230 York Avenue, New York, N.Y. 10021, USA,*

Eisenberg, R. J. (40)   *Dept. of Microbiology, School of Dental Medicine, University of Pennsylvania, Philadelphia, Penn. 19104, USA.*

Franke, Werner W. (24)   *Institut für Exp. Pathologie, Deutsches Krebsforschungszentrum, Im Neuenheimer Feld 280, D-69 Heidelberg 1, BRD.*

Furmanski, Philip (19, 22, 34)   *Dept. of Biology, Michigan Cancer Foundation, 110 East Warren Avenue, Detroit, Mich. 48201, USA.*

Goldstein, Irwin J. (4, 16)   *Dept. of Biological Chemistry, The University of Michigan, Medical School, Ann Arbor, Mich. 48104, USA.*

Gombos, Giorgio (41, 42, 43)   *Centre de Neurochimie de CNRS, 11 rue Humann, F-67 Strasbourg, France.*

Gunther, Gary R. (57)   *Section of Cell Biology, Yale University School of Medicine, 333, Cedar St., New Haven, Conn. 06510, (address whilst preparing manuscript — Rockefeller University).*

Hessle, Helena (40)   *Dept. of Microbiology, School of Dental Medicine, University of Pennsylvania, Philadelphia, Penn. 19104, USA.*

Karsenti, E. (7)

Unité d'Immunocytochimie, Dépt. de Biologie Moléculaire, Institut Pasteur, F-75015 Paris, France.

Keenan, Tom. W. (24)

Dept. of Animal Sciences, Purdue University, West Lafayette, Indiana 47907, USA.

Keller, Robert (56)

Immunobiology Research Group, University of Zürich, Schönleinstrasse 22, CH-8032 Zürich, Switzerland.

Kinzel, Volker (49)

Institut für Exp. Pathologie, Deutsches Krebsforschungszentrum, Im Neuenheimer Feld 280, D-69 Heidelberg 1, BRD.

Kübler, Dieter (49)

Institut für Exp. Pathologie, Deutsches Krebsforschungszentrum, Im Neuenheimer Feld 280, D-69 Heidelberg 1, BRD.

Liener, Irvin E. (2)

Dept. of Biochemistry, University of Minnesota, 140 Gortner Laboratory, St. Paul, Minn. 55101, USA.

Lloyd, Kenneth O. (36, 37)

Department of Biochemistry, Texas Tech University School of Medicine, Lubbock, Texas 79409, USA.

Loor, Francis (6, 29)

Basel Institute for Immunology, Grenzacherstrasse 487, CH-4058 Basel, Switzerland.

Mallucci, Livio (5)

Dept. of Microbiology, Guy's Hospital Med. School, London Bridge, SE1 9RT, England.

Martin, B. J. (12)

Department of Biology, University of Southern Mississippi, Southern Station, Box 477, Hattiesburg, Mississippi 39401, USA.

Martin, W. John (54)

Immunology Section VLLB, National Cancer Institute, Building 41, Suite 300, NIH, Bethesda, Md. 20014, USA.

Mather, IH. (24)

Dept. of Animal Sciences, Purdue University, West Lafayette, Indiana 47907, USA.

Melchers, Fritz (51)

Basel Institue for Immunology, Grenzacherstrasse 487, CH-4058 Basel, Switzerland.

Miras-Portugal, Maria-Teresa (39)

Centre de Neurochimie du CNRS, 11 rue Humann, F-67085 Strasbourg, France.

Neri, Anthony (25)   *Dept. of Biology, Calif. State
                                           University, 18111 Nordhoff Street,
                                           Northridge, Calif. 91324, USA.*

Nicolson, Garth L. (1)   *Department of Cancer Biology, The Salk
                                           Institute for Biological Studies, P.O. Box
                                           1809, San Diego, California, 92112 USA,
                                           and Dept. of Developmental and Cell
                                           Biology, University of California, Irvine,
                                           California 92664, USA.*

Noonan, Kenneth D. (21)   *Dept. of Biochemistry, Box 724, J. Hillis
                                           Miller Health Center, University of
                                           Florida, Gainesville, Florida 32610, USA.*

Oppenheimer, Steven B. (25, 33)   *Dept. of Biology, Calif. State University,
                                           18111 Nordhoff Street, Northridge, Calif.
                                           91324, USA.*

Pelley, Ronald P. (53)   *Dept. of Medicine, Case Western Reserve
                                           University and University Hospitals,
                                           Cleveland, Ohio 44106, USA.*

de Petris, Stefanello (8, 9)   *Basel Institute for Immunology,
                                           Grenzacherstrasse 487, CH-4058 Basel,
                                           Switzerland.*

Phillips, Philip G. (19, 22, 34)   *Trudeau Institute, Inc., P.O. Box 59,
                                           Saranac Lake, N.Y. 12983, USA.*

Pober, Jordan S. (46)   *Dept. of Molecular Biophysics and
                                           Biochemistry, Yale University, New
                                           Haven, Conn. 06520, USA.*

Ponce de Leon, M. (40)   *Dept. of Microbiology, School of Dental
                                           Medicine, University of Pennsylvania,
                                           Philadelphia, Penn. 19104, USA.*

Ralph, Peter (60)   *Sloan-Kettering Institute for Cancer
                                           Research, Donald S. Walker Laboratory,
                                           145 Boston Post Road, Rye, N.Y. 10580,
                                           USA.*

Reeber, André (43)   *Centre de Neurochimie du CNRS, 11 rue
                                           Humann, F-67085 Strasbourg, France.*

Reeke, George N., Jr. (3)   *The Rockefeller University, 1230
                                           York Avenue, New York, N.Y. 10021,
                                           USA.*

Renthal, Robert (46)   *The University of Texas at San Antonio,
                                           Department of Biochemistry, San
                                           Antonio, Texas 78285, USA.*

Richards, James (49) — Institut für Exp. Pathologie, Deutsches Krebsforschungszentrum, Im Neuenheimer Feld 280, D-69 Heidelberg 1, BRD.

Roberson, Marie (25) — Dept. of Biology, Calif. State University, 18111 Nordhoff Street, Northridge, Calif. 91324, USA.

Rosenblith-Borysenko, Joan Z. (11) — Dept. of Anatomy, Tufts University, 136 Harrison Avenue, Boston, Mass. 02111, USA.

Rottmann, Warren L. (35) — Dept. of Zoology, University of Minnesota, Minneapolis, Minn. 55455, USA.

Rutishauser, Urs (48) — The Rockefeller University, 1230 York Avenue, New York, N.Y. 10021, USA.

Schnebli, Hans Peter (17, 27, 31) — Friedrich Miescher-Institut, P.O. Box 273, CH-4002 Basel, Switzerland.

Schwartz, Howard J. (53) — Dept. of Medicine, Case Western Reserve University and University Hospitals, Cleveland, Ohio 44106, USA.

Shier, W. Thomas (55, 59) — The Salk Institute, Cell Biology Laboratory, P.O. Box 1809, San Diego, Calif. 92112, USA.

Smith, David F. (30) — Biochemisches Institut der Universität Freiburg, Hermann Herder-Strasse 7, D-78 Freiburg, BRD.

Spicer, S. S. (12) — Dept. of Pathology, Medical University of South Carolina, 37 Mill Street, Charleston, S.C. 29401, USA.

Stadler, J. (24) — Institut für Exp. Pathologie, Deutsches Krebsforschungszentrum, Im Neuenheimer Feld 280, D-69, Heidelberg 1, BRD.

Steinemann, Adrian (46) — Dept. of Biophys. Chemistry, Biocenter of the University of Basel, Klingelbergstrasse 70, CH-4056 Basel, Switzerland.

Stewart, Margaret L. (50) — Laboratory of Molecular Genetics, National Institue of Child Health and Human Development, Bldg. 6, Room 140, NIH, Bethesda, Md. 20014, USA.

Stöhr, M. (49)

Institut für Exp. Pathologie, Deutsches Krebsforschungszentrum, Im Neuenheimer Feld 280, D-69 Heidelberg 1, BRD.

Stryer, Lubert (46)

Dept. of Molecular Biophysics and Biochemistry, Yale University, New Haven, Conn. 06520, USA.

Walborg, Earl F., Jr. (30)

Biochemistry Department, The University of Texas System, Cancer Center, M.D. Anderson Hospital and Tumor Inst., 6723 Bertner Avenue, Houston, Texas 77025, USA.

Walther, Bernt T. (26, 35)

Dept. of Biology, The Johns Hopkins University, Baltimore, Md. 21218, USA.

Wang, John L. (3, 57)

The Rockefeller University, 1230 York Avenue, New York, N.Y. 10021, USA.

Willingham, Mark C. (32)

Laboratory of Molecular Biology, National Cancer Institute, Bldg. 37, Room 4B27, NIH, Bethesda, Md. 20014, USA.

Zanetta, Jean-Pierre (42)

Centre de Neurochimie du CNRS, 11 rue Humann, F-67 Strasbourg, France.

CONTENTS

Section:1

INTRODUCTION

CHAPTER 1

Concanavalin A: The Tool, the Techniques and the Problems

GARTH L. NICOLSON

The existence of substances called plant haemagglutinins was reported by Stillmark approximately one hundred years ago (1,2), although there is some controversy as to whether Dixson (3) or Stillmark (1) was the first to make these observations. Haemagglutinin(s) and toxin(s) of the jack bean were described by Assmann (4), but the first isolation and purification took place some time later (5,6). The crystalline globin that Sumner and Howell purified was called 'A' or 'concanavalin A' (Con A), and its agglutinating and precipitating activities against erythrocytes, starches, mucins and glycoproteins indicated that the responsible ligand bound to Con A 'may be a carbohydrate group' (6).

Some forty years have now passed since Con A was first isolated and partly purified. We not only know quite a lot about the physical and chemical properties of Con A (ref. 7 and Chapters 2–4), but Con A has also become an invaluable tool in the hands of the cell biologist, biochemist, immunologist, virologist and haematologist. Its uses and potential are astounding and will be discussed briefly here; however, the reader is directed to the appropriate chapters for more detailed information.

Con A is a globular protein composed of identical 25,500 mol. wt. asymmetric subunits (protomers of 237 amino acids each) arranged in dimeric, tetrameric and higher-order forms (8,9). The arrangement of these protomers is dependent on solution characteristics and properties (10). Normally Con A exists as a pseudo-tetrahedral molecule with four saccharide-binding sites, one in each subunit, and the Con A (tetramer) appears from X-ray crystallography to be a union of four ellipsoidal domes paired across a twofold axis of symmetry (Figure 1, Chapter 3). The Con A molecule contains bound calcium and manganese ions (one each per protomer [11]) that are necessary for carbohydrate binding activity (12), but are not part of the actual saccharide binding site (7,13,14). Con A generally binds to saccharides containing α-D-mannose or α-D-glucose residues (refs. 15, 16 and Chapter 4), but it may recognize oligosaccharide sequences lacking mannose or glucose (17). It appears to recognize terminal as well as internal saccharide residues (18).

The saccharide-binding specificities of lectins such as Con A can be utilized in the purification, characterization and sequencing of polysaccharides, glycopeptides and glycoproteins. Polysaccharides isolated using Con A include: D-glucans such as

3

glycogens, amylopectins and dextrans, yeast mannans, phosphomannans, synthetic D-mannans, D-fuctans such as bacterial and other plant levans (15,16,19—24). In these isolations precipitation (25) or affinity chromatography on Con A-containing affinity columns (ref. 26 and Chapter 36) can yield high specific activity preparations of interesting saccharide-containing molecules. Glycoproteins and glycopeptides have been isolated and purified utilizing Con A lectin affinity procedures. This has proven to be a valuable technique for purifying even trace quantities of serum or secreted glycopolypeptides and glycoproteins (for example, hormones [ref. 27 and Chapter 38] ; interferons [28] ; and blood group substances [ref. 29 and Chapter 37]). Serum immunoglobulins (30) and one of the major human lipoproteins (31) have been isolated and purified on Con A affinity columns. Enzymes such as dopamine hydroxylase (Chapter 39) and even other lectins such as soy bean agglutinin (32) have also been purified via their Con A-binding determinants. These procedures will see increased use in the future as more sophisticated methodologies are utilized to purify serum and cellular glycoproteins and glycopeptides. Ultimately the sequential use of different lectin affinity columns with different saccharide specificities will probably constitute a rather general approach to separating and purifying glycocomponents.

Several membrane glucoproteins or their fragments have been the most recent glycocomponents successfully purified on Con A columns. Con A—Sepharose (now available as a commercial product) has been used to purify Con A receptors from detergent-solubilized plasma membranes of human and pig lymphocytes (ref. 33, 34 and Chapter 44), human platelets (35), rat adipocytes (36), rat brain cells (ref. 37 and Chapter 42), hamster fibroblasts (38), mouse ascites tumours (39), rat (40) and human erythrocytes (41) and several animal viruses (e.g. ref. 42). An excellent example is the purification of the fat cell membrane insulin receptor by Con A affinity chromatography (Chapter 45). Proteolytic enzymes have also been used to remove cell surface glycopeptide Con A receptors from Novikoff tumour (39) and rat hepatoma cells (43,44), and in certain cases cell glycolipids interact with lectins (45,46).

Intracellular membrane glycoproteins have also been identified, isolated and purified by Con A affinity chromatography such as rat liver microsome Con A-binding proteins (47) and retinal disc membrane rhodopsin (48). Since a variety of intracellular organelles and membranes contain lectin receptors (49—55), this approach may be as rewarding as those dealing with plasma membrane lectin receptor isolation and characterization (see Chapter 47). It could be that intracellular membrane glycoproteins serve as cytoskeletal or enzyme attachment points or identification molecules for organelle compartmentalization.

Particular attention must be paid during isolation of receptors and during the subsequent assay for activity. Requirements for assaying receptors during their purification by Con A and other lectins and their characterization are discussed in (56) and in Chapter 30. The purified receptors must act as powerful inhibitors either in lectin agglutination or quantitative precipitation assays, and must remain monodispersed during the isolation and chromatography procedures.

Special methods have been developed for purifying macromolecular structures containing Con A-binding components. Viruses and bacteria that bind to Con A can be purified by affinity chromatography or direct Con A precipitation (ref. 57 and Chapter 50), but larger structures such as cells are trapped in the usual

Con A-Sepharose columns or their viability significantly reduced after saccharide reversal of cell agglutination. Cells can be purified by attachment to Con A-derivatized surfaces such as nylon fibres or plastic petri dishes (refs. 58—60 and Chapter 48). The immobilization of glycopeptides, glycoproteins, viruses, bacteria and cells on Con A-derivatized surfaces will be most useful in biological studies on ligand—cell and cell—cell interactions. By having the ligand or one type of cell attached to a Con A-derivatized surface, either the immobilized or the free partner can be manipulated and treated independently (61).

Con A has proven to be a useful tool in the elucidation of cell surface and plasma membrane structural organization and dynamics. Techniques in use include cell agglutination, quantitative labelling, ultrastructural localization and perturbation of receptor function. Cell agglutination is perhaps the most studied and least understood of these approaches (62). Many types of bacteria and cells agglutinate well with Con A: bacterial spheroplasts (63), algae (64), amoeba (65,66), erythrocytes (67—70), embryonic cells (72—74), spermatozoa (71,75), eggs (76,77) and a variety of other cell types (reviewed in 62). Unfortunately, there do not seem to be simple reasons for susceptibility to agglutination by Con A; agglutination appears to be controlled and modified by a variety of cell and environmental factors (ref. 62 and Chapter 26), such as the number, distribution, mobility, membrane location and transmembrane restraints on the cell surface lectin receptor; the physical and biological properties of the cell (surface charge density, deformability, cytoskeletal interactions, stage of the cell cycle, etc.) and solution parameters (temperature, pH, ionic strength, etc.) (62,78 and Chapter 27). Radioisotope-labelled Con A (^{125}I-, ^{3}H-, ^{14}C- and ^{63}Ni-Con A, reviewed in Section III) has been used by a number of investigators to monitor changes in lectin receptors during the cell cycle (79,80), at various development stages (81), at cell contact (82,83), after transformation (84—91), or during non-oncogenic virus infection (92,93). Since agglutination and quantitative labelling data by themselves cannot determine the dynamics of cell surface receptors, receptor localization studies with fluorescence u.v. microscopy and/or electron microscopy utilizing the appropriate techniques must be used. Fluorescent-Con A (71,80,90,92,94—99), ferritin—Con A (62,83,100—109), peroxidase—Con A (65,110—114), haemocyanin—Con A (115—118) and other techniques (see Section II) are currently in use for localizing Con A cell surface receptors. Using these techniques surface properties such as receptor mobility (62,83,90—92,94,97—99,104,106,107,109,112—114,116—118) and restraint of mobility (90,92,98,99,103,104,116,119), membrane oligosaccharide asymmetry (100,101), segregation of receptors into membrane domains (71,75,119), membrane biogenesis (53), endocytosis (120) and others (62) have been accessible problems. Techniques such as these were indispensable in the formulation of new theories on plasma membrane organization and dynamics (2,121—123).

Important changes in the structure, organization and dynamics of cell surface molecules during development, non-oncogenic virus infection, oncogenic virus transformation, cell growth and division, cell movement and cell—cell interactions have been studied using the Con A techniques described above. The Con A agglutinability of chick embryonic neural retinal cells (72,73) and sea urchin eggs (74) was found to decrease during development. Conversely, Con A agglutination of different mating types of *Chlamydomonas* increased during gametic differentiation (64). When normal cells are infected with non-oncogenic viruses such as vaccinia (124), herpes simplex (125), Newcastle (126), influenza (42,93) and others (127), their agglutinability with Con A increases shortly after infection. Cells transformed by oncogenic viruses, chemical carcinogens, X-Rays, etc. are generally much more agglutinable with Con A than their normal counterparts (see reviews 24,62,83,128). Con A differentially inhibits tumour cell transport systems for certain amino acids (129,130) and monosaccharides (129) compared to normal cells, although Con A seems to have the opposite effect on D-glucose transport in fat cells (36,131) and in mitogenically stimulated thymic lymphocytes (see below). Treatment of tumour cells with Con A can enhance their immunogenicity and ability to be rejected (ref. 132 and Chapter 54). The use of lectins like Con A to probe normal and tumour cell surfaces has been the basis of a large and continuing research effort; and studies involving the isolation, characterization, quantitation, distribution, mobility and turnover of cell surface lectin receptors have furthered our knowledge of normal and tumour cell surface components. Hopefully, a better understanding of the cell surface components and processes and their possible modifications after transformation may contribute to a better understanding of normal and tumour cell growth properties, escape from immune attack by surface receptor modulation and the processes by which tumours metastasize.

Con A has been used to kill tumour cells *in vitro* (133—135) and *in vivo* (135,136). In certain cases the sensitivity to Con A may be due to selective killing during certain cell cycle or growth stages (137,138). The toxicity of Con A can be used to select Con A-resistant tumour variants (refs. 139—142 and Chapter 61) which usually have properties more similar to their normal cell counterparts than their transformed parents: Burger and Noonan (143) found that protease-treated Con A inhibited cell growth of polyoma-transformed cells, although Trowbridge and Hilborn (144) failed to duplicate this feat with a succinylated dimeric form of Con A. Con A inhibits cell movement (145), phagocytosis (146) and fertilization (147—149), but mimics insulin stimulation of hexose transport in fat cells (refs. 36,131 and Chapter 58), stimulates platelet release (150), secretion of migration inhibitory factor (151) and histamine (152) and triggers mitogenesis in thymic lymphocytes (153—155). This last property will be discussed further.

Con A is considered to be a specific mitogen because of its ability to stimulate thymic-derived (T) lymphocytes to undergo blastogenesis and division (155—158). This occurs when ∿5% of the Con A surface receptors are occupied (155,158), and it is reversible by the appropriate Con A inhibitors (154). Since Con A can stimulate bone marrow-derived (B) lymphocytes to undergo blastogenesis when attached to a solid support, it is thought that the Con A need not enter cells to activate them toward division (159). The mechanism of Con A action in the mitogenic triggering of lymphocytes is largely unknown, but several important theories have evolved over the last few years (159—164). One theory advanced by Edelman and his collaborators is based on their work with Con A derivatives (165,166). Gunther and coworkers (166) found that a divalent, succinylated derivative of Con A in contrast to native Con A could not immobilize the cell surface receptors for anti-immunoglobulin, does not cap Con A surface receptors or agglutinate lymphocytes and shows a different dose-response curve for mitogenic stimulation (ref. 122 and Chapter 57). Mitogenic stimulation is proposed to occur when Con A receptors are aggregated to form small clusters on the lymphocyte cell surface. These receptors are proposed to be transmembrane linked to the cell cytoskeletal sytem (98,99,122,167) which transduces the 'state' of receptor display at the cell surface to the cytoplasm where further changes occur (perhaps through the messenger molecules cyclic AMP and cyclic GMP [168—170]) affecting permeability, transport, metabolism, phosphorylation, acetylation, RNA synthesis, cell morphological changes, DNA synthesis and finally, mitosis. These changes are reviewed in (62,161—163,165,171).

Con A has a variety of effects on lymphoid cell functions. Perlmann and coworkers (172) reported that although Con A is mitogenic for human lymphocytes, it inhibited cell-mediated killing. However, Stavy and coworkers (173) found that appropriate Con A concentrations can stimulate cell-mediated immune responses, perhaps by stimulation of lymphotoxin release (174). These properties may depend on the type and species of activated lymphocytes and target cells (175,176). The use of Con A to identify and selectively activate particular subpopulations of lympocyte cell mixtures could lead to unique immune activation, but research in this area is plagued by the use of unclonable cell populations.

Con A has pronounced effects on delayed hypersensitivity. Intraperitoneal injection of Con A markedly suppresses delayed hypersensitivity (177). Factors (lymphotoxins?) released from a Con A-stimulated lymphoid subpopulation may produce these symptoms (178,179). Con A treatment also prolongs skin allograft survival (180), and produces an intense inflammatory response upon subcutaneous injection (ref. 181 and Chapter 55).

Con A is a very useful tool, indeed, for exploring the mainstream or even the byways of biology and biochemistry. A testimonial to that fact has been the recent upsurge in the number of publications utilizing Con A. Many of us are searching for other well-characterized molecules like Con A to pursue our own paths to understanding the chemical and biological processes of life.

ACKNOWLEDGEMENTS

The author is supported by a contract from the Tumor Immunology Program of the National Cancer Institute (CB—33879) and grants from the National Cancer Institute (CA—15122) and the National Science Foundation, Human Cell Biology Program (GB—34178).

REFERENCES

1. Stillmark, H. (1888) 'Uber Rizin, ein giftiges Ferment aus dem Samen von *Ricinus communis* L. und einigen anderen Euphorbiaceen, *Inaug. Dis. Dorpat.*
2. Stillmark, H. (1889) in *Arbeiten des Pharmakologischen Institutes zu Dorpat*, Enke, F. (Stuttgart), Vol. 3, p. 59.
3. Dixson (1887) *Australian Med. Gaz.*, p. 156. As quoted in Sumner, J. B. & Howell, S. F. (1936) *J. Bacteriol.*, 34 227—37.
4. Assmann, F. (1911) 'Beitraege zur Kenntnis pflanzlicher Agglutinine', *Arch. für die Gesamte Physiol.*, 137, 489—510.
5. Sumner, J. B. and Howell, S. F. (1935) 'The non-identity of jack bean agglutinin with crystalline urease', *J. Immunol.*, 29, 133—4.
6. Sumner, J. B. and Howell, S. F. (1936) 'The identification of the hemagglutinin of the jack bean with concanavalin A', *J. Bacteriol.*, 32, 227—37.

7. Edelman, G. M., Cunningham, B. A., Reeke, G. N., Jr., Becker, J. W., Waxdal, M. J. and Wang, J. L. (1972) 'The covalent and three dimensional structure of concanavalin A', *Proc. Nat. Acad. Sci. U.S.A.*, **69**, 2580—4.

8. Agrawal, B. B. L. and Goldstein, I. J. (1968) 'Protein—carbohydrate interaction. VII. Physical and chemical studies on concanavalin A, the hemagglutinin of the jack bean', *Arch. Biochem. Biophys.*, **124**, 218—29.

9. Kalb, A. J. and Lustig, A. (1968) 'The molecular weight of concanavalin A', *Biochim. Biophys. Acta*, **168**, 366—7.

10. McKenzie, G. H., Sawyer, W. H. and Nichol, L. W. (1972) 'The molecular weight and stability of concanavalin A', *Biochim. Biophys. Acta*, **263**, 283—93.

11. Yariv, J., Kalb, A. J. and Levitzki, A. (1968) 'The interaction of concanavalin A with methyl α-D-glucopyranoside', *Biochim. Biophys. Acta*, **165**, 303—5.

12. Kalb, A. J. and Levitzki, A. (1968) 'Metal binding sites of concanavalin A and their role in the binding of α-methyl-D-glucopyranoside', *Biochem. J.*, **109**, 669—72.

13. Becker, J. W., Reeke, G. N. and Edelman, G. M. (1971) 'The location of the saccharide binding site of concanavalin A', *J. Biol. Chem.*, **246**, 6123—5.

14. Brewer, C. R., Sternlicht, H., Marcus, D. M. and Grollman, A. P. (1973) 'Interactions of specific carbohydrates with concanavalin A. Mechanism of binding of α- and β-methyl-D-glucopyranoside to concanavalin A as determined by ^{13}C nuclear magnetic resonance', *Biochemistry*, **12** 4448—57.

15. So, L. L. and Goldstein, I. J. (1967) 'Protein—carbohydrate interaction. IX. Application of the quantitative hapten inhibition technique to polysaccharide—concanavalin A interaction. Some comments on the forces involved in concanavalin A—polysaccharide interaction', *J. Immunol.*, **999**, 158—63.

16. So, L. L. and Goldstein, I. J. (1967) 'Protein—carbohydrate interaction. IV. Application of the quantitative precipitin method to polysaccharide—concanavalin A interaction', *J. Biol. Chem.*, **242**, 1617—22.

17. Toyoshima, S., Fukuda, M. and Osawa, T. (1972) 'Chemical nature of the receptor sites for various phytomitogens', *Biochemistry*, **11**, 4000—5.

18. Goldstein, I. J., Reichert, C. M., Misaki, A. and Gorin, P. A. J. (1973) 'An "extension" of the carbohydrate binding specificity of concanavalin A', *Biochim. Biophys. Acta*, **317**, 500—4.

19. Goldstein, I. J., Hollerman, C. E. and Smith, E. E. (1965) 'Protein—carbohydrate interaction. II. Inhibition studies on the interaction of concanavalin A with polysaccharides', *Biochemistry*, **4**, 876—83.

20. Smith, E. E. and Goldstein, I. J. (1967) 'Protein—carbohydrate interaction. V. Further inhibition studies directed toward defining the stereochemical requirements of the reactive sites of concanavalin A', *Arch. Biochem. Biophys.*, **121**, 88—95.

21. Goldstein, I. J. and Iyer, R. N. (1966) 'Interaction of concanavalin A, a phytohemagglutinin, with model substrates', *Biochim. Biophys. Acta*, **121**, 197—200.

22. Cifonelli, J. A., Montogomery, R. and Smith, F. (1956) 'The reaction of concanavalin A with mucopolysaccharides', *J. Amer. Chem. Soc.*, **78**, 2488—9.

23. Reeder, W. J. and Ekstedt, R. D. (1971) 'Study of the interaction of concanavalin A with staphylococcal teichoic acids', *J. Immunol.*, **106**, 334—40.

24. Lis, H. and Sharon, N. (1973) 'The biochemistry of plant lectins (Phyotohemagglutinins)', *Ann. Rev. Biochem.*, **43**, 541—74.

25. Lloyd, K. O. and Biton, M. A. (1971) 'Isolation and purification of a peptido-rhamnomannan from the yeast form of *Sporothrix schenckii*. Structural and immunochemical studies', *J. Immunol.*, **107**, 663—71.

26. Lloyd, K. O. (1970) 'The preparation of two insoluble forms of the phytohemagglutinin, concanavalin-A and their interaction with polysaccharides and glycoproteins', *Arch. Biochem. Biophys.*, **137**, 460—8.

27. Dufau, M. L., Tsuruhara, T. and Catt, K. J. (1972) 'Interaction of glycoprotein hormones with agarose-concanavalin A', *Biochim. Biophys. Acta*, **278**, 281—92.

28. Davey, M. W., Huang, J. W., Sulkowski, E. and Carter, W. A. (1975) 'Hydrophobic interaction of human interferon', *J. Biol. Chem.*, **250**, 348—9.

29. Clarke, A. E. and Denborough, M. A. (1971) 'The interaction of concanavalin A with blood-group-substance glycoproteins from human secretions', *Biochem. J.*, **121**, 811—6.

30. Weinstein, D. B., Marsh, J. B., Glick, M. C. and Warren, L. (1972) 'Membranes of animal cells. VI. The glycolipids of the L cell and its surface membrane', *J. Biol. Chem.*, **245**, 3928—37.
31. McConathy, W. J. and Alaupovic, P. (1974) 'Studies on the interaction of concanavalin A with major density classes of human plasma lipoproteins. Evidence for the specific binding of lipoprotein B in its associated and free forms', *FEBS Lett.*, **41**, 174—8.
32. Lis, H., Fridman, C., Sharon, N. and Katchalski, E. (1966) 'Multiple hemagglutinins in soybean', *Arch. Biochem. Biophys.*, **117**, 301—9.
33. Hayman, M. J. and Crumpton, M. J. (1972) 'Isolation of glycoproteins from pig lymphocyte plasma membrane using *Lens culinaris* phytohemagglutinin', *Biochem. Biophys. Res. Commun.*, **47**, 923—30.
34. Allan, D., Auger, J. and Crumpton, M. J. (1972) 'Glycoprotein receptors for concanavalin A isolated from pig lymphocyte plasma membrane by affinity chromatography in sodium deoxycholate', *Nature (New Biol.)*, **236**, 23—5.
35. Nachman, R. L., Hubbard, A. and Ferris, B. (1973) 'Iodination of the platelet membrane. Studies of the major surface glycoprotein', *J. Biol. Chem.*, **248**, 2928—36.
36. Cuatrecasas, P. and Tell, G. P. E. (1973) 'Insulin-like activity of concanavalin A and wheat germ agglutinin — direct interactions with insulin receptors', *Proc. Nat. Acad. Sci. U.S.A.*, **70**, 485—9.
37. Susz, J. P., Hof, H. I. and Brunngraber, E. G. (1973) 'Isolation of concanavalin A-binding glycoproteins from rat brain', *FEBS Lett.*, **32**, 259—92.
38. Ponce de Leon, M., Hessle, H. and Cohen, G. H. (1973) 'Separation of herpes simplex virus-induced antigens by concanavalin A affinity chromatography', *J. Virol.*, **12**, 766—74.
39. Wray, V. P. and Walborg, E. F., Jr.,. (1971) 'Isolation of tumor cell surface binding sites for concanavalin A and wheat germ agglutinin', *Cancer Res.*, **31**, 2072—9.
40. Akedo, H., Mori, Y., Tanigaki, Y., Shinkai, K. and Morita, K. (1972) 'Isolation of concanavalin A binding protein(s) from rat erythrocyte stroma', *Biochim. Biophys. Acta*, **271**, 378—87.
41. Findlay, J. B. C. (1974) 'The receptor proteins for concanavalin A and *Lens culinaris* phytohemagglutinin in the membrane of the human erythrocyte', *J. Biol. Chem.*, **249**, 4398—403.
42. Rott, R., Becht, H., Klenk, H. -D. and Scholtissek, C. (1972) 'Interactions of concanavalin A with the membrane of influenza virus infected cells and with envelope components of the virus particle', *Z. Naturforsch.*, **27b**, 227—33.
43. Smith, D. F., Neri, G. and Walborg, E. F., Jr. (1973) 'Isolation and partial chemical characterization of cell-surface glycopeptides from AS-30D rat hepatoma which possess binding sites for wheat germ agglutinin and concanavalin A', *Biochemistry*, **12**, 2111—8.
44. Smith, D. F. and Walborg, E. F., Jr. (1972) 'Isolation and chemical characterization of cell surface sialoglycopeptide fractions during progression of rat ascites hepatoma AS-30D', *Cancer Res.*, **32**, 543—9.
45. Hakomori, S-I., Koscielak, J., Bloch, K. J. and Jeanloz, R. W. (1967) 'Immunologic relationship between blood group substances and a fucose-containing glycolipid of human adenocarcinoma', *J. Immunol.*, **98**, 31—38.
46. Gahmberg, C. G. and Hakomori, S-I. (1974) 'Organization of glycolipids and glycoproteins in surface membranes: Dependency on cell cycle and on transformation', *Biochem. Biophys. Res. Commun.*, **59**. 283—91.
47. Winqvist, L., Eriksson, L. C. and Dallner, G. (1974) 'Binding of concanavalin A-"Sepharose" to glycoproteins of liver microsomal membranes', *FEBS Lett.*, **42**, 27—31.
48. Steinemann, A. and Stryer, L. (1973) 'Accessibility of the carbohydrate moiety of rhodopsin', *Biochemistry*, **12**, 1499—502.
49. Nicolson, G. L., Lacorbiere, M. and Delmonte, P. (1972) 'The outer membrane terminal saccharides of bovine liver nuclei and mitochondria', *Exp. Cell Res.*, **71**, 468—73.
50. Henning, R. and Uhlenbruck, G. (1973) 'Detection of carbohydrate structure on isolated subcellular organelles of rat liver by heterophile agglutinins', *Nature (New Biol.)*, **242**, 120—2.
51. Keenan, T. W., Franke, W. W. and Kartenbeck, J. (1974) 'Concanavalin A binding by

isolated plasma membranes and endomembranes from liver and mammary gland', *FEBS Lett.*, **44**, 274—8.

52. Glew, R. H., Kayman, S. C. and Kuhlenschmidt, M. S. (1973) 'Studies on the binding of concanavalin A to rat liver mitochondria', *J. Biol. Chem.*, **218**, 3137—45.

53. Hirano, H., Parkhouse, B., Nicolson, G. L., Lennox, E. S. and Singer, S. J. (1972) 'Distribution of saccharide residues on membrane fragments from a myeloma-cell homogenate: Its implications for membrane biogenesis', *Proc. Nat. Acad. Sci. U.S.A.*, **69**, 2945—9.

54. Monneron, A. and Segretain, D. (1974) 'Extensive binding of concanavalin A to the nuclear membrane', *FEBS Lett.*, **42**, 209—13.

55. Uhlenbruck, G. and Radunz, A. (1972) 'Use of heterophilic agglutinins in plant serology', *Z. Naturforsch.*, **27b**, 1113—6.

56. Sharon, N. and Lis, H. (1975) 'Use of lectins for the study of membranes', *Meth. Memb. Biol.*, **3**, 147—200.

57. Stewart, M. L., Summers, D. F., Soeiro, R., Fields, B. N. and Maizez, J. V., Jr. (1973) 'Purification of oncornaviruses by agglutination with concanavalin A', *Proc. Nat. Acad. Sci. U.S.A.*, **70**, 1308—12.

58. Edelman, G. M., Rutishauser, U. and Millette, C. F. (1971) 'Cell fractionation and arrangement on fibers, beads, and surfaces', *Proc. Nat. Acad. Sci. U.S.A.*, **68**, 2153—7.

59. Anderson, J. and Melchers, F. (1973) 'Induction of immunoglobulin M synthesis and secretion in bone marrow-derived lymphocytes by locally concentrated concanavalin A', *Proc. Nat. Acad. Sci. U.S.A.*, **70**, 416—20.

60. Podolsky, D. K. and Weiser, M. M. (1973) 'Specific selection of mitotically active intestinal cells by concanavalin A-derivatized fibers', *J. Cell Biol.*, **58**, 497—500.

61. Rutishauser, U. and Sachs, L. (1974) 'Receptor mobility and the mechanism of cell—cell binding induced by concanavalin A', *Proc. Nat. Acad. Sci. U.S.A.*, **71**, 2456—60.

62. Nicolson, G. L. (1974) 'The interactions of lectins with animal cell surfaces', *Int. Rev. Cytol.*, **39**, 89—190.

63. Arisawa, M. and Mariyama, H. B. (1974) 'Cell-surface changes of phosphate-deficient *Escherichia coli* as characterized by increased stability and decreased concanavalin A-agglutinability of the spheroplast, *J. Biochem.*, **76**, 65—72.

64. McLean, R. J. and Brown, R. M. (1974) 'Cell surface differentiation of *chlamydomonas* during gametogenesis. I. Mating and concanavalin A agglutinability', *Develop. Biol.*, **36**, 279—85.

65. Martinez-Palomo, A., Wicker, R. and Bernhard, W. (1972) 'Ultrastructural detection of concanavalin A surface receptors in normal and in polyoma-transformed cells', *Int. J. Cancer*, **9**, 676—83.

66. Stevens, A. R. and Kaufman, A. E. (1974) 'Concanavalin A-induced agglutination of *Acanthamoeba*', *Nature*, **252**, 43—5.

67. Mäkelä, O. (1957) 'Studies in hemagglutinins of leguminosae seeds', *Ann. Med. Exp. Fenn.*, **35**, (suppl. 11), 1—133.

68. Boyd, W. C. (1963) 'The lectins: Their present status', *Vox Sang.*, **8**, 1—32.

69. Bird, G. W. G. (1959) 'Anti-A haemagglutinin from a non-leguminans plant *Hyptis snaveolens* poit', *Nature*, **184**, 109.

70. Toms, G. C. and Western, A. (1971) 'Phytohaemagglutinins', in *Chemotaxonomy of the Leguminosae*, eds. Boulter, D and Turner, B. L. (Academic Press, New York), pp. 367—462.

71. Edelman, G. M. and Millette, C. F. (1971) 'Molecular probes of spermatozoan structures', *Proc. Nat. Acad. Sci. U.S.A.*, **68**, 2436—40.

72. Moscona, A. A. (1971) 'Embryonic and neoplastic cell surfaces: Availability of receptors for concanavalin A and wheat germ agglutinin', *Science*, **171**, 905—7.

73. Kleinschuster, S. J. and Moscona, A. A. (1972) 'Interactions of embryonic and fetal neural retina cells with carbohydrate binding phytoagglutinins: Cell surface changes with differentiation', *Exp. Cell. Res.*, **70**, 397—410.

74. Krach, S. W., Green, A., Nicolson, G. L. and Oppenheimer, S. B. (1974) 'Cell surface changes occurring during sea urchin embryonic development monitored by quantitative agglutination with plant lectins', *Exp. Cell Res.*, **84**, 191—8.

75. Nicolson, G. L. and Yanagimachi, R. (1972) 'Terminal saccharides on sperm plasma membranes: Identification by specific agglutinins', *Science*, **177**, 276—9.

76. Oikawa, T., Nicolson, G. L. and Yanagimachi, R. (1975) 'Trypsin-mediated modification of the zone pellucida glycopeptide structure of hamster eggs', *J. Reprod. Fert.*, 43, 133—6.
77. Monroy, A., Ortolani, G., O'Dell, D. and Millonig, G. (1973) 'Binding of concanavalin A to the surface of unfertilized and fertilized ascidian eggs', *Nature*, 242, 409—10.
78. Huet, Ch., Lonchampt, M., Huet, M. and Bernadec, A. (1974) 'Temperature effects on the concanavalin A molecule and on concanavalin A binding', *Biochim. Biophys. Acta*, 365, 28—39.
79. Noonan, K. D. and Burger, M. M. (1973) 'The relationship of concanavalin A binding to lectin-initiated cell agglutination', *J. Cell Biol.*, 59, 134—42.
80. Shoham, J. and Sachs, L. (1974) 'Different cyclic changes in the surface membrane of normal and malignant transformed cells', *Exp. Cell Res.*, 85, 8—14.
81. O'Dell, D. S., Ortolani, G. and Monroy, A. (1973) 'Increased binding of radioactive concanavalin A during maturation of *Ascidia* eggs', *Exp. Cell Res.*, 83, 408—11.
82. Nicolson, G. L. and Lacorbiere, M. (1973) 'Cell contact-dependent increase in membrane D-galactopyranosyl-like residues on normal, but not virus- or spontaneously-transformed murine fibroblasts', *Proc. Nat. Acad. Sci. U.S.A.*, 70, 1672—6.
83. Nicolson, G. L. (1974) 'Factors influencing the dynamic display of lectin-binding sites on normal and transformed cell surfaces', in *Control of Proliferation in Animal Cell Surfaces*, eds. Clarkson, B. and Baserga, R. (Cold Spring Harbor Laboratory, New York), pp. 251—70.
84. Inbar, M. and Sachs, L. (1969) 'Interaction of the carbohydrate-binding protein concanavalin A with normal and transformed cells', *Proc. Nat. Acad. Sci. U.S.A.*, 63, 1418—25.
85. Ozanne, B. and Sambrook, J. (1971) 'Binding of radioactively labeled concanavalin A and wheat germ agglutinin to normal and virus-transformed cells', *Nature (New Biol.)* 232, 156—60.
86. Arndt-Jovin, D. J. and Berg. P. (1971) 'Quantitative binding of ^{125}I-concanavalin A to normal and transformed cells', *J. Virol.*, 8, 716—21.
87. Inbar, M., Ben-Bassat, H. and Sachs, L. (1971) 'A specific metabolic activity on the surface membrane in malignant cell-transformation', *Proc. Nat. Acad. Sci. U.S.A.*, 68, 2748—51.
88. Noonan, K. D. and Burger, M. M. (1973) 'Binding of ^{3}H concanavalin A to normal and transformed cells', *J. Biol. Chem.*, 248, 4286—92.
89. Phillips, P. G. Furmanski, P. and Lubin, M. (1974) 'Cell surface interactions with concanavalin A: Location of bound radiolabeled lectin', *Exp. Cell Res.*, 86, 301—8.
90. Nicolson, G. L. (1973) 'Temperature-dependent mobility of concanavalin A sites on tumour cell surfaces', *Nature (New Biol.)* 243, 218—20.
91. Nicolson, G. L., Lacorbiere, M. and Eckhart, W. (1975) 'Qualitative and quantitative interactions of lectins with untreated and neuraminidase-treated normal, wild-type and temperature-sensitive polyoma-transformed fibroblasts', *Biochemistry*, 14, 172—9.
92. Poste, G. and Reeve, P. (1974) 'Increased mobility and redistribution of concanavalin A receptors on cells infected with Newcastle disease virus', *Nature*, 247 469—71.
93. Penhoet, E., Olsen, C., Carlson, S., Lacorbiere, M. and Nicolson, G. L. (1974) 'Quantitative interaction of *Ricinus communis* agglutinin and concanavalin A with influenza and vesicular stomatitis viruses and virus-infected normal and polyoma-transformed cells', *Biochemistry*, 13, 3561—6.
94. Inbar, M. and Sachs, L. (1973) 'Mobility of carbohydrate containing sites on the surface membrane in relation to the control of cell growth', *FEBS Lett.*, 32, 124—8.
95. Mallucci, L. (1971) 'Binding of concanavalin A to normal and transformed cells as detected by immunofluorescence', *Nature (New Biol.)*, 233, 241—4.
96. Smith, C. W. and Hollers, J. C. (1970) 'The pattern of binding of fluorescein-labeled concanavalin A to the motile lymphocyte', *J. Reticuloendothel. Soc.*, 8, 458—464.
97. Comoglio, P. M. and Guglielmone, R. (1972) 'Two dimensional distribution of concanavalin A receptor molecules on fibroblast and lymphocyte plasma membranes', *FEBS Lett.*, 27, 256—8.
98. Yahara, I. and Edelman, G. M. (1973) 'Modulation of lymphocyte receptor redistribution by concanavalin A, anti-mitotic agents and alterations of pH', *Nature*, 236, 152—5.
99. Yahara, I. and Edelman, G. M. (1973) 'The effects of concanavalin A on the mobility of lymphocyte surface receptors', *Exp. Cell Res.*, 81, 143—55.

100. Nicolson, G. L. and Singer, S. J. (1971) 'Ferritin-conjugated plant agglutinins as specific saccharide stains for electron microscopy: Application to saccharides bound to cell membranes', *Proc. Nat. Acad. Sci. U.S.A.*, **68**, 942—5.
101. Nicolson, G. L. and Singer, S. J. (1974) 'The distribution and asymmetry of saccharides on mammalian cell membrane surfaces utilizing ferritin-conjugated plant agglutinins as specific saccharide stains', *J. Cell Biol.*, **60**, 236—48.
102. Nicolson, G. L. (1971) 'Difference in the topology of normal and tumor cell membranes as shown by different distributions of ferritin-conjugated concanavalin A on their surfaces', *Nature (New Biol.)*, **233**, 244—6.
103. Nicolson, G. L. (1972) 'Topography of cell membrane concanavalin A-sites modified by proteolysis', *Nature (New Biol.)*, **239**, 193—7.
104. Nicolson, G. L. (1975) 'Concanavalin A as a quantitative and ultrastructural probe for normal and neoplastic cell surfaces', in *Concanavalin A*, eds. Chowdhury, T. K. and Weiss, A. K. (Plenum Press, New York), pp. 153—72.
105. Klein, P. A. and Adams, W. R. (1972) 'Location of ferritin-labeled concanavalin A binding to influenza virus and tumor cell surfaces', *J. Virol.*, **10**, 844—54.
106. Barbarese, E., Sauerwein, H. and Simkins, H. (1973) 'Alteration in the surface glycoproteins of chick erythrocytes following transformation with erythroblastosis strain R virus', *J. Memb. Biol.*, **13**, 129—42.
107. de Petris, S., Raff, M. C. and Mallucci, L. (1973) 'Ligand-induced redistribution of concanavalin A receptors on normal, trypsinized and transformed fibroblasts', *Nature (New Biol.)*, **244**, 275—8.
108. Pinto da Silva, P. and Nicolson, G. L. (1974) 'Freeze-etch localization of concanavalin A receptors to the membrane intercalated particles on human erythrocyte membranes', *Biochim. Biophys. Acta*, **363**, 311—9.
109. Matus, A., de Petris, S. and Raff, M. C. (1973) 'Mobility of concanavalin A receptors in myelin and synaptic membranes', *Nature (New Biol.)*, **244**, 278—9.
110. Bernhard, W. and Avrameas, S. (1971) 'Ultrastructural visualization of cellular carbohydrate components by means of concanavalin A', *Exp. Cell Res.*, **64**, 232—6.
111. Barat, N. and Avrameas, S. (1973) 'Surface and intracellular localization of concanavalin A in human lymphocytes', *Exp. Cell Res.*, **76**, 451—5.
112. Huet, Ch. and Garrido, J. (1972) 'Ultrastructural visualization of cell coat components by means of wheat germ agglutinin', *Exp. Cell Res.*, **75**, 523—7.
113. Huet, Ch. and Bernhard, W. (1974) 'Differences in the surfaces mobility between normal and SV40-, polyoma- and adenovirus-transformed hamster cells', *Int. J. Cancer*, **13**, 227—39.
114. Garrido, J., Burglen, M-J, Samolyk, D., Wicker, R. and Bernhard, W. (1974) 'Ultrastructural comparison between the distribution of concanavalin A and wheat germ agglutinin cell surface receptors of normal and transformed hamster and rat cell lines', *Cancer Res.*, **34**, 230—43.
115. Smith, S. B. and Revel, J-P. (1972) 'Mapping of concanavalin A binding sites on the surface of several cell types', *Develop. Biol.*, **27**, 434—41.
116. Rosenblith, J. Z., Ukena, T. E., Yin, H. H., Berlin, R. D. and Karnovsky, M. J. (1973) 'A comparative evaluation of the distribution of concanavalin A-binding sites on the surfaces of normal, virally-transformed, and protease-treated fibroblasts', *Proc. Nat. Acad. Sci. U.S.A.*, **70**, 1625—9.
117. Ryan, G. B., Unanue, E. R. and Karnovsky, M. J. (1974) 'Inhibition of surface capping of macro-molecules by local anaesthetics and tranquillizers', *Nature*, **250**, 56—7.
118. Ukena, T. E., Borysenko, J. Z., Karnovsky, M. J. and Berlin, R. D. (1974) 'Effects of colchicine, cytochalasin B and 2-deoxyglucose on the topographical organization of surface-bound concanavalin A in normal and transformed fibroblasts', *J. Cell Biol.*, **61**, 70—82.
119. Nicolson, G. L. and Yanagimachi, R. (1974) 'Mobility and the restriction of mobility of plasma membrane lectin-binding components', *Science*, **184**, 1294—6.
120. Oliver, J. M., Ukena, T. E. and Berlin, R. D. (1974) 'Effects of phagocytosis and colchicine on the distribution of lectin-binding sites on cell surfaces', *Proc. Nat. Acad. Sci. U.S.A.*, **71**, 394—8.
121. Singer, S. J. and Nicolson, G. L. (1972) 'The fluid mosaic model of the structure of cell membranes', *Science*, **175**, 720—31.

122. Edelman, G. M., Yahara, I. and Wang, J. L. (1973) 'Receptor mobility and receptor—cytoplasmic interactions in lymphocytes', *Proc. Nat. Acad. Sci. U.S.A.,* **70,** 1442—6.
123. Berlin, R. D., Oliver, J. M., Ukena, T. E. and Yin, H. H. (1974) 'Control of cell surface topography', *Nature,* **247,** 45—6.
124. Zarling, J. and Tevethia, S. (1971) 'Expression of concanavalin A binding sites in rabbit kidney cells infected with vaccinia virus', *Virology,* **45,** 313—6.
125. Tevethia, S., Lowry, S., Rawls, W., Melnick, J. and McMillan, V. (1972) 'Detection of early cell surface changes in herpes simplex virus by agglutination with concanavalin A, *J. Gen. Virol.,* **15,** 93—7.
126. Poste, G. and Reeve, P. (1972) 'Agglutination of normal cells by plant lectins following infection with non-oncogenic viruses', *Nature (New Biol.),* **237,** 113—4.
127. Becht, H., Rott, R. and Klenk, H. D. (1972) 'Effect of concanavalin A on cells infected with enveloped RNA viruses', *J. Gen. Virol.,* **14,** 1—8.
128. Burger, M. M. (1973) 'Surface changes in transformed cells detected by lectins', *Fed. Proc.,* **32,** 91—101.
129. Inbar, M., Ben-Bassat, H. and Sachs, L. (1974) 'Location of amino acid and carbohydrate transport sites in the surface membrane of normal and transformed mammalian cells', *J. Memb. Biol.,* **6,** 195—209.
130. Isselbacher, K. J. (1972) 'Increased uptake of amino acids and 2-deoxy-D-glucose by virus-transformed cells in culture', *Proc. Nat. Acad. Sci. U.S.A.,* **69,** 585—9.
131. Czech, M. P. and Lynn, W. S. (1973) 'Stimulation of glucose metabolism by lectins in isolated white fat cells', *Biochim. Biophys. Acta,* **297,** 368—77.
132. Martin, G. S., Venuta, S., Weber, M. and Rubin, H. (1971) 'Temperature dependent alterations in sugar transport in cells infected by a temperature sensitive mutant of Rous sarcoma virus', *Proc. Nat. Acad. Sci. U.S.A.,* **68,** 2739—41.
133. Dent, P. B. (1971) 'Inhibition by phytohemagglutinin of DNA synthesis in cultured mouse lymphomas', *J. Nat. Cancer Inst.,* **46,** 763—73.
134. Ralph, P. and Nakoinz, I. (1975) 'Inhibitory effects of lectins and lymphocyte mitogens on murine lymphomas and myelomas', *J. Nat. Cancer Inst.,* **51,** 883—90.
135. Inbar, M., Ben-Bassat, H. and Sachs, L. (1972) 'Inhibition on ascites tumor development by concanavalin A', *Int. J. Cancer,* **9,** 143—9.
136. Shoham, J., Inbar, M. and Sachs, L. (1970) 'Differential toxicity on normal and transformed cells *in vitro* and inhibition of tumour development *in vivo* by concanavalin A', *Nature,* **227,** 1244—46.
137. Aubery, M. and Bourrillon, R. (1973) 'Physiologie cellulaire', *C.R. Acad. Sci. (D) (Paris),* **276,** 3187—90.
138. Treska-Ciesielski, J., Gombos, G. and Morgan, I. G. (1971) 'Effect of concanavalin A on the neurons of spinal ganglia of chick embryos in culture', *C.R. Acad. Sci. (D) (Paris),* **273,** 1041—43.
139. Ozanne, B. and Sambrook, J. (1971) 'Isolation of lines of cells resistant to agglutination by concanavalin A from 3T3 cells transformed by SV40', in *Biology of Oncogenic Viruses,* ed. Silvestri, L. (North Holland, Amsterdam), pp. 248—57.
140. Wright, J. A. (1973) 'Evidence for pleiotropic changes in lines of Chinese hamster ovary cells resistant to concanavalin A and phytohemagglutinin-P', *J. Cell Biol.,* **56,** 666—75.
141. Schultz, A. R. and Culp, L. A. (1973) 'Contact-inhibited revertant cell lines isolated from SV40-transformed cells. V. Contact inhibition of sugar transport', *Exp. Cell Res.,* **81,** 95—103.
142. Culp, L. A. and Black, P. H. (1972) 'Contact inhibited revertant cell lines isolated from Simian Virus 40-transformed cells. III. Concanavalin A-selected revertant cells', *J. Virol.,* **9,** 611—20.
143. Burger, M. M. and Noonan, K. D. (1970) 'Restoration of normal growth by covering of agglutinin sites on tumour cell surfaces', *Nature,* **228,** 512—5.
144. Trowbridge, I. S. and Hilborn, D. A. (1974) 'Effects of succinyl-concanavalin A on the growth of normal and transformed cells', *Nature,* **250,** 304—7.
145. Friberg, S., Jr., Golub, S. H., Lilliehöök, B. and Cochran, A. J. (1972) 'Assessment of concanavalin A reactivity to murine ascites tumours by inhibition of tumour cell migration', *Exp. Cell Res.,* **73,** 100—6.
146. Berlin, R. D. (1972) 'Effect of concanavalin A on phagocytosis', *Nature (New Biol.),* **235,** 44—5.

147. Lallier, R. (1972) 'Effects of concanavalin A on the development of sea urchin eggs', *Exp. Cell Res.*, **72** 157—63.
148. Oikawa, T., Yanagimachi, R. and Nicolson, G. L. (1973) 'Wheat germ agglutinin blocks mammalian fertilization', *Nature*, **241**, 256—9.
149. Oikawa, T. Nicolson, G. L. and Yanagimachi, R. (1974) 'Inhibition of hamster egg fertilization by phytoagglutinins', *Exp. Cell Res.*, **83**, 239—46.
150. Majerus, P. W. and Brodie, G. N. (1972) 'The binding of phytohemagglutinins to human platelet plasma membranes', *J. Biol. Chem.*, **247**, 4253—73.
151. Schwartz, H. J. Pelley, R. P. and Leon, M. A. (1970) 'Release of a migration inhibitory factor from non-immune lymphoid cells by concanavalin A', *Fed. Proc.*, **29**, 360.
152. Siraganian, P. A. and Siraganian, R. P. (1974) 'Basophil activation by concanavalin A: Characteristics of the reaction, *J. Immunol.*, **112**, 2117—25.
153. Nowell, P. C. (1960) 'Phytohemagglutinin: An initiator of mitosis in cultures of normal human leukocytes', *Cancer Res.*, **20**, 462—6.
154. Powell, A. E. and Leon, M. A. (1970) 'Reversible interaction of human lymphocytes with the mitogen concanavalin A', *Exp. Cell Res.*, **62**, 315—25.
155. Andersson, J., Möller, G. and Sjöberg, O. (1972) 'Selective induction of DNA synthesis in T and B lymphocytes', *Cell Immunol.*, **4**, 381—93.
156. Blomgren, H. and Svedmyr, E. (1971) '*In vitro* stimulation of mouse thymus cells by PHA and allogeneic cells', *Cell Immunol.*, **2**, 285—99.
157. Janossy, G. and Greaves, M. F. (1971) 'Lymphocyte activation. I. Response of T and B lymphocyte to phytomitogens', *Clin. Exp. Immunol.*, **9**, 483—98.
158. Stobo, J. D., Rosenthal, A. S. and Paul, W. E. (1972) 'Functional heterogeneity of murine lymphoid cells. I. Responsiveness to and surface binding of concanavalin A and phytohemagglutinin', *J. Immunol.*, **108**, 1—17.
159. Andersson, J., Edelman, G. M., Möller, G. and Sjöberg, O. (1972) 'Activation of B lymphocytes by locally concentrated concanavalin A', *Euro. J. Immunol.*, **2**, 233—5.
160. Möller, G. (1970) 'Immunocyte triggering', *Cell Immunol.*, **1**, 573—82.
161. Dutton, R. W., Falkoff, R., Hirst, J. A., Hoffmann, M., Kappler, J. W., Kettman, J. R., Lesley, J. F. and Vann, D. (1971) 'Is there evidence for a non-antigen specific diffusable chemical mediator from the thymus-derived cell in the initiation of the immune response', in *Progress in Immunology*, 1st International Congress of Immunology, ed. Amos, B. (Academic Press, New York), pp. 355—68.
162. Greaves, M. and Janossy, G. (1972) 'Elicitation of selective T and B lymphocyte responses by cell surface binding ligands', *Transplant. Rev.*, **11**, 87—130.
163. Andersson, J., Sjöberg, O. and Möller, G. (1972) 'Mitogens as probes for immunocyte activation and cellular cooperation', *Transplant. Rev.*, **11**, 131—77.
164. Bretscher, P. (1972) 'The control of humoral and associative antibody synthesis', *Transplant. Rev.*, **11**, 217—67.
165. Edelman, G. M. (1973) 'Antibody structure and molecular immunology', *Science*, **180**, 830—40.
166. Gunther, G. R., Wang, J. L., Yahara, I., Cunningham, B. A. and Edelman, G. M. (1973) 'Concanavalin A derivatives with altered biological activities', *Proc. Nat. Acad. Sci. U.S.A.*, **70**, 1012—6.
167. Edelman, G. M. (1974) 'Surface changes and mitogenesis in lymphocytes', in *Control of Proliferation in Animal Cells*, eds. Clarkson B. and Baserga, R. (Cold Spring Harbor Laboratory, New York), pp. 357—77.
168. Smith, J. W., Steiner, A. L., Newberry, W. M., Jr. and Parker, C. W. (1971) 'Cyclic adenosine 3', 5'-monophosphate in human lymphocytes. Alterations after phytohemagglutinin stimulation', *J. Clin. Invest.*, **50**, 432—41.
169. Kram, R. and Tomkins, G. M. (1973) 'Pleiotypic control by cyclic AMP: Interaction with cyclic GMP and possible role of microtubules', *Proc. Nat. Acad. Sci. U.S.A.*, **70**, 1659—63.
170. Watson, L., Knox, R. B. and Creaser, E. H. (1974) 'Con A differentiates among grass pollens by binding specifically to wall glycoproteins and carbohydrates', *Nature*, **249**, 574—5.
171. Douglas, S. D. (1972) 'Electron microscopic and functional aspects of human lymphocyte response to mitogens', *Transplant. Rev.*, **11**, 39—59.

172. Perlmann, P., Nilsson, H. and Leon, M. A. (1970) 'Inhibition of cytotoxicity of lymphocytes by concanavalin A *in vitro*', *Science*, **168**, 1112–5.
173. Stavy, L., Treves, A. J. and Feldman, M. (1971) 'Effect of concanavalin A on lymphocyte-mediated cytotoxicity', *Nature*, **232**, 56–8.
174. Schwartz, H. J. and Wilson, F. (1971) 'Target cell destruction by concanavalin A-stimulated lymphoid cells', *Amer. J. Pathol.*, **64**, 295–301.
175. Kirchner, H. and Blaese, R. M. (1973) 'Pokeweed mitogen, concanavalin A and phytohemagglutinin-induced development of cytotoxic effector lymphocytes', *J. Exp. Med.*, **138**, 812–24.
176. Dutton, R. W. (1972) 'Inhibitory and stimulatory effects of concanavalin A on the response of mouse spleen cell suspensions to antigen. I: Characterization of the inhibitory cell activity', *J. Exp. Med.*, **136**, 1445–60.
177. Leon, M. A. and Schwartz, H. J. (1969) 'Inhibition of delayed hypersensitivity to tuberculin by concanavalin A', *Proc. Exp. Biol. Med.*, **131**, 735–6.
178. Schwartz, H. J., Leon, M. A. and Pelley, R. P. (1970) 'Concanavalin A-induced release of skin-reactive factor from lymphoid cells', *J. Immunol.*, **104**, 265–8.
179. Pick, E., Krejĩ, J. and Turk, J. L. (1970) 'Release of skin reactive factor from guinea-pig lymphocytes by mitogens', *Nature*, **225**, 236–8.
180. Markowitz, H., Person, D. A., Gitnick, G. L. and Ritts, R. A. (1969) 'Immunosuppressive activity of concanavalin A', *Science*, **163**, 476.
181. Shier, W. T., Trotter, J. T. and Reading, C. L. (1974) 'Inflammation induced by concanavalin A and other lectins', *Proc. Soc. Exp. Biol. Med.*, **146**, 590–3.

CHAPTER 2

Isolation and Properties of Concanavalin A

I. E. LIENER

1. ISOLATION AND ASSAYS

1.1 Isolation by Direct Crystallization

A systematic study of the proteins of the jack beam (*Canavalia ensiformis*) was first reported by Jones and Johns in 1916 (1). The globulin fraction, which precipitated upon dialysis of a salt extract of the bean, could be separated into two further fractions based on their solubility in ammonium sulphate. The major fraction, which was given the name *canavalin*, could only be precipitated by saturation with ammonium sulphate, whereas a smaller quantity of a fraction called *concanavalin* could be precipitated with 0.6 saturated ammonium sulphate. Sumner (2) subsequently succeeded in separating the concanavalin fraction into two crystallizable components, one of which, concanavalin B was sparingly soluble in 10% NaCl, while the other, concanavalin A was soluble only in concentrated salt solution. The latter was shown to be a haemagglutinin which, in addition to being able to agglutinate erythrocytes of various species of animals, precipitated glycogen and mucoproteins (3—5). In 1936 Sumner and Howell (5) described a simple technique for crystallizing Con A directly from an aqueous acetone extract of jack bean meal with a yield of about 3 g per 100 g of the meal. A further refinement involving the separation of Con A from Con B by serial crystallization was reported in abstract form by Howell in 1953 (6)

Nakamura and Suzuno (7), however, were unable to crystallize Con A from a Japanese variety of jack bean using the techniques of Sumner (2) and Howell (6). These workers found it necessary to remove the canavalin fraction with bentonite treatment before Con A could be crystallized. Once crystallized, however, this preparation of Con A did not differ in crystal form nor in its reactivity towards the sera of various animals from the Con A crystallized from an American source of jack bean by the original procedures of Sumner (2) or Howell (6).

1.2 Isolation by Affinity Chromatography

By taking advantage of the fact that Con A reacts with dextrans, affinity chromatography with commercially available cross-linked dextrans ('Sephadex') has afforded a convenient technique for the direct isolation of Con A from crude extracts of jack bean meal or partially purified fractions (8—10). Because of its simplicity and the high yield which it affords, this method for isolating Con A will be described in detail*.

Defatted, finely ground jack bean meal (General Biochemicals, Chagrin Falls, Ohio; Sigma Chemical Co., St. Louis, Mo), 300 g, is suspended in 0.15 M NaCl (1.5 l), and stirred overnight in the cold (ca. 4°). The resulting suspension is filtered through cheesecloth, the filtrate being saved, and the residue is reextracted for 8 hours with an additional 1.5 litre portion of 0.15M NaCl. The suspension is filtered as before, and the residue is discarded. The combined filtrates are centrifuged for 1 hour at 9500 r.p.m. (14,600 g, using the GSA rotor of a Sorvall RC-2 refrigerated centrifuge), the residue being discarded. The clear yellow supernatant solution is made 30% saturated with $(NH_4)_2SO_4$ by gradual addition of the solid salt (176 g/l). The pH is adjusted to 7.0 with concentrated NH_4OH, and the mixture is stirred gently at room temperature (ca. 25°) for 4 hours. Precipitated proteins are removed by centrifugation (GSA rotor, 9500 r.p.m, 1 hour) and discarded. The bulk of the proteins including Con A is now precipitated by adding $(NH_4)_2SO_4$ (356 g/l) to 80% saturation. The precipitate is collected by centrifugation as above (the supernatant solution being discarded), dissolved in H_2O

*Reprinted in part from Agrawal and Goldstein (11) by permission of the authors and Academic Press, N.Y.

(500 ml) and dialysed in the cold against several changes of water, and finally against 1.0 M NaCl. The dialysed protein solution is centrifuged, and the clear solution is used for dextran gel adsorption.

In lieu of fractionation with ammonium sulphate which serves to remove glycoproteins which may interfere with the subsequent adsorption of Con A to 'Sephadex', Surolia and coworkers (12) found that a crude extract of jack bean meal can be chromatographed directly provided one employs pretreatment with 1 M acetic acid as originally described by Olson and Liener (10). This treatment is as follows: The beans, 50 g, are soaked for 5 h in water, and the seed coats peeled off. The seeds are then further soaked for 2 h in 200 ml 1 M NaCl containing 0.01 M Tris buffer, pH 7.4. The seeds are homogenized in a Waring blender and the homogenate stirred for 6 h at 5°. The suspension is filtered through glass wool and the filtrate brought to 1 M acetic acid by the addition of glacial acetic acid. After stirring for 20 min at room temperature, the solution is clarified by centrifugation, and the supernate dialysed against 1 M NaCl containing 0.001 M of $MgCl_2$ $MnCl_2$ and $CaCl_2$.

A column (4 x 60 cm) of 'Sephadex' G-50 (Pharmacia, fine or medium mesh) is prepared and equilibrated with 1.0 M NaCl. The incorporation of 0.001 M metal ions (Mn^{2+} and Ca^{2+}) is recommended in order to ensure the binding of all of the Con A to the column (see 2.2).

The protein solution from above, in 1 M NaCl, is applied to the column at a rate of 30—40 ml/h. After addition of the sample the column is connected to a reservoir of 1 M NaCl. Fractions (ca. 10 ml) are collected every 15 minutes and monitored for protein by absorbance at 280 nm. Washing the column with 1 M NaCl to elute inert protein normally requires 24—48 hours. After the absorbance at 280 nm ≤ 0.1, a 0.10 M solution of glucose in 1 M NaCl is added to elute Con A from the dextran gel. In place of glucose, other sugars for which Con A has some affinity may be used; these include D-fructose, D-mannose, or the methyl-α-D-glycosides of glucose or mannose. Fractions having an absorbance at 280 nm of 0.20 or greater are combined.

The glucose can be removed by exhaustive dialysis against large volumes of 1.0 M NaCl or by gel filtration on a 'Sephadex' G-25 column equilibrated with 1.0 M NaCl. ('Sephadex' G-25 does not bind Con A presumably because of its high degree of cross-linking.) Except for small quantities of Con A, the latter procedure is not recommended as it requires repetitive cycling to remove all of the glucose bound to Con A. If the product is to be freeze dried the pooled fractions from the original 'Sephadex' G-50 column should be dialysed directly against distilled water. If acid pretreatment is not used, the freeze-dried product is apt to be soluble only at high salt concentrations. Bishayee and coworkers (13) recommend dialysis against 1 M acetic acid followed by distilled water prior to lyophilization in order to obtain a preparation which is readily soluble in solutions of low salt concentrations. In general affinity chromatography may be expected to produce a yield of 2 to 3 g of Con A per 100 g of jack bean meal.

1.3 Con A Isolated from Different Varieties of Jack Bean

Most of the studies described have been conducted with preparations of Con A derived from *Canavalia ensiformis*. In a few instances, the Con A prepared from other varieties of jack bean have been isolated and their properties compared with that isolated from *C. ensiformis*. Reference has already been made to the fact that the Con A crystallized from a Japanese source of jack bean (variety not specified) appeared to be identical to that isolated from *C. ensiformis*, even though a different method of crystallization had to be used (see 1.1). Con A has also been purified from *C. gladiata* or nata bean, and its biological properties were found to be very similar to those of Con A isolated from *C. ensiformis* (12,14). There were some minor differences in physicochemical properties, however. Although gel electrophoresis in SDS revealed the presence of the intact subunit and three minor lower molecular weight fragments, these were slightly different in their molecular weights and in their relative proportions as compared with the pattern obtained with Con A from *C. ensiformis*. Con A from both varieties of beans showed eight components in isoelectric focusing (14); again there were minor differences in the isoelectric points of these bands, the bands from *C. gladiata* showing a uniform shift in the directions of increased positive charge.

Hague (15) found that, although Con A isolated from *C. maritima* could not be distinguished from the Con A derived from *C. ensiformis* or *C. gladiata* on the basis of gel electrophoresis, amino acid analysis revealed it to contain only one methionine residue compared to the two present in the Con A from the other two species. Accordingly, cleavage with cyanogen bromide yielded two fragments compared with the three fragments produced by cleavage of Con A from *C. ensiformis* with this reagent (16). It was concluded that the methionine residue located at position 130 in the Con A derived from *C. ensiformis* is absent in the lectin from *C. maritima*.

1.4 Assay for Biological Activity

Advantage may be taken of the many varied biological properties of Con A for the purpose of evaluating its activity. For the rapid qualitative assessment of activity, the method of double-diffusion precipitation in agar gel (17) affords a very sensitive probe for the specificity of Con A. The most commonly used procedure for obtaining semi-quantitative data is to measure the haemagglutinating activity towards erythrocytes by serial twofold dilution with visual estimation of the end point (18). A more precise estimate of haemagglutinating activity may be obtained by employing a photometric technique which measures the decrease in absorbancy at 620 nm of a suspension of erythrocytes as lectin-induced aggregates sediment under controlled conditions (19,20). The ability of Con A to precipitate specific polysaccharides such as dextran also forms the basis of a quantitative test described by So and Goldstein (21). Abe and coworkers (22) have described a simple technique for measuring the Con A-induced precipitation of glycogen covalently labelled with a dye. Rothman and coworkers (23) have recently published an assay which is based on the ability of Con A to mediate the attachment of single cells in suspension to a confluent cell layer (see also Chapter 35). The assay method of choice will depend on the type of activity most relevant to the study at hand, and for this purpose further details may be obtained from the references cited here.

2. PHYSICOCHEMICAL PROPERTIES

2.1 Molecular Weight and Subunit Composition

It is apparent from Table I that the values reported for the sedimentation behaviour and molecular weight of Con A have in the past been quite variable, but this situation is readily explainable on the basis of our present knowledge that Con A

TABLE I.   Physicochemical properties of Con A

Physicochemical measurement	Remarks	References
Sedimentation constant ($S_{20,w} \times 10^{13}$)		
3.5	At pH 2.1	32
3.8	At pH 5.5	10, 27, 31
3.9	At pH 5.6	35
6.0	In 5% NaCl	25, 26
6.1	At pH 7.0—8.9	10, 32, 35
Diffusion constant ($D_{20,w} \times 10^7$)		
5.43	At pH 5.0	27
5.60	In 5% NaCl	25, 26
Specific volume ($\bar{v}$)		
0.73	In saturated NaCl	25, 26
Molecular weight		
40,000	Gel filtration, pH 2.2	30
55,000	SE^a, pH 3.5 to 5.8	29, 31, 32
68,000	SV^b, pH 5.0	27, 28
71,000	Gel filtration, pH 7.0	30
96,000	SV, in 5% NaCl	25, 26
112,000	SV, at pH 8.9	32
Isoelectric point (pI)		
7.1	Cellulose acetate	27, 28
5.5	Moving boundary electrophoresis	5
	Solubility	33
4.5—5.5 (4 components)	Isoelectric focusing (pH 4 to 6)	24
5.0—8.0 (4 components)	Isoelectric focusing (pH 2 to 9)	14
$E_{1\,cm}^{1\%}$,		
11.4 ± 0.1	At 280 nm in 1.0 M NaCl	27, 28
12.4	At 280 nm, pH 5	49
13.7	At 280 nm, pH 7	49
Metal content (Mn^{2+})		
0.029%	Activation analysis	27, 28
0.036%	Emission spectroscopy	28
No. of amino acid residues		
237	Per subunit having a molecular weight of 27,000	34
Terminal groups	Intact subunit	
N-terminal, alanine		10, 28, 34
C-terminal, asparagine		34

[a]SE, sedimentation equilibrium.
[b]SV, sedimentation velocity.

is a polymeric molecule whose degree of association is dependent on a number of variables, the most important of which is the pH. Evidence that Con A must be comprised of identical subunits, or fragments thereof, has come from experiments which have shown that, in the presence of such denaturants as urea, guanidine and sodium dodecylsulphate (SDS), Con A exists as a monomeric unit having a molecular weight of about 27,000 (30,36,37). Between pH 2 and 5.5, Con A exists as a dimer of two subunits and therefore has a molecular weight of approximately 55,000, whereas at pH values above 5.5 a tetrameric structure with a molecular weight of 110,000 predominates (29,32,34). The effect of pH on the structure of Con A is discussed in more detail below (3.1).

The above picture is somewhat complicated, however, by the observation that, upon gel electrophoresis in the presence of SDS, in addition to an intact subunit having a molecular weight of 27,000, as many as three to four lower molecular weight fragments have been noted, even in preparations of Con A which are presumably homogeneous with respect to activity (22,31,36,37). These 'extraneous' fragments have been reported to have molecular weights ranging from 13,000 to 18,000. Wang and coworkers (36) postulated that the dimer of Con A could exist as (i) a dimer of two intact subunits, (ii) a dimer of one intact and one fragmented subunit and (iii) a hybrid dimer of two fragmented units; the tetramer could then exist as combinations of these various possibilities. McKenzie and Sawyer (38) however, concluded on the basis of binding studies with methyl-α-D-glucopyranoside, concluded that hybrid dimers are probably not present to any appreciable extent. Cunningham and coworkers (39) have described a simple procedure for isolating the intact subunit which may be recovered to the extent of 50% in the supernatant of a solution of Con A which has been incubated in 1% NH_4HCO_3 at $37°$ for 12 to 16 h. The exact cause for the presence of these fragments is not known; presumably they are already present in the intact seed or else they are produced during the course of the isolation of Con A.

2.2 The Role of Metal Ions

Sumner and Howell (66) were the first to establish the essential role which metal ions play in the activity of Con A. They found that dialysis against dilute acid led to a loss in activity which could be restored by the addition of salts of Ca^{2+}, Mn^{2+}, Zn^{2+}, or Mg^{2+}. The activating effect of metal salts was subsequently confirmed and studied in more detail by Agrawal and Goldstein (40) and Kalb and Levitzski (41). The latter workers found that Con A binds bivalent metals at two different sites — an S_1 site which binds transition metals such as Mn^{2+}, Ni^{2+} and Co^{2+}, and an S_2 site which preferentially binds Ca^{2+}. They concluded that the occupation of the S_1 site by transition metals was necessary before the S_2 site could bind Ca^{2+}, and that occupation of both of these sites was necessary for the binding of carbohydrate. A more detailed study (42) of the specificity of the metal requirement of Con A has revealed that only divalent metal ions which have an affinity for nitrogen ligands are bound to the S_1 site (see also 4.4); these include the transition metals already referred to as well as Zn^{2+} and Cd^{2+} which are bound less strongly. The interaction with the S_2 site is restricted to divalent metals whose radii are very close to 1 Å; Ca^{2+} conforms most closely to this requirement. Present evidence, based on studies involving electron spin resonance (43,44), circular

dichroism (45), spin—lattice relaxation (46) and electron paramagnetic resonance (47), support the view that the saccharide binding site does not depend
directly on calcium, but that the binding of Ca^{2+} to the S_2 site causes an alteration
in the environment of the S_1 site which in turn is important for the creation of the
saccharide binding site. Each of the subunits of Con A contains one of a trio of
binding sites — the S_1 and S_2 metal binding sites just described, and the sugar
binding site (34,48—51). The location of these sites in relation to the overall structure
of Con A will be considered in more detail in Chapter 3.

Preparations of Con A which have been demetallized by exposure to acid
(pH < 5) lose their dextran-precipitating ability and will not bind to 'Sephadex' in
the absence of metal salts (40). Commercial preparations of Con A which appear
homogeneous on electrophoresis have been found to contain variable proportions
of demetallized forms of Con A which are bound weakly, or not at all, to
'Sephadex' (52,53). Treatment with Mn^{2+} and Ca^{2+} will restore the ability of such
preparations to bind completely to 'Sephadex' as a single component.

2.3 Isoelectric Point

It will be noted from the data shown in Table I that the values reported for the
isoelectric point of Con A range from a value as low as 4.5 to one as high as 8.0.
This descrepancy presumably arises from the fact that preparations of Con A which
appear homogeneous by most other criteria may be resolved by isoelectric focusing
into as many as four (32) and even eight (14) components. The exact reason for the
existence of multiple forms of Con A is not known, but these diffferences in charge
are most likely due to differing amounts of amides in a single gene product,
although the possibility that these differences in charge may be due to multiple
gene products cannot be excluded at the present time.

2.4 Other Properties

Unlike most lectins, Con A is distinguished by the fact that it does not contain
any covalently bound carbohydrate (10,28). It is also unique in the fact that it is
completely devoid of disulphide bonds (10,28,34). Jaffé and Palozzo (54) have
reported that Con A contains approximately 3% lipid material. It is not likely,
however, that this lipid component is covalently attached to the Con A molecule; it
may simply represent lipid material which is strongly bound in a non-covalent
manner to a hydrophobic site which has been postulated to exist in the Con A
molecule (34,51,55,56).

3. FACTORS INFLUENCING THE STRUCTURE AND ACTIVITY OF CON A

3.1 Effect of pH

The effect of pH on the ability of Con A to interact with various polysaccharides, glycosides, glycogen and erythrocytes is summarized in Table II. It is
evident that the pH of maximum activity of those reactions which involve
agglutination or precipitation is between 6 and 7. It is significant to note that this
pH range also favours the existence of the tetrameric form of Con A (32),

TABLE II.   The effect of pH on the activity of Con A

Activity observed	pH range of activity	Optimum pH	Reference
Agglutination of erythrocytes	5.2—7.5		5
Precipitation of glycogen	5—9	6	10
Precipitation of dextran	4—9	7	21
Precipitation of mannan	4—8	6	57
Reaction with p-nitrophenyl-α-D-gluco- and manno-pyranosides[a]	3—8	4—8	58

[a]Measured by ultraviolet difference spectroscopy.

suggesting that those reactions which involve aggregation are mediated by the tetrameric form of Con A (see also 3.2 and 4.2 for further discussion of this point). However, the dimeric form of Con A must still retain the ability to interact with low molecular weight sugar derivatives as shown by the fact that Con A binds the p-nitrophenylglycoside derivatives of α-D-glucose and α-D-mannose at a pH as low as 2.4 (58), where Con A exists exclusively as a dimer. Even yeast mannan is bound to Con A at pH 2.4 despite the fact that precipitation does not occur under these conditions.

In the pH range where Con A manifests maximum activity, the molecule exists largely in the β-conformation and is virtually devoid of α-helical regions (44,59). At higher pH's (60), however, where Con A is inactive, the conformation of Con A undergoes a drastic and irreversible change and approaches a random coil (44,59—61). Under alkaline conditions an aggregation of Con A must also occur as evidenced by the fact that a sedimentation coefficient of 11.35 is observed at pH 9 (61).

3.2 Effect of Temperature

The biological activity of Con A is markedly influenced by temperature. Reducing the temperature from 20° or 37° to 0° or 4° has been found to cause a marked impairment in the ability of Con A to precipitate dextran (21), to agglutinate human erythrocytes (62,63) and to agglutinate transformed cells (64,65). In the latter instance, although little agglutination of transformed cells is observed at the lower temperature, electron microscopic studies have shown that the binding of Con A had occurred (65). The effect of temperature can be readily explained by the observation that lower temperatures favour the existence of the dimer of Con A, whereas higher temperatures favour the tetramer (32,62, 65). Again it appears that the tetramer is necessary for those reactions which involve precipitation of agglutination. If the latter event is thought of as the consequence of a two-step mechanism, the first step may involve a simple monovalent binding process in which both the dimeric and tetrameric forms of Con A are equally effective, and the second step the precipitation or agglutination process itself. This second stage would occur because of the polyvalent property of the tetramer which endows it with the ability to cross-link molecules (as in the precipitation of dextran) or cells via their receptor sites (as in agglutination).

Unrelated to the effects of temperature just described which occur at temperatures below $40°$ are the effects observed at higher temperatures. Temperatures within the range of $30°$ to $50°$ cause a dissociation of the glycogen—Con A complex (67), suggesting that the binding sites in the Con A molecule are heat sensitive and that the binding process involves hydrogen bonding. Temperatures in excess of $50°$ cause denaturation of Con A itself (67); some protection from thermal denaturation, however, was observed in the presence of methyl-α-D-mannopyranoside (68). This would indicate that perhaps the binding of a ligand by Con A stabilizes a particular conformation which is more resistant to thermal disruption.

3.3 Effect of Ionic Strength and Salts

The stability of Con A, as evidenced by the absence of turbidity, is influenced by the ionic strength of its environment (32). An ionic strength (I) of at least 0.3 is necessary to stabilize Con A; at this low ionic strength Con A is largely in the form of a dimer. High ionic strength ($I = 1.0$), on the other hand, stabilizes Con A and favours the formation of the tetramer. Concentrations of NaCl as high as 3.75 M have no effect on the ability of Con A to precipitate dextran when the system is buffered at pH 7.0 with phosphate (21). Precipitate formation, however, is inhibited by anions with inhibition paralleling the lyotropic series: $CNS^- > I^- > Br^- > Cl^- > F^-$. The reason for the unique inhibitory effect of these anions is not readily apparent.

3.4 Effect of Denaturants

The saccharide-binding properties of Con A are lost upon exposure to high concentrations of such denaturing agents as urea, guanidine hydrochloride, formamide and 1,1,3,3-tetramethylurea (40,67,69). Since Con A activity is destroyed by concentrations of urea and guanidine which are less than that required for its dissociation into subunits, this loss in activity is most likely due to a drastic irreversible change in conformation (61), rather than dissociation.

Based on circular dichroism studies, Jirgensons (71) found that the conformation of Con A was not appreciably affected by concentrations of sodium dodecyl-sulphate (SDS) up to 0.65 g per g protein: this stability was attributed to the high content of the β-pleated sheet structure of Con A which has been estimated to be about 50% to 60% (34). It should be noted that this critical concentration of SDS is far below that generally employed for gel electrophoresis wherein the dissociation of Con A into subunits has been observed (36,37).

3.5 Effect of Saccharides on Conformation of Con A

Doyle and coworkers (67) observed that D-glucose induced an ultraviolet difference spectrum in Con A which was interpreted as a reflection of a change in the environment of the aromatic amino acids. Similar spectral changes were noted when Con A interacted with the p-nitrophenylglycosides of glucose and mannose (58). Bessler and Goldstein (50) have taken advantage of this spectral change to measure the ability of various ligands to displace the chromogenic ligands, and were thus able to calculate association constants for a wide variety of saccharides.

Akedo and coworkers (70) observed that when Con A was preincubated with D-mannose, α-methyl-D-glucopyranoside or D-glucose, there was an extensive shift (an increase in net positive charge) in the isoelectric points of the various components revealed by isoelectric focusing (see also 2.3). No such shift was observed with sugars such as galactose which are poorly bound to Con A. The authors believe that the alteration in charge in the presence of certain sugars is a reflection of a conformational change induced by the binding of specific carbohydrates. The exact significance of the conformational change induced by sugars is not known, but it may represent an essential step in the mechanism whereby Con A manifests its unique biological activity.

4. MODIFICATIONS OF CON A

4.1 Limited Proteolytic digestion

Con A is readily inactivated by digestion with pepsin, chymotrypsin, papain and pronase, but is only very slowly attacked by trypsin (5,27). Burger and Noonan (72) reported that Con A which had been exposed to limited digestion with trypsin or chymotrypsin was capable of restoring the growth pattern of transformed cells to that of normal cells. The exact nature of the change in the Con A molecule resulting from this treatment was not ascertained in this study. Cunningham and coworkers (39) found that limited digestion of the intact subunit of Con A with trypsin or chymotrypsin yielded high molecular weight fragments, none of which, however, were identical to the naturally occurring fragments which accompany Con A during its isolation (see 2.1). One of these fragments which had a moelcular weight of 20,000 did retain the ability to bind to 'Sephadex' and may well be identical to the trypsin-modified Con A described earlier by Burger and Noonan (72).

Doyle and coworkers (68) noted that the digestion of Con A by pronase was reduced by almost one half in the presence of methyl-α-D-mannopyranoside. As in the case of thermal denaturation, the ligand may be stabilizing a conformation which is more resistant to proteolytic attack as well as heat.

4.2 Chemical Modification of Amino Groups

The ε-amino groups of the lysine residues of Con A can be extensively acetylated without any significant effect on its reactivity towards dextran, yeast mannan and glycogen, or its ability to cross react with antibodies to the native protein (73). Acetylated Con A also retains its capacity to induce blast formation, albeit with diminished efficiency (74). Acetylated Con A, however, is no longer capable of binding to 'Sephadex' (73). There are conflicting reports (35,73) as to whether acetylation causes a dissociation of the Con A molecule, although Kind and coworkers (75) found that the acetylation of Con A eliminated its ability to stimulate immune spleen lymphocytes to undergo DNA synthesis unless acetylated Con A was coupled to erythrocyte stroma. This would suggest that acetylated Con A may no longer be a tetramer since it lacks the capacity for multipoint attachment necessary for DNA stimulation, a deficiency which is overcome when acetylated Con A is bound to stroma.

The succinylation of the amino groups of Con A does not affect its ability to bind methyl-α-D-glucopyranoside or 'Sephadex', its mutagenic activity or the extent to which it is bound to sheep erythrocytes or splenic lymphocytes (35), Significant, however, is the fact that those biological properties which involve aggregation are very much affected by succinylation. Those properties which are diminished as a result of succinylation include the agglutination of erythrocytes, cap formation by glycoprotein acceptors and the inhibition of cap formation by immunoglobulin acceptors on spleen cells. Accompanying these changes in biological activities is the dissociation of the tetrameric form of Con A to the dimeric state. This alteration in physical behaviour most likely accounts for the fact that succinylated Con A, although still capable of binding saccharides, has lost its capacity to carry out reactions involving aggregation; the dimer, however, because of its reduced valence, can no longer provide the degree of cross-linking necessary for aggregation. The maleylation of Con A appears to cause changes in its biological and physical behaviour similar to those produced by succinylation (76).

Trowbridge and Hilborn (77) have taken advantage of the fact that succinylated Con A does not cause cellular aggregation by testing its effect on transformed cells, to see if its was capable of restoring normal growth in the same fashion as had been reported for trypsinated Con A (72). Unlike trypsinated Con A, however, these workers found that succinylated Con A was incapable of restoring normal growth to transformed cells. This discrepancy thus far remains unresolved.

4.3 Chemical Modification of Carboxyl Groups

Hassing and Goldstein (78) reported that approximately eight carboxyl groups (per tetramer) were not available for titration in the presence of α-methyl-D-mannopyranoside. The modification of carboxyl groups by condensation with glycine methyl ester resulted in a loss of activity which, however, could be partially prevented in the presence of this sugar. Evidence for the involvement of carboxyl groups in the activity of Con A is also provided by potentiometric titration data which indicate that two carboxyl groups are involved in the binding of Ca^{2+} to the S_2 site of Con A (79). The specific carboxyl groups so implicated are discussed in connexion with the molecular structure of Con A (Chapter 3).

4.4 Modification of Tyrosine, Trytophan and Histidine

The acetylation of the amino groups of Con A with acetic anhydride resulted in the concomitant acetylation of about 30% of the phenolic groups of tyrosine, but this had little effect on its ability to precipitate dextran (73). Neither did the acetylation of tyrosyl residues with N-acetylimidazole, to the extent of about 40%, have any effect on the activity of Con A (80). More extensive acetylation of the tyrosyl residues (about 70%) with N-acetylimidazole, however, did produce a marked decrease in the extent to which Con A precipitated glycogen (81). McCubbin and coworkers (82) observed alterations in the aromatic bands of the circular dichroic spectrum of Con A in which the tyrosyl groups had been modified with tetranitromethane, but they were not certain whether these changes were the direct result of chemical modification or an indirect consequence of an alteration in the immediate environment of the aromatic residues. Thus, the precise role that tyrosyl residues play in the activity of Con A must be considered somewhat ambiguous at the present time.

Up to 40% of the tryptophan residues of Con A could be modified with 2-hydroxy-5-nitrobenzyl bromide without any detectable loss in activity (80). More extensive modification with two other reagents which are presumably specific for tryptophan residues in proteins, namely 2-nitrophenyl sulphenyl chloride and N-bromosuccinimide, was accompanied by losses in activity which paralleled the degree to which the tryptophan residues had been modified.

The carboxymethylation of Con A with bromoacetate caused inactivation in a pH-dependent manner and was accompanied by corresponding losses in recoverable histidine (80). The ethoxyformylation of the histidine residues of demetallized Con A resulted in an impairment in its ability to bind Ni^{2+} ions at the transition metal binding (S_1) site (83). When the S_1 site was occupied with Ni^{2+} ions, however, two otherwise ethoxyformylated histidines were rendered inert to chemical modification. The destruction of histidine by photooxidation in the presence of methylene blue causes almost complete inactivation (27). Similarly, the capacity to bind Ni^{2+} ions was also lost when Con A was photooxidized in the presence of rose bengal, and protection of histidines and of the S_1 binding site was observed in the presence of Ni^{2+} ions (83).

5. REFERENCES

1. Jones, D. B. and Johns, C. O. (1916) 'Some proteins from the jack bean, *Canavalia ensiformis*', *J. Biol. Chem.*, **28**, 67–75.
2. Sumner, J. B.(1919) 'The globulins of the jack bean, *Canavalia ensiformis*', *J. Biol. Chem.*, **37**, 137–44.
3. Sumner, J. B. and Howell, S. F. (1935) 'The non-identity of jack bean agglutinin with crystalline urease', *J. Immunol.*, **29**, 133–4.
4. Sumner, J. B., Howell, S. F. and Zeissig, A. (1935) 'Concanavalin A and hemagglutination', *Science*, **82**, 65–6.
5. Sumner, J. B. and Howell, S. F. (1936) 'The identification of the hemagglutinin of the jack bean with concanavalin A', *J. Bacteriol.*, **32**, 227–37.
6. Howell, S. F. (1953) 'Serial crystallization of concanavalin A and concanavalin B', *Fed. Proc.*, **12**, 220–1.
7. Nakamura, S. and Suzuno, R. (1965) 'Crystallization of concanavalin A and B and concanavalin from Japanese jack beans', *Arch. Biochem. Biophys.*, **111**, 499–505.
8. Agrawal, B. B. L. and Goldstein, I. J. (1965) 'Specific binding of concanavalin A to cross-linked gel', *Biochem. J.*, **96**, 23c–25c.
9. Agrawal, B. B. L. and Goldstein, I. J. (1967) 'Protein–carbohydrate interaction. VI. Isolation of concanavalin A by specific adsorption on cross-linked dextran gels', *Biochim. Biophys. Acta*, **147**, 262–71.
10. Olson, M. O. J. and Liener, I. E. (1967) 'Some physical and chemical properties of concanavalin A, the phytohemagglutinin of the jack bean', *Biochemistry*, **6**, 105–11.
11. Agrawal, B. B. L. and Goldstein, I. J. (1972) 'Concanavalin A, the jack bean (*Canavalia ensiformis*) phytohemagglutinin', in *Methods in Enzymology*, Vol. 28, ed. Ginsberg, V. (Academic Press, New York), pp. 314–6.
12. Surolia, A., Prakash, N., Bishayee, S. and Bachawat, B. K. (1973) 'Isolation and comparative physicochemical studies of concanavalin A from *Canavalia ensiformis* and *Canavalia gladiata*', *Indian J. Biochem. Biophys.*, **10**, 145–8.
13. Bishayee, S., Farooqui, A. A. and Bachawat, B. K. (1973) 'Purification of brain lysosomal arylsulfatases by concanavalin A–"Sepharose" column chromatography', *Indian J. Biochem. Biophys.*, **10**, 1–2.

14. Akedo, H., Mori, Y., Tanigaki, Y., Shinkai, K. and Morita, K. (1972) 'Isolation of concanavalin A binding protein from rat erythrocyte stroma', *Biochim. Biophys. Acta*, **271**, 378–87.
15. Hague, D. R. (1974) 'Concanavalin A from three species of *Canavalia*', personal communication.
16. Waxdal, M. J., Wang, J. L., Pflumm, M. N. and Edelman, G. M. (1971) 'Isolation and order of the cyanogen bromide fragments of concanavalin A', *Biochemistry*, **10**, 3343–7.
17. Goldstein, I. J. and So, L. L. (1965) 'Protein–carbohydrate interaction. III. Agar-gel diffusion studies on the interaction of concanavalin A, a lectin isolated from jack bean, with polysaccharides', *Arch. Biochem. Biophys.*, **111**, 407–14.
18. Kabat, E. A. and Mayer, M. M. (1961) *Experimental Immunochemistry* (Thomas & Co., Springfield, Ill.), pp. 114–6.
19. Liener, I. E. (1955) 'The photometric determination of the hemagglutinating activity of soyin and crude soybean extracts', *Arch. Biochem. Biophys.*, **54**, 223–31.
20. Lis, H. and Sharon, N. (1972) 'Soybean (*Glycine max*) agglutinin', in *Methods in Enzymology*, Vol. 28, ed. Ginsberg, V. (Academic Press, New York), pp. 360–8.
21. So, L. L. and Goldstein, I. J. (1967) 'Protein–carbohydrate interaction. IV. Application of the quantitative precipitation method to polysaccharide–concanavalin A interaction', *J. Biol. Chem.*, **242**, 1617–22.
22. Abe, Y., Iwabuchi, M. and Ishii, I. I. (1971) 'Multiple forms in the subunit structure of concanavalin A', *Biochem. Biophys. Res. Commun.*, **45**, 1271–8.
23. Rottmann, W. L., Walther, B. T., Hellerqvist, C. G., Umbreit, J. and Roseman, S. (1974) 'A quantitative assay for concanavalin A-mediated cell agglutination', *J. Biol. Chem.*, **249**, 373–80.
24. Entlicher, G., Kostir, J. V. and Kocourek, J. (1971) 'Studies on phytohemagglutinins. VIII. Isoelectric point and multiplicity of purified concanavalin A', *Biochim. Biophys. Acta*, **236**, 795–7.
25. Sumner, J. B., Gralén, N. and Ericksson-Quensel, I.-B. (1938) 'The molecular weights of canavalin, concanavalin A, and concanavalin B', *J. Biol. Chem.*, **125**, 45–8.
26. Sumner, J. B., Gralén, N. and Eriksson-Quensel, I.-B. (1938) 'The molecular weights of urease, canavalin, concanavalin A and concanavalin B', *Science*, **87**, 395–6.
27. Agrawal, B. B. L. and Goldstein, I. J. (1967) 'Physical and chemical characterization of concanavalin A, the hemagglutinin from jack bean (*Canavalia ensiformis*)', *Biochim. Biophys. Acta*, **133**, 376–9.
28. Agrawal, B. B. L. and Goldstein, I. J. (1968) 'Protein–carbohydrate interaction. VII. Physical and chemical studies on concanavalin A, the hemagglutinin of the jack bean', *Arch. Biochem. Biophys.*, **124**, 218–9.
29. Kalb, A. J. and Lustig, A. (1968) 'The molecular weight of concanavalin A', *Biochim. Biophys. Acta*, **168**, 366–7.
30. Olson, M. O. J. and Liener, I. E. (1967) 'The association and dissociation of concanavalin A, the phytohemagglutinin of the jack bean', *Biochemistry*, **6**, 3801–8.
31. McCubbin, W. D. and Kay, C. M. (1971) 'Molecular weight studies on concanavalin A', *Biochem. Biophys. Res. Commun.*, **44**, 101–9.
32. McKenzie, G. H., Sawyer, W. H. and Nichol, L. W. (1972) 'The molecular weight and stability of concanavalin A', *Biochim. Biophys. Acta*, **263**, 283–93.
33. Csonka, F. A., Murphy, J. C. and Jones, D. B. (1926) 'The isoelectric points of various proteins', *J. Amer. Chem. Soc.*, **48**, 763–8.
34. Edelman, G. M., Cunningham, B. A., Reeke, G. N., Jr., Becker, J. W., Waxdal, M. J. and Wang, J. L. (1972) 'The covalent and three-dimensional structure of concanavalin A', *Proc. Nat. Acad. Sci. U.S.A.*, **69**, 2580–4.
35. Gunther, G. R., Wang, J. L., Yahara, I., Cunningham, B. A. and Edelman, G. M., 'Concanavalin A derivatives with altered biological activities', *Proc. Nat. Acad. Sci. U.S.A.*, **70**, 1012–6.
36. Wang, J. L., Cunningham, B. A. and Edelman, G. M. (1971) 'Unusual fragments in the subunit structure of concanavalin A', *Proc. Nat. Acad. Sci. U.S.A.*, **68**, 1130–4.
37. Edmundson, A. B., Ely, K. R., Sly, D. A., Westholm, F. A., Powers, D. A. and Liener, I. E. (1971) 'Isolation and characterization of concanavalin A polypeptide chains', *Biochemistry*, **10**, 3554–9.

38. McKenzie, G. H. and Sawyer, W. H. (1973) 'The binding properties of dimeric and tetrameric concanavalin A. Binding of ligands to non-interacting monomolecular acceptors', *J. Biol. Chem.*, **248**, 549–56.

39. Cunningham, B. A., Wang, J. L., Pflumm, M. N. and Edelman, G. M. (1972) 'Isolation and proteolytic cleavage of intact subunit of concanavalin A', *Biochemistry*, **11**, 3233–9.

40. Agrawal, B. B. L. and Goldstein, I. J. (1968) 'Protein–carbohydrate interaction. XV. The role of bivalent ions in concanavalin A–polysaccharide interaction', *Can. J. Biochem.*, **46**, 1147–50.

41. Kalb, A. J. and Levitzski, A. (1968) 'Metal binding sites of concanavalin A and their role in binding of α-D-glucopyranoside', *Biochem. J.*, **109**, 669–72.

42. Shoham, M., Kalb, A. J. and Pecht, I. (1973) 'Specificity of metal ion interaction with concanavalin A', *Biochemistry*, **12**, 1914–7.

43. Nicolau, C., Kalb, A. J. and Yariv, J. (1969) 'Electron spin resonance of the transition metal-binding site of concanavalin A', *Biochim. Biophys. Acta.*, **194**, 71–3.

44. Brewer, C. F., Marcus, D. M. and Grollman, G. P. (1974) 'Interactions of saccharide with concanavalin A. Relation between calcium ions and the binding of saccharides to concanavalin A', *J. Biol. Chem.*, **249**, 4614–9.

45. McCubbin, W. D., Oikawa, K. and Kay, C. M. (1971) 'Circular dichroism studies on concanavalin A', *Biochem. Biophys. Res. Commun.*, **43**, 666–74.

46. Barber, B. H. and Carver, J. P. (1973) 'The proton relaxation enhancement properties of concanavalin A', *J. Biol. Chem.*, **248**, 3353–5.

47. Von Goldhammer, E. and Zorn, H. (1974) 'Electron-paramagnetic-resonance study of manganese ions bound to concanavalin A', *Eur. J. Biochem.*, **44**, 195–9.

48. So, L. L. and Goldstein, I. J. (1968) 'Protein–carbohydrate interaction. XX. On the number of binding sites on concanavalin A, the phytohemagglutinin of the jack bean', *Biochim. Biophys. Acta.*, **165**, 398–404.

49. Yariv, J., Kalb, A. J. and Levitzski, A. (1968) 'The interaction of concanavalin A with methyl-α-D-glucopyranoside', *Biochim. Biophys. Acta.*, **165**, 303–5.

50. Bessler, W., Shafer, J. A. and Goldstein, I. J. (1974) 'A spectrophotometric study of the carbohydrate binding site of concanavalin A', *J. Biol. Chem.*, **249**, 2819–22.

51. Hardman, K. D. and Ainsworth, C. F. (1972) 'Structure of concanavalin A at 2.4 Å resolution', *Biochemistry*, **11**, 4910–9.

52. Uchida, T. and Matsumoto, T. (1972) 'Heterogeneity of commercially available concanavalin A with respect to carbohydrate-binding ability', *Biochim. Biophys. Acta.*, **257**, 230–4.

53. Karlstram, B. (1973) 'Evidence of the requirement of a full complement of manganese and calcium ions for optimal binding of carbohydrates to electrophoretically homogeneous concanavalin A', *Biochim. Biophys. Acta.*, **329**, 295–304.

54. Jaffé, W. G. and Polozzo, A. (1971) 'Concanavalin A, a lipo-protein', *Acta Ciént, Venezolana*, **22**, 102–5.

55. Hardman, K. D. and Ainsworth, C. F. (1973) 'Binding of non-polar molecules by crystalline concanavalin A', *Biochemistry*, **12**, 4442–8.

56. Poretz, R. D. and Goldstein, I. J. (1968) 'The hydrophobic character of phenyl glycosides and its relation to the binding of saccharides to concanavalin A', *Arch. Biochem. Biophys.*, **125**, 1034–5.

57. So, L. L. and Goldstein, I. J. (1968) 'Protein–carbohydrate interaction. XIII. The interaction of concanavalin A with α-mannans from a variety of microorganisms', *J. Biol. Chem.*, **243**, 2003–7.

58. Hassing, G. S. and Goldstein, I. J. (1970) 'Ultraviolet difference spectral studies on concanavalin A', *Eur. J. Biochem.*, **16**, 549–56.

59. Zand, R., Agrawal, B. B. L. and Goldstein, I. J. (1971) 'pH-dependent conformational changes of concanavalin A', *Proc. Nat. Acad. Sci. U.S.A.*, **68**, 2173–6.

60. Pflumm, M. N., Wang, J. L. and Edelman, G. M. (1971) 'Conformational changes in concanavalin A', *J. Biol. Chem.*, **246**, 4369–75.

61. Pflumm, M. N. and Beychok, S. (1974) 'Alkali and urea induced conformation changes in concanavalin A', *Biochemistry*, **13**, 4982–6.

62. Gordon, J. A. and Marquardt, M. D. (1974) 'Factors affecting hemagglutination by concanavalin A and soybean hemagglutinin', *Biochim. Biophys. Acta.*, **332**, 136–44.

63. Vlodavsky, I., Inbar, M. and Sachs, L. (1973) 'Membrane changes and adenosine triphosphate contents in normal and malignant transformed cells', *Proc. Nat. Acad. Sci. U.S.A.*, **70**, 1780—4.

64. Inbar, M., Ben-Bassat, H. and Sachs, L. (1971) 'A specific metabolic activity on the surface membrane in malignant cell transformation', *Proc. Nat. Acad. Sci. U.S.A.*, **68**, 2748—52.

65. Huet, C., Lonchampt, M., Huet, M. and Bernadac, A. (1974) 'Temperature effects on the concanavalin A molecule and on concanavalin A binding', *Biochim. Biophys. Acta.*, **365**, 28—39.

66. Sumner, J. B. and Howell, S. F. (1936) 'The role of divalent metals in the reversible inactivation of jack bean hemagglutinin', *J. Biol. Chem.*, **115**, 583—8.

67. Doyle, R. J., Pittz, E. P. and Woodside, E. E. (1968) 'Carbohydrate—protein complex formation. Some factors affecting the interaction of D-glucose and polysaccharides with concanavalin A', *Carbohydrate Res.*, **8**, 89—100.

68. Doyle, R. J., Nicholson, S. K., Gray, R. D. and Glew, R. H. (1973) 'Carbohydrate-induced conformational change in concanavalin A', *Carbohydr. Res.*, **29**, 265—70.

69. Doyle, R. J., Woodside, E. E. and Fishel, C. W. (1968) 'Protein—polyelectrolyte interactions. The concanavalin A precipitin reaction with polyelectrolytes and polysaccharide derivatives', *Biochem. J.*, **106**, 35—40.

70. Akedo, H., Mori, Y., Kobayaski, M. and Okada, M. (1972) 'Changes of isoelectric points of concanavalin A induced by the binding of carbohydrates', *Biochem. Biophys. Res. Commun.*, **49**, 107—13.

71. Jirgensons, B. (1973) 'The sensitivity of some non-helical proteins to structural modification by sodium dodecyl sulfate and its homologues. Circular dichroism studies on Bence—Jones protein, concanavalin A, soybean trypsin inhibitor, and trypsin', *Biochim. Biophys. Acta.*, **328**, 314—22.

72. Burger, M. M. and Noonan, K. D. (1970) 'Restoration of normal growth by covering of agglutinin sites on tumor cell surface', *Nature*, **228**, 512—5.

73. Agrawal, B. B. L., Goldstein, I. J., Hassing, G. S. and So, L. L. (1968) 'Protein—carbohydrate interaction. XVIII. The preparation and properties of acetylated concanavalin A, the hemagglutinin of the jack bean', *Biochemistry*, **7**, 4211—8.

74. Reichert, C. F., Pan, P. M., Mathews, K. P. and Goldstein, I. J. (1973) 'Lectin-induced blast transformation of human lymphocytes', *Nature New Biol.*, **242**, 146—8.

75. Kind, L. S., Wang, H. and Lee, S. H. S. (1973) 'Activation of mouse spleen cells with acetylated Con A coupled to red blood cell stroma', *Proc. Soc. Exp. Biol. Med.*, **142**, 680—2.

76. Young, N. M. (1974) 'Effects of maleylation on the properties of concanavalin A', *Biochim. Biophys. Acta.*, **336**, 46—52.

77. Trowbridge, I. S. and Hilborn, D. A. (1974) 'Effects of succinyl-Con A on the growth of normal and transformed cells', *Nature*, **250**, 304—7.

78. Hassing, G. S., Goldstein, I. J. and Marini, M. (1971) 'The role of protein carboxyl groups in carbohydrate—concanavalin A interaction', *Biochim. biophys. Acta.*, **243**, 90—7.

79. Gachelin, G. and Goldstein, L. (1972) 'Characterization of the amino acids involved in calcium binding in concanavalin A', *FEBS Lett.*, **26**, 264—6.

80. Hassing, G. S. and Goldstein, I. J. (1972) 'Further chemical modification studies on concanavalin A, the carbohydrate binding protein of the jack bean', *Biochim. Biophys. Acta.*, **271**, 388—99.

81. Doyle, R. J. and Roholt, O. A. (1968) 'Tyrosyl involvement in the concanavalin A—polysaccharide precipitin reaction', *Life Sciences*, **7**, 841—6.

82. McCubbin, W. D., Oikawa, K. and Kay, C. M. (1972) 'Circular dichroism studies on chemically modified derivatives of concanavalin A', *FEBS Lett.*, **23**, 100—4.

83. Gachelin, G., Goldstein, L., Hofnung, D. and Kalb, A. J. (1972) 'Implication of two histidines in transition-metal binding in concanavalin A', *Eur. J. Biochem.*, **30**, 155—62.

CHAPTER 3

The Molecular Structure of
Concanavalin A

JOSEPH W. BECKER
BRUCE A. CUNNINGHAM
GEORGE N. REEKE, JR.
JOHN L. WANG
GERALD M. EDELMAN

Lectins interact with specific carbohydrate structures on cell surfaces and provide useful tools for studying alterations in the number, distribution and mobility of the cell surface receptors associated with the control of cell proliferation and cell—cell interactions (1—3). The use of concanavalin A (Con A) in the analysis of the role of cell surface saccharides in these complex phenomena has been particularly fruitful because this lectin can be obtained in homogeneous form (4,5), because it has been characterized in terms of its chemical and three-dimensional structure (6—11), and because its binding specificity has been extensively studied (12—14).

Con A possesses haemagglutinating activity (15) and the ability to precipitate various polysaccharides (12—14). Alterations in the ability of cells to be agglutinated by Con A have been correlated with cellular transformation and escape from density-dependent growth control (16). Particularly important is the fact that Con A is mitogenic for lymphocytes (17,18), and the binding of the lectin to the lymphocyte surface results in modulation of the mobility and distribution of a variety of cell surface molecules (19,20).

Con A is the first lectin and macromolecular mitogen for which the chemical and three-dimensional structure have been established (6—11). This structure, therefore,

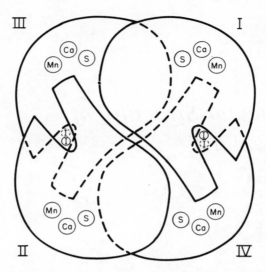

Figure 1. Schematic diagram of the Con A tetramer. The four subunits, indicated in heavy outline, are related by D_2 point symmetry. Binding sites for Ca^{2+}, Mn^{2+}, β-IPGlc, and specific saccharides are indicated by *Ca*, *Mn*, *I*, and *S*, respectively. Roman numerals designate the individual subunits referred to in the text and in Figure 6.

serves as a reference to which those of other lectins can be compared. In addition, correlation of the structural results with biological assays have suggested several mechanisms to explain the various activities of the lectin.

Structural studies have demonstrated a number of features of the molecule that may be significant in understanding its biological activities. At physiological pH, Con A is a tetramer of identical subunits each of which contains 237 amino acid residues (5,21,22), one calcium ion, one manganese ion and one binding site for glucose-, mannose- or fructose- like saccharides (23,24) (Figure 1). The polypeptide chain is folded into a compact ellipsoidal dome, approximately 40 x 39 Å in cross section and 42 Å high (Figure 2). Near the base, the protomer is somewhat thinner, being about 40 x 25 Å at the narrowest point. The molecular surface is generally smooth and uninterrupted with the exception of a large cavity which extends deep into each protomer at the lower right in Figure 2. Individual subunits are paired base-to-base across an axis of twofold symmetry to form ellipsoidal dimers approximately 84 x 40 x 40 Å in size. The dimers are paired back-to-back across additional twofold axes to form tetramers of exact 222 (D_2) symmetry.

1. CHEMICAL STRUCTURE OF CON A

Unlike most other lectins, Con A contains no carbohydrate. The protein also contains no lipid, nucleic acid or other prosthetic groups. The 237 amino acids and

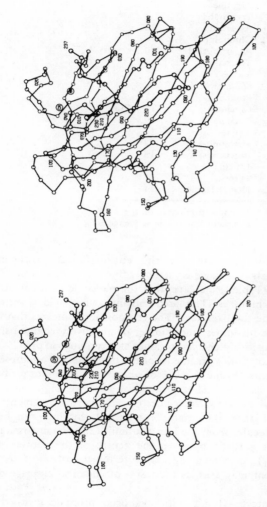

Figure 2. Stereoscopic representation of the polypeptide backbone of a single Con A protomer, oriented approximately the same as subunit I in Figure 1. Circles represent α-carbons and lines, peptide groups. The Ca²⁺, and Mn²⁺ ions are indicated by Ca and Mn, respectively. The site of the natural cleavage between residues 118 and 119 is at the bottom right, and the β-1PGlc binding cavity is just above it at the far right. (Reproduced by permission of the copyright owner from J. Biol. Chem., 250, (1975).)

TABLE I.   Amino acid
composition of Con A[a]

Residue	Number
Lysine	12
Histidine	6
Arginine	6
Tryptophan	4
Asparagine	12
Glutamine	5
Aspartic Acid	20
Threonine	19
Serine	31
Glutamic Acid	7
Proline	11
Glycine	16
Alanine	19
Valine	16
Methionine	2
Isoleucine	15
Leucine	18
Tyrosine	7
Phenylalanine	11

[a]Composition based on the
complete amino acid sequence.

two metal ions therefore account for the entire Con A protomer, which corresponds to the contents of a single crystallographic asymmetric unit.

Amino acid analysis (Table I) indicates that the molecule contains no half-cystines and only two methionines. The predominant amino acid is serine, which accounts for 13.1% of the total residues. The composition indicates that regardless of the ionic state of the histidines, the number of negatively charged side chains exceeds the number of positively charged residues at physiological pH.

In most preparations of Con A fragments of the basic subunit are present (5,21) (Figure 3). These fragments, A_1 and A_2, are of unknown origin, but probably arise as the result of proteolysis (5). Such a proteolytic cleavage might serve as a first step in the catabolism of the lectin. Molecules containing such fragments are more sensitive to denaturation than the intact protein and this fact can be used to separate fragmented molecules from the intact subunit (25). Fragmented molecules appear to be cleaved at a single point in the centre of the peptide chain. An additional fragment (A'_1) containing the amino terminal portion of A_1 has been isolated but this fragment may arise as the result of chemical cleavage during the separation of A_1 and A_2.

Treatment of the intact subunit with cyanogen bromide gives three fragments (26), F_1, F_2 and F_3 (Figure 3), which were used to establish the primary structure of Con A. The Met-Glu bond between fragments F_1 and F_2 is not completely cleaved and this results in the appearance of a fourth fragment ($F_{1,2}$) in small but detectable amounts. Although similar in size, F_1 and F_2 are readily separated by gel filtration because F_2 (and A_2) aggregates in solution.

The complete amino acid sequence of Con A is shown in Figure 4. Protein

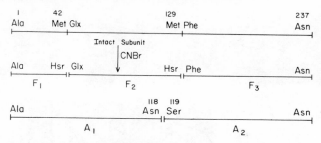

Figure 3. Size and arrangement of the CNBr fragments, F_1, F_2 and F_3 and the naturally occurring fragments A_1 and A_2 of the intact subunit of Con A. (Reproduced by permission of the International Union of Pure and Applied Chemistry from *Pure Appl. Chem.*, **41**, Nos. 1–2 (1975).)

Figure 4. Amino acid sequence of Con A. Residues interacting directly (solid lines) or through water molecules (dashed lines) with the metal ions of Con A are enclosed in boxes. The natural cleavage occurring in some molecules of Con A is indicated by a wavy arrow, and sites of cleavage by CNBr are indicated by solid arrows. (Reproduced by permission of Plenum Press.)

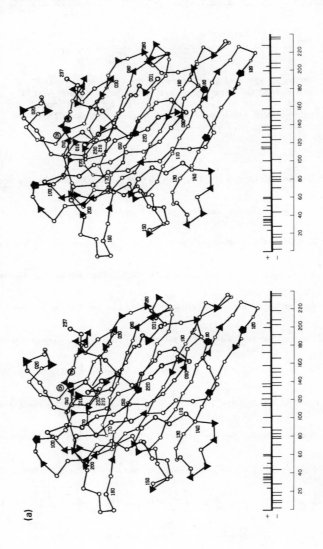

(a)

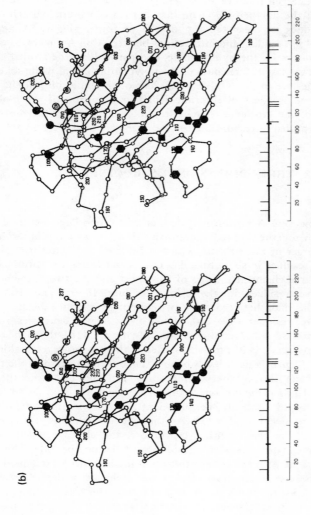

Figure 5. Distribution of charged and aromatic amino acid residues in the Con A protomer. (a) The locations of charged residues are indicated by the following symbols: ▼, acidic side chain; ▲, basic side chain; ⬣, histidine. The line drawing indicates the locations of these residues in the linear sequence: ⌐, acidic side chain; T, basic side chain; ⌐, histidine. (b) The locations of aromatic amino acid residues: ●, tyrosine; ■, tryptophan; ⬟, phenylalanine. In the line drawing: ⌐, tyrosine; +, tryptophan; T, phenylalanine.

ligands for the metals, methionyl residues and the point of cleavage in fragmented molecules are indicated. In the three-dimensional structure the distribution of amino acid side chains within the Con A subunit is like that normally seen in globular proteins, with hydrophobic side chains generally in the interior of the molecule and hydrophilic groups on the surface. In terms of the linear sequence, however, there is an irregularity in the distribution of certain types of amino acid residues (Figure 5). Charged residues are generally more densely distributed in the NH_2-terminal half of the polypeptide chain than in the COOH-terminal half. Many of the charged residues are near the top of the protomer (Figure 5), and are associated with metal ion binding and interactions between Con A dimers, as will be discussed below. Conversely, the hydrophobic, and particularly the aromatic, side chains tend to be in the last 127 residues of the polypeptide chain. While six of the seven tyrosine residues are in the NH_2-terminal half of the molecule, all eleven of the phenylalanines are in the COOH-terminal half in a portion of the molecule toward the bottom of the protomer. The hydrophobic character of this portion of the peptide chain is probably responsible for the aggregation of fragments F_2 and A_2.

2. THREE-DIMENSIONAL STRUCTURE

2.1 Polypeptide Chains

The folding of the Con A protomer is dominated by two large β-structures or pleated sheets, which together contain more than half of the amino acid residues in the molecule. The larger of these sheets, which is depicted in Figures 6, 7 and 9a, comprises about 64 residues in six antiparallel chains and forms the entire back of the molecule. The three chains at the top of this sheet and the two at the bottom comprise consecutive residues in the chain, residues 48—79 and 103—130, respectively, while the chain between these two groupings is from a distant part of the primary structure, residues 188—200. Unlike most large β-structures, such as those of carboxypeptidase A (27) and carbonic anhydrase C (28), the plane of this sheet is relatively untwisted. It is somewhat curled at the top, however, making the chains at the top of Figure 7a lie at levels behind those at the bottom of the figure. The site of the natural cleavage between residues 118 and 119 is on the loop connecting the two chains at the bottom of this sheet. This loop is on the surface of the molecule and would probably be sufficiently accessible for the proposed enzymatic cleavage to occur readily (5). As the structure in this region of the molecule is stabilized by the extensive β-structure, the cleavage might not be expected to have any major effect on the folding of the polypeptide chain. This expectation is consistent with the observations that the presence of fragments does not affect either the crystal structure (10) or the biological activity (29) of the molecule. Once this β-structure is perturbed, however, the cleavage might allow rapid unfolding of the pleated sheet, consistent with circular dichroism measurements and the observation that fragmented molecules are denatured more readily than intact subunits.

The second pleated sheet (Figures 6 and 7b) passes through the centre of the molecule and consists of about 57 residues arranged in 7 antiparallel chains. Unlike the β-structure at the back of the molecule, the plane of this sheet is twisted, with the chains at opposite ends of the sheet nearly perpendicular to each other. As in the other pleated sheet, the three strands of the sheet nearest the top of the protomer are from a continuous region of the primary structure, residues 4—40. The four lower strands, however, are from widely separated parts of the polypeptide chain.

The remainder of the molecule displays no regular secondary structures. To the left of the twisted β-structure, residues 131—168 are arranged in three loosely organized turns. Between these chains and the back β-structure is a large region containing mostly hydrophobic side chains, and no main chain atoms. To the right of the twisted β-structure there is another irregularly folded region containing the NH_2- and COOH-termini and the metal ion binding region.

2.2 Metal Ion Binding Sites

Each Con A protomer contains two metal ions, Mn^{2+} and Ca^{2+}, which are bound near the top of the molecule and about 5 Å from each other (Figures 2 and 8). The top two strands of the twisted β-structure form one side of the metal-binding region, and a loop between two of these chains is folded over the top of the metal ions. Two additional parts of the chain around residues 208 and 228 pass near this site. Both ions are approximately octahedrally coordinated, and both have four protein groups and two water molecules as ligands (Figure 8). The Mn^{2+} ion coordination is quite symmetrical, and the ligands are the side chains of Glu 8, Asp 10, Asp 19 and His 24 and two water molecules. One of the water ligands is apparently joined by a hydrogen-bonding network to the side chain of Ser 34 and the carbonyl oxygen of Val 32, while the other is at the bottom of a shallow solvent-filled depression in the surface of the molecule. This site is capable of binding several transition metal ions other than Mn^{2+}. Chemical (24,30) and crystallographic (31) studies have shown that Co^{2+}, Ni^{2+} and Cd^{2+} can replace Mn^{2+} at this site.

The Ca^{2+} ion is 4.6 Å away from the Mn^{2+} ion, and its ligands are the side chains of Asp 10, Asn 14 and Asp 19, and the carbonyl oxygen of Tyr 12. There are also two water ligands, one hydrogen bonded to the side chain of Asp 208, and the other to the carbonyl oxygen of Arg 228. The coordination of the Ca^{2+} ion is somewhat less symmetrical than that of the Mn^{2+} ion, but it possesses a clearly octahedral organization. This site, unlike the Ca^{2+} binding sites of trypsin and trypsinogen (32), staphlococcal nuclease (33) and thermolysin (34), shows a considerable metal ion specificity. Cd^{2+} can replace Ca^{2+} (30) but studies conducted both in solution (30) and in the crystal (10) have failed to reveal any substitution by Ba^{2+}, Sm^{3+}, Gd^{3+}, Mg^{2+}, Mn^{2+} or Ni^{2+}. Two of the protein ligands, Asp 10 and Asp 19 are shared by both metal ions, so the metal-binding region of Con A can be described as a binuclear complex of two octahedra sharing a common edge.

The structure of this region is consistent with much of the solution chemical data concerning metal ion interactions with Con A. The presence of four carboxylic acid groups near to the metals accounts for the fact that acid treatment removes the metal ions from the protein (24). The observations that Mn^{2+} ion must be bound before Ca^{2+} ion (35) and that Ca^{2+} ion has a strong influence on the rate of Mn^{2+} binding (36) are consistent with the linked nature of the two metal ion binding sites. Binding of Mn^{2+} ion may bring the NH_2-terminal portion of the chain and the

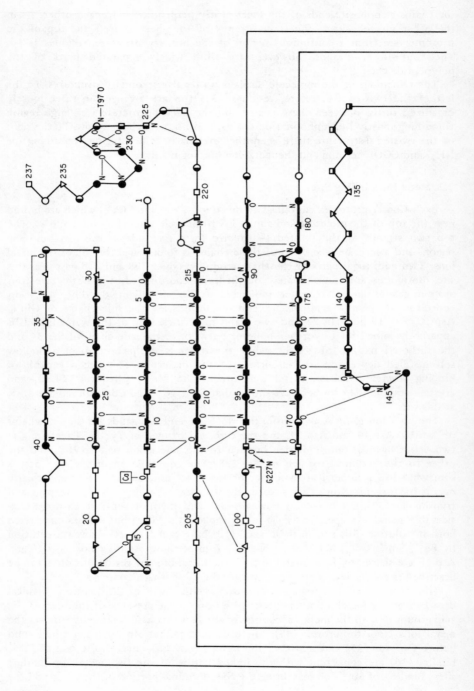

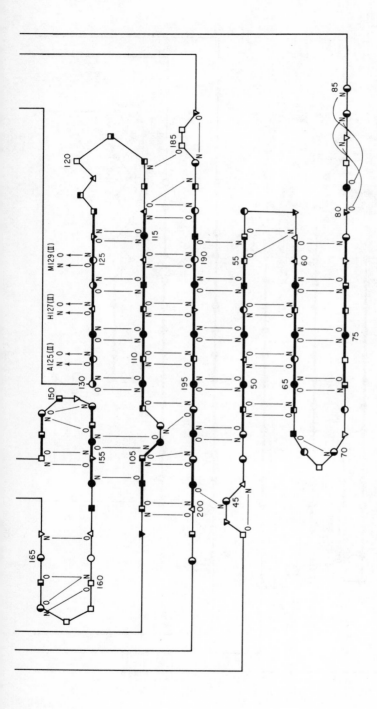

Figure 6. (a) Hydrogen bonding in Con A. Hydrogen bonds involving only main-chain atoms. The solid lines represent the polypeptide backbone and heavy lines represent residues involved in β-structure. Thin lines represent hydrogen bonds. Symbols for amino acid residues are: ○, hydrophobic side chain; △, basic side chain; ▽, acidic side chain; □, uncharged polar side chain. The location of each side chain is depicted by the shading of the symbols: ●, interior; ◐, surface; ○, exposed; ◑, involved in monomer–monomer contact; ◒, involved in dimer–dimer contact. Roman numerals refer to different subunits as indicated in Figure 1. (Reproduced by permission of the copyright owner from *J. Biol. Chem.*, **250**, (1975).)

43

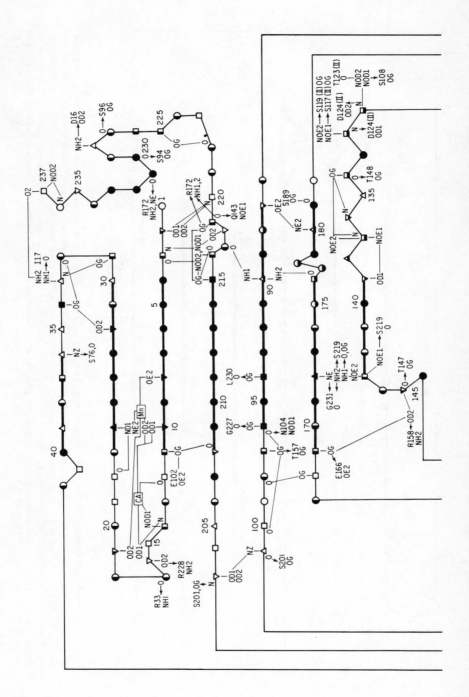

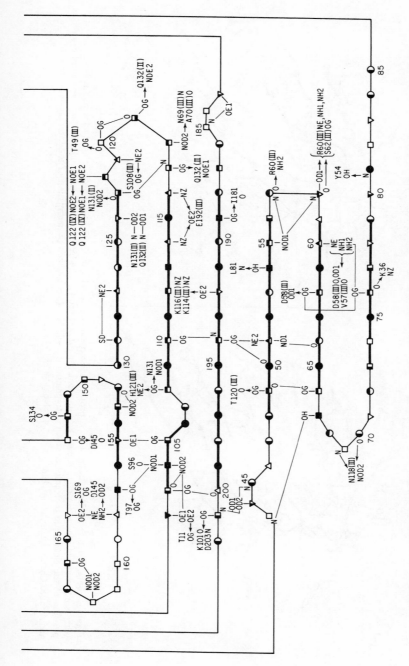

Figure 6. (b) Hydrogen bonding in Con A. Hydrogen bonds involving side-chain atoms. The notation is the same as in Fig. 6(a).

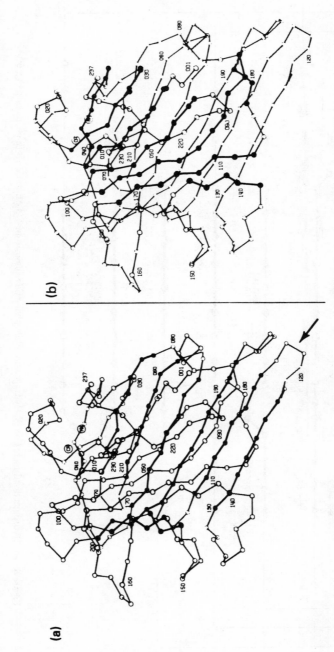

Figure 7. Beta structures in Con A. The Con A protomer is represented as in Figure 2, with residues involved in β-structures emphasized (this is not a stereo illustration). (a) The back (non-twisted) pleated sheet. (b) The front (twisted) pleated sheet. The arrow indicates the position of the cleavage in molecules containing fragments A_1 and A_2. (Reproduced by permission of the International Union of Pure and Applied Chemistry from *Pure Appl. Chem.*, **41**, Nos. 1–2 (1975).)

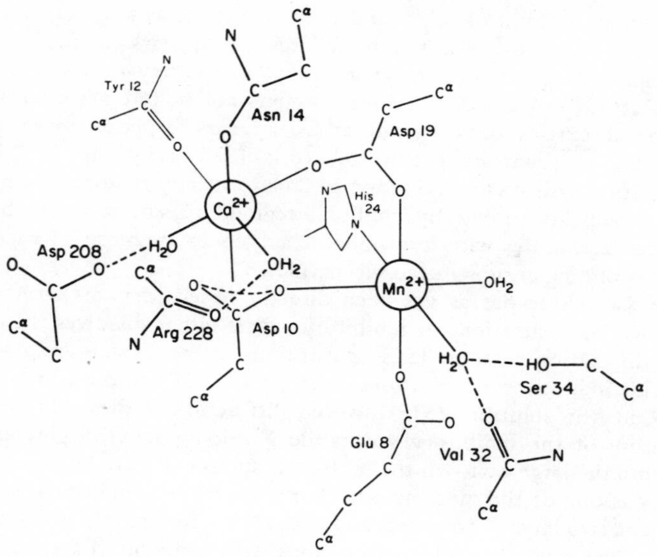

Figure 8. Schematic illustration of the metal-binding region of Con A. Both the Mn^{2+} and Ca^{2+} ions are octahedrally coordinated by four protein ligands and two water molecules. (Reproduced by permission of the New York Academy of Sciences from *Ann. N. Y. Acad. Sci.*, **234**, 369–82 (1974).)

shared ligands into the proper orientation to create the Ca^{2+} binding site. The Ca^{2+} ion, in turn, participates in a network of hydrogen bonds and coordination interactions linking COOH-terminal regions of the chain to the metal-coordinating groups near the NH_2-terminus. The possibility of a conformational change upon metal ion binding is indicated by circular dichroism studies (37,38) which show changes associated with aromatic residues, as well as magnetic resonance studies (36,39) which indicate conformational changes on metal binding and a strong effect of Ca^{2+} on the rate of Mn^{2+} binding. Similarly, a crystallographic study (40) suggests an alteration of the geometrical relationships among the Con A subunits on metal binding, with the general structure of the protomer remaining the same.

2.3 Saccharide Binding Sites

The detailed manner in which Con A interacts with saccharides is of considerable interest, inasmuch as the biological activities of Con A appear to require binding to the cell surface via glycoprotein or polysaccharide receptors. Specific monosaccharides, such as D-glucose and D-mannose, have been shown to be potent inhibitors of the biological activities of Con A, while other closely related carbohydrates, such as D-galactose, are not. Extensive solution studies on the ability of various carbohydrates to inhibit the precipitation of specific polysaccharides (12–14) have indicated that Con A can bind glucosyl and mannosyl residues at the non-reducing termini of oligo- and polysaccharides as well as certain specific non-terminal mannosyl residues. The bound residue appears to interact with the protein through hydrogen bonds involving the anomeric oxygen and the hydroxyl groups at C-2, C-3, C-4 and C-6. Alpha anomers are bound more strongly than β-anomers, and an axial hydroxyl group at C-2 appears to permit more favourable interactions than an equatorial group. Equatorial hydroxyl groups at C-3 and C-4 and an unmodified hydroxymethyl group at C-6 are strictly required for binding. In addition, there appears to be a hydrophobic region adjacent to the carbohydrate-binding site, as saccharides with hydrophobic aglycones are more potent inhibitors than the corresponding unsubstituted saccharides.

No bound saccharide has as yet been directly observed in a Con A crystal, as addition of high concentrations of inhibitory sugars either dissolves the crystals or destroys the diffraction pattern. These disruptions of the crystals are probably due to a saccharide-induced conformational change (37—39) and an increase in the stability of Con A in solution (25). However, diffraction studies (41) have located the iodine atom of the inhibitory saccharide β-(*o*-iodophenyl)-D-glucopyranoside (β-IPGlc) within the large cavity in the molecular surface (Figure 1).

This cavity is one of the most noticeable features of the molecular structure. It lies between the two large β-structures and extends approximately half way through the molecule. The cavity includes two regions with different characteristics. Near the molecular surface the cavity is a narrow, tunnel-like structure, 3.5 to 6 Å wide, and about 7 Å high and lined with predominantly hydrophilic side chains and main-chain groups. In the interior of the molecule the cavity is somewhat larger and is lined almost exclusively with hydrophobic groups. It is in this inner, hydrophobic region that the iodine atom of β-IPGlc is located, surrounded by the side chains of Tyr 54, Leu 81, Leu 85, Val 89, Val 91, Phe 111, Ser 113, Val 179, Ile 181, Phe 191, Phe 212 and Ile 214. The saccharide moiety of β-IPGlc is probably located nearer the molecular surface in the hydrophilic region, where it could form specific hydrogen bonds with the side chains of Tyr 54, Ser 56, Asn 82, Ser 113 and Ser 189 and the main-chain nitrogen and oxygen of Lys 114 and Ile 181.

This site is probably not the specific carbohydrate site observed in solution, as suggested by the observation that crystalline Con A binds compounds such as *o*-iodophenol and β-(*o*-iodophenyl-D-galactopyranoside (β-IPGal) within the hydrophobic subsite (10,42). It is therefore possible that the binding of β-IPGlc may be due to the aglycone rather than the saccharide moiety. In addition, magnetic resonance experiments indicate that the bound carbohydrate is 10 Å from the Mn^{2+} ion (43—46) rather than the 20 Å distance suggested by the diffraction studies. O-iodophenal and β-IPGal are apparently not bound to Con A in solution (29), suggesting that the binding specificity of the cavity may change on crystallization. Recent diffraction studies (47) on a cross-linked Con A-saccharide complex have indicated that the specific saccharide binding site is in a shallow pocket at the top of the molecule, approximately 13 Å from the Mn^{2+} (Fig. 1). Although the specific contact residues cannot as yet be identified, the site is near residues 15, 98, 168, 208, 226, 236 (cf. Fig. 2). This assignment is supported by parallel studies on inactive demetallized Con A, which showed this region to be the only part of Con A where the structure changes significantly on removal of the metals. Detailed characterization of this site must await high resolution crystallographic studies on the Con A-saccharide complex.

2.4 Quaternary Structure

At physiological pH, Con A exists predominantly as the approximately tetrahedral tetramers represented in Figure 1. Dimeric Con A is the dominant form below pH 6 (22) and chemical derivatives of Con A have been prepared which have unmodified carbohydrate binding properties, but which are dimers even under physiological conditions (ref. 48 and Chapter 57). As these derivatives have different biological activities from those of native Con A, the quaternary structure of Con A may play a major role in these activities.

The arrangement of the four protomers of the Con A tetramer (Figure 1) brings the large β-structures at the back of each protomer into contact with one another, and it is primarily non-covalent interactions among these pleated sheets which stabilize the dimeric and tetrameric structures.

The Con A dimer is formed by joining two dome-shaped monomers base-to-base across a twofold rotational axis to form an ellipsoid about 84 x 40 x 40 Å in size (Figure 9). In this assembly, the large β-structures at the back of each protomer are

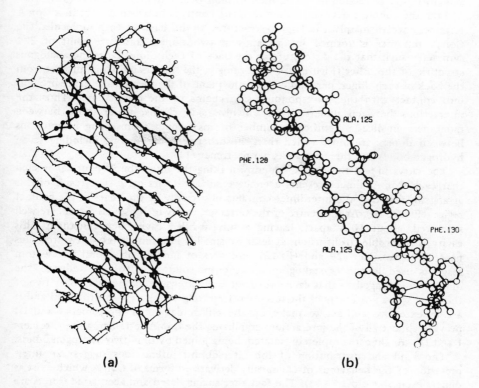

(a)   **(b)**

Figure 9. Structure of the Con A dimer. (a) Two protomers related by a twofold axis at the centre of the figure. Residues belonging to the back β-structures of the two protomers are emphasized, illustrating the formation of a single 12-stranded pleated sheet in the dimer. (b) Details of the interaction between β-structures in the region of contact at the centre of the dimer. Hydrogen bonds between the subunits are represented by thin lines.

joined edge-to-edge in such a manner tat three residues at the bottom of one monomer, Ala 125, His 127 and Met 129 can form β-structure-type hydrogen bonds with the same residues in the other monomer. This extension of the β-structure across the monomer—monomer interface means that the entire back of the Con A dimer is a single pleated sheet made up of 12 chains, six from each protomer. There are many other non-covalent interactions between the two members of a Con A dimer, includin eight additional hydrogen bonds and a total of about 194 atoms in van der Waals contact. As was noted above, the monomers are somewhat constricted in the area of contact, which is about 40 x 25 Å in cross-section, so the dimer has an hour-glass shape when viewed from the side of Figure 9.

The formation of a protein dimer by intersubunit β-structure formation has been observed previously in the structure of insulin (49). It is worth noting that the regions of monomer—monomer contact in these two proteins appear to be quite similar, in that both proteins have a grouping of large uncharged side chains near the intersubunit twofold axis: Phe B24, Phe B25 and Tyr B26 in insulin, and Leu 126, His 127, Phe 128, Met 129 and Phe 130 in Con A. This structural similarity may be related to the insulin-like biological activities of Con A (50).

Like the contacts between monomers, the contacts between dimers in a Con A tetramer involve mainly the large β-structures at the back of each molecule. The Con A tetramer is formed by juxtaposing two dimers across a point of 222 symmetry such that the β-structure at the back of each dimer faces the analogous structure of the other (Figure 10). Accordingly, the side chains which project from the back of each dimer, perpendicular to the plane of the pleated sheet, are brought into contact with the corresponding side chains of the other dimer. Thus, the interactions between dimers mainly involve side chains, while those between monomers involve a significant number of main-chain atoms. The interactions between dimers are more varied than those between monomers, including seven hydrogen bonds and three salt links per protomer (11).

The curvature of the large β-structure making the intersubunit contacts is such that each Con A dimer presents a concave surface to its mate. The result of this arrangement is that the interdimer contacts occur only at the edges of the contact region (Figure 10). At the centre of the tetramer, the two ellipsoidal dimers are well separated, about 10 Å apart, leaving a solvent-filled cavity between them. This cavity is accessible to diffusion, at least by small solute molecules, as two residues near its centre, Met 129 and His 127, are sites of binding of several heavy-atom reagents used in the crystallographic structure solution (6,10). In addition, the dimers are arranged so that each protomer is in extensive contact with only two of the other three protomers of the tetramer. Referring to Figure 1, protomers I and II are in extensive contact to make up the ellipsoidal dimer, protomers I and III provide almost all of the interactions stabilizing the Con A tetramer, and protomers I and IV are almost completely isolated, being joined by only two hydrogen bonds.

The kind and distribution of the intersubunit interactions suggest an interpretation of the behaviour of chemically derivatized forms of Con A which exist as dimers even above pH 7 (48). The fact that succinylated and acetylated Con A are dimeric under conditions where the underivatized protein is a tetramer may be due to disruption of the salt links between lysines 114 and 116 on one dimer and Glu 192 on the other. The interactions between Con A monomers do not include

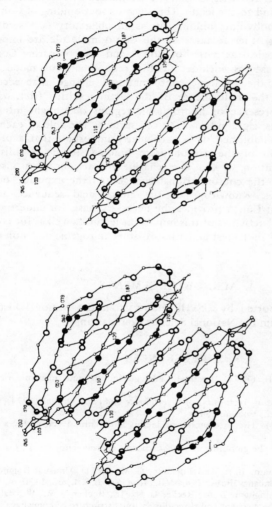

Figure 10. Contacts between the two dimers of a Con A tetramer. Only α-carbons belonging to the back β-structures and connecting loops of two Con A dimers are shown. The large circles indicate interacting residues as follows: ○, hydrophobic interactions; ●, hydrogen bonds; ●, salt links. The point of D_2 symmetry is at the centre of the figure. Residue numbers are shown only for protomer I. Compare with Figure 1. (Reproduced by permission of the copyright owner from *J. Biol. Chem.*, **250**, (1975).)

any which would be expected to be sensitive to this derivatization, consistent with the observation that dimers and not monomers are formed.

3. CONCLUDING REMARKS

Most of the biological activities of Con A are inhibited by low molecular weight saccharides known to bind to the lectin. This suggests that binding of Con A to the cell surface via its carbohydrate binding site is an obligatory first step in these activities. At least four other features of the Con A molecule are important in understanding the biological activities of Con A and for comparing Con A with other lectins. These features include: (1) the requirement of metal ions for saccharide binding; (2) the apparent conformational change that accompanies saccharide binding; (3) the multivalence of the lectin resulting from its subunit structure and (4) the presence of hydrophobic regions that might interact with other types of molecules, e.g. lipids, on the cell surface. The role of each of these factors remains to be established. It is particularly important that a precise and detailed characterization of the carbohydrate binding site be established. The details, at atomic resolution, of Con A-saccharide interactions might also provide clues to the nature of the conformational change and the exposure of possible secondary binding sites. Knowledge of the amino acid sequence and three-dimensional structure of Con A provides the necessary basis for understanding the chemistry and biological activities of this important lectin, as well as for comparison with the structures and biological activities of other mitogenic and non-mitogenic lectins (51—53).

4. ACKNOWLEDGEMENTS

This work was supported by USPHS grants from the National Institutes of Health and by grants from the National Science Foundation.

5. REFERENCES

1. Sharon, N. and Lis, H. (1972) 'Lectins: Cell-agglutinating and sugar-specific proteins', *Science*, 177, 949—59.
2. *Biomedical Perspectives of Agglutinins of Invertebrate and Plant Origins* (1974), ed. Cohen, E., *Ann. N.Y. Acad. Sci.*, Volume 234.
3. Nicolson, G. L. (1974) 'The interaction of lectins with animal cell surfaces', *Int. Rev. Cytol.*, 39, 89—190.
4. Sumner, J. B. (1919) 'The globulins of the jack bean *Canavalia ensiformis*', *J. Biol. Chem.*, 37, 137—42.
5. Wang, J. L., Cunningham, B. A. and Edelman, G. M. (1971) 'Unusual fragments in the subunit structure of concanavalin A', *Proc. Nat. Acad. Sci. U.S.A.*, 68, 1130—4.
6. Edelman, G. M., Cunningham, B. A., Reeke, G. N., Jr., Becker, J. W., Waxdal, M. J. and Wang, J. L. (1972) 'The covalent and three-dimensional structure of concanavalin A', *Proc. Nat. Acad. Sci. U.S.A.*, 69, 2580—4.
7. Hardman, K. D. and Ainsworth, C. F. (1972) 'Structure of concanavalin A at 2.4 Å resolution', *Biochemistry*, 11, 4910—9.
8. Wang, J. L., Cunningham, B. A., Waxdal, M. J. and Edelman, G. M. (1975) 'The covalent and three-dimensional structure of concanavalin A. I. Amino acid sequence of cyanogen bromide fragments F_1 and F_2', *J. Biol. Chem.*, 250, 1490—502.
9. Cunningham, B. A., Wang, J. L., Waxdal, M. J. and Edelman, G. M. (1975) 'The covalent

and three-dimensional structure of concanavalin A. II. Amino acid sequence of cyanogen bromide fragment F_3', J. Biol. Chem., 250, 1503—12.

10. Becker, J. W., Reeke, G. N., Jr., Wang, J. L., Cunningham, B. A. and Edelman, G. M. (1975) 'The covalent and three-dimensional structure of concanavalin A. III. Structure of the monomer and its interactions with metals and saccharides', J. Biol. Chem., 250, 1513—24.

11. Reeke, G. N., Jr., Becker, J. W. and Edelman, G. M. (1975) 'The covalent and three-dimensional structure of concanavalin A. IV. Atomic coordinates, hydrogen bonding, and quaternary structure', J. Biol. Chem., 250, 1525—47.

12. Goldstein, I. J., Hollerman, C. E. and Smith, E. E. (1965) 'Protein-carbohydrate interactions. II. Inhibition studies on the interaction of concanavalin A with polysaccharides', Biochemistry, 4, 876—83.

13. Poretz, R. D. and Goldstein, I. J. (1970) 'An examination of the topography of the saccharide binding sites of concanavalin A and of the forces involved in complexation', Biochemistry, 9, 2890—6.

14. Goldstein, I. J., Reichert, C. M. and Misaki, A. (1974) 'Interaction of concanavalin A with model substrates', Ann. N.Y. Acad. Sci., 234, 283—96.

15. Sumner, J. B. and Howell, S. F. (1935) 'The non-identity of jack bean agglutinin with crystalline urease', J. Immunol., 29, 133—4.

16. Inbar, M. and Sachs, L. (1969) 'Interaction of the carbohydrate-binding protein concanavalin A with normal and transformed cells', Proc. Nat. Acad. Sci. U.S.A., 63, 1418—25.

17. Powell, A. E. and Leon, M. A. (1970) 'Reversible interaction of human lymphocytes with the mitogen concanavalin A', Exp. Cell Res., 62, 315—25.

18. Beckert, W. H. and Sharkey, M. M. (1970) 'Mitogenic activity of the jack bean (Canavalia ensiformis) with rabbit peripheral blood lymphocytes', Int. Arch. Allergy Appl. Immunol., 39, 337—41.

19. Yahara, I. and Edelman, G. M. (1972) 'Restriction of the mobility of lymphocyte immunoglobulin receptors by concanavalin A', Proc. Nat. Acad. Sci. U.S.A., 69, 608—12.

20. Yahara, I. and Edelman, G. M. (1973) 'The effects of concanavalin A on the mobility of lymphocyte surface receptors', Exp. Cell Res., 81, 143—5.

21. Edmundson, A. B., Ely, K. R., Sly, D. A., Westholm, F. A., Powers, D. A. and Liener, I. E. (1971) 'Isolation and characterization of concanavalin A polypeptide chains', Biochemistry, 10, 3554—9.

22. Kalb, A. J. and Lustig, A. (1968) 'The molecular weight of concanavalin A', Biochem. Biophys. Acta, 168, 366—7.

23. Sumner, J. B. and Howell, S. F. (1936) 'The identification of the hemagglutinin of the jack bean with concanavalin A', J. Bacteriol., 32, 227—37.

24. Kalb, A. J. and Levitzki, A. (1968) 'Metal-binding sites of concanavalin A and their role in the binding of α-methyl-D-glucopyranoside', Biochem. J., 109, 669—72.

25. Cunningham, B. A., Wang, J. L., Pflumm, M. N. and Edelman, G. M. (1972) 'Isolation and proteolytic cleavage of the intact subunit of concanavalin A', Biochemistry, 11, 3233—9.

26. Waxdal, M. J., Wang, J. L., Pflumm, M. N. and Edelman, G. M. (1971) 'Isolation and order of the cyanogen bromide fragments of concanavalin A', Biochemistry, 10, 3343—7.

27. Lipscomb, W. N., Reeke, G. N., Jr., Hartsuck, J. A., Quiocho, F. A. and Bethge, P. H. (1970) 'The structure of carboxypeptidase A. VIII. Atomic interpretations at 0.2 nm resolution, a new study of the complex of glycyl-L-tyrosine with CPA, and mechanistic deductions', Phil. Trans. Roy. Soc. Lond. B, 257, 177—214.

28. Liljas, A., Kannan, K. K., Bergstén, P.-C., Waara, I., Fridborg, K., Strandberg, B., Carlbom, U., Järup, L., Lövgren, S. and Petef M. (1972) 'Crystal structure of human carbonic anhydrase C', Nature New Biol., 235, 131—7.

29. Wang, J. L. and Edelman, G. M., unpublished observations.

30. Shoham, M., Kalb, A. J. and Pecht, I. (1973) 'Specificity of metal ion interactions with concanavalin A', Biochemistry, 12, 1914—7.

31. Weinzierl, J. and Kalb, A. J. (1971) 'The transition metal-binding site of concanavalin A at 2.8 Å resolution', Fed. Eur. Biochem. Soc. Lett., 18, 268—70.

32. Darnall, D. W. and Birnbaum, E. R. (1970) 'Rare earth metal ions as probes of calcium ion binding sites in proteins', J. Biol. Chem., 245, 6484—8.

33. Arnone, A., Bier, C. J., Cotton, F. A., Day, V. W., Hazen, E. E., Jr., Richardson, D. C., Richardson, J. S. and Yonath, A. (1971) 'A high resolution structure of an inhibitor complex of the extracellular nuclease of *Staphlococcus aureus*. I. Experimental procedures and chain tracing', *J. Biol. Chem.*, **246**, 2302–16.
34. Colman, P. M., Weaver, L. H. and Matthews, B. W. (1972) 'Rare earths as isomorphous calcium replacements for protein crystallography', *Biochem. Biophys. Res. Commun.*, **46**, 1999–2005.
35. Yariv, J., Kalb, A. J. and Levitzki, A. (1968) 'The interaction of concanavalin A with methyl-α-D-glucopyranoside', *Biochim. Biophys. Acta*, **165**, 303–5.
36. Barber, B. H. and Carver, J. P. (1973) 'The proton relaxation enhancement properties of concanavalin A', *J. Biol. Chem.*, **248**, 3353–5.
37. Pflumm, M. N., Wang, J. L. and Edelman, G. M. (1971) 'Conformational changes in concanavalin A', *J. Biol. Chem.*, **246**, 4369–70.
38. McCubbin, W. D., Oikawa, K. and Kay, C. M. (1971) 'Circular dichroism studies on concanavalin A', *Biochem. Biophys. Res. Commun.*, **43**, 666–74.
39. Grimaldi, J. J. and Sykes, B. D. (1975) 'Concanavalin A: stopped flow nuclear magnetic resonance study of conformational changes induced by Mn^{++}, Ca^{++}, and α-methyl-D-mannopyranoside', *J. Biol. Chem.*, **250**, 1618–24.
40. Jack, A., Weinzierl, J. and Kalb, A. J. (1971) 'An X-ray study of demetallized concanavalin A', *J. Mol. Biol.*, **58**, 389–95.
41. Becker, J. W., Reeke, G. N., Jr. and Edelman, G. M. (1971) 'Location of the saccharide binding site of concanavalin A', *J. Biol. Chem.*, **246**, 6123–5.
42. Hardman, K. D. and Ainsworth, C. F. (1973) 'Binding of nonpolar molecules by crystalline concanavalin A', *Biochemistry*, **12**, 4442–7.
43. Brewer, C. F., Sternlicht, H., Marcus, D. M. and Grollman, A. P. (1973) 'Binding of ^{13}C-enriched α-methyl-D-glucopyranoside to concanavalin A as studied by carbon magnetic resonance', *Proc. Nat. Acad. Sci. U.S.A.*, **70**, 1007–11.
44. Brewer, C. F., Sternlicht, H., Marcus, D. M. and Grollman, A. P. (1973) 'Interactions of saccharides with concanavalin A. Mechanisms of binding of α- and β-methyl-D-glucopyranosides to concanavalin A as determined by ^{13}C nuclear magnetic resonance', *Biochemistry*, **12**, 4448–57.
45. Villafranca, J. J. and Viola, R. E. (1974) 'The use of ^{13}C spin lattice relaxation times to study the interaction of α-methyl-D-glucopyranoside with concanavalin A', *Arch. Biochem. Biophys.*, **160**, 465–8.
46. Alter, G. M. and Magnuson, J. A. (1974) 'Characterization of concanavalin A sugar binding site by ^{19}F nuclear magnetic resonance', *Biochemistry*, **13**, 4038–45.
47. Becker J. W., Reke, G. N., Jr., Cunningham, B. A. and Edelman, G. M. (1976) 'New evidence on the location of the saccharide-binding site of concanavalin A', *Nature*, **259**, 406–9.
48. Gunther, G. R., Wang, J. L., Yahara, I., Cunningham, B. A. and Edelman, G. M. (1973) 'Concanavalin A derivatives with altered biological activities', *Proc. Nat. Acad. Sci. U.S.A.*, **70**, 1012–6.
49. Adams, M. J., Blundell, T. L., Dodson, E. J., Dodson, G. G., Vijayan, M., Baker, E. N., Harding, M. M., Hodgkin, D. C., Rimmer, B. and Sheat, S. (1969) 'Structure of rhombohedral 2 zinc insulin crystals', *Nature*, **224**, 491–5.
50. Cuatrecasas, P. and Tell, G. P. E. (1973) 'Insulin-like activity of concanavalin A and wheat germ agglutinin – direct interactions with insulin receptors', *Proc. Nat. Acad. Sci. U.S.A.*, **70**, 485–9.
51. Wang, J. L., Becker, J. W., Reeke, G. N., Jr. and Edelman, G. M. (1974) 'Favin, a crystalline lectin from *Vicia faba*', *J. Mol. Biol.*, **88**, 259–62.
52. Wei, C. H. (1973) 'Two phytotoxic anti-tumor proteins: Ricin and abrin. Isolation, crystallization, and preliminary X-ray study', *J. Biol. Chem.*, **248**, 3745–7.
53. Wright, C. S., Keith, C., Langridge, R., Nagata, Y. and Burger, M. M. (1974) 'A preliminary crystallographic study of wheat germ agglutinin', *J. Mol. Biol.*, **87**, 843–6.

CHAPTER 4

Carbohydrate Binding Specificity of Concanavalin A

I.J. GOLDSTEIN

The purpose of this chapter is to discuss the carbohydrate binding specificity of concanavalin A (Con A) in terms of its usefulness as a structural probe for the detection and preliminary characterization of specific structural features in simple sugars and the carbohydrate portion of complex carbohydrates, both is solution and as components of cells and subcellular organelles (1).

The dissolution of specific precipitates between Con A and carbohydrate-containing macromolecules, the displacement of polysaccharides and glycoproteins from immobilized Con A and the inhibition or reversal of the biological effects of Con A all depend on a knowledge of the lectins's carbohydrate binding specificity.

1. TECHNIQUES FOR THE STUDY OF SUGAR BINDING SPECIFICITY

The sugar binding specificity of the jack bean lectin has been studied using a wide variety of techniques. These include precipitation with carbohydrate-containing macromolecules (2—28), elution of Con A bound to 'Sephadex' columns by a variety of different sugars (29), quantitative hapten inhibition of precipitation (11,12,15,25—27,30—38), equilibrium dialysis (37,39—41) and ultraviolet difference spectroscopy (42—44).

Although the techniques of equilibrium dialysis and ultraviolet difference spectroscopy are capable of yielding precise thermodynamic parameters such as association constants and free energies of binding of sugar ligands, the method of hapten inhibition of polysaccharide precipitation has been the procedure most widely employed to obtain information regarding the carbohydrate binding specificity of Con A. Through the use of simple sugars and their derivatives as inhibitors (11,12,15,25—27,30—38) of the Con A—polysaccharide (e.g. Con A—dextran) interaction, it has been possible to ascertain the configurational features that a molecule must possess in order to bind Con A.

It is of considerable interest, that despite many years of intensive study employing the most advanced techniques of x-ray crystallographic analysis (45—48) and n.m.r. spectroscopy (49—52), the precise location of the carbohydrate binding site of the jack bean lectin is still a matter of controversy. What has emerged from these studies is the fact that the Con A molecule displays different carbohydrate binding properties in the crystalline state and in solution (44,45,49—53).

2. BINDING OF SIMPLE SUGARS

2.1 *Alpha*-D-mannopyranoside and Derivatives

Studies of the kind described above have indicated that the concanavalin A binding site is most complementary to α-D-mannopyranosyl residues (Figure 1) (11,12,26,27,30,31).

SPECIFICITY OF CONCANAVALIN A

Figure 1. Structure of an α-D-mannopyranosyl unit. Hydroxyl groups which are essential for binding to concanavalin A are underlined.

Although there appear to be loci in the protein binding site for each of the hydroxyl groups of the α-D-mannopyranosyl ring structure (27,33), the hydroxyl groups which appear to be most critical for binding to the lectin are situated at positions C-3, C-4 and C-6 of the pyranosyl ring system. Any modification of these hydroxyl groups, for example a change in configuration or a conversion to deoxy or O-methyl ether derivative, drastically reduces or abolishes completely interaction with the protein (27,31—33).

On the other hand, Con A will tolerate considerable variation at the C-2 position. It has been shown that 2-deoxy-D-glucose (2-deoxy-D-*arabino*hexose) itself binds to concanavalin A indicating that a C-2 hydroxyl group is not required for binding to the protein. However, an *axial* hydroxyl group at the C-2 position (the D-*manno* configuration) enhances the affinity of the saccharide for Con A threefold whereas an *equatorial* C-2 hydroxyl froup (D-*gluco* configuration) diminishes the ability of the resulting sugar to bind to the protein (27,33). It is important to point out that N-acetyl-D-glucosamine also binds to the jack bean lectin, but with approximately one half the affinity of D-glucose, whereas N-acetyl-D-mannosamine does not bind (27,32,33). In all cases it is the α-pyranosyl form of the above sugars which binds to the protein. The specificity of Con A is also reflected in the precipitation of several types of polysaccharides as discussed below.

Subsequent studies indicated that 2-O-substituted α-D-mannopyranosyl units as occur in animal glycoproteins also provide receptor sites for Con A (26,27,53—57).

2.2 Five-membered Sugar Rings

Of great interest is the fact that certain sugars in the 5-membered furanoid ring may also bind to Con A (9,12,58). These furanoid sugars possess configurational features in common with the α-D-hexopyranoses alluded to above. Thus, the disposition of the hydroxyl groups at C-3, C-4 and C-6 of the D-fructofuranosyl residue and the C-2, C-3 and C-5 of the D-arabinofuranosyl residue is similar to the orientation of the hydroxyl groups at C-3, C-4 and C-6 of the D-glucopyranosyl (or D-mannopyranosyl) residue (12). For these reasons α- and β-D-arabinofuranose and α- and β-D-fructofuranose will bind to Con A (Figure 2).

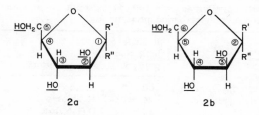

Figure 2. Structural formulae of furanosyl sugars which interact with concanavalin A. Note the disposition of the underlined hydroxyl groups and their relationship to the C-3, C-4 and C-6 hydroxyl groups of the α-D-mannopyranosyl residue in Figure 1. (a) R″ = OH, R′ = H—α-D-arabinofuranose; R′ = OH, R″ = H—β-D-arabinofuranose. (b) R″ = OH, R′ = CH₂OH—α-D-fructofuranose; R′ = OH, R′ = OH, R″ = CH₂OH—β-D-fructofuranose.

These findings have special significance because they suggest that Con A should interact in a specific fashion with polyaccharides or glycoproteins containing sugars with these structural features. Such is indeed the case. For example, Con A will form precipitates with plant β-D-fructans and with bacterial levans (9,12,58). Both of these polysaccharides are branched and contain multiple, terminal non-reducing β-D-fructofuranosyl residues. Similarly, Con A will precipitate arabino-galactans isolated from the cell walls of the genus *Mycobacterium* (17,59). These carbohydrate polymers possess D-arabinofuranosyl end groups (probably α-linked).

2.3 Assessment of the Specificity of Concanavalin A-mediated Biological Activity

In all studies of lectin—substrate interaction, especially those that relate to lectin-potentiated biological effects, it is important to establish whether a specific interaction has occurred between lectin and binding substrate. This is generally determined by attempting to inhibit or reverse the binding phenomenon or biological activity promoted by lectin binding. Advantage is taken of the known sugar binding specificity of the lectin. Several sugars which are known to interact with the lectin combining site and a negative control (non-binding sugar) should be employed. A list of the more common sugars with their relative inhibition potency in the Con A—polysaccharide precipitation system is presented in Table I. The sugar of choice is methyl α-D-mannopyranoside (α-methyl mannoside). If this is not available, D-mannose itself, D-glucose or sucrose may be used. A number of non-inhibitors are also listed in Table I. Lactose or D-galactose are obvious choices. A word of caution: The use of very high concentrations of any sugar could lead to spurious results. Except under very unusual conditions, it is recommended that concentrations greater than 0.1 M should not be used.

TABLE I[a].   List of some readily available inhibitors of concanavalin A—substrate interaction along with the relative potency compared to methyl α-D-mannopyranoside set at 100 (the most potent common inhibitor of Con A—polysaccharide interaction).

Sugar inhibitors	Relative inhibition potency	Non-inhibitors
Methyl α-D-mannopyranoside	100	Lactose
Methyl α-D-glucopyranoside	24	Melibiose
D-Mannose	13.6	D-Galactose
D-Glucose	2.9	D-Ribose
Sucrose	2.6	D-Xylose
		N-Acetyl-D-galactosamine

[a]Taken from 12. 31. 33.
[b]Some common non-inhibitor sugars are also tabulated. p-Notrophenyl-α-D-mannopyranoside is twice as potent an inhibitor as methyl α-D-mannopyranoside (25,33). It may be synthesized or purchased.

In an excellent example of how inhibitor sugars were employed to demonstrate that the primary event in the Con A—stimulated mitogenesis of human lymphocytes involves binding to cell surface carbohydrate-containing residues, Powell and Leon (60) showed that the incorporation of thymidine into DNA was inhibited by sugars in the order methyl α-D-mannopyranoside > methyl α-D-glucopyranoside > D-mannose; this order parallels the known affinities of these sugars for Con A (30—33). Conversely, D-galactose, a non-inhibitor of the Con A system was without effect. Furthermore, inhibition of mitogenesis by α-methyl mannoside was shown to be proportional to its concentration.

Similarly, Kaneko and coworkers (61) showed that the binding of Con A to rat liver and rat ascites hepatoma nuclei was inhibited by D-glucose, D-mannose, methyl α-D-glucopyranoside, sucrose and N-acetyl-D-glucosamine but not by D-galactose, lactose or N-acetyl-D-galactosamine; Noonan and Burger (62) demonstrated inhibition of the binding of acetylated and succinylated Con A to Py3T3 cells by α-methyl mannoside (cf. Agrawal and coworkers, 63). Numerous other examples could also be cited: for example, demonstration of the specificity of binding of ferritin-conjugated concanavalin A to cell membranes (64,65), binding of haemocyanin labelled concanavalin A to various cell types (66) and the inhibition of concanavalin A—specific agglutination of spheroplasts by methyl α-D-gluco-pyranoside (67).

The specificity of precipitin formation between Con A and carbohydrate-containing macromolecules may also be examined in Ouchterlony plates (1,9,11,18,25,68). After washing with saline, the gel diffusion plates may be layered over with solutions of inhibitor (and non-inhibitor) sugars. Precipitin bands formed upon specific interaction of Con A and polysaccharide and glycoprotein should disappear in the presence of inhibitor sugars. Furthermore, the time required for precipitin band disappearance serves as a useful indicator of the affinity of Con A for carbohydrate-containing macromolecules (9,25). Yeast mannan, for example, gives a very vigorous precipitation reaction with Con A, and a longer period of time is required for dissolution of the precipitin band than is required for the disappearance of a Con A—levan or Con A-dextran band of precipitation (9).

A particularly interesting case in which Con A was disqualified from use as a plasma membrane marker involved isolation by centrifugation in isotonic sucrose; rapid dissociation of Con A and plasma membrane occurred in the presence of sucrose, a Con A inhibitor (69).

Another situation in which it is imperative to distinguish between two types of specific interactions involves the use of antibodies against Con A. The jack bean lectin is known to interact with a variety of plasma proteins (16,54,70—73). For this reason it is necessary to add a sugar inhibitor (e.g. α-methyl mannoside) to any system (fluid or gel) in which Con A interacts with antiserum. In this manner, the formation of precipitates between Con A and plasma proteins other than with specific antibodies is prevented (63).

3. PRECIPITATION OF POLYSACCHARIDES AND GLYCOPROTEINS BY CONCANAVALIN A

The specific precipitation reaction between Con A and polysaccharides and glycoproteins has been studied by turbidimetry and by the quantitative precipitin procedure. The latter method involves nitrogen or protein determinations on the precipitated complex. Branched α-D-glucans (glycogens, amylopectins and dextrans) (2—10,13,74), α-D-mannans (yeast mannans and phosphomannans) (2,5,8,9,11,75,76) and β-D-fructans (plant and bacterial levans) (9,12,58), are precipitated by the jack bean phytohaemagglutinin whereas linear α-D-glucans (amylose and a synthetic, linear α-(1→6)-D-glucan), α-D-mannan (synthetic α-(1→6)-D-mannan) and β-D-fructan (inulin) did not form a precipitate with Con A. Certain pneumococcal polysaccharides (28,77,78), lipopolysaccharides (18,78,79), teichoic acids (19—23) and arabinogalactans (17,59) also afford specific precipitates with the protein. These findings in conjunction with the inhibition studies discussed above prompted the formulation of a 'chain-end mechanism' (8,9,11,25,32,53) for Con A—polysaccharide interaction in which it was suggested that Con A (a multivalent protein) interacted with terminal non-reducing α-D-gluco-pyranosyl, α-D-mannopyranosyl or β-D-fructofuranosyl residues at polysaccharide chain ends.

Reports (54—57) that certain glycopeptides which lack terminal non-reducing α-D-gluco- and α-D-mannopyranosyl residues interacted with Con A prompted further hapten inhibition studies on a series of oligosaccharides containing terminal non-reducing α-D-galactopyranosyl units and an investigation of the precipitation of Con A by linear polysaccharides containing *internal* α-D-mannopyranosyl residues (26,27,53). The conclusion which emerged from these studies was that in addition to binding to saccharide chain ends of the appropriate configuration, Con A will also interact with internal 2-O-substituted α-D-mannopyranosyl (and 2-O-substituted α-D-glucopyranosyl) (6) units situated in the interior or core region of glycoproteins and polysaccharides, provided of course, that these determinant sugars are accessible to the protein. *Alpha*-D-mannopyranosyl units substituted with a glycosyl residue at the C-2 position (Figure 3) are of rather common occurrence in animal glycoproteins (80). The fact that these units interact with Con A is consonant with the specificity of the protein for the C-3, C-4 and C-6 hydroxyl groups of D-mannose (27,30—33,53).

Figure 3. Structure of a 2-O-substituted α-D-mannopyranosyl residue. Note the free hydroxyl groups at C-3, C-4 and C-6

A related finding which could be of significant importance with regard to the size or extent of the Con A—binding site is the demonstration that the binding affinity of *sequences* of α-(1→2) -linked α-D-mannopyranosyl units increases as the series is extended, the mannobiose and mannotriose being 5 and 20 times, respectively, more potent inhibitors than α-methyl mannoside (11,26,27,53; cf. 32).

For many years it has been known that crude extracts of the jack bean lectin react with the components of various animal secretions. Recently the precipitation reaction between Con A and certain purified glycoproteins has been studied. For example, it has been shown that Con A will form specific precipitates with several other purified lectins which are glycoproteins (16,27,81,82), with certain myeloma proteins and immunoglobulins (16,71—73,83) and high concentrations of ovalbumin (84).

An exciting new procedure for the isolation of glycopeptides and proteins which interact with Con A depends on their adsorption to Con A linked to insoluble supports, e.g. Con A—Sepharose (see also section V). In this manner a number of enzymes (e.g. brain lysosomal arylsulphatases [85] and α-glutamyltransferase from rat kidney [86]), glycoprotein hormones (human chorionic gonadotrophin, luteinizing hormone and follicle-stimulating hormone [87] , plasma components (α-antitrypsin protein [88], immunoglobulins [83,89—91] and antihaemophilic factor (factor VIII) [92])and other lectins (81) and cell surface receptors from a variety of cells have been isolated (93,94). Polysaccharides that interact with the jack bean phytohaemagglutinin may also be purified by this technique (89,95).

4. INTERACTION OF CONCANAVALIN A WITH GLYCOLIPIDS

The interaction of Con A with glycolipids has not received much attention. This probably derives from the fact that glycolipids are not easily dissolved in water and that animal glycolipids that have been described do not possess receptors for the jack bean lectin. N-Acetyl-D-glucosamine and D-glucose when present in glycolipids occur in the β-anomeric form which will not react with Con A, and D-mannose has not been reported as a constituent of animal glycolipids. However, D-mannose is present as a component of a glycolipid from *M. lysodeikticus,* and the mannosyl–glycerol obtained by alkaline hydrolysis was shown to inhibit the dextran–Con A precipitation reaction to the same extent as methyl α-D-mannopyranoside (32). The α-configuration was, therefore, assigned to the mannosyl–glycerol linkage, a result which was subsequently supported by chemical and enzymatic analysis (96).

5. REFERENCES

1. Goldstein, I. J. (1972) 'Use of concanavalin A for structural studies', in *Methods in Carbohydrate Chemistry,* eds. Whistler, R. L. and BeMiller, J. N. (Academic Press, New York and London), Vol 6, pp. 106–19.
2. Sumner, J. B. and Howell, S. F. (1936) 'The identification of the hemagglutinin of the jack bean with concanavalin A', *J. Bacteriol.* 32, 227–37.
3. Cifonelli, J. A., Montgomery, R. and Smith, F. (1956) The reaction between concanavalin A and glycogen', *J. Am. Chem. Soc.,* 78, 2485–8.
4. Cifonelli, J. A. and Smith, F. (1957) 'The interaction of bacterial polyglucosans with concanavalin A', *J. Am. Chem. Soc.,* 79, 5055–7.
5. Cifonelli, J. A. and Smith, F. (1955) 'Turbidimetric method for the determination of yeast mannan and glycogen', *Anal. Chem.,* 27, 1639–41.
6. Hehre, E. J. (1960) 'Contribution of classical immunology to the development of knowledge of dextran structures', *Bull. Soc. Chim.'Biol.,* 42, 1581–90.
7. Manners, D. J. and Wright, A. (1962) 'α-1,4-Glucosans. Part XIV. The interaction of concanavalin A with glycogens', *J. Chem. Soc.,* 4592–5.
8. Goldstein, I. J., Hollerman, C. E. and Merrick, J. M. (1965) 'Protein–carbohydrate interaction. I. The interaction of polysaccharides with concanavalin A', *Biochim. Biophys. Acta,* 97, 68–76.
9. Goldstein, I. J. and So, L. L. (1965) 'Protein–carbohydrate interaction. III. Agar gel-diffusion studies on the interaction of concanavalin A, a lectin isolated from jack bean, with polysaccharides', *Arch. Biochem. Biophys.,* 111, 407–14.
10. So, L. L. and Goldstein, I. J. (1967) 'Protein–carbohydrate interaction. IV. Application of the quantitative precipitin method to polysaccharide–concanavalin A interaction', *J. Biol. Chem.,* 242, 1617–22.
11. So, L. L and Goldstein, I. J. (1968) 'Protein–carbohydrate interaction. XIII. The interaction of concanavalin A with α-mannans from a variety of microorganisms', *J. Biol. Chem.,* 243, 2003–7.
12. So, L. L. and Goldstein, I. J. (1969) 'Protein–carbohydrate interactions. XXI. Interaction of concanavalin A with D-fructans', *Carbohydr. Res.,* 10, 231–44.
13. Goldstein, I. J., Poretz, R. D., So, L. L. and Yang, Y. (1968) 'Protein–carbohydrate interaction. XVI. The interaction of concanavalin A with dextrans from *L. mesenteroides* B-512-F, *L. mesenteroides* (Birmingham), *Streptococcus bovis,* and a synthetic α-(1 → 6)-D-glucan', *Arch. Biochem. Biophys.,* 127, 787–94.
14. Markowitz, H. (1969) 'Interaction of concanavalin A with polysaccharides of *Histoplasma capsulatum',* *J. Immunol.,* 103, 308–18.
15. Lloyd, K. O., Kabat, E. A. and Beychok, S. (1969) 'Immunochemical studies on blood groups. XLIII. The interaction of blood group substances from various sources with a plant lectin, concanavalin A', *J. Immunol.,* 102, 1354–62.
16. Goldstein, I. J., So, L. L., Yang, Y. and Callies, Q. C. (1969) 'Protein–carbohydrate

interaction. XIX. The interaction of concanavalin A with IgM and the glycoprotein phytòhemagglutinins of the waxbean and the soybean', *J. Immunol*, 103, 695–8.

17. Goldstein, I. J. and Misaki, A. (1970) 'Interaction of concanavalin A with an arabinogalactan from the cell wall of *Mycobacterium bovis*', *J. Bacteriol*, 103, 422–5.

18. Goldstein, I. J. and Staub, A. M. (1970) 'Interaction of concanavalin A with polysaccharides of *Salmonellae*', *Immunochem.*, 7, 315–9.

19. Reeder, W. J. and Ekstedt, R. D. (1971) 'Study of the interaction of concanavalin A with staphylococcal teichoic acids', *J. Immunol.*, 106, 334–40.

20. Archibald, A. R. and Coopes, H. E. (1971) 'The interaction of concanavalin A with teichoic acids and bacterial walls', *Biochem. J.*, 123, 665–7.

21. Tze-Jou, K., Doyle, R. J. and Birdsell, D. C. (1973) 'Mechanism of the formation of concanavalin A–teichoic acid complexes', *Carbohyd. Res.*, 31, 401–4.

22. Birdsell, D. C. and Doyle, R. J. (1973) 'Modification of bacteriophage ϕ25 adsorption to *Bacillus subtilis* by concanavalin A', *J. Bacteriol.*, 113, 198–202.

23. Doyle, R. J., Birdsell, D. C. and Young, F. E. (1973) 'Isolation of the teichoic acid of *Bacillus subtilis* 168 by affinity chromatography', *Prep. Biochem.* 3, 13–18.

24. Svensson, S., Hammarström, S. G. and Kabat, E. A. (1970) 'The effect of borate on polysaccharide–protein and antigen–antibody reactions and its use for the purification and fractionation of cross reacting antibodies', *Immunochem.*, 7, 413–22.

25. Iyer, R. N. and Goldstein, I. J. (1973) 'Quantitative studies on the interaction of concanavalin A, the carbohydrate-binding protein of the jack bean with model carbohydrate–protein conjugates', *Immunochem.*, 10, 313–22.

26. Goldstein, I. J., Reichert, C. M., Misaki, A. and Gorin, P. A. J. (1973) 'An "extension" of the carbohydrate binding specificity of concanavalin A', *Biochim. Biophys. Acta*, 317, 500–4.

27. Goldstein, I. J., Reichert, C. M. and Misaki, A. (1974) 'Interaction of concanavalin A with model substrates', *Ann. N. Y. Acad. Sci.*, 234, 283–96.

28. Goldstein, I. J., Cifonelli, J. A. and Duke, J. (1974) 'Interaction of concanavalin A with the capsular polysaccharide of pneumococcus type XII and the isolation of kojibiose from the polysaccharide', *Biochemistry*, 13, 867–70.

29. Agrawal, B. B. L. and Goldstein, I. J. (1967) 'Protein–carbohydrate interaction. VI. Isolation of concanavalin A by specific adsorption on cross-linked dextran gels', *Biochim. Biophys. Acta*, 147, 262–71.

30. Goldstein, I. J., Hollerman, C. E. and Smith, E. E. (1965) 'Protein–carbohydrate interaction. II. Inhibition studies on the interaction of concanavalin A with polysaccharides', *Biochemistry*, 4, 876–83.

31. So, L. L. and Goldstein, I. J. (1967) 'Protein–carbohydrate interaction. IX. Application of the quantitative hapten inhibition technique to polysaccharide–concanavalin A interaction. Some comments on the forces involved in concanavalin A–polysaccharide interaction', *J. Immunol.*, 99, 158–63.

32. Smith, E. E. and Goldstein, I. J. (1967) 'Protein–carbohydrate interaction. V. Further inhibition studies directed toward defining the stereochemical requirements of the reactive sites of concanavalin A', *Arch. Biochem. Biophys*, 121, 88–95.

33. Poretz, R. D. and Goldstein, I. J. (1970) 'An examination of the topography of the saccharide binding sites of concanavalin A and of the forces involved in complexation', *Biochemistry*, 9, 2890–6.

34. Plow, E. F. and Resnick, H. (1970) 'Effects of hydroxyl compounds on the interaction of concanavalin A with polysaccharides', *Biochim. Biophys. Acta*, 221, 657–61.

35. Poretz, R. D. and Goldstein, I. J. (1971) 'Protein–carbohydrate interaction. On the mode of binding of aromatic moieties to concanavalin A, the phytohemagglutinin of the jack bean', *Biochem. Pharm.*, 20, 2727–39.

36. Duke, J., Goldstein, I. J. and Misaki, A. (1972) '*D*-glucuronic acid: A noninhibitor of the concanavalin A system', *Biochim. Biophys. Acta*, 271, 237–41.

37. Loontiens, F. G., Van Wauwe, J. P., DeGussem, R. and DeBruyne, C. K. (1973) 'Binding of *para*-substituted phenyl glycosides to concanavalin A', *Carbohydr. Res.*, 30, 51–62.

38. Goldstein, I. J., Iyer, R. N., Smith, E. E. and So, L. L. (1967) 'Protein–carbohydrate interaction. X. The interaction of concanavalin A with sophorose and some of its derivatives', *Biochemistry*, 6, 2373–7.

39. So, L. L. and Goldstein, I. J. (1968) 'Protein—carbohydrate interaction. XX. On the number of combining sites on concanavalin A, the phytohemagglutinin of the jack bean', *Biochim. Biophys. Acta*, 165, 398—404.

40. Yariv, J., Kalb, A. J. and Levitzki, A. (1968) 'The interaction of concanavalin A with methyl α-D-glucopyranoside', *Biochim. Biophys. Acta*, 165, 303—5.

41. McKenzie, G. H. and Sawyer, W. H. (1973) 'The binding properties of dimeric and tetrameric concanavalin A. Binding of ligands to no-interating macromolecular acceptors', *J. Biol. Chem.* 248, 549—56.

42. Doyle, R. J., Pittz, E. P. and Woodside, E. E. (1968) 'Carbohydrate—protein complex formation. Some factors affecting the interaction of D -glucose and polysaccharide with concanavalin A', *Carbohydr. Res*, 8, 89—100.

43. Hassing, G. S. and Goldstein, I. J. (1970) 'Ultraviolet difference spectral studies on concanavalin A—carbohydrate interaction', *Eur. J. Biochem.*, 16, 549—56.

44. Bessler, W., Shafer, J. A. and Goldstein, I. J. (1974) 'A spectrophotometric study of the carbohydrate binding site of concanavalin A', *J. Biol. Chem.*, 249, 2819—22.

45. Hardman, K. D. and Ainsworth, C. F. (1973) 'Binding of nonpolar molecules by crystalline concanavalin A', *Biochemistry*, 12, 4442—8.

46. Becker, J. W., Reeke, G. N. Jr. and Edelman, G. M. (1971) 'Location of the saccharide binding site of concanavalin A', *J. Biol. Chem.*, 246, 6123—5.

47. Edelman, G. M., Cunningham, B. A., Reeke, G. N., Jr., Becker, J. W., Waxdal, M. J. and Wang, J. L. (1972) 'The covalent and three-dimensional structure of concanavalin A', *Proc. Nat. Acad. Sci. U.S.A.*, 69, 2580—4.

48. Hardman, K. D., Wood, M. K., Schiffer, M., Edmundson, A. B. and Ainsworth, C. F. (1971) 'Structure of concanavalin A at 4.25 Ångström resolution', *Proc. Nat. Acad. Sci. U.S.A.*, 68 1393—7.

49. Brewer, C. F., Sternlicht, H., Marcus, D. M. and Grollman, A. P. (1973) 'Binding of ^{13}C-enriched α-methyl-D-glucopyranoside to concanavalin A as studied by carbon magnetic resonance', *Proc. Nat. Acad. Sci. U.S.A.*, 70, 1007—11.

50. Brewer, C. F., Sternlicht, H., Marcus, D. M. and Grollman, A. P. (1973) 'Interaction of saccharides with concanavalin A. Mechanism of binding of α- and β-methyl D-glucopyranoside to concanavalin A as determined by ^{13}C nuclear magnetic resonance', *Biochemistry*, 12, 4448—57.

51. Villafranca, J. J. and R. E. Viola (1974) 'The use of ^{13}C spin lattice relaxation times to study the interaction of α-methyl-D-glucopyranoside with concanavalin A', *Arch. Biochem. Biophys.*, 160, 465—8.

52. Alter, G. M. and Magnuson, J. A. (1974) 'Characterization of concanavalin A sugar binding site by ^{19}F nuclear magnetic resonance', *Biochemistry*, 13, 4038—45.

53. Goldstein, I. J. (1975) 'Studies on the combining sites of concanavalin A', in *Concanavalin A*, eds. Chowdhury, T. K. and Weiss, A. K. (Plenum Publishing Corporation, New York), pp. 35—53.

54. Andersen, B. R. (1969) 'Studies on the structure of the carbohydrate moiety of rabbit γG-globulin. I. degradation with glycosidases', *Immunochem.* 6, 739—49.

55. Chase, G. M. W. and Miller, F. (1973) 'Preliminary evidence for the structure of the concanavalin A binding site on human lymphocytes that induce mitogenesis', *Cell. Immunol.*, 6, 132—9.

56. Toyoshima, S., Fukuda, M. and Osawa, T. (1972) 'Chemical nature of the receptor site for various phytomitogens', *Biochemistry*, 11, 4000—5.

57. Kornfeld, R. and Ferris, C. (1975) 'Interaction of immunoglobulin glycopeptides with Concanavalin A', *J. Biol. Chem.*, 250, 2614—9.

58. Lewis, B. A., St. Cyr, M. J. and Smith, F. (1967) 'The constitution of the fructan produced by *Leuconostoc mesenteroides* Strain C', *Carbohydr. Res.* 5, 194—201.

59. Daniel, T. M. and Wisnieski, J. J. (1970) 'The reaction of concanavalin A with mycobacterial culture filtrates', *Amer. Rev. Resp. Disease*, 101, 762—4.

60. Powell, A. E. and Leon, M. A. (1970) 'Reversible interaction of human lymphocytes with the mitogen concanavalin A', *Exp. Cell Res.*, 62, 315—25.

61. Kaneko, I., Satoh, H. and Ukita, T. (1972) 'Binding of radioactively labeled concanavalin A and *Ricinus communis* agglutinin to rat liver and rat ascites hepatoma-nuclei', *Biochem. Biophys. Res. Commun.*, 48, 1504—10.

62. Noonan, K. D. and Burger, M. M. (1973) 'Binding of [^{3}H]concanavalin A to normal and transformed cells', *J. Biol. Chem.*, **248**, 4286—92.
63. Agrawal, B. B. L., Goldstein, I. J., Hassing, G. S. and So, L. L. (1968) 'Protein—carbohydrate interaction. XVIII. The preparation and properties of acetylated concanavalin A, the hemagglutinin of the jack bean', *Biochemistry*, **7**, 4211—8.
64. Nicolson, G. L. (1971) 'Difference in topology of normal and tumor cell membranes shown by different surface distributions of ferritin-conjugated concanavalin A', *Nature (New Biol)*, **233**, 244—6.
65. Nicolson, G. L. and Singer, S. J. (1971) 'Ferritin-conjugated plant agglutinins as specific saccharide stains for electron microscopy: application to saccharides bound to cell membranes', *Proc. Nat. Acad. Sci. U.S.A.*, **68**, 942—5.
66. Smith, S. B. and Revel, J.-P. (1972) 'Mapping of concanavalin A binding sites on the surface of several cell types', *Developmental Biol.*, **27**, 434—41.
67. Maruyama, H. B. (1972) 'Agglutination of bacterial spheroplasts. I. Effects of concanavalin A', *Biochim. Biophys. Acta*, **274**, 499—504.
68. So, L. L. and Goldstein, I. J. (1969) 'Protein—carbohydrate interaction. XVII. The effect of polysaccharide molecular weight on the concanavalin A—polysaccharide precipitation reaction', *J. Immunol.*, **102**, 53—7.
69. Chang, K.-J., Bennett, V. and Cutrecasas, P. (1975) 'Membrane receptors as general markers for plasma membrane isolation procedures. The use of ^{125}I-labeled wheat germ agglutinin, insulin and cholera toxin', *J. Biol. Chem.*, **250**, 488—500.
70. Nakamura, S., Tanaka, K. and Murakawa, S. (1960) 'Specific protein of legumes which reacts with animal proteins', *Nature*, **188**, 144—5.
71. Harris, H. and Robson, E. B. (1963) 'Precipitin reactions between extracts of seeds of *Canavalia ensiformis* (jack bean) and normal and pathological serum proteins', *Vox Sang.*, **8**, 348—55.
72. Nakamura, S., Tominaga, S., Katsuno, A. and Murakawa, S. (1965) 'Specific reaction of concanavalin A with sera of various animals', *Comp. Biochem. Physiol*, **15**, 435—44.
73. Leon, M. A. (1967) 'Concanavalin A reaction with human normal immunoglobulin G and myeloma immunoglobulin G', *Science*, **158**, 1325—6.
74. Smith, E. E., Smith, Z. H. G. and Goldstein, I. J. (1968) 'Protein—carbohydrate interaction. A turbidimetric study of the interaction of concanavalin A with amylopectin and glycogen and some of their enzymic and chemical degradation products', *Biochem. J.*, **107**, 715—24.
75. Robinson, R. and Goldstein, I. J. (1970) 'Protein—carbohydrate interaction. Part XXII. A chemically-synthesized D-mannan and the interaction of some synthetic D-mannans with concanavalin A', *Carbohydr. Res.*, **13**, 425—31.
76. Slodki, M. E., Ward, R. M. and Boundy, J. A. (1973) 'Concanavalin A as a probe of phosphomannan molecular structure', *Biochim. Biophys. Acta*, **304**, 449—56.
77. Cifonelli, J. A., Rebers, P., Perry, M. B. and Jones, J. K. N. (1966) 'The capsular polysaccharide of pneumococcus type XII, SXII', *Biochemistry*, **5**, 3066—72.
78. Goldstein, I. J. and Iyer, R. N. (1966) 'Interaction of concanavalin A, a phytohemagglutinin, with model substrates', *Biochim. Biophys. Acta*, **121**, 197—200.
79. Doyle, R. J., Woodside, E. E. and Fishel, C. W. (1968) 'Protein—polyelectrolyte interactions. The concanavalin A precipitin reaction with polyelectrolytes and polysaccharide derivatives', *Biochem. J.*, **106**, 35—40.
80. Marshal, R. D. (1972) 'Glycoproteins', *Ann. Rev. Biochem.*, **41**, 673—702.
81. Bessler, W. and Goldstein, I. J. (1973) 'Phytohemagglutinin purification: a general method involving affinity and gel chromatography', *FEBS Lett.*, **34**, 58—62.
82. Jaffé, W. G., Levy, A. and González, D. I. (1974) 'Islolation and partial characterization of bean phytohemagglutinins', *Phytochem.*, **13**, 2685—93.
83. Errson, B. and Porath, J. (1974) 'Immobilized lectins as adsorbents for serum proteins', *FEBS Lett.*, **48**, 126—9.
84. Young, N. M. and Leon, M. A. (1974) 'The affinity of concanavalin A and *Lens culinaris* hemagglutinin for glycopeptides', *Biochim. Biophys. Acta*, **365**, 418—24.
85. Bishayee, S., Farooqui, A. A. and Backhawat, B. K. (1973) 'Purification of brain lysosomal arylsulphatases by concanavalin A—'Sepharose' column chromatography', *Ind. J. Biochem. Biophys.*, **10**, 1—2.

86. Takahashi, S., Pollack, J. and Seifter, S. (1974) 'Purification of α-glutamyltransferase of rat kidney by affinity chromatography using concanavalin A conjugated with "Sepharose" 4B', *Biochim. Biophys. Acta,* **371,** 71—5.
87. Dufau, M. L., Tsuruhara, T. and Catt, K. J. (1972) 'Interaction of glycoprotein hormones with agarose—concanavalin A', *Biochim. Biophys. Acta,* **278,** 281—92.
88. Liener, I. E., Garrison, O. R. and Pravda, Z. (1973) 'The purification of human serum α-antitrypsin by affinity chromatography on concanavalin A', *Biochem. Biophys. Res. Commun.,* **51,** 436—43.
89. Donnelly, E. H. and Goldstein, I. J. (1970) 'Glutaraldehyde-insolubilized concanavalin A: an adsorbent for the specific isolation of polysaccharides and glycoproteins', *Biochem. J.,* **118,** 679—80.
90. Aspberg, K. and Porath, J. (1970) 'Group-specific adsorption of glycoproteins', Acta *Chem. Scand.,* **24,** 1839—41.
91. Weinstein, Y., Givol, D. and Strausbauch, (1972) 'The fractionation of immunoglobulins with insolublized concanavalin A', *J. Immunol,* **109,** 1402—4.
92. Kass, L., Ratnoff, O. D. and Leon, M. A. (1969) 'Studies on the purification of antihemophilic factor (factor VIII). I. Precipitation of antihemophilic factor by concanavalin A', *J. Clin. Invest.,* **48,** 351—8.
93. Allan, D., Auger, J. and Crumpton, M. (1972) 'Glycoprotein receptors for concanavalin A isolated from pig lymphocyte plasma membrane by affinity chromatography in sodium deoxycholate', *Nature (New Biol.),* **236.** 23—5.
94. Hunt, R. C., Bullis, C. M. and Brown, J. C. (1975) 'Isolation of a concanavalin A receptor from mouse L cells', *Biochemistry,* **14,** 100—15.
95. Lloyd, K. O. (1970) 'The preparation of two insoluble forms of the phytohemagglutinin, concanavalin A, and their interaction with polysaccharides and glycoproteins', *Arch. Biochem. Biophys.,* **137,** 460—8.
96. Lennarz, W. J. and Talamo, B. (1966) 'The chemical characterization and enzymatic synthesis of mannolipid in *Micrococcus lysodeikticus',* *J. Biol. Chem.,* **241,** 2707—19.

Section II:

MICROSCOPY

CHAPTER 5

Preparation and Use of Fluorescent Concanavalin A Derivatives

LIVIO MALLUCCI

Various optical techniques can be employed to detect concanavalin A (Con A) on biological structures but fluorescent microscopy is the only one that can be applied to the study of living cells.

Concanavalin A binds to the cell surface and fluorescent tracers can be used to locate the lectin receptors. Thus fluorescent Con A techniques represent a suitable tool to investigate the topographic organization of the surface membrane. These techniques have been employed as a structural probe in the study of the surface in a variety of cell types (1–6), during development (7), during the cell cycle (8,9) and in malignant transformation (2,8–11). They have also been used to observe the distribution and the dynamic rearrangement of the lectin-specific determinants in a number of different conditions such as lymphocyte capping, patching, their relationship to other surface antigens and their dependence on association with cytokinetic structures (12–15). They have also been used to examine the significance of some of these events in lectin-induced agglutination (9,10). These studies have been reviewed recently (16) and they have been discussed in some detail in recent works (17,18).

This section describes the preparation and the use of reagents for immuno-fluorescent detection of Con A and methods for the preparation and for direct detection of fluorescent Con A on the cell surface. Basic and well-established techniques will only be mentioned briefly.

For more detailed information on immunofluorescence the reader is advised to refer to the books by Nairn (19) and Holborow (20).

A list of commercially available Con A preparations and reagents is given in the appendix to this book.

1. INDIRECT IMMUNOFLUORESCENT STAINING. THREE-STEP PROCEDURE (SANDWICH TECHNIQUE)

1.1 Principle of the Method

The principle of this method is that of the multiple-layer technique first described by Weller and Coons (21). Cells are treated with Con A, then with antibodies to Con A and are finally stained with fluorescein-labelled antiglobulins (antibodies prepared against the immunoglobulins of the animal species in which the anti-Con A sera have been raised).

1.2 Con A Preparations for Cell Surface Binding

Unless directly prepared from jack bean meal (see appropriate section) Con A from commercial preparations can be readily used. Two- or three-times crystallized Con A or reconstituted lyophilized Con A is dissolved in 1.0 M NaCl (50–100 mg/ml) and kept at 4°. Immediately before use it is diluted further in physiologic phosphate buffer solution (PBS) containing Ca^{2+} and Mg^{2+} adjusted to pH 7.0.

1.3 Antisera to Con A

Antisera to concanavalin A are available commercially and they are acceptable if the purity of the immunogen (Con A) is guaranteed by the manufacturer. When a large stock of antiserum is required it is cheaper and quite simple to make one's own.

1.3.1 Purification of Con A for immunization

Unless the absolute purity is guaranteed, commercial preparations of Con A need to be further purified when employed to raise antibodies. This is easily done by means of affinity chromatography using specific adsorption on and elution from 'Sephadex' dextran beads as described by Agrawal and Goldstein (22). Concanavalin A is dissolved in 1.0 M NaCl and applied on a 'Sephadex' G-50 column equilibrated with the same molar concentration of NaCl. After washing the column with 1.0 M NaCl for about 24 hours, Con A is displaced by a solution of 0.1 M D-glucose in 1.0 M NaCl. An alternative way to displace Con A from the column is to use a buffer of low pH as described by Olson and Liener (23). Fractions with optical adsorbance greater than 0.1 at 280 nm can be pooled and dialysed against 1.0 M NaCl. In the author's laboratory the major peak from the eluted fractions is purified a second time and concentrated by means of vacuum dialysis against 1.0 M NaCl, 2×10^{-4} M $CaCl_2$, 1×10^{-4} M $MnCl_2$, 20 mM phosphate buffer pH 7.0. After purification Con A is tested for specific biological properties such as haemagglutination or agglutination of either transformed or trypsinized normal cells (see appropriate sections), before being stored at $4°$.

1.3.2 Induction of Con A Antibodies

A potent antiserum is the first requirement for success. A simple and efficient method is to inject rabbits at weekly intervals with three or four doses of purified Con A (2–4 mg per dose) homogenized 1:1 with complete Freund's adjuvant and injected partly subcutaneously and partly intraperitoneally. The animals are bled 7–10 days after the last injection. When very small amounts of purified Con A are available (0.5 mg or less) good antibody titres can be obtained in hamsters. The amount of serum collected from each of these animals is not large but intraperitoneal injections of Freund's complete adjuvant and saline (0.5 ml of 1:1 emulsion) repeated daily at the time when the animals are about to be bled will elicit a peritoneal exudate rich in antibodies.

1.3.3 Specificity of Antisera

Each antiserum should be tested for presence of anti-Con A antibodies. A rapid and sensitive method is that of double immunodiffusion (Ouchterlony method) using 0.7% agar gel in 0.25 M NaCl. Serial twofold dilutions of serum (outer wells) are tested against 200–300 µg/ml Con A in 0.25 M NaCl (central well). Con A dilutions are tested in the same way against each antiserum. A single band of precipitation whose size is roughly proportional to the concentration of the reagent should be obtained. Sera with satisfactory anti-body content (visible band at dilution of 1:8 or more) are stored at $-20°$ and are used without separation of the IgG fraction.

1.4 Fluorescein-conjugated Immunoglobulin

This reagent is prepared by conjugating with fluorescein isothiocyanate (FITC) the IgG fraction of an antiserum to the immunoglobulins of the species in which anti-Con A antibodies have been raised. Purified IgG fractions and purified crystalline FITC (isomer I) both ready for conjugation can be purchased commercially. The procedures involved in the various conjugation processes (see 3.2) are based on well-established techniques, but good quality FITC—IgG conjugates are readily available commercially and are quite satisfactory. It is necessary to ascertain from the manufacturer the degree of purity of the conjugate (e.g. gel filtration with 'Sephadex' and fractionation on DEAE—cellulose) as the presence of unreacted fluorescent material can lead to very troublesome non-specific fluorescence.

1.5 Immunofluorescent Staining

1.5.1 General Considerations

Before exposing cells to Con A the following must be remembered: 1) serum should not be present when Con A is added because this lectin binds to some of the serum constituents; 2) cross-linking caused by Con A increases with temperature and is further accentuated by the addition of antibodies, hence according to the aims of the experiment the temperature at which incubation is carried out might have to be preselected for each single step of the immunoassay. This is illustrated by studies on temperature-dependent mobility of Con A receptors on the cell surface (10); 3) variations of pH may affect the distribution of macromolecular components capable of translational movement in the plane of the membrane, therefore it is advisable not to use media based on CO_2-bicarbonate buffers for cell washing and incubation during the assay.

1.5.2. Cell Preparations

Tissue culture monolayers are highly suitable for fluorescent studies. Cells are usually grown on 'flying' coverslips in culture tubes or in petri dishes. When accurate temperature-dependent kinetic studies are required or when rapid changes of temperature are necessary a better (but more elaborate) alternative is the use of vials made from 'Sterilin' universal containers. These are cut to a height of about one inch below the screw cap and a 25 mm round everslip is placed at the bottom of the vial, held in place by a smear of silicone grease and a rubber sleeve which allows observation under an inverted microscope. These vials are easily sterilized with u.v. light and they can be incubated in a water bath.

If the experiment allows it, it is preferable to use cells at low density in order to have a better definition of the surface staining and to reduce possible occurrence of cell clustering (agglutination) by Con A.

Isolated cells in suspension (e.g. lymphocytes) are washed by centrifugation and maintained in suspension throughout the various incubation steps.

1.5.3 Con A Binding

Growth medium is removed and cells are washed three or four times either with PBS plus Ca^{2+} and Mg^{2+} pH 7.0—7.2 or with serum-free medium of the same pH. If cells are on coverslips. these are transferred in a humidified chamber and exposed to Con A diluted in PBS or medium (50—200 μg/ml for most purposes). An incubation period of 15—20 minutes is usually adequate. When cells are kept at low temperature longer periods of incubation cause only moderate increase of Con A binding. Excess of Con A is removed by repeated PBS washings.

1.5.4 Binding of Anti-Con A Antibodies

It is of the utmost importance that non-specific antibodies be removed from the antiserum. This is achieved by preadsorbing the antiserum onto a PBS-washed monolayer of the same cell type that it will be reacted with in the test (0.5—1.0 ml of antiserum/1—5 x 10^7 cells). Even when high dilutions of antiserum are used, cell preadsorption must still be carried out. Cells are exposed to the highest dilution of antiserum giving optimal specific staining for 15 minutes to one hour, depending on the temperature at which the reaction is allowed to occur, and then washed several times in PBS.

1.5.5 Binding of Fluorescent Conjugate

Preparations of FITC—IgG conjugate even when highly purified may sometimes require adsorption with cells. We have found that when using macrophages preadsorption with these cells is necessary to prevent non-specific fluorescence during the immunoassay (2).

Cells are reacted with FITC—IgG (ca. 1:30—1:50 dilutions) applied for periods varying from 15 minutes to one hour according to the temperature of incubation. Lower temperatures require longer exposures.

Specificity of staining is ascertained by absence of fluorescence when Con A is omitted and when the antiserum is substituted with cell-preadsorbed non-immune serum.

2. INDIRECT IMMUNOFLUORESCENT STAINING. TWO-STEP PROCEDURE

Fluorescein isothiocyanate can be conjugated directly to anti-Con A antibodies (see 3.2 for conjugation procedures). FITC-conjugated anti-Con A antibodies are not at present available commercially. Anti-Con A γ-globulins are isolated by means of ammonium sulphate precipitation and then separated by fractionation on a DEAE—cellulose chromatography column. Conjugation of FITC to the IgG fraction is routinely followed by filtration on a G-25 'Sephadex' column in order to remove unreacted fluorescent material. Further purification on DEAE—cellulose and cell preadsorption may be necessary as described above in order to avoid non-specific staining. Apart from the shortened staining procedure, the principles to be followed in this method are as in the previous one. Specificity of staining is ascertained by treating cells that have not been exposed to Con A with cell-preadsorbed FITC—conjugated anti-Con A antibodies or Con A-exposed cells with equally preadsorbed FITC—conjugated non-immune serum. Staining should be absent in either instance.

3. DIRECT FLUORESCENT STAINING

3.1. Principle of the Method

This is a rapid one-step technique where the fluorescent tracer is conjugated to the lectin allowing Con A to be visualized directly after its binding on the cell surface.

3.2 Preparation of FITC—Con A Conjugates

The procedures for the conjugation of FITC to Con A are similar to those employed for the conjugation of FITC to immunoglobulins and they are based on two main methods: direct mixing and dialysis. It is worth remembering that as in the case of globulins purified fluorescein-labelled Con A free of unreacted material can be purchased commercially, a fact of considerable convenience to the occasional user and the less specialized laboratory.

3.2.1 Conjugation by Direct Mixing (adapted and modified from Cebra and Goldstein method [24])

A commercial preparation of lyophilized Con A is dissolved at a concentration of 10 mg/ml in a cold standard salt solution (SS) made of 1.0 M NaCl, 0.25 mM $CaCl_2$, 0.25 mM $MnCl_2$, 0.1 M glucose and brought to pH 7.0 with 10 mM of Na phosphate buffer. Any undissolved material is removed by centrifugation at 10,000 g for twenty minutes. After 4—6 hours at 4° to allow formation of tetramers, the pH of the solution is raised to 9.0 with 1/10 volume of 0.5 M carbonate—bicarbonate buffer. Fluorescein isothiocyanate isomer I dissolved in carbonate—bicarbonate buffer 0.05 M pH 9.0 (15 μg FITC/mg Con A in 0.1 ml volume) is added slowly while the mixture is agitated by means of a magnetic stirrer. The conjugation is carried out for two hours maintaining the temperature at 4° and stirring throughout. The reaction mixture is then left overnight in the cold. The high salt concentration used in this procedure prevents aggregation of Con A that would otherwise occur at high pH. If conjugation is carried out at lower pH the ratio of FITC to Con A is increased (ca. 40—50 μg FITC/mg Con A at pH 8.5). Conjugation can be also carried out maintaining the pH at neutral values by using buffers of lower ionic strength (PBS) in the presence of 0.1 M glucose and using lower concentrations of Con A.

3.2.2 Rapid Conjugation on Inert Powder (adapted and modified from Rinderknecht's method [25])

This is a modification of the direct mixing method that allows considerable shortening of the time required for conjugation by using FITC dispersed on the large reactive surface of an inert material ('Celite'). FITC (10%) on 'Celite' can be purchased commercially or easily prepared by heating under continuous reflux for 30 min 1 g of FITC and 2 g of phosphorus pentachloride in 20 ml of ethylenglycoldimethylether as described by Rinderknecht (26). FITC—conjugated Con A is prepared by mixing 100 mg of commercial FITC on 'Celite' with 200 mg of Con A dissolved in 1—2 ml of standard salt solution (see 3.2.1) whose pH has been raised to 9.0 with 0.5 M carbonate—bicarbonate buffer. The mixture is shaken for 30 min and then centrifuged at 10,000 g for 15 min.

3.2.3 Conjugation by Dialysis (modified from Clark and Shepard method [27])

This procedure for the labelling of Con A has recently been described in detail (15). Con A is dissolved in cold standard salt solution (10 mg/ml) and freed of undissolved material by centrifugation (same steps as in 3.2.1). After sixteen hours in the cold at pH 7.0 the solution is dialysed for six hours against standard salt solution adjusted to pH 9.0—9.2 with 0.1 M carbonate—bicarbonate buffer and then dialysed overnight at 4° against 10 volumes of the same solution containing 300—350 μg/ml of FITC (ca. 0.25 μg/ml Con A inside the dialysis bag).

3.3 Purification of Conjugates

Con A—FITC conjugates are extensively dialysed in standard salt solution pH 7.2 in the absence of glucose and purified by affinity chromatography on a column of 'Sephadex' G—50. The column is washed overnight with cold standard salt solution buffer and the conjugate is then eluted with two subsequent concentrations of glucose in buffer, 0.01 M and 0.1 M. The more strongly bound fraction which is eluted with 0.1 M glucose and contains most of the conjugate is concentrated by means of vacuum dialysis against 1.0 M NaCl, 2×10^{-4} M $CaCl_2$, 1×10^{-4} M $MnCl_2$, 20 mM Na phosphate buffer pH 7.0 and stored at —20° or at 4° under sterile conditions ('Millipore' filtration). It can also be lyophilized and reconstituted in PBS as when supplied by the manufacturer. Assessment of the degree of conjugation is given by the ratio of the optical densities at 495 and 280 nm. Well-conjugated preparations of Con A have 495/280 O.D. ratio of 0.75—1.00. Purification of FITC—Con A by affinity column chromatography cannot be followed, obviously, by preadsorption on cell monolayers.

3.4 Preparation of TRITC—Con A Conjugates

Tetramethylrhodamine isothiocyanate (TRITC) is a fluorescent tracer sometimes used to replace FITC. The procedure for the conjugation of TRITC to Con A are the same as those described in 3.2 except that larger amounts of TRITC are used (60—80 μg TRITC/mg Con A), because the conjugation process is less efficient (de Petris, personal communication). This is probably due to the fact that the substance that is commercially available is partially inactive, consisting of a mixture of crystalline and amorphous TRITC. Purification of conjugate preparations is carried out by affinity chromatography as described in 3.3 and the degree of conjugation is given by the ratio of optical densities at 555 and 280 nm.

3.5 Staining Procedure

The same general consideration discussed in 1.5 apply here, however, this is a single-step procedure where the only cross-linking that can occur, and that the operator may wish to control, is that induced by Con A itself. Washing of the cells, incubation with labelled Con A for 20—30 minutes and further extensive washing is all that is required for staining to be successful. Specificity of staining can be tested by carrying out washings and incubation in the presence of a competing sugar (see appropriate section). For inhibition to be most efficient the sugar must be added to Con A before this is added to the cells. It must be noted, however, that in the author's experience inhibitors are not always totally effective in suppressing fluorescence and that their effectiveness varies with different cell types.

4. FIXATION, MOUNTING, MICROSCOPY

Fixation is by no means a necessary step. Living cells mounted on a drop of PBS or medium (most serum-free media have negligible fluorescence) can manifest changes of the staining pattern during microscopic observation. A microscope with temperature-controlled stage is highly suitable for this type of study. On the other hand, it may be preferable to arrest the movement of Con A receptors either before they interact with Con A or in between the different incubation steps which involve shifting of temperature. This is done by fixation with 2% formaldehyde for 30—60 minutes or preferably with 2% glutaraldehyde in phosphate buffer for 10—30 minutes. The fixation is most efficient with glutaraldehyde, but there is some autofluorescence that is negligible after formaldehyde fixation. A detailed study on the effect of fixation with both formaldehyde and glutaraldehyde in membrane immunofluorescent tests has been reported recently (28). Fixed cells can be mounted in a drop of fixative or better still in phosphate-buffered glycerol that allows storage in a deep freezer and preservation of fluorescence. Microscopic observations of the fluorescent-stained preparations are carried out using wavelength in the blue light range preferably by epi-illumination.

5. COMPARISON OF METHODS

5.1 FITC versus TRITC

A first distinction between methods can be made on the basis of the emission spectra of the labelling tracers and of their resistance to fading under excitation. The orange—yellow emission of TRITC is slower to fade and can create a better contrast with the blue—green autofluorescence of living cells than can the green fluorescence of FITC. Fluorescein, on the other hand, is a more sensitive and versatile tracer for Con A studies as it gives a very strong fluorescence. It allows a better evaluation of Con A receptor sites when the degree of Con A binding is limited (as in the case of fixed cells or cells kept at low temperature) and when the experiment demands the use of low concentrations of Con A.

5.2 Direct versus Indirect Method

Of the methods described the one based on direct fluorescence is by far the simplest but it is at the same time the least sensitive. Of the two indirect immuno-stainings both of high sensitivity the sandwich technique is estimated to be several-fold more sensitive (29). The choice of the method will very much depend upon the aims of the experiment. Translational movement of Con A receptors in lymphocytes has been optimally investigated by the direct staining technique. Despite its poorer sensitivity fluorescein-labelled Con A has proved to be the most suitable reagent for the study of lymphocyte capping inhibition of capping and related phenomena (12—15). On the other hand, indirect methods are best used when detection of Con A sites is the primary object (especially if they are present at low density on the membrane) and when it is desired to investigate translational movement of clustered receptor sites (patching) under conditions of increased cross-lining by addition of antibody molecules. Clustering of the fluorescent probe may also increase the efficiency of visual recognition by increasing the number of multiple emissions from smaller surface regions.

6. REFERENCES

1. Smith, C. W. and Hollers, J. C. (1970) 'The pattern of binding of fluorescent labelled concanavalin A to the motile lymphocyte', *J. Ret. Endothel. Soc.*, **8**, 458—64.
2. Mallucci, L. (1971) 'Binding of concanavalin A to normal and transformed cells as detected by immunofluorescence', *Nature (New Biol.)*, **233**, 241—4.
3. Edelman, G. M. and Millette, C. F. (1971) 'Molecular probes of spermatozoa structures', *Proc. Nat. Acad. Sci. U.S.A.*, **68**, 2436—40.
4. Inbar, M., Huet, C., Oseroff, A. R., Ben-Bassat, H. and Sachs, L. (1973) 'Inhibition of lectin agglutinability by fixation of the cell surface membrane', *Biochem. Biophys. Acta.*, **311**, 594—9.
5. Comoglio, P. M. and Filogamo, G. (1973) 'Plasma membrane fluidity and surface motility of mouse C-1300 neuroblastoma cells', *J. Cell Sci.*, **13**, 415—20.
6. Steinemann, A. and Stryer, L. (1973) 'Accessibility of the carbohydrate moiety to rhodopsin', *Biochemistry*, **12**, 1499—502.
7. Monroy, A., Ortolani, G., O'Dell, D. and Millonig, G. (1973) 'Binding of concanavalin A to the surface of unfertilized and fertilized eggs:, *Nature*, **242**, 409—10.
8. Shohan, J. and Sachs, L. (1972) 'Differences in the binding of fluorescent concanavalin A to the surface membrane of normal and transformed cells', *Proc. Nat. Acad. Sci. U.S.A.*, **69**, 2479—82.
9. Imbar, M. and Sachs, L. (1973) 'Mobility of carbohydrate containing sites on the surface membrane in relation to the control of cell growth', *FEBS Lett.*, **32**, 124—8.
10. Nicolson, G. L. (1973) 'Temperature-dependent mobility of concanavalin A sites on tumour cell surfaces', *Nature (New Biol.)*, **243**, 218—20.
11. Robbins, J. C. and Nicolson, G. L. (1975) 'Surfaces of normal and transformed cells', in *Cancer: A Comprehensive Treatise*, Vol. III, section III, Chapter 21. ed. Becker, F. F. (Plenum Press, New York).
12. Yahaza, I. and Edelman, G. M. (1973) 'The effects of concanavalin A on the mobility of lymphocyte surface receptors', *Exp. Cell Res.*, **81**, 143—55.
13. De Petris, S. (1974) 'Inhibition of capping by cytochalasin B, vinblastine and colchicine', *Nature*, **250**, 54—6.
14. Loor, F. (1974) 'Binding and redistribution of lectins on lymphocyte membrane', *Eur. J. Immunol.*, **4**, 210—20.
15. De Petris, S. (1975) 'Concanavalin A receptors, immunoglobulins and θ antigen of the

lymphocyte surface. Interaction with concanavalin A and with cytoplasmic structures', *J. Cell Biol.*, **65**, 123–46.

16. Nicolson, G. L. (1974) 'The interaction of lectins with animal cell surfaces', in *International Review of Cytology*, Vol. 30, eds. Bourne, G. H. and Danielli, J. F. (Acad. Press, New York & London).

17. Nicolson, G. L. (1974) 'Factors influencing the dynamic display of lectin binding sites on normal and transformed cell surfaces', in *Control of Proliferation in Animal Cells*, eds. Clarkson, B. and Baserga, R. (Cold Spring Harbor and Laboratory, New York), pp. 251–70. pp. 251–70.

18. Raff, M. C., De Petris, S. and Mallucci, L. (1974) 'Distribution and mobility of membrane macromolecules: ligand-induced redistribution of concanavalin A receptors and its relationship to cell agglutination' in *Control of Proliferation in Animal Cells,* eds. Clarkson, B. and Baserga, R. (Cold prng Harbor Laboratory, New York), pp. 271–81.

19. *Fluorescent Protein Tracing* (1969), ed. Nairn, R. C. (E. & S. Livingston Ltd., Edinburgh and London).

20. *Standardization in Immunofluorescence* (1970), ed. Holborow, E. J. (Blackwell Scientific Publications, Oxford and Edinburgh).

21. Weller, T. S. and Coons, A. H. (1954) 'Fluorescent antibody studies with agents of varicella and herpes zoster propagated *in vitro, Proc. Soc. Exp. Biol. (N.Y.),* **86**, 789–94.

22. Agrawal, B. B. L. and Goldstein, I. J. (1967) 'Protein–carbohydrate interaction. VI. Isolation of concanavalin A by specific adsorption on cross-linked dextran gels', *Biochim. Biophys. Acta*, **147**, 262–71.

23. Olson, M. O. J. and Liener, I. E. (1967) 'Some physical and chemical properties of concanavalin A, the phytohaemagglutinin of the jack bean', *Biochemistry*, **6**, 105–11.

24. Cebra J. J. and Goldstein, G. (1965) 'Chromatographic purification of tetramethyl-rhodamine-immune globulin conjugates and their use in the cellular localization of rabbit γ-globulin polypeptide chorius', *J. Immunol.*, **95**, 230–45.

25. Rinderknecht, H. (1962) 'Ultra-rapid fluorescent labelling of proteins', *Nature*, **193**, 167.

26. Rinderknecht, H. (1960) 'A new technique for fluorescent labelling of proteins', *Experientia*, **16**, 430–1.

27. Clark, H. F. and Shepard, C. C. (1963) 'A dialysis technique for preparing fluorescent antibody', *Virology*, **20**, 642–4.

28. Biberfeld, P., Biberfeld, G., Molnar, Z. and Fagraeus, A. (1974) 'Fixation of cell-bound antibody in the membrane immunofluorescent test', *J. Immunol. Methods*, **4**, 135–48.

29. Pressman, D., Yagi, Y. and Hiramoto, R. (1958) 'A comparison of fluorescein and I^{131} as labels for determining the *in vivo* localization of anti-tissue antibodies', *Int. Arch. Allergy*, **12**, 125–36.

CHAPTER 6

Detection of Microprojections on Lymphocytes by Fluorescence Microscopy after Concanavalin A Binding

FRANCIS LOOR

Optical microscopy has a maximum resolution of about 100—200 nm which is not very much lower than the resolution of scanning electron microscopy (about 20 nm) (1). Though it does not provide as nice a picture of microvilli-bearing cells as scanning electron microscopy, membrane immunofluorescence may be used to detect lymphocyte plasma membrane microextensions of various sizes on viable cells in suspension. Microvilli are small extensions of the cell membrane having a uniform diameter of about 100 nm and variable lengths (2). Various ligands which bind to the cell membrane are found equally distributed on the whole cell surface including the microvilli (3). When the ligands are labelled with fluorochromes these microvilli then fluoresce brightly upon appropriate u.v. illumination and can be detected though being very thin (4). Fluorescence microscopy has one important advantage over scanning electron microscopy: it is that one can work with viable cells kept in suspension and follow the fate of microvilli in response to various stimuli or microenvironmental changes and study their nature, their function, their modulation, etc. Fluorescent concanavalin A can be used directly as a cell membrane ligand allowing visualization of microvilli on the lymphocytes by fluorescence microscopy. Con A can also be used as an agent which modulates the expression of lymphocyte microvilli that would be detected by use of another fluorescent membrane ligand, for instance a fluorescent antibody bound to the immunoglobulins of the lymphocyte membrane.

Not all the essential factors which influence the expression of microextensions on viable lymphocytes in suspension are yet under control. For instance with some batches of medium the cells do not show microprojections, but on the contrary there is pynocytosis of part of the membrane-bound ligand, which is then found into small vesicles under the plasma membrane (4). The actual reasons for such differences are still unknown. This fairly recent methodology will thus probably require much improvement. Its present principle and some sample experiments are described hereafter, and a complete report has been made elsewhere (4). It should still be pointed out that the detection of microvilli by fluorescence microscopy has recently been obtained also by Dr. de Petris (personal communication) by labelling, with fluorescent Con A, cells that had been prefixed with glutaraldehyde. This shows that microvilli were present before Con A binding and not — or not only — induced by the binding of Con A to the cell surface.

1. MATERIALS AND EQUIPMENT

1.1 Fluorochrome Labelling of Con A

Concanavalin (Miles-Yeda, Rehovot, Israel or Sigma, St. Louis, Missouri, U.S.A.) is further purified by affinity chromatography on 'Sephadex' G—50 (Pharmacia, Uppsala, Sweden), according to the method of Agrawal and Goldstein (5). Fluorescein isothiocyanate (FITC) and tetramethylrhodamine isothiocyanate (TRITC) (purchased from Baltimore Biological Laboratories, Becton Dickinson and Co., Cockeysville, Maryland), are allowed to react with the purified Con A in solution (5—10 mg of Con A/ml 9‰ NaCl) at the following ratios: 12.5 μg FITC/mg Con A or 30 μg TRITC/mg Con A. The reaction is continued for 1 h at 0°C at pH 9.0 (adjusted by addition of 0.1—0.01 N NaOH and maintained by 0.1—0.01 M carbonate—bicarbonate), in the presence of 0.1 M glucose in order to protect the Con A-active sites. The excess of unbound fluorochrome and the glucose are removed by continuous flow dialysis. The fluorescent Con A is fractionated again by affinity chromatography on 'Sephadex' G-50 and three peaks are obtained. A first fraction of fluorescent Con A passes through the column with the void volume; when checked on cells, it does not show any detectable binding by fluorescence microscopy and is probably mostly inactivated. A second fraction of fluorescent Con A is only delayed in the 'Sephadex' column and is eluted without use of glucose; this fraction can label cell membrane. A third fraction, considered a fully active fluorescent Con A, firmly binds to the 'Sephadex' and can only be eluted after addition of glucose (0.1 M); this fraction is also able to stain cell membrane after removal of the glucose by continuous flow dialysis. Fluorochrome-conjugated Con A is kept lyophylized. Before use it is dissolved in physiological saline (9‰ NaCl) supplemented with 1 mM $MnCl_2$ and 1 mm $CaCl_2$. It can be kept as a concentrated Con A solution (> 10 mg/ml) for a few weeks at $+4^\circ$C with no detectable loss of cell membrane binding activity.

The use of FITC—Con A is recommended if fluorescence microscopy has to be performed with ordinary fluorescence microscopes, which usually have a poorer sensitivity for TRITC; both FITC—and TRITC—Con A can be used with modern fluorescence microscopes equipped with an epi-illuminator (Ploemopak).

1.2 Preparation of Cell Suspensions

The expression of microvilli on lymphocytes strongly depends on the temperature and is very much decreased when the lymphocytes are kept at or near 0°C. After such cooling, the lymphocytes do not immediately recover their ability to express microvilli, which now require incubation for a few hours in culture medium at 37°C. Thus one has to avoid preparing the cell suspension by the usual methods performed at or near 0°C. Suspension of lymphocytes can be obtained at room temperature or at 37°C with good viability, provided the cells are prepared in a good culture medium, e.g. RPMI 1640, (Microbiological Associates Inc., Bethesda) equilibrated to pH 7.2 by 0.03 M HEPES (N-2-hydroxyethylpiper-azone-N-2-ethanesulphonic acid, Calbiochem., Los Angeles) and diluted NaOH, and containing 0.5% BSA (bovine serum albumin, Sigma, St. Louis, Missouri). Apart from the fact that the cells are never cooled down, the procedure is identical to that described in another section (6).

1.3 Microscopy

Not all optical systems, especially high-power lens systems, allow detection of microvilli. We used a Wild-Leitz Orthoplan microscope equipped with the Ploemopak 2.1 vertical illuminator or a Wild-Leitz Epivert microscope equipped with the Ploemopak 2.2. The filter combination for optimal detection of FITC and TRITC are those recommended by Ploem (7). In our opinion, the lymphocyte microvilli are best detected by fluorescence microscopy using a Wild-Leitz achromate oil immersion objective 40X/NA 1.30 and H6.3 x M eyepieces, or a Wild-Leitz oil immersion objective 63 x /1.30 and the new wide-angle low-power 4 x eyepieces (7). These lens combinations give a good depth of field which allows one to see a rather large — but not total — area of the round cell surface. These indications refer to Wild-Leitz objectives and eyepieces, but similar equipment is of course also available from other sources.

2. METHOD

2.1 Incubation of Cells with Fluorescent Ligands

One mixes 0.1 ml of the lymphocyte suspension at 25×10^6 cells/ml with 0.1 ml of the fluorescent ligand solution (fluorescent Con A or antibody) in the bottom of a small plastic tube. This mixture is kept for some 10—15 min at room temperature (or 37°C). The excess or unbound fluorescent ligand is washed off at room temperature or 37°C ,. The cell pellet is gently resuspended in 50—100 μl medium, one droplet is put on a microscope slide, covered with a coverslip which is immobilized by a few spots of nail polish. The small chamber is *not sealed* by the nail polish; in fully sealed chambers microvilli formation seems to be less. Microvilli are also quite sensitive to cell resuspension by use of a Pasteur pipette.

Two types of assay for detection of microvilli can be performed in this general way depending on whether the cells are agglutinated or not. Examples are given hereafter.

2.1.1 Free Cells and Small Clumps

Metabolic inhibitors, including NaN_3, were found to increase the percentage of lymphocytes showing detectable microvilli to virtually 100% of the viable cells (4). Con A at low doses also seems to have a similar effect and to allow a faster reexpression of microvilli after lymphocyte chilling (4). On the other hand, NaN_3 inhibits the lectin-induced aggregation of cells when low doses of Con A are used (8). As a trick to keep separated cells showing nice microvilli, one can thus work with 10 mM NaN_3 in the medium throughout the experiment, which is otherwise run exactly as described above. Figure 1 shows lymphocytes from mouse lymph nodes treated in that way with 10 μg/ml of TRITC—Con A: they are separated and show either a fuzzy ring made of multiple short microvilli or longer microprojections. The whole cell surface cannot be in one single focus, the cells continuously move, and photographic exposure requires 30 sec. The photograph is thus poorly representative of the actual microscope appearance of the microvilli which are much more distinct and longer.

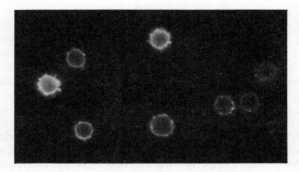

Figure 1. Lymphocytes stained with TRITC—Con A: microextensions of various aspects can easily be detected on these cells, from the numerous short microvilli giving a fuzzy ring appearance to the cell, to the long microextensions giving the aspect of a star

2.1.2 Agglutinated Cells

High doses of Con A (e.g. 100 μg/ml) agglutinate the cells and no microvilli can be detected: the plasma membrane gets flatter (4). There is, however, a range of doses where cells are agglutinated and show long microvilli (normal medium without NaN_3). In this last case the definite observation of the individual microvilli from any one cell is very difficult with a fluorescent lectin since practically all the agglutinated cells have fixed the lectin and all have fluorescent microvilli. To visualize single-cell microvilli, the trick consists of introducing into the system an independent fluorescent reagent which will bind to some cells only. For instance with lymphocytes one can use a fluorescent reagent detecting the membrane-bound immunoglobulins or any other membrane antigen which is characteristic of only a fraction of the total lymphocyte population. These immunoglobulin-bearing lymphocytes are agglutinated by Con A as well as the others and the agglutinates are actually a perfect mixture of the two cell types. Thus if one uses FITC—Con A to agglutinate mouse lymphocytes (and label their membrane), one can at the same time use a TRITC-labelled anti-mouse immunoglobulin antibody to label the immunoglobulin-bearing lymphocytes. To avoid the occurrence of spotting and capping one will preferably use a monovalent antibody (9). This labelling of the immunoglobulin-bearing lymphocytes can be performed before, during or after incubation with the Con A, following the usual procedures for membrane immunofluorescence (10) which are basically similar to what is written above as experimental design. Thus when the agglutinated cells are examined for FITC fluorescence, the strong green fluorescence of the cells will make the distinction of microvilli practically impossible. However, when the cells are examined for TRITC fluorescence only the immunoglobulin-bearing cells present in the agglutinate will fluoresce while all the other cells will remain black (provided the excess of the TRITC-antibody has been well washed off); if immunoglobulin-bearing cells are not too frequent in the clump, one can easily follow their microvilli by fluorescence microscopy; they frequently extend in the clump far from the original cell, frequently up to 20—30 μm. Figure 2 shows an example of mouse lymph node lymphocytes whose immunoglobulin-bearing cells had been prestained with a TRITC-labelled monovalent antibody against mouse immunoglobulin; the cells were then agglutinated by 20 μg/ml FITC—Con A. Though the cells did not significantly move in the clump during the exposure time, still the photograph is poorly representative of the actual aspect of the microvilli on the cells due to the low depth of field which does not allow one to obtain the whole picture on one single focus.

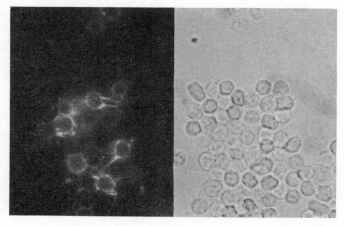

Figure 2. Lymphocytes agglutinated with Con A, some of them being stained by a TRITC-labelled monovalent antibody against membrane immunoglobulins: long microextensions are visible going far from the cells in the clump

3. REFERENCES

1. Kessel, R. G. and Shick, C. Y. (1974) in *Scanning Electron Microscopy in Biology*, Introduction. (Springer-Verlag, Berlin, Heidelberg, New York), pp. 1—18.
2. Kessel, R. G. and Shick, C. Y. (1974) in *Scanning Electron Microscopy in Biology*, Chapter 3, 'Cells in culture' (Springer-Verlag, Berlin, Heidelberg, New York), pp. 63—85.
3. Molday, R. S., Dreyer, W. J., Rembaum, A. and Yen, S. P. S. (1975) 'New immunolatex spheres: visual markers of antigens on lymphocytes for scanning electron microscopy', *J. Cell Biol.*, 64, 75—88.
4. Loor, F. and Hägg, L. -B. (1975) 'The modulation of microprojections on the lymphocyte membrane and the redistribution of membrane-bound ligands, a correlation', *Eur. J. Immunol.*, 5, 854—65.
5. Agrawal, B. B. L. and Goldstein, I. J. (1965) 'Specific binding of concanavalin A to cross-linked dextran gels', *Biochem. J.*, 96, 23c—25c.
6. Loor, F. (1976) 'Quantitation of the Con A-induced lymphocyte agglutination by optical microscopy', in *Concanavalin A as a Tool*, eds. Bittiger, H. and Schnebli, H. P. (John Wiley and Sons Ltd., International Publishers) p. 267.
7. Ploem, J. S. (1975) 'General introduction', *Ann. N.Y. Acad. Sci.*, 254, 4—20.
8. Loor, F. (1973) 'Lectin-induced lymphocyte agglutination. An active cellular process?', *Exp. Cell Res.*, 82, 415—25.
9. Loor, F., Forni, L. and Pernis, B. (1972) 'The dynamic state of the lymphocyte membrane. Factors affecting the distribution and turnover of surface immunoglobulins', *Eur. J. Immunol.*, 2, 203—12.
10. Loor, F. (1974) 'Binding and redistribution of lectins on lymphocyte membrane', *Eur. J. Immunol.*, 4, 210—20.

CHAPTER 7

Peroxidase–Concanavalin A Method: Application in Light and Electron Microscopy

S. AVRAMEAS
E. KARSENTI
M. BORNENS

1. PRINCIPLE

At neutral pH, Concanavalin A (Con A) has four equivalent active sites (1), all of which can react with sugars or glycoproteins which bear terminal non-reducing α-D-glucopyranosyl, α-D-mannopyranosyl or β-D-fructofuranosyl residues (2). If Con A interacts with a sugar in solution, saturation of the four binding sites can be expected. However, when sugar residues are present in an insoluble form, one might expect that due to steric hindrance, Con A might not engage all its four active sites. This is the case when Con A binds to glycoproteins or mucopolysacch-arides present on a cell membrane (Figure 1). The remaining free active sites can then operate as acceptors of another sugar, secondarily added to the system (3). Horse-radish peroxidase (HRP), which is a glycoprotein containing 18% carbo-hydrate, can be used for this purpose. The catalytic activity of the peroxidase molecules thus associated with the cell can then be used either for the light and

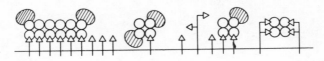

Figure 1. Various possibilities of interaction of Con A with cell receptors and HRP

△   Con A receptors.

88 Tetravalent Con A.

🖉 HRP.

electron microscopic visualization of Con A molecules using the diaminobenzidine method of Graham and Karnovsky (4), or for the quantitation of these molecules by using the *o*-dianisidine method (5,6). The interaction of Con A with cellular carbohydrate and with HRP can be inhibited by an excess of a sugar such as α-D-methyl mannoside (0.1 to 0.2 M) which competes for the binding sites of Con A.

It is evident that in addition to peroxidase, other glycoprotein enzymes such as glucose oxidase, as well as sugar-containing molecules like polydextrans which possess at the light and/or electron microscope level a characteristic shape, can be employed as markers in such procedures (3).

It is likely that the method does not allow visualization of all cell-bound Con A molecules (see Figure 1 and paragraph 3.3).

2. TECHNICAL PROCEDURE

2.1 General procedure

The procedure described below refers to the staining of carbohydrates specifically interacting with Con A on the surface of viable lymphocytes in suspension. Lymphoid cells (5×10^7) in 1 ml of medium* are incubated for 15 minutes at 4°C with 100 μg/ml of Con A and then washed three times with 10 ml of medium, by centrifugation at 4°C (300 *g* for 7 min). The cells are then incubated for 15 min at 4°C in 5 ml of medium containing 50 μg/ml HRP, and then washed three times with medium.

The same procedure is employed for attached cells. In that case washing is carried out by successive decantations.

2.2 Staining of Cells for Examination by Light Microscopy

The cells treated as described in 2.1 are fixed for 30 min at 4°C in 1.25% glutaraldehyde in 0.1 M phosphate buffer, pH 7.5, and then washed twice with

---

*Any cell medium can be used provided glucose or any inhibitory Con A-specific sugar is avoided. Generaly Earle's or Eagle's medium buffered with 25 mM HEPES without bicarbonate and containing galactose instead of glucose is used.

phosphate-buffered saline. Enzyme activity is revealed by incubating the cells for 10 min at room temperature, in 10 ml of 0.1 M Tris HCl buffer, pH 7.4, containing 5 mg of diaminobenzidine tetrahydrochloride (Sigma) and 0.03% H_2O_2. The cells are finally washed twice with 10 ml of phosphate-buffered saline, resuspended in 1 ml of the same medium and examined under coverslips with a light microscope. The enzymatic reaction produces a dense brown precipitate, visible on the surface. An example is given in Figure 2. The same staining procedure is used for attached cells.

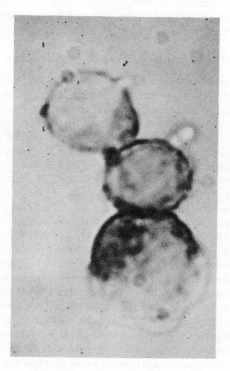

Figure 2. Example of staining observed in light microscopy. Rat thymocytes were incubated at 4°C with Con A and HRP as described in the text

The procedure described above can also be employed for the intracellular detection of carbohydrate constituents. In that case, cells fixed in suspension, frozen tissue sections, or tissues fixed (for example 4% formaldehyde in 0.1 M phosphate buffer, pH 7.5, for 30 minutes at room temperature), embedded in paraffin, cut into sections and then deparaffinized, are employed. After peroxidase staining, if desired, any conventional histochemical staining which does not impair the observation of the brownish peroxidase reaction product can be employed (toluidine blue–methyl green pyronin).

In order to assess the specificity of the staining obtained, controls performed in parallel are indispensable. The control preparations are treated in the same way as the experiments but 0.2 M α-D-methyl mannoside is incorporated in the solutions of Con A and HRP. Generally these controls are negative except when cells possess endogeneous peroxidase activity. For the procedure, the purest commercially available peroxidase* (RZ = 3) should be used; with less pure peroxidase preparations (RZ = 0.6; RZ = 1.5) a diffuse background staining might sometimes be observed. Con A is prepared according to Agrawal and Goldstein (7).

The procedure in general permits specific detection of cell-bound Con A. However it must be stressed again that only cell-bound Con A molecules which still possess free binding sites are stained by this procedure.

2.3 Staining of Cells for Examination by Electron Microscopy

Cells are processed in the same way as for visualization of Con A by light microscopy. However, staining with diaminobenzidine may be prolonged for up to 30 minutes. After enzymatic staining, the cells are washed twice in 10 ml of phosphate-buffered saline and then post-fixed for 1 hour at room temperature in 2% osmium tetroxide. After post-fixation, the cells are dehydrated in ascending concentrations of alcohol and embedded in 'Epon'. The embedded cell suspensions are cut with an ultramicrotome and mounted on 'Formvar' and carbon-coated 200 mesh copper grids. Sections are examined unstained or slightly counterstained for 30 sec with lead citrate (8). An example of cell surface staining is shown in Figure 3.

As for light microscopy, the procedure can also be employed for the intracellular detection of carbohydrate constituents. In that case, tissue blocks are fixed for 24 hours at 4°C with 4% formaldehyde in 0.2 M cacodylate buffer at pH 7.2. After fixation, tissues are washed at 4°C for at least 24 hours with several changes of the buffer solution. The tissue blocks are reduced to small fragments with a razor blade and from these fragments, frozen sections (10−40 μ) are cut and incubated with Con A and subsequently with HRP. Incubation times from 2 to 4 hours with either Con A or HRP are optimal. It should be emphasized however that penetration of Con A molecules into tissues or cells fixed for electron microscopy is not always possible (8). Thus, negative results are not conclusive. Sections counterstained with lead citrate should always be examined in parallel with unstained ones because with the former, negative staining might possibly be interpreted as a faintly positive reaction.

Again, one has to assess the specificity of the staining obtained and for this reason the same controls as those described for light microscopy must be performed.

3. MOLAR RATIO OF PEROXIDASE TO CON A

The HRP-Con A method is obviously an indirect method for assaying Con A receptors. It involves at least two binding mechanisms, which most probably depend on many parameters as accessibility and mobility of binding sites, properties of the cell membrane, fixation etc. In order to evaluate these effects we developed a method for independent determination of bound Con A and peroxidase.

*Grade I ref. 15629 Boehringer Mannheim, Germany Type VI ref. P8375 Sigma.

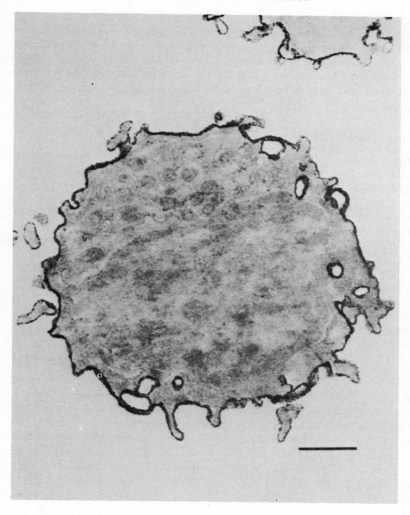

Figure 3. Example of staining observed in electron microscopy. Normal human lymphocyte treated at 4°C with Con A and HRP as described in the text

3.1 Binding of ^{125}I-labelled Con A and Horse-radish peroxidase (HRP) to Cells

Cells (5×10^7) are incubated for 15 minutes at 4°C in 1 ml of the appropriate medium, with ^{125}I-labelled Con A at various concentrations. Cells are then washed (centrifugation at 300 g for 7 min at 4°C) three times with medium. After the third wash, cells are incubated in 5 ml of Eagle galactose medium containing 50 μg/ml of HRP, for 15 min at 4°C. Cells are then washed three times with 10 ml of medium. After the third wash, cells are incubated for 15 min at room temperature in 1 ml of 0.2 M α-D-methyl mannoside in medium and the final cell concentration is determined in a haemocytometer. Radiocativity and peroxidase activity are determined in tubes on the same duplicate samples of 0.1 ml of the cell suspension and of the supernatant after cell sedimentation. Peroxidase activity is allowed to develop as described below and the tubes are then used directly for measuring radioactivity in a gamma counter and subsequently for reading the absorption at 400 nm.

3.2 Measurement of HRP Activity

The following solutions are used; solution A: 1% o-dianisidine in methanol, solution B: 0.3% H_2O_2 in distilled water, and the procedure is as follows.

Add 1 ml of B to 1 ml of 1 M phosphate buffer, pH 6. Complete to 100 ml with distilled water and add, with stirring, 0.83 ml of solution A. HRP activity is assayed by adding 2.9 ml of the reaction mixture (A,B) to 0.1 ml of sample. The reaction is developed during 5 min. It is stopped by one drop of 6 N HCl. Absorption at 400 nm is recorded and referred to a standard peroxidase activity curve established with known concentrations of HRP.

Depending on the amount of HRP released from the cells by α-D-methyl mannoside, the time allowed for the enzymatic reaction may be varied to facilitate recording of the OD_{400}. However, the linear range of the standard peroxidase activity curve must be determined for each set of experiments. The product of the enzymatic reaction is stable for at least 24 hours at room temperature, after addition of 6 N HCl.

3.3 Effect of Fixation on HRP/Con A Ratio

We will give in some detail the example which served to develop the technique. We will deal only with HRP fixation. Other aspects of this example, like the maximum number of lectin molecules bound to cells in the different cases (see Figure 4) will not be discussed here. The cells employed are rat thymocytes. Using ^{125}I-labelled Con A or succinyl-Con A, we have measured the HRP fixation to cell-bound lectins on living cells, or on cells whose receptors were immobilized by fixation with glutaraldehyde.

To fix the cells the following procedure was employed. 10 ml of cell suspension (5×10^7 cells ml) are incubated for 15 min at $4°C$ with 1% glutaraldehyde in Earle's medium buffered with HEPES (25 mM) at pH 7.4. Forty ml of 0.1 M lysine in the same medium (pH 7.4) is then added to the cell suspension and the cells are sedimented (7 min at 500 g). The cell pellet is suspended in fifty m l of 0.1 M lysine in the same medium and centrifuged. Two supplementary washes are performed with the medium alone, before any binding experiment with Con A.

Experiments performed on living cells, have been described in paragraph 3.1. Typical results are shown in Figure 4.

HRP fixation to cell-bound Con A or succinyl-Con A parallels lectin binding when cells are prefixed with glutaraldehyde (Figure 4, C,D).

In living cells HRP fixation to cell-bound Con A does not parallel Con A binding to the cell (Figure 4, A); HRP fixation is low at low doses of Con A. It increases considerably for high concentrations of Con A.

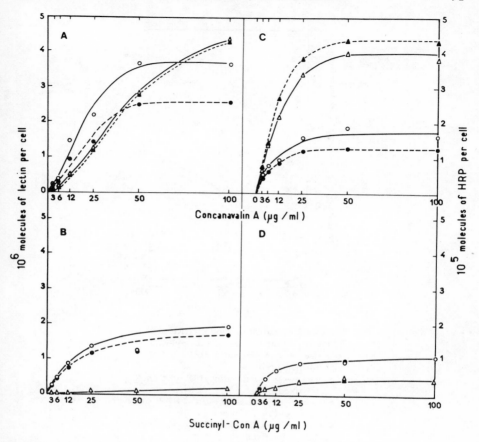

Figure 4. Binding of Con A or succinyl-Con A and HRP on fixed or living cells. ○————————○ Number of lectin molecules bound per cell; ●————————● α-D-methyl-mannoside-eluted Con A molecules; △————————△ number of HRP molecules bound per cell; ▲————————▲ α-D-methyl-mannopyranoside-eluted HRP molecules. A, Binding of Con A and HRP on living cells; B, binding of succinyl-Con A and HRP on living cells; C, binding of Con A and HRP on fixed cells; D, binding of succinyl-Con A and HRP on fixed cells

In living cells, HRP fixation to cell-bound succinyl-Con A is low for any dose of succinyl-Con A (Figure 2, B).

Figure 5 compares the values of the molar ratio HRP to cell-bound lectin in the different cases. One can see that it varies between 0.06 (1 HRP/17 Con A) for living cells incubated with low concentrations of Con A, to 0.32 (1 HRP/3 Con A) for any dose of Con A on fixed cells. Intermediary values are observed with living cells incubate with increasing doses of Con A.

For succinyl-Con A, which is divalent, values are lower than for native Con A. When cells are fixed, constant ratio of 1 HRP/20 succinyl-Con A is observed. For

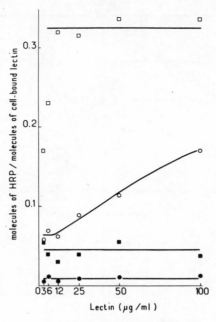

Figure 5. Relative binding of HRP to cell-bound   Con A   or   succinyl-Con A.
○———————○ Molar ratio HRP to Con A for living cells;
□———————□ molar ratio HRP to Con A for fixed cells;
●———————● molar   ratio   HRP   to succinyl-Con A for living cells;
■———————■ molar ratio HRP to succinyl-Con A for fixed cells

living cells, the ratio is also constant for any dose of succinyl-Con A and corresponds to 1 HRP/100 succinyl-Con A. The above results demonstrate that HRP fixation may vary considerably depending on the conditions of the experiments. The significance of these results, particularly the correlation beteen HRP fixation and receptor immobilization are discussed elsewhere (10).

4. GENERAL COMMENTS

As already stressed, the peroxidase—Con A technique can detect only cell-bound Con A molecules which still have free sugar binding sites. This has to be kept in mind for the interpretation of results obtained by light or electron microscopy. If for example, any heterogeneity exists among the Con A receptors of the cell, leading to the involvement of more or less sugar binding sites for each Con A molecule the peroxidase labelling could be different for the two cell-bound Con A molecules.

Although the specificity of the technique is satisfactory, absence of label is not sufficient to completely exclude the presence of Con A. Similarly comparisons between two differents cell types must be interpreted with caution.

For quantitative considerations it is important to note that the ratio HRP/Con A may vary considerably depending on the conditions of the experiments (paragraph 3.3), most notably fixation (see also 9 and 10). In order to circumvent these limitations, one needs an independent labelling of Con A, beside HRP. This was accomplished by developing the present peroxidase–[125]I-labelled Con A procedure.

The reliability of the technique must be assessed as well; one has, in each case, to make sure that HRP does not compete with membrane receptors for Con A binding. This is done by measuring the radioactivity that might be released after incubation with HRP. No such a release has been found for the cells we have tested (rat and mouse lymphocytes and thymocytes). Similarly, we have tested the ability of HRP to compete with 'Sephadex' for the binding of Con A. ^{125}I-Labelled Con A was adsorbed on a 'Sephadex' column and an excess HRP was applied to the column (5-fold the amount of HRP retained by the adsorbed Con A). Only 0.5% of the 'Sephadex'-bound Con A was displaced by HRP. The molar ratio HRP/'Sephadex'-bound Con A was 0.5, i.e. 1 molecule of HRP per 2 molecules of 'Sephadex'-bound Con A.

5. REFERENCES

1. Edelman, G. M. Cunningham, B. A., Reeke, G. N., Becker, J. W., Waxdal, M. J. and Wang, J. L. (1972) 'The covalent and three-dimensional structure of concanavalin A'. *Proc. Nat. Acad. Sci. U.S.A.*, **69**, No. 9, 2580–4.
2. Goldstein, I. J., So, L. L., Young, Y. and Callies, Q. C. (1969) 'Protein–carbohydrate interaction. XIX. The interaction of concanavalin A with IgM and the glycoprotein phytohemagglutinins of the waxbean and the soybean', *J. Immunol.*, **103**, 695–8.
3. Avrameas, S. (1970) 'Empode la concanavaline-A pour l'isolement, la détection et la mesure des glycoprotéines et glucides extra- ou endo-cellulaires', *Compte Rend. Acad. Sci.*, **270**, 2205–8.
4. Graham, R. C. and Karnovsky, M. J. (1966) 'The early stages of absorption of injected horseradish peroxidase in the proximal tubules of mouse kidney: ultrastructural cytochemistry by a new technique', *J. Histochem. Cytochem.*, **14**, 291–302.
5. Bergmeyer, H. U. (1963) 'Inorganic peroxides', in *Methods of Enzymatic Analysis* (Verlag Chemie, Weinheim and Acad. Press, N.Y.), pp. 633–5.
6. Avrameas, S. and Guilbert, B. (1972) 'Enzyme-immunoassay for the measurement of antigens using peroxidase conjugates', *Biochimie*, **54**, 837–42.
7. Agrawal, B. B. L. and Goldstein, I. J. (1967) 'Protein–carbohydrate interaction. VI. Isolation of Con A by specific adsorption on cross linked dextran gels', *Biochim. Biophys. Acta*, **147**, 262–71.
8. Bernhard, W. and Avrameas, S. (1971) 'Ultrastructural visualization of cellular carbohydrate components by means of concanaval in A', *Exp. Cell. Res.*, **64**, 232–6.
9. Collard, J. G. and Temmink, J. H. (1974) 'Binding and cytochemical detection of cell-bound concanavalin A', *Exp. Cell. Res.*, **86**, 81–6.
10. Bornens, M., Karsenti, E. and Avrameas, S. (1976) 'Cooperative binding of concanavalin A to thymocytes at $6°$C and microredistribution of concanavalin A receptors', *Eur. J. Biochem.* in press.

CHAPTER 8

Preparation of Concanavalin A–Ferritin

STEFANELLO DE PETRIS

Several interesting biological phenomena are elicited by the binding of concanavalin A (Con A) to living cells. As discussed in several articles in this volume, the occurrence of these phenomena, such as the differential agglutination of normal and virus-transformed cells, or the stimulation of certain classes of lymphocytes to proliferate, depends on the interaction of the lectin with specific receptors on the surface of the cells. It appears, therefore, clearly important to determine whether any correlation can be established between the pattern of distribution of these receptors on the surface of the cells types in the various experimental situations and the different biological effects of Con A binding.

The pattern of distribution and of redistribution (see next chapter) of the Con A receptors can be studied by immunofluorescence and by electron microscopy. The former (see Mallucci, Chapter 5 this volume) can only give information about the gross distribution of the Con A receptors, namely the distribution at a level of resolution of the order of 0.2–0.4 μm. The finer distribution can be studied only by electron microscopic labelling techniques which can reach a level of spatial resolution of the order of 10–30 nm, these values being essentially determined by the size of the lectin itself, or by the size or other characteristics of the marker added to increase its visibility, 'Native', i.e. unmodified, Con A can be detected on freeze-etched surfaces (see Bächi, Chapter 14, this volume) or in conventional thin sections (1). The first technique, however, is difficult to use routinely, and in the second the low contrast of the Con A molecules stained with heavy atoms allows the detection of the molecules only on surfaces which are cut more or less perpendicular to the plane of the section (1). For general applications, therefore, it is common to tag the Con A molecule with a suitable marker of high electron density. The markers commonly used are ferritin, enzyme markers (horse-radish peroxidase, cytochrome c), iron dextran and haemocyanin. All these markers with the exception of ferritin and cytochrome c (2) are normally used with indirect labelling methods (see this volume, Chapters 7,11,12,13) in which no conjugation of Con A to the marker is involved. For direct labelling methods ferritin directly conjugated to Con A is almost universally used. The main theoretical advantage of using ferritin-conjugated Con A consists in the fact that each Con A molecule bound to the cell surface becomes detectable, ideally with each ferritin molecule corresponding to one Con A molecule. In practice the degree to which this condition can be achieved is determined by the quality of the conjugate.

1. PRINCIPLE

Conjugation of Con A to ferritin is generally carried out by using glutaraldehyde as a coupling agent according to the general method introduced by Avrameas (3). Other coupling agents, such as the diisocyanates used by Singer and Schick (4) to couple ferritin to antibody, could also be used, but in the case of Con A conjugation they do not appear to offer any practical advantage over glutaraldehyde. In the following section only the glutaraldehyde conjugation method will be considered in detail.

2. COUPLING OF FERRITIN TO CON A

2.1 One-step Method

Usually Con A is covalently coupled to ferritin with glutaraldehyde by a one-step procedure (2,5—9). In this procedure the two proteins and the coupling agent are mixed together and the latter is allowed to react simultaneously with the two proteins. Under these conditions if the reaction proceeds too far large branched polymers of mixed composition are formed. To limit the formation of these polymers glutaraldehyde is added in limiting amount, and the polymerization reaction is usually stopped before completion. This results in a relatively poor yield of useful conjugate, as usually not more than 20%, or less, of the Con A initially present is recovered in the purified conjugate. This disadvantage is not as serious in the case of Con A conjugation as in the case of antibody conjugation, as purified Con A is commercially available and relatively inexpensive, and is constituted by a population of molecules all with the same specificity. A relatively high concentration of active Con A—ferritin complexes can therefore be obtained even if the total yield is low.

Crystallized Con A, as supplied by commercial sources (see Appendix) is in general suitable for conjugation. The author has mainly used Con A from Miles-Yeda (Kankakee, Illinois), supplied either as a lyophilized powder or as a concentrated solution in saturated NaCl. Con A dissolved or diluted in the conjugation buffer (see below) may be used directly for conjugation, or, preferably, may be further purified by affinity chromatography on a column of 'Sephadex' G-50 or G-75 according to the method of Agrawal and Goldstein (10; see Liener and Goldstein, this volume). The active Con A, suspended in 0.5—1.0 M NaCl buffered at pH 6.5 or 7.0 with 20—50 mM phosphate or Tris buffer is bound to the column, and is eluted by 0.1 M glucose or 0.2 M sucrose in the same buffer (cf. ref. 6) and then reconcentrated to 10—30 mg/ml. Before conjugation any aggregated material is removed by centrifugation (e.g. 15,000 g x 20 min) or 'Millipore' filtration (0.22 or 0.45 μm diameter methylcellulose filters) to reduce formation of aggregates during conjugation.

The *ferritin* normally used for conjugation is horse-spleen ferritin which can be obtained as an aqueous solution from several commercial sources. The author has mainly used 3-times crystallized Schwarz-Mann (Orangeburg, New York) ferritin without further purification. Any aggregate, if present, is removed from the ferritin solution as indicated for Con A. Nicolson and Singer (6) have used filtration through a 0.05 μm diameter 'Millipore' filter. The ferritin used by the author normally did not contain sufficient aggregates to make these purification steps necessary. It is essential, however, to use aqueous ferritin solutions which have never been exposed to temperature below $0°C$, as freezing causes extensive and irreversible denaturation and aggregation of ferritin (cf. for instance ref. 6). If required, ferritin can be recrystallized using cadmium sulphate as described by several authors (e.g. 6,11,12). Ferritin consists of a hollow protein shell (m.wt. 445,000) of about 12 nm diameter, formed by 24 identical subunits, surrounding a cavity of about 70 Å where a variable amount (up to 100% of the protein weight) of ferric oxide hydrate is contained (13). This iron core, which may contain up to 5000 Fe atoms (usually 2—3000) gives the molecule its typical electron-dense appearance. Absolute ferritin concentration cannot be determined by u.v. absorption at 280 nm, as the strong absorption of the variable amount of the iron component masks that of the protein. Protein concentration can be measured by a micro-Kjeldahl analytical method (6), but for practical conjugation purposes for which the ferritin concentration is not very critical, it is often sufficient to use the concentration as given by the supplier. For relative concentration determinations, the optical density at 325 or 330 nm, or at higher wavelength, where the absorption is lower, can be used taking the commercial preparation as a standard. Extinction coefficients (optical density of a 1 mg/ml solution in 1 cm cell) of commercial preparations at 325—330 nm may vary between 3 and 9 according to the iron content of ferritin. Ferritin preparations may contain some apoferritin which is not visible in thin sections, but can be detected by negative staining of ferritin molecules with uranyl acetate or sodium phosphotungstate, but usually its amount is negligible for all practical purposes. If this is not the case, apoferritin can be separated from ferritin by high-speed differential centrifugation (12).

Glutaraldehyde ($OHC(CH_2)_3CHO$; m.wt. 100.12) as supplied by the manufacturers may contain either pure monomer (absorption peak at 280 nm; usually supplied under N_2) or monomer plus various amounts of polymeric derivatives (absorption peak at about 235 nm). Both types of glutaraldehyde have been used for conjugation in several laboratories with comparable results. Until recently most of the conjugates made in the author's laboratory were prepared with satisfactory results using polymer-containing glutaraldehyde (Taab, Reading, England). At nominally identical glutaraldehyde concentrations a smaller amount of large protein aggregates is formed when polymer-containing glutaraldehyde is used, possibly because the actual number of aldehyde groups available for reaction is lower. Secondary reactions of the condensation products of glutaraldehyde (14,16) are expected to be negligible during the short time of the conjugation if the starting material is monomeric. It is not known, however, if secondary reactions occur subsequently to the binding of glutaraldehyde to the protein, contributing to the stability of the coupling. The molar ratio of glutaraldehyde to protein (Con A +

ferritin) in the conjugation mixture is of the order of 15 to 70 according to the conditions of conjugation (time, protein concentration, temperature, etc.), which is insufficient to modify all the amino groups of the proteins.

The conjugation procedures used by different authors are all substantially identical with only minor modifications. An important parameter is the pH of the reaction, as Con A is present essentially as a tetramer at pH 7 or above, and as a dimer at pH below 6. Formation of tetramers and higher aggregates is also favoured by higher temperatures (see Liener and Goldstein, this volume). According to the conjugation conditions (pH, temperature), conjugates containing either mainly dimers or mainly tetramers (or higher aggregates) of Con A can be obtained. In the following procedure, employed by Nicolson and Singer (6), Con A aggregation is minimized by using a relatively low pH, at which, however, some dimers could still be present.

Ferritin (30—45 mg/ml) is mixed at room temperature with 15 mg/ml of Con A (final concentrations) in 0.5 M NaCl 0.050 M phosphate buffer, pH 6.5, containing 0.2 M sucrose to protect the active sites of Con A. To this mixture monomeric glutaraldehyde is added slowly with gentle stirring to give a final concentration of 0.02—0.04%. After 30—45 min at room temperature, the reaction is stopped by adding a drop of 0.1 M NH_4Cl, and the conjugate is fractionated by filtration on agarose gel as described in paragraph 2.2.

In his preparations the author has used slightly lower protein concentrations and a higher nominal polymer-containing glutaraldehyde concentration, and the reaction has been carried out at low temperature for a longer time at either pH 5.9—6.0 (mainly dimeric Con A) or pH 7.0—7.2 (mainly tetrameric Con A) (8,9). For instance: 4—6 mg of Con A and 30—35 mg of ferritin in one ml are gently stirred with 0.05—0.075% of glutaraldehyde for 2—6 h at 4°C. The unreacted aldehyde groups are neutralized by adding 25 mg/ml of lysine HCl in PBS (pH 7.2) and the conjugate is then fractionated by sucrose gradient centrifugation (see paragraph 2.2).

2.2 Purification of the Conjugate

'Purification' of the conjugate is required in order to free the conjugate from the unconjugated Con A molecules which would otherwise interfere with labelling, and from large aggregates formed during conjugation. The fractionation is carried out either by sucrose gradient centrifugation taking advantage of the several-fold difference between the sedimentation coefficients of ferritin and ferritin conjugates and that of unconjugated Con A (8,9), or by gel filtration through an agarose column taking advantage of differences in size and shape of the various molecules and complexes (5,6).

As a preliminary step the conjugate mixture is centrifuged at about 15,000g for 30 min to remove the largest aggregates. This step may be substituted or supplemented by a precipitation step in which cold saturated ammonium sulphate is added to the conjugation mixture at 4°C to a final 15% volume, in order to precipitate all the poorly soluble molecules or aggregates in the suspension (9,17). These poorly soluble aggregates (corresponding mainly to large aggregates) are usually responsible for non-specific staining and other artifacts. Conjugates prepared at pH ≤ 6.0 usually contain little or no precipitable material, whereas an increasing amount of precipitable material may be present in conjugates prepared at higher pH. The insoluble material is removed by centrifugation as indicated above, the supernatant is diluted to 15—20 mg/ml of protein with phosphate-buffered saline (PBS) and layered on the sucrose gradient.

The sucrose gradient centrifugation is carried out using a swinging bucket rotor with transparent or semi-transparent centrifuge tubes of 8—9 cm length. For instance, if a Spinco SW41 rotor is used, a discontinuous gradient of sucrose in PBS is prepared by layering 2.0 ml of 2.0 M sucrose, 3.0 ml of 1.0 M sucrose, 3.0 ml of 0.5 M sucrose, 0.5 ml of 0.4 M sucrose (this layer has been introduced to reduce 'droplet formation'). On top of the gradient about 3.5 ml of the conjugate are layered. The tubes are centrifuged for 3 h at about $150,000g$ (average) at $4°C$. If different rotors are used, equivalent conditions have to be employed (e.g. 10—12 h are required at a speed of 22,000—24,000 r.p.m. with SW26 rotors). The yellowish upper layer is completely removed and discarded together with the 0.5 M sucrose layer: these layers contain practically all the unconjugated Con A. The ferritin-containing 1.0 M layer and the upper 1 ml of the 2.0 M sucrose layer are collected and dialysed overnight against PBS. The conjugate is concentrated by vacuum dialysis without stirring to 10—15 mg of ferritin/ml. The conjugate can be concentrated also by precipitation with 50% ammonium sulphate and by resuspending the precipitate in a small volume of 0.5 M NaCl/0.05 M phosphate buffer or PBS (higher buffer molarity is recommended for higher protein concentrations), but some aggregation may result if poorly soluble material is still present. The conjugate is thoroughly dialysed against PBS or conjugation buffer and sterilized by filtration through 'Millipore' filter (0.22 or 0.45 μm diameter). Loss of iron during extensive dialysis can be avoided by including 1—5 mM Fe^{III} salts in the second to last change of dialysis buffer. Further purification to obtain predominantly active Con A—ferritin conjugate can be achieved by affinity chromatography on a 'Sephadex' G-200 column (2.5 x 10 cm) (8,15) or 'Sephadex' G-75 column (4 x 40 cm) (6) prior to or after the density gradient step. Only active Con A—ferritin is bound to the column, which is eluted with α-methyl mannoside (15), 0.1 M glucose or 0.2 M sucrose (6). This step is not absolutely necessary, as unconjugated ferritin or inactive conjugate molecules cause negligible non-specific binding. Purified material is, however, necessary to determine the molar ratio of ferritin: Con A in the conjugate. Ferritin concentration can be estimated by spectrophotometry (see above) or by iron analysis (6). The lectin content can be determined by radioactivity measurements if the lectin used for conjugation had been previously labelled with ^{3}H, ^{14}C or ^{125}I, or estimated from the total protein content of the conjugate determined by micro-Kjeldahl analysis (6). Alternatively an estimate of the molar ratio of Con A to ferritin conjugates can be obtained entirely by u.v. absorption measurements (H. P. Schnebli, personal communication). The absorption curves of Con A and ferritin differ significantly between 240 and 340 nm, and the molar extinction coefficient of Con A at 330 nm ($< 0.4 \times 10^3$) is at least 3 orders of magnitude lower than that of ferritin. The amount of ferritin in the purified conjugates can, therefore, be determined directly at 330 nm. Absorption of the conjugate at 280 nm however, is composed of contributions from both Con A and ferritin. Optimally the OD at 280 and 330 nm of both the conjugate and a small sample of native ferritin (dialysed parallel to the conjugate!) are measured. Depending on the batch and previous handling, the

OD_{280} of ferritin is higher than the OD_{330} by a factor (k) of 1.5 to 2. The contribution of Con A to the extinction of the conjugate can therefore be calculated according to the following equation:

$$OD_{280} \text{ (Con A)} = OD_{280} \text{ (conjugate)} - k \times OD_{330} \text{ (conjugate)}$$

The concentration of Con A can now be calculated from the molar extinction coefficient $\epsilon = 1.51 \times 10^5$. The Con A—ferritin ratio in conjugates is usually between 1 and 5.

The alternative method of purification of the conjugate is by filtration through a column of agarose gel (5,6). After centrifugation at 15,000g for 15—30 min, the conjugation mixture is applied to a 1.5 x 120 cm 'Bio-Gel' A-1.5 m column (Biorad) equilibrated with 0.5 M NaCl/0.05 M phosphate buffer. The fractions (0.5 ml) are monitored by absorbance at 280 nm (which corresponds almost exclusively to the absorbance of ferritin) and by agglutinability of (rabbit) erythrocytes. The first fractions containing high molecular weight aggregates are discarded and the peak of agglutinating material which follows these fractions in the excluded volume is collected and concentrated to 1—5 mg/ml protein (6). The resolution of the gel filtration method for relatively high protein loads is probably less than that of the centrifugation method. Monomeric conjugate molecules are also present in the delayed fractions, and some of the later fractions containing ferritin and monomeric conjugates are usually contaminated by some unconjugated Con A, possibly Con A oligomers. For optimal yield and purification of the conjugate the two methods may be combined by removing first the excluded large aggregates on an agarose column and then subjecting all the delayed ferritin-containing fractions, pooled and reconcentrated, to sucrose gradient purification; the procedure, however, is rather lengthy. For application to living cells the molarity of the conjugate buffer (if not isotonic) has to be adjusted by dilution with water, or by dialysis.

The values given above for the final concentration of the conjugate are only indicative as they depend on the efficiency of the conjugation and purification procedures. In order to insure saturation of the Con A receptors in any cell sample, the final concentration of active Con A in the conjugate should not be less than 100 μg/ml, which in preparations not purified on 'Sephadex' may often correspond to concentrations of up to 10—15 mg/ml of ferritin. In these preparations the actual amount of active Con A in the conjugate can be approximately estimated, as mentioned above, by comparing the haemagglutinating titre of the conjugate with the titre of an unlabelled Con A solution of known concentration (see also Mallucci, Chapter 5 this volume); the test is however, only roughly indicative of the real Con A concentration, as the agglutinating ability of the conjugated Con A molecules is not necessarily identical to that of the native Con A, and may also be affected by the size distribution of the conjugate molecules in the conjugate mixture. The relative titre of active Con A in different conjugates can also be estimated by immunofluorescence, by applying appropriate dilutions of the conjugate to a fixed number of cells (e.g. lymphocytes) and labelling the cell. with a rhodamine-(or fluorescein-)conjugated anti-Con A antibody.

Examples of labelling using conjugates prepared according to the methods described above are presented in the next chapter.

2.3 Alternative Coupling Methods and Possible Improvements

The main problem in the preparation of Con A—ferritin conjugates by one-step conjugation with glutaraldehyde consists in the unavoidable formation of aggregates. Although these can be removed to a larger or lesser extent by the various purification procedures, it would be of great advantage to be able to prepare conjugates containing little or no polymer. Theoretically it should be possible to obtain some improvement by using two-stage conjugation procedures, in which the ferritin is first 'activated' by reaction with a suitable coupling agent, and after removal of the excess of unreacted reagent, is allowed to react with the unmodified Con A molecule (e.g. as in the original method of Singer and Schick (4) for the coupling of antibody to ferritin). Two-stage coupling methods using glutaraldehyde have been developed for ferritin—antibody coupling (16,18) taking advantage of the fact that ferritin is polymerized by glutaraldehyde much less readily than many other proteins (e.g. antibodies) (16,18) and therefore can be activated by it without gross formation of polymers. When applied to the coupling of Con A to ferritin, however, these methods do not seem to offer any advantage over the simpler one-stage method, as some ferritin aggregates are formed anyway during the first stage of the reaction and the yield of useful conjugate is generally lower than that obtained with the direct one-step method (de Petris, unpublished data). A theoretically more satisfactory two-stage procedure has been used by Kishida and coworkers, (16) for the conjugation of antibody and antibody fragments to ferritin and could in principle be applied also to the coupling of Con A to ferritin. According to this method, polymerization of ferritin by the coupling agent is prevented by a preliminary reaction step in which the ferritin amino groups are blocked by succinylation. The succinylated ferritin is activated by reaction with a water-soluble carbodiimide and N-hydroxysuccinimide and after removal of excess of the latter reagents is allowed to react with the protein to be conjugated. It has been reported that with this method monomeric conjugate complexes are almost exclusively produced (16). If the same result could also be obtained in the preparation of Con A conjugates, the extra complication of the conjugation procedure would be compensated by the production of a homogeneous conjugate.

The advantage of using monomeric conjugates would not only reside in the possibility of obtaining a cleaner and more precise localization of the labelled receptors, but also in the reduction of possible biological 'artifacts' in living cells, such as pinocytosis and other secondary metabolically dependent redistribution phenomena which may be enhanced by the binding to the membrane of larger complexes. It should be noted, however, that when a reversibly dissociable molecule such as Con A is conjugated to ferritin by a two-stage method, some dissociation into dimers of the relatively unmodified tetrameric Con A molecule could occur during storage, or during cell labelling, causing some loss of activity and sensitivity of the conjugate. To avoid this potential difficulty the tetrameric Con A molecules can be stabilized before conjugation or immediately after conjugation by treating them with a very low concentration (0.0025—0.005%) of glutaraldehyde for 1—2 h at room temperature, which may be sufficient to form some intramolecular bridges without causing appreciable intermolecular cross-linking.

3. REFERENCES

1. de Petris, S. (1975). 'Concanavalin A receptors, immunoglobulins and θ antigen on the lymphocyte surface: interactions with concanavalin A and with cytoplasmic structures', *J. Cell. Biol.*, **65**, 123—46.
2. Stobo, J. D. and Rosenthal, A. S. (1972) 'Biologically active concanavalin A complexes suitable for light and electron microscopy', *Exp. Cell. Res.*, **70**, 443—7.
3. Avrameas, S. (1969). 'Coupling of enzymes to proteins with glutaraldehyde. Use of the conjugates for the detection of antigens and antibodies', *Immunochem.*, **6**, 43—52.
4. Singer, S. J. and Schick, A. F. (1961) 'The properties of specific stains for electron microscopy prepared by the conjugation of antibody molecules with ferritin', *J. Biophys. Biochem. Cytol.*, **9**, 519.
5. Nicolson, G. L. and Singer, S. J. (1971) 'Ferritin-conjugated plant agglutinins as specific saccharide stains for electron microscopy: application to saccharides bound to cell membranes', *Proc. Nat. Acad. Sci. U.S.A.*, **68**, 942—5.
6. Nicolson, G. L. and Singer, S. J. (1974). 'The distribution and asymmetry of mammalian cell surface saccharides utilizing ferritin-conjugated plant agglutinins as specific saccharide stains', *J. Cell. Biol.*, **60**, 236—48.
7. Nicolson, G. L. (1971). 'Difference in topology of normal and tumor cell membranes as shown by different surface distributions of ferritin-conjugated concanavalin A', *Nature (New Biol.)*, **233**, 244—6.
8. De Petris, S., Raff, M. C. and Mallucci, L. (1973) 'Ligand-induced redistribution of concanavalin A receptors on normal, trypsinized and transformed fibroblasts', *Nature (New Biol.)*, **244**, 275—8.
9. De Petris, S. and Raff, M. C. (1974) 'Ultrastructural distribution and redistribution of alloantigens and concanavalin A receptors on the surface of mouse lymphocytes', *Eur. J. Immunol.*, **4**, 130—7.
10. Agrawal, B. B. and Goldstein, I. J. (1967) 'Physical and chemical characterization of concanavalin A, the hemagglutinin from jack bean (*Canavalia ensiforms*)', *Biochim. Biophys. Acta*, **133**, 376—9.
11. Granick, S. (1942) 'Ferritin. I. Physical and chemical properties of horse spleen ferritin', *J. Biol. Chem.*, **146**, 451—61.
12. Andres, G. A., Hsu, C. and Seegal, B. C. (1973) 'Immunoferritin technique for the identification of antigens by electron microscopy', in *Handbook of Experimental Immunology*, ed. Weir, D. M. (Blackwell Scientific Publication, Oxford and Edinburgh), 2nd ed., Vol. 2, pp. 34.1—34.45.
13. Hoare, R. J., Harrison, P. M. and Hoy, T. G. (1975) 'Structure of horse apoferritin at 6 Å resolution', *Nature*, **255**, 653—4.
14. Richards, F. M. and Knowles, J. R. (1968) 'Glutaraldehyde as a protein cross-linking reagent', *J. Mol. Biol.*, **37**, 231—3.
15. Bittiger, H. and Schnebli, H. P. (1974) 'Binding of concanavalin A and ricin to synaptic junctions of rat brain', *Nature*, **249**, 370—1.
16. Kishida, Y., Olsen, B. R., Berg, R. A. and Prockop, D. J. (1975) 'Two improved methods for preparing ferritin—protein conjugates for electron microscopy', *J. Cell. Biol.*, **64**, 331—9.
17. de Petris, S. and Raff, M. C. (1972) 'Distribution of immunoglobulin on the surface of mouse lymphoid cells as determined by immunoferritin electron microscopy. Antibody-induced, temperature-dependent redistribution and its implications for membrane structure', *Eur. J. Immunol.*, **2**, 523—35.
18. Otto, H., Takamiya, H. and Vogt, A. (1973) 'A two-stage method for cross-linking antibody globulin to ferritin by glutaraldehyde. Comparison between the one-stage and the two-stage method', *J. Immunol. Methods*, **3**, 137—46.

CHAPTER 9

Application of Concanavalin A–Ferritin to Whole Cells

STEFANELLO DE PETRIS

1. LIGAND-INDUCED REDISTRIBUTION OF CON A RECEPTORS

The main point to be taken into consideration when labelling a membrane component with a multivalent ligand is that the observed distribution will not in general represent the original distribution of that component in the membrane, but a redistribution pattern generated by the interaction of the component with the ligand (1). This is due to the fact that cellular membranes, and in particular the plasma membrane, are essentially fluid structures (2–6) and that membrane components which are mobile in the plane of the membrane can be cross-linked by multivalent ligands to form two-dimensional aggregates (clusters and 'patches') which remain segregated from the other components in discrete areas of the membrane (1,4,7). This phenomenon is general and occurs in various degrees in different cells and under different experimental conditions also when Con A receptors are labelled and cross-linked by Con A or by a Con A–ferritin conjugate.

The primary process of clustering, based on diffusion of the membrane component, appears to be essentially metabolically independent. In several types of cells this process may be followed by more dramatic rearrangements and displacements of the cross-linked membrane complexes, which require metabolic energy, and which may lead to the accumulation of the complexes over one particular area of the cell, e.g. over one pole of the cell as in the phenomenon of 'capping' in lymphocytes (1,4,7).

The main factors determining the extent and the characteristics of the redistribution are the valence of the ligand and that of the surface receptor-carrying molecule, as these determine the extent of cross-linking of the reacting molecules. The metabolically dependent phenomena, however, are also largely influenced by direct or indirect interactions of some of the surface components with cytoplasmic structures, such as microfilaments and microtubules, which are considered to be actively involved in the displacement of surface molecules (1,4,8—10).

The contribution of metabolically dependent and metabolically independent factors in determining the pattern of redistribution is relatively clear in gross-redistribution phenomena, such as capping, but the distinction is more difficult to make in the case of less dramatic redistribution, e.g. in the formation of large patches. In the case of Con A receptors the analysis of the phenomena following the binding of the ligand (Con A or Con A conjugates) is complicated by the fact that various subpopulations of receptors with different valence and possibly different interaction with cytoplasmic structures may be present in different cell types, and by the fact that the valence of the ligand itself (Con A or Con A—ferritin conjugate) may vary under different experimental conditions and in different conjugate mixtures. Moreover, in some cells (e.g. lymphocytes), the interaction of surface molecules with cytoplasmic structures under different conditions may either enhance or inhibit Con A receptor redistribtion (e.g. capping) by a mechanism which is still largely unknown, but which depends on the extent of cross-linking of the Con A receptors by the ligand (10, 11).

In general, both metabolically independent and metabolically dependent redistribution phenomena may occur at relatively high temperatures, i.e. at temperatures of $15-20°C$ or above. Metabolically dependent redistribution may be inhibited or strongly reduced at these temperatures if the cells are incubated in the presence of metabolic inhibitors, such as $10-50$ mM sodium azide in the absence of glucose (1, 7), and some other drugs which affect membrane or cytoplasmic components (1, 7, 11, 12). Metabolic inhibitors, however, do not as a rule prevent passive formation of clusters and patches (1, 7, 8).

Active redistribution is also largely or completely inhibited if the cells are incubated at low temperature $(0°-4°C)$. The extent also to which metabolically independent phenomena are inhibited at these temperatures seems to vary in different cells. In lymphocytes some clustering of Con A receptors has been observed at low temperatures, in agreement with the observation made in other cell surface systems (8,9). In other cells, such as fibroblasts, diffusion seems to be more restricted and Con A receptor clustering seems to be negligible or greatly reduced at $0°-4°C$ (13—15). However, in this type of cell also, some clustering has been observed in some experimental situations (13), indicating that incubation at low temperature cannot *a priori* ensure a complete absence of redistribution (cf. Figure 7). Redistribution of Con A receptors can be effectively prevented and the receptors can be immobilized in their 'original' distribution if the cells (or the membranes) are prefixed with glutaraldehyde or formaldehyde before labelling them with Con A or with a Con A conjugate. These fixatives are able to cross-link membrane proteins without altering the ability of the carbohydrate receptor sites to react with Con A (13—16).

Different patterns of distribution will therefore, be observed on the cell surface under different experimental conditions (Figures 1—8). The application of Con A—ferritin to whole cells or whole cell membranes under these different conditions is described in the following sections.

2. LABELLING PROCEDURES

2.1 Labelling of Unfixed Cells

Unfixed cells are labelled with ferritin—Con A conjugates either in suspension [e.g. lymphocytes (Figures 1 and 4), fibroblasts in agglutination studies (Figures 5—7)], or attached to a substrate [e.g. fibroblasts (Figure 8)]. The pattern of redistribution may not be necessarily the same for the same cells in the two situations.

Suspended cells are usually labelled in tissue culture medium or in a balanced salt solution (Hank's, Dulbecco's PBS), preferably not containing glucose and serum, some components of which may be adsorbed on the membranes and bind Con A—ferritin, but containing some carbohydrate-free protein such as 0.1—1% bovine serum albumin to keep the cells in good conditions. A pellet of washed cells $(3—5 \times 10^6$ cells or more) is resuspended in 0.2—0.4 ml of conjugate at a suitable concentration. As a rough indicative value it can be estimated that conjugates containing 50—100 $\mu g/ml$ of active Con A (i.e. Con A capable of binding to a 'Sephadex' column) are adequate for staining 10^7 cells/ml carrying 10^7 receptors per cell. For agglutination studies a cell concentration of $1—2 \times 10^6$ cells/ml would be more suitable, as it would correspond more closely to the concentrations normally used in light microscopy studies, but such dilutions are often impracticable, as they require the use of an excessive amount of conjugate to stain an adequate number of cells. In many systems incubation of concentrated cell suspensions with concentrated conjugate at room temperature or at $37°C$ often has the undesirable effect of causing gross cell agglutination. Even more severe agglutination, however, is caused by centrifugation during the subsequent washing steps. Agglutination is largely avoided if labelling and washing is carried out at $0°C$ (17), but generally less Con A is bound to the cell at $0°C$ (18,19), and the pattern of distribution cannot be considered equivalent to that at higher temperatures (cf. Figures 5 and 6).

After incubation for the established time (e.g. 10—30 min at room temperature or at $37°C$; 60 min at $0°—4°C$) with occasional stirring, the cells are diluted to 5—10 ml with fresh medium and washed twice by centrifugation. To avoid the excessive agglutination and clumping caused by repeated centrifugations in samples incubated at 20—37°C, washing can be done in one step by layering in a centrifuge tube the cell sample diluted to 2—4 ml over 2—5 ml of a medium of higher density (e.g. 5—10% bovine serum albumin in PBS), and spinning the cells through this medium. The supernatant with the top layer containing the conjugate is carefully removed and discarded, the tube walls are dried, and the cell pellet is resuspended in 2 ml of PBS or 0.1 M phosphate buffer pH 7.2—7.4 and fixed either by gradual dropwise addition with gentle and continuous agitation over a period of 1—2 min of 2 ml of 6% glutaraldehyde in 0.12 M sodium phosphate buffer pH

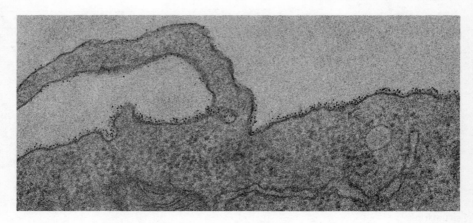

Figure 1. Mouse lymphocyte labelled with Con A—ferritin for 20 min at 21°C. Conjugate purified by gel filtration (see Chapter 8): first peak of the delayed fractions. The label is distributed in large, almost continuous patches, but is almost completely absent from a microvillus (x 52,000). All figures in this chapter represent unstained sections from samples block stained with uranyl acetate

Figure 2. Same as in Figure 1, but the cells were prefixed for 2 h at 21°C with 3% glutaraldehyde before being incubated (90 min, 21°C) with Con A—ferritin. The label is uniformly distributed also over the microvilli (a microvillus in this section is cut tangentially) (x 52,000)

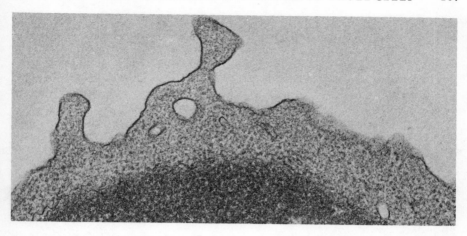

Figure 3. Same as in Figure 2, but the prefixed cells were incubated with Con A—ferritin in the presence of 0.1 M α-methyl mannoside. Only an occasional ferritin molecule is attached to the cell surface (x 52,000)

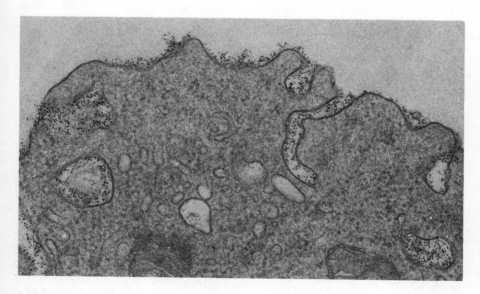

Figure 4. Same as in Figure 1. Conjugate purified by gel filtration: excluded volume fractions. This conjugate contained some aggregates. Marked clustering and pinocytosis. Pinocytosis, capping and agglutination were very marked in this sample (x 52,000)

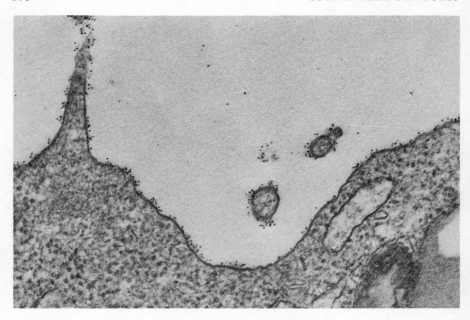

Figure 5. Mouse embryo fibroblast (secondary culture) labelled in suspension for 20 min at 0°–4° C. Conjugate purified by density gradient centrifugation. Cells washed fixed by centrifugation through a gradient of PBS/3% glutaraldehyde/6% glutaraldehyde (13). The label is distributed over the whole surface, although the distribution is not completely uniform (x 60,000). Fig. 5 to 8 are from work of de Petris, Raff and Mallucci (ref. 13 and unpublished).

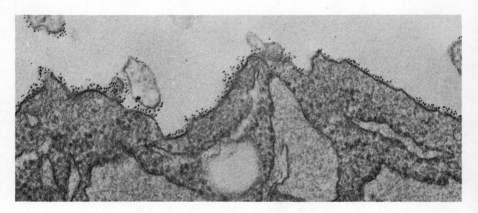

Figure 6. Same as in Figure 4, but the cells were labelled at 22° C (in the presence of 5 μg/ml of cytochalasin B). The labelling is more intense and 'patchy' (x 60,000)

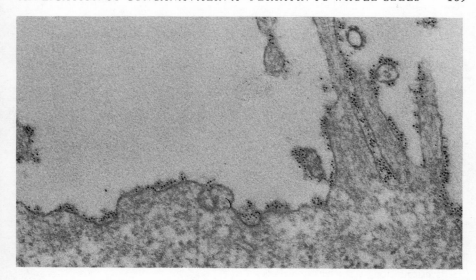

Figure 7. Polytransformed 3T3 fibroblast labelled for 40 min at 0°C with Con A—ferritin, washed (4°C), and incubated (45 min, 0°C) with anti-ferritin antiserum (diluted 1:100). The ferritin molecules are collected in small clusters and patches and are embedded in a material of moderate density corresponding to the antibody. Incipient self-agglutination of the microvilli (× 60,000)

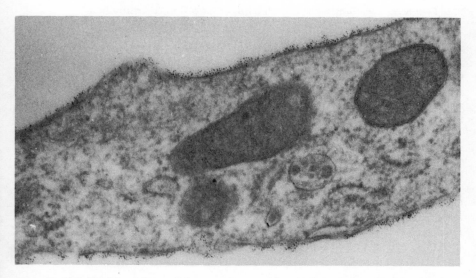

Figure 8. 3T3 fibroblast of a monolayer grown on a glass surface. Cells stained for 20 min at room temperature, washed and fixed *in situ*. Cells detached with a rubber policeman after glutaraldehyde fixation. The upper side of the cell corresponds to the exposed surface of the cell, the lower side probably corresponds to the surface in contact with the substrate. The ferritin is distributed over most of the exposed surface, but is more concentrated in some cases (× 50,000)

7.2—7.4, or by resuspending directly the pellet in buffered 2.5—3.0% glutaralde-hyde. After 5—10 min the cells are pelleted by centrifugation and left in the fixative for 1—48 h. They are then processed according to standard electron microscopic techniques [e.g. the pellets are washed for 30—60 min (2 changes) with saline, post-fixed in OsO$_4$ (1—2 h at 0°C), washed once with distilled water, washed (1 h) with 0.5% uranyl acetate in distilled water (0°C), rinsed with water, dehydrated and embedded].

Cells attached to a substrate (e.g. fibroblasts) can be labelled and treated in a similar way. It is convenient to grow the cell in small culture dishes (or small vials, like scintillation counter vials) to limit the volume of conjugate solution required. Samples of 1—2 x 10^6 cells are usually adequate. After each step the cells can be washed rapidly by rinsing the monolayers 3—4 times with medium. The cells can be fixed *in situ* with glutaraldehyde and then detached using a rubber policeman, pelleted, post-fixed and embedded as indicated above (Figure 8). Alternatively, they may be embedded *in situ* [e.g. on a thick coverslip using inverted gelatin capsule filled with embedding resin; or by flat embedding a monolayer grown on 'Araldite' (20)]. The first method is much more practical for examination of a large number of cells, but the second allows the orientation of the monolayer for cutting cell sections along a chosen direction.

The same labelling procedures are applied to cells to be used for freeze etching or surface replication. After washing and fixation the cells are treated according to the appropriate techniques (see for instance, Bächi, Chapter 14 this volume).

2.2 Two-layer Labelling using Anti-ferritin or Anti-Con A Antibodies

In some cases, to increase the degree of cross-linking (which often induces a more marked redistribution of the receptors, a tighter packing of the Con A—ferrit-in/receptor complexes and usually increased agglutination [13,15,21]) the washed Con A—ferritin-labelled cells are further incubated for 15—30 min at the chosen temperature with an antiserum (or antibody-containing IgG fraction) directed against Con A or against ferritin. Anti-ferritin antibodies are perhaps preferable as they do not disturb the binding of Con A and react with the most accessible part of the complex. Relatively diluted antibody concentrations are used (e.g. good titre anti-ferritin antisera diluted 1:100 or 1:200, which usually corresponds to $\leqslant 50$ μg/ml of specific antibody); it has been noted by the authors that sometimes high antibody concentrations can cause some detachment and loss of Con A —ferritin molecules from the membrane, for reasons which are unclear but that may be somehow connected with the fact that the binding of Con A to individual Con A receptor sites is probably relatively weak and that the rigid constraint imposed by the antibody may hamper reassociation when the bond dissociates reversibly.

At the end of the incubation with the antibody, the cells are washed once and fixed, or fixed directly without washing (Figure 7).

2.3 Labelling of Fixed Cells

This procedure has been used to study the 'normal' distribution of the Con A receptors in various fibroblastic cell lines (13—15), in lymphocytes and thymocytes, liver-derived cell lines, subcellular fractions, etc. (e.g. 22; de Petris and Raff, unpublished data). In all these cases the distribution of the receptors was found to be uniformly disperse (i.e. essentially random) within the limits of resolution of the technique (Figure 2), although some long-range variation of the receptor density in different areas of the cell has been observed in particular instances (de Petris; unpublished observation). Using glutaraldehyde fixation it has been possible to establish that the differences in the aggregation of receptors observed on membranes of normal trypsinized and transformed cells (23,24) do not exist prior to the binding of Con A but are actually due to the secondary phenomena of redistribution (13—15).

The fixative, almost universally used to immobilize Con A receptors is glutaraldehyde. Immobilization is probably very rapid at $\geqslant 20°C$ and using 2—3% glutaraldehyde is probably completed within a few minutes. At $4°C$, as judged by the inhibition of agglutination, fixation by glutaraldehyde is completed within 30 min (16). If the cross-linking by the aldehyde causes some displacements and clustering of membrane proteins, these are probably below the limit of resolution of the Con A—ferritin labelling technique and in fact they are not apparently reflected in the observed distribution of the Con A—ferritin molecules. Formaldehyde (e.g. 4% in phosphate buffer) could also be used to immobilize receptors, although it is a less efficient protein cross-linking reagent (16), but is has not been extensively employed for electron microscopy. Immobilization is probably more or less complete after 1—2 h, but no detailed data is available. Some reduction of Con A binding occurs after glutaraldehyde fixation, but practically no reduction occurs after formaldehyde fixation (16).

The cells, either in suspension or in a monolayer, are fixed with 3% glutaraldehyde as indicated above for 60—120 min at room temperature or at $37°C$. They are washed once with PBS and then (20—30 min) with 0.05 M NH_4Cl, made isotonic with phosphate-buffered saline. The cells are washed two additional times with PBS and then labelled with the conjugate (e.g. for 60 min at room temperature) as in the case of unfixed cells. The washing step with NH_4Cl has the purpose of blocking free glutaraldehyde sites on the cell surface which might bind non-specifically conjugate molecules. If the cells are well fixed, however, and the conjugate is free of aggregates, this step does not seem strictly necessary and prolonged washing with PBS alone may be adequate as shown by the fact that very few ferritin molecules remain attached to the cells if the conjugate is subsequently added to them in the presence of 0.1 M α-methyl mannoside (13) (Figure 3). After incubation the cells are quickly washed twice with PBS by centrifugation, resuspended and fixed in the cold with 1% OsO_4 in veronal acetate buffer, which precipitates onto the membrane the Con A—ferritin which is bound to it. The cells are immediately pelleted and then processed as usual. If the cell pellet tends to disaggregate during processing, as is sometimes the case with prefixed cells, the cells may be embedded in a small drop of 1% agar at $37°—40°C$ after block staining with uranyl acetate and before dehydration.

In some cases pre-fixation may also be employed to study the distribution of Con A receptors resulting from the binding of *native Con A*. Membrane-bound unlabelled Con A can be visualized as a fuzzy material covering the cell surface (11). To increase its detection without causing further alterations in the distribution, the Con A—labelled cells can be fixed with glutaraldehyde or formaldehyde as indicated above and then stained with an anti-Con A antibody—ferritin conjugate.

2.4 Control for the Specificity of Labelling

A control for the specificity of labelling is usually done by exposing the cells to Con A—ferritin in the presence of a sugar which competitively inhibits the binding. The sugar inhibitor commonly used is α-methyl-mannoside which binds to Con A with relatively high affinity (cf. Liener and Goldstein, this volume). Glucose also binds to Con A, but much less strongly, thus the amount of glucose normally present in tissue culture media (about 10 mM) is insufficient to prevent appreciable binding. The isotonic inhibitor is mixed with the Con A—ferritin conjugate (e.g. one volume of 0.3 M sugar and two volumes of conjugate) before the latter is added to the cells, and the incubation is carried out as usual. If the sample is diluted with fresh medium for the first wash, the sugar should be present also in the diluting medium. The inhibition of labelling is virtually complete (13, 23) (Figure 3). If the inhibitor is added when the cells have been in contact with the conjugate already, the bound Con A—ferritin is usually only incompletely removed, in particular from the areas of contact between agglutinated cells.

2.5 Labelling of Isolated Cell Membranes

Con A—ferritin conjugates are also used to label cell surface receptors on membranes isolated and mounted on collodion-carbon-coated grids according to the method of Nicolson and Singer (23—26). Membrane ghosts 'strengthened', but not fixed, by brief exposure (2 min) of the intact cell to 0.1% buffered formaldehyde are prepared by lysis of a droplet of concentrated cell suspension on the surface of distilled water (25,26), or 2 mM Tris buffer (21). The ghosts remaining on the water surface [which may correspond to only a small fraction (< 1%) of the lysed cells (21)] are collected on the coated grids, exposed to a 5% bovine serum albumin solution to reduce non-specific adsorption of the conjugate molecules, stained for 5—10 min with a droplet of the conjugate, washed with several drops of buffer and water and air dried. According to this technique only the exposed face of the flattened ghost is labelled. If the cells are labelled before lysis (21), both faces of the flattened ghosts are stained and the number of ferritin molecules counted per unit area of the ghost is twice the number per unit area of the cell surface. Topographical distortions may take place during lysis and flattening of the ghost, but typical membrane characteristics such as fluidity are still maintained.

3. EVALUATION OF THE CON A—FERRITIN METHOD

3.1 Comparison of Con A—Ferritin with Unconjugated Con A

The primary factors determining the changes in distribution of the Con A receptors upon the binding of the ligand appears to be the degree of cross-linking of the receptors by the multivalent ligand, which in turn depends on the valence of the reacting molecules. It is clear that when using Con A—ferritin conjugates as a ligand the analysis of the experimental results will be greatly facilitated if the conjugate is homogeneous, free of large aggregates and possibly consisting of monomeric complexes (see previous chapter), since polymeric complexes and large aggregates may drastically modify the pattern of the redistribution. As regards to its cross-linking properties, however, ferritin-conjugated Con A may not be completely equivalent to native Con A even in preparations from which large aggregates have been removed, as a certain heterogeneity in the effective valence of the ferritin-bound Con A is certainly introduced as a result of the conjugation. Inactivation or steric blockade of one or more binding sites may lower the effective valence of the complex, whereas the presence of more than one Con A molecule on some of the conjugate complexes may actually increase it. This second factor may be more important as it is likely to enhance clustering and local secondary redistribution phenomena, such as pinocytosis (Figure 4). Moreover, Con A and Con A—ferritin differ in size, since a conjugate complex, even a monomeric (1:1) one, has a volume at least 3—4 times as large as that of the unconjugated Con A molecule. The larger size might increase the hindrance to the binding of more than one molecule to receptors located very closely to each other, e.g. on different carbohydrate chains of the same molecule, thus reducing the total number of occupied receptors. The combined effect of all these factors is difficult to predict, but some differences in the redistribution effects and probably a comparatively lower quantitative binding of the Con A conjugates should be expected.

Some of these factors may be responsible for the reported observations that the number of bound Con A—ferritin molecules counted in thin sections of lymphocytes (21) or on surface replicas of erythrocytes (Bächi, this volume) is considerably lower than the number of unconjugated Con A molecules bound under comparable conditions. In some cases the lower binding may be simply due to the use of a conjugate containing too low a concentration of active Con A molecules (cf. previous chapter). Using thin sections the number of ferritin molecules present over the cell membrane may be underestimated as a result of the overlapping of the image of molecules present at different levels in the section (21). This effect, however, seems unlikely to account completely for the large discrepancies observed*, and it cannot explain the lower binding observed on surface replicas. Higher density of labelling with ferritin—Con A has been obtained on isolated membrane ghosts (21,23), although in some of these cases (21) the amount of bound ferritin might have been actually overestimated as the cells were labelled before being lysed and ferritin was present on both faces of the ghost (cf. section 2.5). It is not known whether fixation for conventional electron microscopy (and subsequent treatments) might fail to prevent detachment of some Con A—ferritin from the membrane which might not occur on the unfixed and quickly prepared membrane ghosts. Although the cause of these discrepancies remains unclear, it seems likely that the lower binding of Con A—ferritin on fixed and

---

*It can be estimated, for instance, that in order to account for the discrepancies observed by Yahara and Edelman (21) roughly 50% of the area of the Con A—ferritin-containing layer above the membrane in the section should have been covered by a dark mass of superimposed or contiguous ferritin molecules.

embedded cells is in fact real, especially at high labelling density, and that Con A—ferritin conjugates may not be suitable for quantitative determinations at a density of labelling exceeding $1-3 \times 10^3$ molecules/μm^2. Under conditions of reduced labelling it is possible that Con A—ferritin binds preferentially to particular subpopulations of Con A receptors (e.g. those of higher valence or affinity).

3.2 Comparison with Other Labelling Methods

Despite the uncertainties and limitations of the Con A—ferritin method for quantitative determination, the method has the intrinsic advantage that the number of ferritin molecules bound to the membrane is directly proportional to the number of the bound Con A molecules. This proportionality does not generally hold for indirect methods which use non-covalently bound glycoprotein markers, such as haemocynin (see Rosenblith-Borysenko, Chapter 11 this volume) and horse-radish peroxidase (see Avrameas, Chapter 7 this volume), which bind to the Con A-binding sites with their own Con A receptors. These methods have the advantage that they utilize the unmodified Con A as specific ligand in the first step of the procedure, which allows a better correlation to be established between the distribution of Con A and its other biological properties (e.g. agglutination), despite the fact that the competitive binding of the marker to the binding sites of the Con A molecules which have reacted with the surface receptors may modify the equilibrium of this reaction. For quantitative purposes, however, these methods have the more serious disadvantage that the extent to which the marker binds to Con A strongly depends on the characteristics (multiple binding, degree of cross-linking) of the Con A—Con A surface receptor reaction, which in general will vary in different cell types and will leave a different fraction of the Con A sites available for the reaction with the marker. A clear example of this effect has been illustrated by Collard and Temmink (27) who have shown that much less peroxidase staining was present on transformed than on normal fibroblasts although the amount of Con A bound to the cells was similar in the two cases.

Also as regards its resolution for the localization of the Con A receptors, the Con A—ferritin method seems to be generally superior to the indirect methods, and, therefore, more suitable for the detection of short-range redistribution and clustering of these receptors. The location of a Con A receptor is expected to be within $10-12$ nm of the position of the ferritin core whereas the resolution of haemocyanin indirect staining is of the order of $40-50$ nm. The variable size of the reaction product of peroxidase under the usual reaction conditions (28) and problems of diffusion of this product over the membrane (29) makes peroxidase indirect labelling even less suitable for the study of the fine distribution of receptors, although the method is probably more practical than ferritin labelling for the rapid scanning of the gross features of the distribution in a large member of cells.

In conclusion, Con A—ferritin has the main drawbacks that it may be unable to give a high density of labelling and it may have cross-linking and redistribution-inducing characteristics somewhat different from those of the unconjugated Con A. It has, however, the advantage of giving a degree of labelling proportional to the actual amount of bound Con A, and of being probably the label with the highest resolution for the localization of surface receptors.

4. REFERENCES

1. Taylor, R. B., Duffus, W. P. H., Raff, M. C. and de Petris, S. (1971) 'Redistribution and pinocytosis of lymphocyte surface immunoglobulin molecules induced by anti-immunoglobulin antibody', *Nature (New Biol.)*, 233, 225—9.
2. Frye, L. D. and Edidin, M. (1970) 'The rapid intermixing of cell surface antigens after formation of mouse—human heterokaryons', *J. Cell Sci.*, 7, 319—35.
3. Raff, M. C. and de Petris, S. (1972) 'Antibody—antigen reactions at the lymphocyte surface: implications for membrane structure, lymphocyte activation and tolerance induction', in *Cell Interactions*, ed. Silvestri, L. G. (North-Holland Publishing Co., Amsterdam—London), pp. 237—43.
4. de Petris, S. and Raff, M. C. (1972) 'Distribution of immunoglobulin on the surface of mouse lymphoid cells as determined by immunoferritin electron microscopy. Antibody-induced, temperature-dependent redistribution and the implications for membrane structure', *Eur. J. Immunol.*, 2, 523—35.
5. Singer, S. J. and Nicolson, G. L. (1972) 'The fluid mosaic model of the structure of cell membranes', *Science*, 175, 720—31.
6. Scandella, C. J., Devaux, P. and McConnell, H. M. (1972) 'Rapid lateral diffusion of phospholipids in rabbit sarcoplasmic reticulum', *Proc. Nat. Acad. Sci. U.S.A.*, 69, 2056—60.
7. Loor, F., Forni, L. and Pernis, B. (1972) 'The dynamic state of the lymphocyte membrane. Factors affecting the distribution and turnover of surface immunoglobulins', *Eur. J. Immunol.*, 2, 203—12.
8. de Petris, S. and Raff, M. C. (1973) 'Normal distribution, patching and capping of lymphocyte surface immunoglobulin studied by electron microscopy', *Nature (New Biol.)*, 241, 257—9.
9. de Petris, S. and Raff, M. C. (1974) 'Ultrastructural distribution and redistribution of alloantigens and concanavalin A receptors on the surface of mouse lymphocytes', *Eur. J. Immunol.*, 4, 130—7.
10. Edelman, G. M., Yahara, I. and Wang, J. L. (1973) 'Receptor mobility and receptor—cytoplasmic interactions in lymphocytes', *Proc. Nat. Acad. Sci. U.S.A.*, 70, 1442—6.
11. de Petris, S. (1975) 'Concanavalin A receptors, immunoglobulins and θ antigens on the lymphocyte surface: interactions with concanavalin A and with cytoplasmic structures', *J. Cell. Biol.*, 65, 123—46.
12. Yahara, I. and Edelman, G. M. (1975) 'Modulation of lymphocyte receptor mobility by locally bound concanavalin A', *Proc. Nat. Acad. Sci. U.S.A.*, 72, 1579—83.
13. de Petris, S., Raff, M. C. and Mallucci, L. (1973) 'Ligand-induced redistribution of concanavalin A receptors on normal, trypsinized and transformed fibroblasts', *Nature (New Biol.)*, 244, 275—8.
14. Rosenblith, J. Z., Ukena, T. E., Yin, H. H., Berlin, R. D. and Karnovsky, M. J. (1973) 'A comparative evaluation of the distribution of concanavalin-binding sites on the surface of normal, virally-transformed, and protease-treated fibroblasts', *Proc. Nat. Acad. Sci. U.S.A.*, 70, 1625—9.
15. Nicolson, G. L. (1973) 'Temperature-dependent mobility of concanavalin A sites on tumour cell surfaces', *Nature (New Biol.)*, 243, 218—220.
16. Inbar, M., Huet, C., Oseroff, A. R., Ben-Bassat, H. and Sachs, L (1973) 'Inhibition of lectin agglutinability by fixation of the cell surface membrane', *Biochim. Biophys. Acta*, 311, 594—9.
17. Inbar, M., Ben-Bassat, H. and Sachs L. (1971) 'A specific metabolic activity on the surface membrane in malignant cell-transformation', *Proc. Nat. Acad. Sci. U.S.A.*, 68, 2748—51.
18. Noonan, K. D. and Burger, M. M. (1973) 'Binding of ^{3}H-concanavalin A to normal and transformed cells', *J. Biol. Chem.*, 248, 4286—92.
19. Yahara, I. and Edelman, G. M. (1973) 'The effects of concanavalin A on the mobility of lymphocyte surface receptors', *Exp. Cell. Res.*, 81, 143—55.
20. Abercrombie, M., Heaysman, J. E. M. and Pegrum, S. M. (1971) 'The locomotion of fibroblasts in culture. IV. Electron microscopy of the leading Lamella', *Exp. Cell. Res.*, 67, 359—67.
21. Yahara, I. and Edelman, G. M. (1975) 'Electron microscopic analysis of the modulation of lymphocyte receptor mobility', *Exp. Cell. Res.*, 91, 125—42.

22. Matus, A., de Petris, S. and Raff, M. C. (1973) 'Mobility of concanavalin A receptors in myelin and synaptic membranes', *Nature (New Biol.)*, **244**, 278–80.
23. Nicolson, G. L. (1971) 'Difference in topology of normal and tumor cell membranes as shown by different surface distributions of ferritin-conjugated concanavalin A', *Nature (New Biol.)*, **233**, 244–6.
24. Nicolson, G. L. (1972) 'Topography of membrane concanavalin A sites modified by proteolysis', *Nature (New Biol.)*, **239**, 193–7.
25. Nicolson, G. L. and Singer, S. J. (1971) 'Ferritin-conjugated plant agglutinins as specific saccharide stains for electron microscopy: application to saccharides bound to cell membranes', *Proc. Nat. Acad. Sci. U.S.A.*, **68**, 942–5.
26. Nicolson, G. L. and Singer, S. J. (1974) 'The distribution and asymmetry of mammalian cell surface saccharides utilizing ferritin-conjugated plant agglutinins as specific saccharide stains', *J. Cell Biol.*, **60**, 236–48.
27. Collard, J. G. and Temmink, J. H. M. (1974) 'Binding and cytochemical detection of cell-bound concanavalin A', *Exp. Cell. Res.*, **86**, 81–6.
28. Sternberger, L. A. (1974) *Immunocytochemistry*, (Prentice-Hall, Inc., Englewood Cliffs, N.J.), p. 172.
29. Matter, A., Lisowska-Bernstein, B., Ryser, J. E., Lamelin, J-P. and Vassalli, P. (1972) 'Mouse thymus-independent and thymus-derived lymphoid cells. II. Ultrastructural studies', *J. Exp. Med.*, **136**, 1008–30.

CHAPTER 10

Application of Concanavalin A–Ferritin to Subcellular Fractions

HELMUT BITTIGER

1. METHOD

Subcellular material (100–250 μg protein) is incubated in 200 to 500 μl phosphate-buffered saline (PBS), pH 7.3, containing Con A–ferritin (1–5 mg/ml protein of active Con A–ferritin conjugates purified on 'Sephadex' G-200) for 30 min at room temperature. Care has to be taken that the subcellular material is well dispersed (e.g. by extensive homogenization with a glass–'Teflon' homogenizer). The incubation is terminated by diluting the suspension about 20-fold with PBS and centrifugation ($2–3 \times 10^5 g \times$ min). The pellet is rinsed several times with PBS. Visual inspection of the pellets showed that this washing procedure is adequate to remove unbound Con A–ferritin in most cases. Concerning the force and the time of centrifugation two effects have to be considered: on the one hand, all subcellular material must be pelleted to obtain a representative picture of the structures, on the other hand, the packing density of the pellets increases with the time and strength of centrifugation and may lead to difficulties in recognizing unambiguously subcellular particles in the electron micrographs. Thus, the details of the centrifugation have to be optimized for each type of subcellular material.

The sedimented material is fixed with 2% glutaraldehyde in Millonig's buffer (1), pH 7.2, for 1 hour at room temperature, followed by extensive washing with PBS and post-fixation with osmium tetroxide at 4°C for 1–2 hours. Embedding and staining are performed according to standard procedure (2). Specificity controls for Con A—ferritin binding are obtained by incubation of the subcellular material with Con A—ferritin in PBS containing 0.1—0.2 M α-methyl mannoside.

Subcellular fractions can also be fixed with glutaraldehyde prior to Con A—ferritin incubation. However, in this case unspecific reactions of Con A—ferritin with reactive aldehyde groups of monovalently membrane-bound bifunctional glutardialdehyde must be excluded by reacting the fixed material with an excess of amines (e.g. ethanolamine, lysine, glycine) or ammonium chloride prior to incubation with Con A—ferritin. In any case, unspecific binding has to be assessed by control incubations done in the presence of hapten inhibitor. Pelleting of the subcellular material is also possible by filtration on 'Millipore' filters (3). Removal of unbound Con A—ferritin is somewhat more difficult with the 'Millipore' method than with the centrifugation method.

2. APPLICATION OF CON A—FERRITIN TO SUBCELLULAR COMPONENTS OF RAT BRAIN

2.1 Preparation of Subcellular Fractions

A crude mitochondrial fraction is obtained according to standard procedures (4). A fraction of synaptic junctions is prepared following the procedure of Cotman and coworkers (5). Synaptosomal membranes, isolated from the crude mitochondrial fraction according to the technique of Levitan and coworkers (6), are treated with the detergent 'Triton' X-100 at a concentration of 4 mg/ml (protein and 'Triton' X-100 in a ratio 1:1 by weight). The resulting suspension is layered on a discontinuous sucrose gradient (5) to isolate the synaptic junctions.

2.2 Binding of Con A—ferritin to Fractions

In Figures 1 to 5 representative examples for the binding of Con A—ferritin to subcellular components of rat brain are shown. The outer membranes of mitochondria (M) are virtually devoid of binding sites (Figure 1); in contrast a very high density of Con A receptors is observed on the postsynaptic membrane (P) of the synaptic junction in the region of the subsynaptic web (Figure 1). This location of Con A receptors is more clearly demonstrated in partially opened clefts (Figure 2). Extensive staining with Con A-ferritin is only observed on those parts of the postsynaptic membrane that are connected with the subsynaptic web. As demonstrated in Figures 2 and 3, Con A binding sites are not accessible, or only partially, for Con A—ferritin in intact synaptic junctions. Staining on the presynaptic membrane has been found to be much less intense than on the postsynaptic membrane. In fractions enriched in synaptic junctional complexes (Figure 4) the binding of Con A—ferritin occurs almost exclusively at the region of contact between the subsynaptic web and the postsynaptic membrane. A similar staining pattern is observed when the postsynaptic membrane has been removed by detergents, suggesting that the Con A receptors are anchored within the subsynaptic web.

With the exception of mitochondria and the postsynaptic membranes subcellular

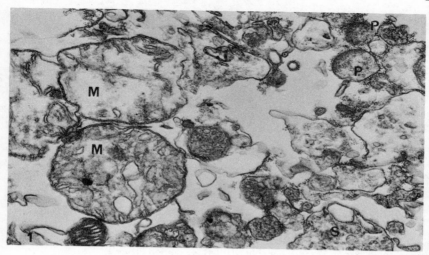

Figure 1. Subcellular fraction from rat brain incubated with Con A—ferritin. Crude mitochondrial fraction. Mitochondria and separated postsynaptic particles (x 32,400). M, Mitochondria; P, postsynaptic part of synaptic junctional complex; S, synaptosomes

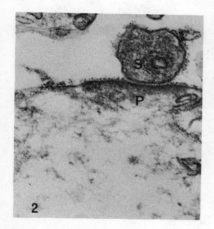

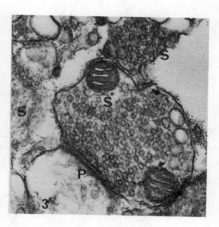

Figure 2. Subcellular fraction from rat brain incubated with Con A—ferritin. Crude mitochondrial fraction. Synaptic junctions (x 50,000). For abbreviation see caption Figure 1

Figure 3. Subcellular fraction from rat brain incubated with Con A—ferritin. Crude mitochondrial fraction. Synaptosomes (x 44,800). For abbreviations see caption Figure 1

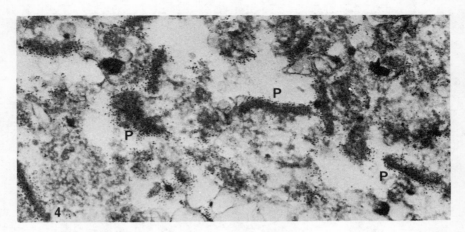

Figure 4. Subcellular fraction from rat brain incubated with Con A—ferritin. Isolated synaptic junctional complexes (x 52,000). For abbreviations see caption Figure 1

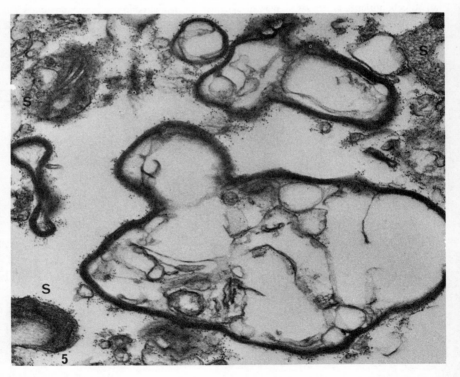

Figure 5. Subcellular fraction from rat brain incubated with Con A—ferritin. Crude mitochondrial fraction. Myelin (x 44,800). For abbreviations see caption Figure 1

components from rat brain do not show a distinctive staining pattern. Synaptosomes, axons and myelin bind much less Con A—ferritin than postsynaptic membranes and it appears in a scattered and dispersed distribution, although clusters of Con A—ferritin are observed in rare instances. In general, synaptosomes bind less Con A—ferritin than myelin. When the subcellular material is prefixed with glutaraldehyde and subsequently reacted with glycine, the staining patterns have been found to be similar as in unfixed preparations; under these conditions, however, intact synaptic junctions are completely inaccessible to Con A—ferritin conjugates.

Our findings pertinent to the subcellular distribution of Con A receptors in rat brain (7) are in general agreement with those reported by other authors (8,9).

The unequal distribution of Con A receptors among subcellular components of rat brain was exploited for the separation of mitochondria from crude mitochondrial fractions and for the enrichment of postsynaptic parts of synaptic junctional complexes (see Chapter 47).

3. PROBLEMS IN CON A—FERRITIN BINDING

3.1 Accessibility of Con A Receptor Sites to Con A—Ferritin

In Con A—ferritin conjugates 1—10 Con A molecules are coupled to one ferritin molecule. Such conjugates thus represent large macromolecular units. The external diameter of the ferritin molecule and the major axis of the Con A molecule measure about 120 Å (10) and 80 Å (see Chapter 3) respectively. Therefore, the diameter of Con A—ferritin conjugates might reach 280 Å. Obviously macromolecules of such size cannot penetrate intact membrane vesicles, and their location is confined to the surface membranes of these structures. Binding of such large macromolecular units to the Con A receptors may be restricted by the presence of a dense glycocalyx of the membranes. In addition, Con A binding sites at specialized cell contacts such as desmosomes, gap junctions or synaptic junctions will not be accessible or at least only partially so (see Figures 2 and 3). The accessibility of Con A—ferritin binding sites might further be restricted because of excessive packing of subcellular components in pellets; careful homogenization of the material is, therefore, a prerequisite for obtaining meaningful results in such binding studies. In general, only damaged subcellular particles reveal internal binding sites.

3.2 Concentration of Con A—Ferritin for Incubation

The dissociation constant for Con A and its binding sites on erythrocytes has been found to be 2.5×10^{-7} M (11). Thus the concentrations in the range of several mg/ml of active Con A—ferritin conjugates, corresponding to about $0.3-2 \times 10^{-5}$ M Con A are sufficient to saturate Con A binding sites on membranes; concentrations of 100 μg/ml (10^{-6} M) of active Con A are generally considered as the lower limit.

3.3 Con A—Ferritin as a Charged Macromolecular Unit

Ferritin (12,13) and Con A (see Chapter 3) are negatively charged at neutral pH. This may lead to repulsion of Con A—ferritin conjugates by highly negatively charged membrane surfaces; alternatively these complexes may be attracted by positively charged surfaces. Therefore, it is important to be aware of possible electrostatic interactions between the binding substrate and the conjugates, and to minimize such charge effects by the use of sufficiently high ionic concentrations.

3.4 Reliability of Con A—Ferritin Staining for Con A Binding Sites

It was discussed above that the accessibility of Con A binding sites for Con A—ferritin is a major problem of this method. A second point of concern is the possible chemical change of Con A induced by the coupling procedure which may lead to a decreased affinity of Con A for its receptors. In addition, the mobility of Con A conjugates is probably reduced compared to that of free Con A. All these effects taken together imply an incomplete staining of Con A binding sites by Con A—ferritin conjugates. In fact, a direct comparison of the number of bound Con A molecules and Con A—ferritin conjugates in freeze-fractured preparations (Chapters 13 and 14) showed that only about 20% of the Con A binding sites are occupied by the conjugates. The method, therefore, gives a qualitative rather than a quantitative picture of Con A binding sites on membranes. The main advantage of the method is the very high electron microscopical resolution of binding sites due to the defined form and the high contrast of the ferritin iron core and the possibility of staining the specimens for high contrast.

4. REFERENCES

1. Millonig, G. (1962) *Proc. Fifth. Int. Congr. Electron Microscopy*, Philadelphia, Aug. 29—Sept. 5, ed., Breese, S. S. Jr. (Academic Press Inc., New York), p. 8.
2. Fraska, J. M. and Parks, V. R. (1965) 'A routine technique for double staining ultrathin sections using uranyl and lead salts', *J. Cell. Biol.*, **25**, 157—61.
3. Baudhuin, P., Evrard, P. and Berthet, J. (1967) 'Electron microscopic examination of subcellular fractions', *J. Cell. Biol.*, **32**, 181—91.
4. Whittaker, V. P. (1969) 'The synaptosome', in *Handbook of Neurochemistry*, Vol. 2, ed. Lajtha, A. (Plenum Press, New York, London), pp. 327—64.
5. Cotman, C. W. and Taylor, D. (1972) 'Isolation and structural studies on synaptic complexes from rat brain', *J. Cell Biol.*, **55**, 696—711.
6. Levitan, I. B., Mushynski, W. E. and Ramirez, G. (1972) 'Highly purified synaptosomal membranes from rat brain', *J. Biol. Chem.*, **247**, 5376—81.
7. Bittiger, H. and Schnebli, H. P. (1974) 'Binding of concanavalin A and ricin to synaptic junctions of rat brain', *Nature*, **249**, 370—1.
8. Matus, A., de Petris, S. and Raff, M. C. (1973) 'Mobility of concanavalin A receptors in myelin and synaptic membranes', *Nature (New Biol.)*, **244**, 278—80.
9. Cotman, C. W. and Taylor, D. (1974) 'Localization and characterization of concanavalin A receptors in the synaptic cleft', *J. Cell. Biol.*, **62**, 236—42.
10. Hoy, T. G., Harrison, P. M. and Hoare, R. J. (1974) 'Quaternary structure of apoferritin: The rotation function at 9 Å resolution', *J. Mol. Biol.*, **84**, 515—22.
11. Schnebli, H. P. and Bächi, T. (1975) 'Reaction of lectins with human erythrocytes', *Exp. Cell. Res.*, **91**, 175—83.
12. Crichton, R. R. (1973) 'Struktur und Funktion von Ferritin', *Angew. Chemie*, **85**, 53—62.
13. Danon, D., Goldstein, L., Marikovsky, H. and Skutelsky, E. (1972) 'Use of cationized ferritin as a label of negative charges on cell surfaces', *J. Ultrastr. Res.*, **38**, 500—10.

CHAPTER 11

Haemocyanin Labelling for Visualization of Concanavalin A on Platinum–Carbon Replicas of the Cell Surface

J. Z. ROSENBLITH-BORYSENKO

1. PRINCIPLE OF THE METHOD

Smith and Revel (1) originally developed this simple technique in which *Busycon canaliculatum* haemocyanin, a large molecule of distinct size and shape, is employed as a morphological marker for Con A previously bound to the cell surface. The method has been used primarily in conjunction with conventional platinum—carbon replicas (1—4). The three-dimensional replica preparations allow mapping of the distribution of Con A binding sites (CABS) on the upper surface of intact cells, using transmission electron microscopy. In addition, the haemocyanin can be visualized both in thin section (3,5) and in the scanning electron microscope (6).

The labelling procedure is conducted in two steps. Intact cells are first treated with Con A and rinsed to remove excess lectin. Haemocyanin, in an appropriate buffer, is then applied to the cells. Free valences remaining on the membrane-bound Con A presumably react with sugar residues on the haemocyanin, binding the marker molecule to the cell. In this respect, the technique is similar to peroxidase labelling of Con A since both can be employed without prior coupling. The resultant Con A—haemocyanin complex (Con A/H) is then stabilized by aldehyde fixation. The haemocyanin molecule (m.wt. $\sim 0.5-9 \times 10^6$) is shaped like a truncated cylinder, 350 Å in diameter. Individual haemocyanin molecules, therefore, are easily resolved both by their constant size and by their characteristic appearance either as circles (end-on view) or squares (side view).

2. PREPARATION OF CELL MONOLAYERS

In order to produce replicas of the cell surface, cells should be prepared as monolayers on glass coverslips. Tissue culture cell lines are seeded directly onto glass coverslips (Corning) and grown to the desired confluency. Consistency in monolayer density will obviate several problems. First, cells that are touching are probably not physiologically equivalent to isolated cells. Monolayers in which some cells are isolated and others in close contact may, therefore, represent a heterogeneous population with respect to physiological parameters. Secondly, in very dense monolayers, details of cell perimeters are often obscured by processes of neighbouring cells. In too sparse a monolayer cells are simply difficult to find.

Blood cells of many types which have been washed by centrifugation can be suspended in an appropriate buffer and pipetted onto a coverslip. A volume of 0.5 ml solution, containing 1×10^6 cells, gives an excellent monolayer when pipetted onto a 22 mm (diameter) round coverslip (Corning) which has been precleaned in 95% ethanol. These coverslips are then incubated at 37°C for 30—45 minutes, during which time a monolayer of cells will settle out and adhere to the coverslip. The coverslip is then rinsed in three serial changes of 37° buffer (holding the coverslip in forceps and gently swishing it in a beaker works well) to remove non-adherent cells and the monolayer is ready for experimentation or labelling. A handy incubation chamber can be made in a Petri dish lined with a piece of wet filter paper by supporting the coverslips on a flat rod (the grooved plastic supports that come in boxes of Clay-Adams Gold Seal microslides are perfect, the grooves giving protection against slipping without forming a tight seal by capillary action). This chamber can be used throughout the labelling procedure as well.

3. LABELLING OF CELLS WITH CON A

3.1 Treatment of Cells with Con A

Cells on coverslips are rinsed serially three times in Dulbecco's phosphate-buffered saline, pH 7.4 (PBS) to remove culture medium. Con A is dissolved in PBS at concentrations ranging from 5 μg/ml to 200 μg/ml depending on cell type and temperature, and 0.5 ml of the solution is pipetted onto the coverslip. With rabbit polymorphonuclear leucocytes, for instance (author's unpublished observations), 5 μg/ml Con A is sufficient for optimal labelling since the same density of labelling is observed as with 50 μg/ml Con A. With Swiss mouse 3T3 cells, however, we used 100 μg/ml Con A to correlate with [^{125}I] Con A-binding studies on these cells (2). This high concentration is, however, toxic to some cells such as lymphocytes. The optimal concentration will vary, then, both with cell type and temperature, and may be influenced by conditions used in simultaneous physiological measurements.

3.2 Effects of Temperature

Cells incubated at 37° apparently bind more Con A than cells incubated at 4°. For example, twice the amount of Con A (200 μg/ml) was required at 4°, to give a density of label on 3T3 cells equivalent to that obtained at 37° (2).

Con A is a multivalent ligand which is capable of cross-linking its own binding sites into clusters on several cell lines (2,3,7—9) if incubation proceeds at 37° (Figure 1). Incubation at 4°, where mobility in the lipid phase of the membrane is greatly restricted, however, prevents ligand-induced clustering of CABS and allows examination of inherent binding site topography. Therefore a temperature of incubation is chosen dependent on whether Con A is able to cross-link its binding sites on a given cell type, and on whether one is planning to investigate inherent or Con A-induced binding site distributions.

3.3 Effects of Fixation

Fixation of cells prior to treatment with Con A can also prevent ligand-induced clustering of binding sites, provided that an appropriate time and temperature of fixation is chosen (Figure 2). The fluid dynamic state (10) of membrane glycoproteins and/or glycolipids at the time of Con A/H treatment is a critical parameter of CABS distribution. It has been reported by Pinto da Silva (11) that fixation of erythrocyte ghosts in 1% glutaraldehyde at temperatures ranging from —16° to 25° is sufficient to restrict the translational mobility of proteinaceous membrane particles revealed by freeze fracture. The conditions of fixation leading to immobilization of other membrane components, however, may vary according to the physical—chemical properties of the particular membrane as well as the

Figure 1. Surface replica of part of an SV3T3 cell labelled with Con A/H at 37°. Note the uniform background binding of haemocyanin to glass (B). Clusters of Con A/H (arrows) present on the upper cell surface are characteristically excluded from pseudopodia (P) and cell edges (E). Isolated haemocyanin molecules (circle) are readily distinguished by size and square or round profile. (× 14,875)

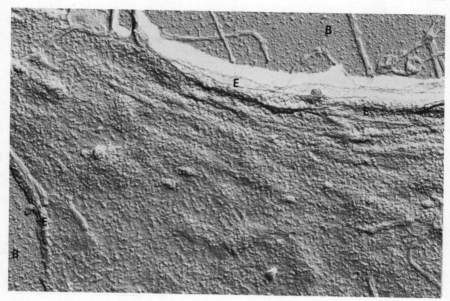

Figure 2. SV3T3 cell labelled with Con A/H at 4° to demonstrate the inherent dispersed distribution of CABS. Note that the label extends to the cell edges (E) and exhibits no spatial hetereogeneity. (x 14,875)

fixative and its molecular substrate. Prior treatment of fibroblast membranes with 0.1—0.3% formaldehyde for two minutes at room temperature does not prevent the movement of CABS within the plane of the membrane, as clusters of sites are observed after ferritin—Con A labelling of such preparations (12). We have likewise demonstrated that whereas prior fixation of SV3T3 or protease-treated 3T3 cells in 1% paraformaldehyde for ten minutes at 37° prevents Con A-induced movement of CABS, fixation at 4° for as long as 30 minutes does not prevent lateral movement of binding sites into clusters (2). Therefore, it is recommended that fixation be carried out at a minimum of room temperature for ten minutes in 1% paraformaldehyde to insure immobilization of binding sites, if their inherent distribution is to be studied in this manner. The effects of temperature and fixation on CABS distribution on 3T3 and SV3T3 cells is summarized in Table I.

4. LABELLING OF CELLS WITH HAEMOCYANIN

4.1 Treatment of Cells with Haemocyanin

Following treatment with Con A, coverslips are rinsed serially three times in PBS at the temperature of Con A incubation. 0.5 ml of a solution of *B. canaliculatum* haemocyanin (500 μg/ml at 37° or 1 mg/ml at 4°) is then applied to the coverslip, which is incubated for ten minutes at the temperature of choice.

TABLE I.   Con A/H treatment of 3T3 and SV3T3 cells

Cells	Con A/H[a] labelling (°C)	Incubation[b] post Con A/H (°C)	CABS[d]
3T3	4	4	Dispersed
	4	37	Dispersed
	37	37	Dispersed
3T3 Pre-fix[c]	37	37	Dispersed
SV3T3	4	4	Dispersed
	4	37	Clustered
	37	37	Clustered
	37	4	Clustered
SV3T3 Prefix[c]	37	37	Dispersed

[a]Cells were treated with Con A (10 min), rinsed three times in PBS, treated with haemocyanin (10 min), and rinsed and
[b]mono-layers were then incubated for an additional 10 min.
[c]Fixation before Con A/H labelling was in 1% paraformaldehyde for 10 min.
[d]CABS = Con A-binding-site topography after Con A/H labelling.

4.2 Effects of Temperature

When Con A treatment has been carried out at 4° to map the inherent CABS distribution, haemocyanin incubation should also be carried out at 4°. Temperatures of both incubations, of course, vary according to experimental design. We have found (2) that both the Con A and haemocyanin contribute to formation of CABS clusters in the following way. SV3T3 cells incubated with Con A at 37°, then cooled to 4° before addition of haemocyanin, show loose clusters of CABS, i.e. the haemocyanin molecules are loosely packed within the clusters. If, however, cells treated as above are warmed again to 37° after haemocyanin treatment, tight clusters of CABS are formed. The lectin itself, therefore, is responsible for the formation of clusters which are then further cross-linked by the haemocyanin.

4.3 Effects of Fixation

When cells have been fixed in 1% paraformaldehyde prior to Con A treatment, haemocyanin labelling can proceed at room temperature.

4.4 Preparation of Haemocyanin

This method for the purification of haemocyanin from whelk haemolymph was developed by Karnovsky (5). One of these large marine whelks, *Busycon canaliculatum* (Woods Hole Biological Laboratories, Woods Hole, Mass.) will yield 20–50 ml of haemolymph. The haemolymph is obtained by drawing the operculum out of the shell and holding it with forceps while cutting deeply into the foot with a scalpel. The animal may then be inverted into a large funnel and the haemolymph collected in an iced beaker. The crude haemolymph should then be purified as follows:

(1) Filter through layer of glass wool.

(2) Centrifuge at 2000g for 20 minutes at 4° to get rid of debris. Discard pellet.

(3) Spin supernatant at 100,000g for 100 minutes at 4°. Discard clear supernatant.

(4) The resultant pellet has two layers, a dark-blue gelatinous layer which is not easily disturbed and a light-blue flocculent layer above it. Carefully resuspend the flocculent upper layer in 1% saline and discard.

(5) Resuspend the gelatinous layer in 1% saline at 4° by carefully rubbing it against the tube with a glass rod. This is somewhat tedious, but should be completely accomplished.

(6) Repeat step 3 twice.

(7) Resuspend pellet in 3% saline at 4°.

(8) Measure protein content by the Lowry method (13) against a standard curve constructed from micro-Kjeldahl estimations. *B. Canaliculatum* haemolymph contains approximately 7 mg/ml of protein, so resuspension of the pellet in 3% NaCl to the original volume will give a fair estimation of haemocyanin concentration.

(9) Sterilize by filtration through a 'Millipore' filter (Falcon Plastics, Oxnard, Calif.), pore size 0.45 μ.

(10) Store in refrigerator. Batches we prepared two years ago are still good.

4.5 Controls for Specificity of Labelling

4.5.1 Application of Haemocyanin without Con A

If coverslips which have been cleaned in dilute nitric acid or 95% ethanol are treated with haemocyanin alone, there is no labelling (1). Experimental mono-layers, however, should be tested for possible haemocyanin binding in case some cellular product should cause non-specific adherence of label. In our experience, however, this has not happened.

4.5.2 Reversal of Con A Binding

Following Con A/H labelling, coverslips are incubated in 50 mM α-methyl-D-glucopyranoside for 60 minutes (2) to reverse Con A binding.

5. POST-LABELLING INCUBATION AND FIXATION

5.1 Incubation

Following the ten-minute incubation in haemocyanin, coverslips are again rinsed serially three times in PBS. 0.5 ml PBS at the appropriate temperature for experimental design is then pipetted onto the coverslip which is incubated for an additional ten minutes or whatever time is deemed necessary to allow the cell to react to the bound Con A/H complex (e.g. the complex may patch, cap and/or be pinocytosed).

5.2 Fixation and Preparation of Tissue

5.2.1 Replicas

Monolayers are fixed in 1% glutaraldehyde in cacodylate buffer (pH 7.4) for ten minutes at room temperature. After brief post-fixation (ten minutes) in 2% aqueous OsO_4, the monolayers are quickly dehydrated in a graded series of ethanols. After two rinses in absolute ethanol, the monolayers are submerged in amyl acetate (mixed isomers) for a minute and dried quickly in a warm air stream from an Oster hair dryer. This drying technique was an expedient innovation of Smith and Revel (1) who found that it gave results comparable to those obtained with critical-point drying. We have tried both methods and agree that there is no detectable difference in quality. Drying from amyl acetate in warm air, however, is clearly the simpler method. Care must be taken to avoid drying between transfer from alcohol to amyl acetate.

5.2.2 Thin Sections

When preparing monolayers for thin sectioning, eventual separation of the embedding medium from the coverslip is facilitated by giving the coverslip a *light* carbon coating before making the monolayer. Too thick a layer will flake off in the various incubation solutions. Fixation and dehydration is as above, except that the tissues are run through propylene oxide and the embedding resin mixtures of choice rather than being dried. The cells can be easily embedded by inverting the coverslips onto an aluminium weighing pan filled with resin, or by filling three or four Beem (Ladd) capsules with resin and inverting them on top of the coverslip.

6. PREPARATION OF SURFACE REPLICA

6.1 Principle of the Method

A shadowed replica is a three-dimensional cast of surface detail produced by the high-vacuum evaporation of an electron-opaque metal (platinum) at a $35°-45°$ angle onto the surface of the dried monolayer. The platinum shadow cast, however, is too fragile to separate from the cells and examine in the microscope. Therefore, another metal (carbon) which is electron lucent but imparts strength to the fragile platinum, is evaporated at $90°$ to the surface of the monolayer. The resultant replica is very strong and is easily separated from the glass, washed and mounted on standard grids for electron microscopy.

The carbon electrode is set at $90°$ to the specimen. The electrode should be degassed by glowing it until the vacuum holds steady. The carbon is then evaporated in several 3–4 second bursts. The vacuum will decrease after firing and should be allowed to recover between bursts. A good carbon coating should impart a medium gray mirror finish to the replica. Too light a coating will fail to strengthen the replica which will then disintegrate in subsequent operations. Too heavy a coating yields a dark, granular appearance.

6.2 Separation of Replicas from Glass Coverslips

(1) The replicas are carefully marked off into 2 mm squares with a diamond pencil. Care must be taken to use a light touch so that the underlying glass is not scored.

(2) The coverslip, replica side up, is tilted into hydrofluoric acid (48%). To do this use a *plastic* (not glass) Petri dish, filled to a depth of about 0.5 cm with the acid. The coverslip is held at the very edge with forceps, and tilted 30° toward the surface of the acid. The edge of the coverslip opposite the forceps is angled into the acid *very* gradually. At first, the coverslip should barely break the surface. In a few seconds, the acid will begin to loosen the replica from the glass. When you see the replica loosening, *slowly* advance the coverslip further into the solution, maintaining the 30° angle. The 2 mm squares of replica will float off the coverslip onto the surface and the bare coverslip can be removed from the acid. If you forget to score the replica, irregular fragments will come off onto the acid, and much may be lost in subsequent steps. If you advance the coverslip into the acid too rapidly, wetting the surface of the replica rather than allowing the acid to work between glass and replica, the wet portion of the replica is ruined. In this case, remove the coverslip from the acid and move the forceps so that the procedure can be reinitiated with a dry leading edge. *NOTE OF CAUTION:* Hydrofluoric acid is *extremely* corrosive. The operator should protect his clothing with a lab jacket and wear plastic gloves. The acid should only be used under a fume hood with good exhaust, and be placed far back in the hood to avoid any danger of inhalation or exposure of the eyes to vapour.

(3) After the pieces of replica are floating on the acid, the surface tension is reduced by squirting 2—3 ml of distilled water into the acid. The replicas will suddenly be attracted to the area into which the water enters, so the operator must watch for signs of this and begin squirting the water elesewhere.

(4) The replicas are then transferred to distilled water, using a wire loop (Ladd Research Industries, Inc., Burlington, Vt.). A drop of ethanol added to the water is a great aid in minimizing surface forces, thus preventing breakage of replicas.

(5) The replicas are next transferred onto 'Clorox' (sodium hypochlorite) for a one hour incubation to remove adherent cellular debris. Spot dishes are excellent for this purpose and other wells can be filled with water for rinses.

(6) The cleaned replicas are looped onto two changes of distilled water, and picked up from the final rinse on 200 mesh uncoated grids. The filled grids should be held at 90° to a piece of filter paper to draw liquid off from the side of the grid. If a wet grid is placed flat on filter paper, much of the replica will be drawn through the grid mesh and ruined.

7. PITFALLS

7.1 Non-specific Adsorption of Con A to Glass

Smith and Revel (1) showed that clean coverslips did not bind haemocyanin. When the glass was pretreated with Con A, however, haemocyanin was bound. This background binding is constant and does not vary with experimental manipulation, nor is it reversed by treatment with α-methyl glucoside (1). We tried to eliminate background binding by using carbon-coated coverslips, changing coverslip brands and cleaning in a variety of solvents. None of these measures influenced the background. Binding to the cell, however, which is reversible with sugars, is specific.

7.2 Steric Considerations

The large size of the haemocyanin molecule (350 Å) imposes a theoretical steric limitation to saturation of CABS, since the Con A molecule itself is considerably smaller. Smith and Revel (1) suggested that 200 molecules per square micron might approach the steric limit for haemocyanin binding. Since the stoichiometry of Con A/H binding has not been worked out, this method is best suited for obtaining qualitative information concerning the topographical distribution of CABS, rather than quantitative data.

7.3 Morphological Limitations

Cells which do not flatten out well on glass pose the difficulty of falling out of the replicas, leaving holes. This may also sometimes occur, when the density of cells is very great. Phillips and Perdue (14) introduced the technique of dipping the replicas in 0.2% 'Formvar' prior to removing them from glass. The 'Formvar' backing, while leaving some dark spots as a result of uneven drying, increases the strength of the replicas and prevents areas of greater relief from falling out. These authors also describe an artifact we have not seen. Using chicken embryo fibroblasts, they report that the cell margins had a melted, amorphous appearance while the background replicated well. They attribute this appearance to collapse and/or volatilization of surface components during evaporation of metals. They found that shadowing on a cold stage, or simply allowing dried specimens to rest in ambient room atmosphere for several days before shadowing, alleviated this problem. This effect, then, may occur on some cell types but not on others.

8. COMPARISON WITH OTHER METHODS

There are several advantages in using the Con A/H replica technique as compared to other methods. First, the technique allows visualization of the *entire* upper surface of the cell. As seen in Figure 3, CABS distribution is often inhomogeneous. If a thin section, or even a large series of thin sections were take from Zone 1, no label would be apparent. In Zone 2, label appears as small clusters associated with long microspikes of membrane. Zone 3 reveals a central cap of CABS. It would be laborious indeed to reconstruct such information from thin sections labelled with Con A—ferritin or Con A—peroxidase. Likewise, techniques including the use of cell ghosts labelled with Con A—ferritin cannot yield information concerning spatial distribution of CABS over the membrane surface as a whole.

Secondly, individual cells may be in somewhat different states in regard to how far toward patching or capping they have gone, or a cell population may be heterogeneous. With replicas one can easily sample hundreds of cells in little time. The task of such a sampling using thin sections would be enormous. Thin sections, of course, are required to demonstrate events occurring in the cytoplasm which

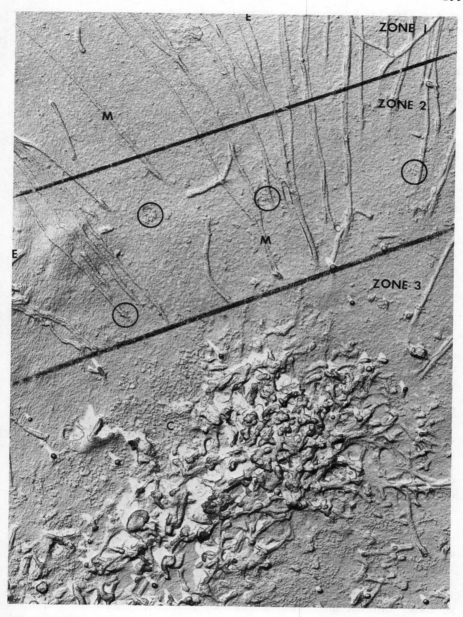

Figure 3. SV3T3 cell treated with 10^{-6} M colchicine before A/H labelling at 37°. Micro-spikes (M) extend from the central area of the cell (C) past the cell edges (E). Zone 1, near the cell periphery, contains only occasional haemocyanin molecules. Zone 2 shows several small clusters of label (circles) which are often associated with the points of origin of the microspikes. Zone 3 consists of a central cap of haemocyanin clusters associated with an area of membrane acitivity. (x 11,900)

may be related to experimental manipulation. Since haemocyanin is visible on thin sections after uranyl acetate staining, the same labelling technique may be used to prepare coverslips in tandem for replication and thin sectioning. It may be of considerable interest that Phillips and Perdue (14), using haemocyanin from the keyhole limpet, stained it in thin section using the polycationic dye, ruthenium red, according to Luft (15). This greatly enhances contrast as compared to conventional uranyl acetate staining.

Another advantage, in addition to those above, is the ability to prepare and examine replicas on equipment standard to any laboratory equipped for electron microscopy. An ordinary vacuum coating unit and transmission electron microscope are the only major pieces of equipment required.

9. EXAMPLE OF APPLICATION: EFFECTS OF COLCHICINE, CYTO-CHALASIN B AND 2-DEOXYGLUCOSE ON THE SURFACE-BOUND CON A IN NORMAL AND TRANSFORMED CELLS

In the course of mapping the distribution of CABS on the surfaces of normal and virus-transformed cells (2), we observed that the Con A-induced clustering typical of virally transformed cells had a particular spatial organization. As described above, when cells were fixed in paraformaldehyde prior to Con A/H labelling, the CABS distribution was dispersed on both normal and transformed fibroblasts (Figure 2). On the other hand, labelling of transformed cells with Con A/H at 37° resulted in the organization of CABS into clusters (primary organization) which were not present on the pseudopodia and other peripheral areas of the membrane (secondary organization) (Figure 1). Clearly, this kind of information can only be obtained if the entire cell surface of the intact cell is visualized.

Treatment of transformed cells with colchicine, cytochalasin B or 2-deoxyglucose did not alter the inherent dispersed distribution of binding sites as determined by fixation before labelling. However, these drugs did produce marked changes in the secondary (but not the primary) organization of CABS on transformed cells labelled at 37° (3). Colchicine treatment resulted in the formation of a cap-like aggregation of binding-site clusters near the centre of the cell (Figure 3), whereas cytochalasin B and 2-deoxyglucose led to the formation of patches over the entire membrane, eliminating the inward displacement of patches observed on untreated cells. The distribution of bound Con A on normal cells at 37° was always dispersed, in both control and drug-treated preparations.

It had previously been reported that colchicine inhibits the Con A-mediated agglutination of erythrocytes to SV3T3 cells (16). It was proposed that this resulted from an inhibition by colchicine of the primary organization of CABS (clustering). Our experiments suggest, however, that the effect of colchicine on agglutination is, instead, probably related to a change in the secondary organization of CABS (distribution of patches). That is, if all clusters are withdrawn into a relatively small area after colchicine treatment, the probability that cell-to-cell contact will occur in areas containing CABS is reduced and less agglutination is observed. Also, the association of CABS on these cells with the central area of filopodial activity may hinder red cell access to the bound Con A, preventing cross-links from forming between the cells. This kind of information is easily yielded by the Con A/H replica technique. The colchicine effect is intrinsically interesting because it promotes movement of CABS, leading to their aggregation and redistribution. Several lines of evidence have suggested that a colchicine-sensitive protein, perhaps in microtubules, is involved in control of binding-site distribution. The use of Con A/H labelling in conjunction with surface replicas has made information concerning CABS distribution on surface domains readily available, while the visibility of the haemocyanin in thin section has allowed correlation of surface phenomena with cytoplasmic events. In Con A/H-labelled SV3T3 cells treated with colchicine, for instance, thin sections revealed a disappearance of most microtubules (3) (Figure 4).

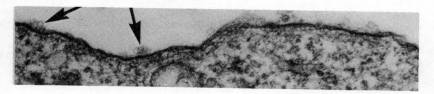

Figure 4. Thin section through an SV3T3 cell treated with 10^{-6} M colchicine before Con A/H labelling at $37°$. The section is presumably through zone 2 as shown on Figure 3. Several small clusters of haemocyanin are visible on the cell surface (arrows). (x 75,000)

In summary, the opportunity to see the whole surface of a large number of intact cells, allowing evaluation of both the short- and long-range topographical distribution of CABS, is unique to the Con A/H replica technique. In addition, the use of replicas in conjunction with thin sections, labelled by the same technique, gives much needed informatin regarding surface/cytoplasmic interactions.

10. REFERENCES

1. Smith, S. B. and Revel, J. P. (1972) 'Mapping of concanavalin A binding-sites on the surface of several cell types', *Dev. Biol.*, 27, 434—6.
2. Rosenblith, J. Z., Ukena, T. E., Yin, H. H., Berlin, R. D. and Karnovsky, M. J. (1973) 'A comparative evaluation of the distribution of concanavalin A-binding sites on the surfaces of normal, virally-transformed, and protease-treated fibroblasts', *Proc. Nat. Acad. Sci. U.S.A.*, 70, 1625—9.
3. Ukena, T. E., Borysenko, J. Z., Karnovsky, M. J. and Berlin, R. D. (1974) 'Effects of colchicine, cytochalasin B, and 2-deoxyglucose on the topographical organization of surface-bound concanavalin A in normal and transformed fibroblasts', *J. Cell. Biol.*, 61, 70—82.
4. Ryan, G. B., Borysenko, J. Z. and Karnovsky, M. J. (1974) 'Factors affecting the redistribution of surface-bound concanavalin A on human polymorphonuclear leukocytes', *J. Cell Biol.*, 62, 351—65.
5. Karnovsky, M. J., Unanue, E. R. and Leventhal, M. (1972) 'Ligand-induced movement of lymphocyte membrane macromolecules. II. Mapping of surface moieties', *J. Exp. Med.*, 136, 907—30.
6. Weller, N. K. (1974) 'Visualization of concanavalin A-binding sites with scanning electron microscopy', *J. Cell Biol.*, 63, 699—707.
7. De Petris, S., Raff, M. C. and Malucci, L. (1973) 'Ligand-induced redistribution of concanavalin A receptors on normal, trypsinized and transformed fibroblasts', *Nature (New Biol.)*, 244, 275—8.

8. Inbar, M., Huet, C., Oseroff, A. R., Ben-Bassat, H. and Sachs, L. (1973) 'Inhibition of lectin agglutinability by fixation of the cell surface membrane', *Biochim. Biophys. Acta.,* **311**, 594−9.

9. Nicolson, G. L. (1973) 'Temperature-dependent mobility of concanavalin A sites on tumor cell surfaces', *Nature (New Biol.),* **243**, 218−20.

10. Singer, S. J. and Nicolson, G. L. (1972) 'The fluid mosaic model of the structure of cell membranes', *Science,* **175**, 720−31.

11. Pinto da Silva, P. (1972) 'Translational mobility of the membrane intercalated particles of human erythrocyte ghosts. PH-dependent, reversible aggregation', *J. Cell Biol.,* **53**, 777−82.

12. Nicolson, G. L. (1972) 'Topography of membrane concanavalin A sites modified by proteolysis', *Nature (New Biol.),* **239**, 193−7.

13. Lowry, O. H., Rosebrough, N. J., Farr, A. L. and Randall, R. J. (1951) 'Protein measurement with the Folin phenol reagent', *J. Biol. Chem.,* **193**, 265−75.

14. Phillips, E. R. and Perdue, J. F. (1974) 'Ultrastructural distribution of cell surface antigens in avian tumor virus-infected chick embryo fibroblasts', *J. Cell Biol.,* **61**, 743−56.

15. Luft, H. J. (1964) 'Electron microscopy of cell extraneous coat as revealed by ruthenium red staining', *J. Cell Biol.,* **23**, 54A−55A.

16. Yin, H. H., Ukena, T. E. and Berlin, R. D. (1972) 'Effect of colchicine, colcemid, and vinblastine on the agglutination, by concanavalin A, of transformed cells', *Science,* **178**, 867−8.

CHAPTER 12

Concanavalin A–Iron Dextran Technique

S. S. SPICER
B. J. MARTIN

Cytochemical staining of mucosubstances at the ultrastructural level was extended to localisation of glycoproteins containing glucose or mannose by the concanavalin A (Con A)–horse-radish peroxidase method (HRP) of Bernhard and Avrameas (1) (see Chapter 7), the Con A–ferritin method of Nicolson and Singer (2) (see Chapters 8–10) and the Con A–haemocyanin method introduced by Smith and Revel (3) (see Chapter 11).

1. PRINCIPLE

We developed another technique, which allows precise and perhaps stoichiometric demonstration of Con A binding sites by transmission electron microscopy. This method employs iron dextran as the electron-opaque labelling moiety (4). The tetravalent Con A binds to glucose and mannose residues of glycoproteins and glycolipids of cell membranes, unoccupied receptor sites of Con A can again bind the glucose units of iron dextran.

2. METHOD

For subsequent Con A–iron dextran staining, tissues may be fixed in 3% glutaraldehyde buffered with 0.1 M cacodylate or phosphate-buffered saline (PBS) at pH 7.3 (5). Initially, small pieces of the freshly obtained tissues are minced to small fragments no greater than a cubic millimetre in size in drops of fixative on dental wax. The finely minced fragments are then transferred to several millilitres of fixative and retained therein for 1 hour at room temperature. After three post-fixation rinses of 15 minutes each in PBS, the fixed tissue fragments are exposed one hour at room temperature to a 1 mg/ml solution of Con A (Sigma

Grade IV) in PBS. Following three more 5 to 15 minute rinses with PBS, the specimen is immersed 30 minutes at room temperature in a 5—50 mg/ml solution of iron dextran (Imferon Lakeside Laboratories, Inc. 1707 East North Avenue, Milwaukee, Wisconsin 53201, USA) in PBS. The tissue is again rinsed three times in PBS and then post-fixed one hour at room temperature ($25°C$) in 1% osmium tetroxide in PBS. Subsequent processing includes routine dehydration through graded alcohols and propylene oxide and embedment in 'Epon'. Thin sections are examined in the electron microscope, preferably without the usual staining by uranyl acetate and lead citrate. Controls are prepared by omitting the Con A solution or by treating the specimen with 0.2 M α-methyl-D-mannoside in PBS prior to the iron dextran step.

3. APPLICATION TO HUMAN PLACENTA

Human placenta stained by the Con A—iron dextran method disclosed discrete, electron-opaque particles of iron dextran along the maternal, or luminal surface of the microvilli of the syncytiotrophoblast (Figure 1). The particles lay periodically

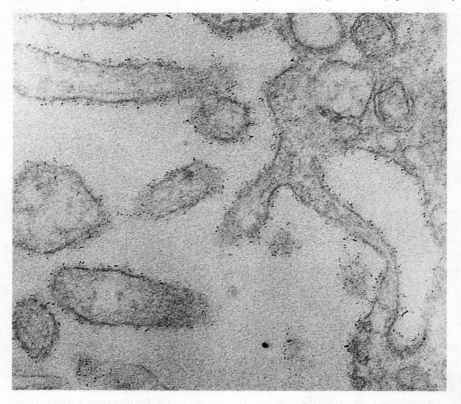

Figure 1. This micrograph illustrates Con A—iron dextran staining of the microvillar surface of 4-month placenta. Note that the individual iron dextran particles are localized within the region of the surface coat external to the plasmalemma. (x 89,370)

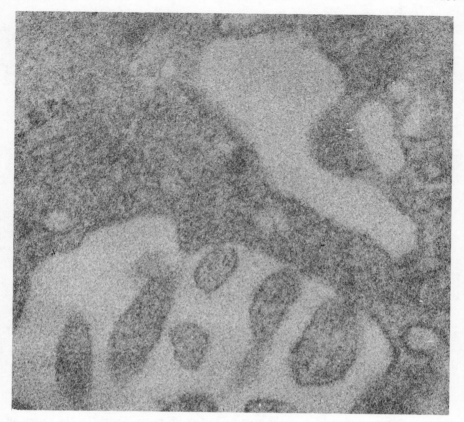

Figure 2. This micrograph, showing a similar region of the same placenta as Figure 1, is taken from a control in which the Con A step of the procedure was omitted. Note that the surface is completely devoid of iron dextran particles. (x 77,580)

aligned in a thin layer closely applied to the surface of the plasmalemma. Since intracellular particles were not encountered, the reagents appear not to penetrate into intact cells. A control specimen lacked such staining (Figure 2). The nature of the Con A reactive moiety on the placental surface remains undetermined, but human chorionic gonadotropin, known from immunostaining to coat the surface of the placenta, contains mannose residues and (6) could account for the Con A binding.

4. COMPARISON WITH OTHER METHODS

Compared with the HRP technique, which visualizes the Con A receptor with enzymatic reaction product which is subject to some diffusion or displacement, the iron dextran technique allows more precise and perhaps stoichiometric demonstration of Con A binding sites. As shown in Figures 1 and 3, the iron dextran particles appear to localize the Con A receptors more precisely than does the

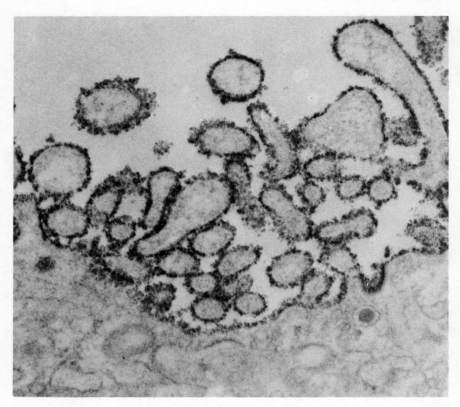

Figure 3. Con A—HRP staining of the maternal surface of the syncytium (1). The pattern of staining appears continuous but periodically intensifed in cross-sections of the plasma membrane. However, on tangential surfaces (arrows) discontinuous foci of reaction product may be observed. There is no staining within the syncytioplasm. Term placenta. (x 48,640)

Con A—HRP technique in this site. Like the Con A—HRP method, the present technique stained only the exposed surface of the placental villi and is generally applicable to staining the luminal surface of epithelial cells or surfaces of cells in suspension. In general the Con A—iron dextran technique has not been compared with the Con A—HRP method extensively as yet. However, it has been found to stain the fuzzy coat on the luminal surface of the feline intestinal epithelium (S. Ito, Harvard University, personal communication). The latter site contains abundant glycoprotein which can be demonstrated by strong periodic acid-Schiff reactivity at the light microscope level and, presumably, accounts for the Con A—iron dextran reactivity. Compared with the high-resolution Con A—ferritin method the technique proposed is much simpler and chemical coupling procedures, which may alter the affinity of Con A to its natural receptor, are not necessary.

5. REFERENCES

1. Bernhard, W. and Avrameas, S. (1971) 'Ultrastructural visualization of cellular carbohydrate components by means of concanavalin A', *Exp. Cell Res.*, **64**, 232—6.
2. Nicolson, G. L. and Singer, S. J. (1971) 'Ferritin-conjugated plant agglutinins as specific saccharide stains for electron microscopy: application to saccharide bound to cell membranes', *Proc. Nat. Acad. Sci. U.S.A.*, **68**, 942—5.
3. Smith, S. B. and Revel, J. P. (1972) 'Mapping of concanavalin A binding sites on the surface of several cell types', *Dev. Biol.*, **27**, 434—41.
4. Martin, B. J. and Spicer, S. S. (1974) 'Concanavalin A—iron dextran technique for staining cell surface mucosubstances', *J. Histochem. Cytochem.*, **22**, 206—7.
5. Mason, T. E., Phifer, R. R., Spicer, S. S., Swallow, R. A. and Dreskin, R. B. (1969) 'An immunoglobulin—enzyme bridge method for localizing tissue antigens', *J. Histochem. Cytochem.*, **17**, 563—9.
6. Bahl, O. P. (1969) 'Human chorionic gonadotropin. II. Nature of the carbohydrate units', *J. Biol. Chem.*, **244**, 5 575—83.

CHAPTER 13

Indirect Visualization of Concanavalin A with Immune Reactions

THOMAS BÄCHI

Substrate-bound Con A cannot be seen directly in thin sections because of its low electron density. The principle of ferritin tagging can be employed for direct detection of individual lectin molecules (Figure 1b: see also Chapter 8). The conjugation of Con A with this macromolecular marker (described in Chapter 8) however, always involves a certain degree of chemical denaturation which may decrease the avidity of the ligand for receptors. Moreover, the size of the ferritin molecule can sterically interfere with binding and mask some of the four binding sites of the Con A tetramer. Freeze-etching replicas allowing direct visualization of native lectin molecules (Figure 1a) avoid this problem (see Chapter 14).

Here we describe an 'indirect visualisation' of Con A in thin sections, involving a secondary reaction with native (Figure 1c) or ferritin-labelled antibody directed against Con A. A further amplification is obtained by a sandwich technique (1) in which the Con A molecules are first reacted with unlabelled anti-Con A antibodies and subsequently with a ferritin conjugated anti-immunoglobulin antibody (Figure 1d). In both methods, the unaltered lectin is allowed to react with the receptor before the indicator reagents are added.

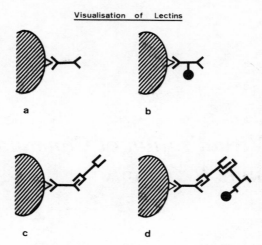

Figure 1. Scheme for visualization of Con A receptors. (a) direct visualization (see Chapter 14); (b) Con A–ferritin method; (c) indirect method with anti-Con A antibody; (d) indirect visualization by sandwich technique: anti-Con A antibody/ferritin-conjugated anti-immunoglobulin antibody

1. MATERIALS

1.1 Preparation of Antibodies

Antisera to Con A, a protein with excellent antigenic properties, are commercially available or can be prepared by hyperimmunization of laboratory animals. It is desirable to use lectin preparations of the purest grade available in order to avoid the formation of antibodies against contaminants. A high-titre rabbit antiserum was obtained by five weekly intramuscular injections (1 ml in each hind leg) of an emulsion (1:1) of complete Freund's adjuvant with PBS containing 1 mg/ml Con A (1). The animals were bled 14 days after the last injection. For controls, we used rabbit serum of unrelated specificity. The sera were inactivated ($56\,^{\circ}$C, 30 min), extensively absorbed with washed human erythrocytes and stored frozen or at $4\,^{\circ}$C after sterilization by filtration.

The presence of specific anti-Con A antibodies was demonstrated by an indirect haemagglutination assay (see Chapter 17) and by gel double diffusion (2). It should be noted that pre-immune sera, by virtue of their content of glycoproteins and glycolipids, can also form precipitin lines. Serial dilutions of hyperimmune sera around a centre well containing a solution of 100 µg/ml Con A, however, will be positive at dilutions above 1/10, whereas unspecific precipitations are obtained with undiluted sera only.

Because of the glycoprotein components in serum which may react with Con A, purified antibody fractions (IgG) have to be used in order to avoid interference with the immune reaction. Antibody fractions can be prepared by salt fractionation (3). With our rabbit hyperimmune serum, fractionation was achieved by dropwise addition of 1 volume of saturated $(NH_4)_2SO_4$ to 1 volume of serum in the cold. This mixture was allowed to stand for at least one hour and was then centrifuged at 10,000 r.p.m. for 30 min. The sediment was redissolved in water to 50% of the original serum volume. (This procedure may be repeated if necessary.) The redissolved serum proteins were extensively dialysed against the appropriate buffer for the subsequent experiments. Further fractionation was achieved by DEAE cellulose chromatography (4). For this, the protein was again precipitated by salt and dialysed against the starting buffer (0.015 M phosphate, pH 8.0) used in the ion exchange chromatography. Pure IgG fractions were obtained by using an ionic strength gradient of 0.015 M to 0.25 M phosphate buffer, pH 8.0. The first protein-containing fraction, which starts to elute at about 0.02 M, consists of IgG without traces of other proteins as judged by immunoelectrophoresis (2). The pooled fractions are concentrated by vacuum dialysis or lyophilization. The relative concentration of IgG reactive with Con A can be determined by indirect haemagglutination (Chapter 17).

The preparation of an anti-IgG sera (guinea pig anti-rabbit gG) has been described elsewhere (5).

1.2 Ferritin Labelling of Antibodies

The widespread use of ferritin-labelled antibodies for the detections of cellular and viral antigens has lead to the development of a variety of established conjugation procedures (6). They usually involve the linkage of ϵ-amino groups of the antibody protein with those of the ferritin protein shell with bifunctional reagents.

For the indirect sandwich technique described in this chapter, the conjugation of an IgG fraction prepared from guinea pig anti-rabbit IgG serum with ferritin was achieved in a single step reaction with p,p'-difluoro-m,m'-dinitrophenyl sulphone (FNPS) (7). The remaining unbound IgG was removed from the preparation by pelleting the ferritin-containing material 3 times at 165,000 g(150 min) from PBS solution.

2. METHOD

2.1 Reaction of Cells with Con A and Antibodies

Some general rules to be observed when cells or membrane material is incubated with Con A have already been proposed in other chapters of this book. In order to reveal the distribution of lectin binding sites and to get quantitatively meaningful results it is important to ascertain that the concentrations of the reagents used are sufficient to saturate the cell receptors. We routinely incubated 10^8 cells per ml with 200 μg of Con A per ml for 30 min, which saturates more than 90% of the available receptor sites (at least in the case of erythrocytes, [1,8]).

After completion of the reaction with the native lectin, aggregates of cells are dispersed. Dilutions of anti-Con A globulins are added to the samples which are kept in suspension by constant agitation for 30 min at 0 °C. After the first few minutes, heavy clumping is frequently observed. Redispersion of the material at this stage is often only possible by trituration. Excess antibodies are removed by washing with cold media. This first immune reaction can be followed by a similar second reaction (30 min at 0°C) with anti-IgG antibodies; this second immune reagent is usually ferritin labelled. The last reaction is terminated by extensive washing in the cold, followed by chemical fixation (glutaraldehyde and osmium tetroxide) and embedding in plastic.

Controls include experiments in which a) the lectin is incubated in the presence of hapten inhibitors (methyl glucoside), b) the complete sequence of specific immune reactions is done without pre-incubation of the material with Con A or c) the anti-Con A IgG is replaced by an IgG fraction prepared from sera of unrelated specificity.

2.2 Interpretation of Electron Micrographs

As an illustration we prepared thin sections for the visualization of the Con A molecules on human erythrocytes by indirect ferritin labelling. For this, the cells were agglutinated with Con A and subsequently incubated at 0°C for 30 min with rabbit anti-Con A serum followed by guinea pig anti-rabbit IgG serum conjugated with ferritin (0°C, 30 min). Thin sections reveal electron-dense cores of the ferritin arranged in discrete clusters along the surface (Figure 2). The ferritin molecules are embedded in an opaque matrix which was also present on cells reacted with Con A and anti-Con A alone (Figure 3). In controls, cells were reacted with anti-Con A and anti-IgG ferritin conjugates without prior reaction with Con A. Sections of these cells (not shown) never displayed any ferritin molecules.

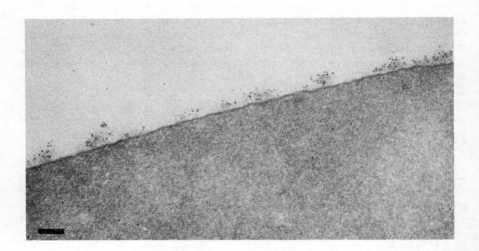

Figure 2. Human erythrocyte after reaction with anti Con A IgG and a ferritin conjugate of guinea pig anti-rabbit IgG (see also ref. 1). The Con A molecules are tagged by discrete clusters of ferritin. The electron-opaque mass embedding the clusters is formed by the antibody complex. Bar indicates 10 nm.

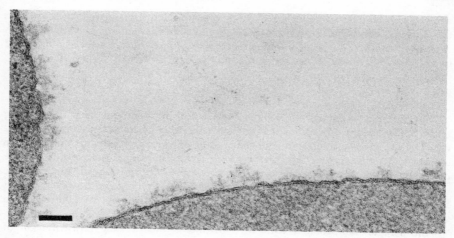

Figure 3. Human erythrocyte after a reaction with Con A and rabbit anti-Con A IgG. The presence of a surface-bound lectin is indicated by an irregular mass coating the membrane surface. Bar indicates 10 nm

3. COMPARISON OF DIRECT AND INDIRECT VISUALIZATION OF CON A BOUND TO ERYTHROCYTES

A variety of techniques are available to study Con A receptors on cells and membranes, but not all produce quantitatively reliable results.

Earlier it was shown that native Con A molecules bound to the intact or lysed erythrocyte membrane can be visualized directly by freeze etching (Chapter 14). This technique reveals 400 to 700 Con A binding sites per μm^2 (1): the number of molecules per cell (60,000–100,000 based on a surface of 145 μm^2) is in close agreement with the values determined by binding experiments using radioactively labelled Con A (8).

On the other hand, procedures which depend on direct coupling of Con A with electron-dense markers showed a much sparser arrangement of lectin binding sites. On freeze-fractured cells reacted with ferritin-labelled Con A we found only 50–100 binding sites per μm^2, a number that is comparable with that seen on previously published pictures of human erythrocyte ghosts reacted with a ferritin conjugate (9). This value, being at least five times smaller than the ones which were obtained by counts of the native molecules, probably reflects the loss of binding affinity by the ferritin conjugation procedure or could be due to a steric interference of lectin binding by the large marker molecule.

However, this problem can be avoided by the indirect technique described here, using ferritin-labelled anti-IgG reacting with the Con A rabbit anti-Con A complexes and thus allowing the receptors to be scored with the same degree of sensitivity as by the direct visualization (Figure 2). Although, an estimate of the number of binding sites based on such one-dimensional views cannot be as accurate as the counts of native Con A molecules revealed by freeze etching, the indirect visualization by immunological tagging of Con A shown here (Figures 2 and 3) allowed the demonstration of a receptor density which is comparable with that of the native lectin. The location of Con A molecules is detectable with the high sensitivity and specificity of immune reactions at a resolution which is only limited by the degree of antibody clustering around the lectin antigen.

4. PROBLEMS

The indirect tagging of Con A molecules by immune reactions using ferritin-labelled anti-Con A, or a sandwich consisting of anti-Con A and anti-IgG ferritin conjugates, usually results in the binding of more than one ferritin per lectin molecule. This clustering, which diminishes the resolution of the method, can be overcome to some degree by lowering of the concentration of the immune reactants. The quantitative information gained by this technique, however, always has to be interpreted with prudence and is further limited by the failure of thin sections to display directly a two-dimensional distribution.

Another general problem with this method is the possible redistribution of Con A by antibody-mediated cross-linking. Such antibody-induced movements are avoided in most systems for limited periods of time by keeping the temperatures of incubations at $0°C$ during all immune reactions and washing procedures.

REFERENCES

1. Bächi, T. and Schnebli, H. P. (1975) 'Reaction of lectins with human erythrocytes. II. Mapping of Con A receptors by freeze-etching electron microscopy', *Exp. Cell Res.*, 91, 285–5.
2. Ouchterlony, O. and Nilsson, L. A. (1973) 'Immunodiffusion and immunoelectrophoresis', in *Handbook of Experimental Immunology*, ed. Weir, D. M. (Blackwell, Oxford), pp. 19.1–19.39.
3. Heide, K. and Schwick, H. G. (1973) 'Salt fractionation of immunoglobulins', in *Handbook of Experimental Immunology*, ed. Weir, D. M. (Blackwell, Oxford), pp. 6.1–6.11.
4. Fahey, J. L. and Terry, E. W. (1973) 'Ion exchange chromatography and gel filtration', in *Handbook of Experimental Immunology*, ed. Weir, D. M. (Blackwell, Oxford), pp. 7.1–7.16.
5. Aguet, M. and Bächi, T. (1973) 'Comparison of the direct and indirect immunoferritin tagging of Sendai virus antigen', *Ann. Immunol. (Inst. Pasteur)*, 124C, 407–16.
6. Andres, G. A., Hsu, K. G and Seegal, B. C. (1973) 'Immunologic techniques for the identification of antigens or antibodies by electron microscopy', in *Handbook of Experimental Immunology*, ed. Weir, D. M. (Blackwell, Oxford), pp. 34.1–33.45.
7. Sri Ram, J., Twade, S. S., Pierce, G. B. and Midgley, A. R. (1963) 'Preparation of antibody—ferritin conjugates for immuno-electron microscopy', *J. Cell Biol.*, 17, 673–5.
8. Schnebli, H. P. and Bächi, T. (1975) 'Reaction of lectins with human erythrocytes. I. Factors governing the agglutination reaction', *Exp. Cell Res.*, 91, 175–83.
9. Nicolson, G. L. and Singer, S. J. (1971) 'Ferritin-conjugated plant agglutinins. as specific saccharide stains for electron microscopy: application to saccharides bound to cell membranes', *Proc. Nat. Acad. Sci. U.S.A.*, 68, 942–5.

CHAPTER 14

Direct Visualization of Concanavalin A Molecules by Freeze Etching

THOMAS BÄCHI

Freeze-etching techniques have been adapted for the demonstration of the outer surface of membranes (1, 2). In conjunction with specific markers, this elegant method shows the two-dimensional distribution of surface receptors of a given surface reactant. Among other markers, viruses, antibodies and lectins have been employed (3—5). Because the replication of structures by platinum—carbon shadowing techniques requires a minimal size of approximately 5 nm, ferritin conjugates have been used for the detection of small markers. Attempts to visualize unlabelled immunoglobulins (IgG) have failed because they seem to become flattened upon binding to membranes. On the other hand, the rigidity, size and shape of native Con A allows direct depiction after its reaction with membrane receptors without further labelling (3). This technique therefore provides the possibility for analysing the topographical distribution of the native lectin molecules on membranes, whereas other methods employing a macromolecular marker such as ferritin are prone to distort the original distribution, to sterically hinder and to influence the binding of the lectin due to altered physicochemical properties. Moreover, the freeze-etch technique which visualizes single molecules, attains optimal resolution, recommending its use in studies of the surface distribution of Con A.

149

1. PRINCIPLE

Native Con A is reacted with cells or membrane material. After chemical fixation, the samples are frozen in distilled water and replicated by freeze-etching techniques. The lectin molecules can be observed as globular particles on the etched surface of membranes.

2. METHOD

2.1 Reaction of Con A with Cells

The reaction of cells or subcellular fractions with Con A does not have to be modified for the use of the samples as freeze-etching specimens. However some general rules always apply: 1) To prevent self-aggregation, the lectin is dissolved in the incubation buffer only shortly before use. 2) Because serum glycoproteins and glycolipids react with Con A, only serum-free incubation media can be used. We routinely used Dulbecco's phosphate-buffered saline (PBS) which is isotonic and fulfills the ionic requirements for Con A binding. 3) For quantitative studies where a saturation of the receptors is required, the necessary lectin concentration has to be estimated by radiolabelling. 4) One lectin molecule may bridge two receptors localized on two different substrates (membranes). In some cases this can be prevented by shaking during incubation. 5) Controls include incubations with Con A in the presence of specific sugar hapten inhibitors (methyl glucoside) as well as incubations without Con A. 6) To prevent the internalization of cell-surface-bound lectin molecules by endocytosis and to minimize redistribution of the receptors, we routinely incubated the samples for 30 min (or less) at $0°C$ (ice bath).

2.2 Preparation of Cells for Freeze Etching

After completion of the lectin reaction the excess Con A is removed by washing with PBS (3 to 4 times). This is followed by a brief fixation with a solution of 3% glutaraldehyde in a physiological buffer or with a mixture of 3% glutaraldehyde and 3% acrolein in 0.05 M cacodylate buffer, pH 7.2, applied for 10 min at $0°C(3)$, which serves the following purposes: The Con A molecules are irreversibly cross-linked by glutaraldhyde to the membrane, and the membrane is stabilized for the subsequent procedures. After fixation, the material is washed three times in distilled water and pelleted. Small droplets of this pellet are mounted on specimen discs and rapidly frozen in the liquid phase of 'Freon' 22. Freeze fracturing of such a specimen is carried out by standard methods, and the specimen is allowed to sublimate at $-100°C$ for 3 min under high vacuum ($10^{-5}-10^{-6}$ torr).

2.3 Interpretation of Electron Micrographs

Within the replica, some variation with regard to structural preservation is usually observed and it is often necessary to select interpretable areas, i.e. where membranes are clearly recognizable. Next to the area where the typical intra-membranous particles are revealed by the fracturing process, the cell surface is exposed through the lowering of the ice table during sublimation. Figures 1 and 2

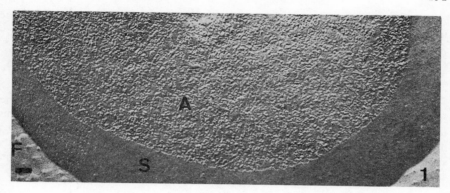

Figure 1. Human erythrocyte after incubation with Con A in presence of methyl glucoside (control experiment). Freeze fracturing of the membrane has exposed the intramembranous particles (A), whereas the membrane surface (S), revealed by sublimation of the frozen water (F) appears smooth. Bar indicates 10 nm

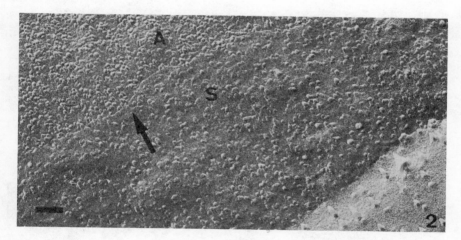

Figure 2. Human erythrocyte incubated with a saturating concentration of Con A (see also ref. 3). The lectin molecules appear as particles with a diameter of 10 nm on the etched surface of the membrane (S). Above the fracture edge (arrow), the intramembranous particles are seen (A). Bar indicates 10 nm

show human erythrocytes as an example. Here the membrane surface is smooth in the native state (Figure 1), but after reaction with the Con A molecules, particles with a diameter of approx. 10 nm can be discerned on this part of the membrane (Figure 2). Since it is conceivable that in certain other systems the Con A globules may be mistaken for membrane particles, a clear demonstration of the fracture edge (Figure 2, arrow) occurring between the etched and the fractured part of the membrane, will be needed to identify the particles. When the lectin molecules are in a clustered rather than a dispersed arrangement, as e.g. on MDBK cells, the resolution of single particles is more difficult; granulated patches covering the membrane indicate receptor-containing surface areas (Figure 4).

3. EXAMPLE OF APPLICATION: RELATIONSHIP OF INTRAMEMBRANOUS PARTICLES AND CON A RECEPTORS IN HUMAN ERYTHROCYTES

Earlier it has been proposed that the intramembranous particles appearing on the fractured part of the membrane represent anchoring structures for the lectin receptor (6). We used the freeze-etching technique to compare the distribution of directly observable native Con A, and the arrangement of immunological reactants superimposed on the lectin with the distribution of intramembranous particles on erythrocytes (3).

Figure 3. Surface of human erythrocyte after reaction with Con A followed by a reaction anti-Con A IgG (30 min at 37°C). The irregular structures on the surface represent Con A–IgG clusters. No redistribution is noted in the arrangement of the membrane particles. Bar indicates 10 nm

Figure 4. MDBK cell after reaction with Con A. The etched surface of the membrane (S) reveals granulated patches of lectin molecules which, by their clustered arrangement, are not always discernible as individual molecules. Bar indicates 10 nm

From freeze-etch electron micrographs (e.g. Figure 2) we estimated that about 10^5 Con A molecules bind to each erythrocyte under saturating condition (3). This figure is in close agreement with the value determined by binding experiments using radioactively labelled Con A (7) but is considerably lower than the number of intramembranous particles (approximately 5×10^5 per cell).

The random arrangement of intramembranous particles found in untreated erythrocytes (Figure 1) and cells reacted with Con A at $0°C$ or $37°C$ (Figure 2) was unaffected by the temperature-dependent redistribution of receptor sites after the addition of anti-Con A antibodies (Figure 3). This, together with the quantitative difference between receptor sites and membrane particles has led us to conclude that the surface Con A receptors are independent of the intramembranous particles revealed by freeze etching (3).

4. PROBLEMS

The essential point in preparing the cells for this technique is the use of distilled water instead of a buffered medium containing cryoprotectants for freezing. The chemical fixation is necessary for the stability of the membrane since distilled water is a highly unfavourable medium for freezing. Damage through ice crystal formation is unavoidable. Whereas the cell interior invariably displays gross disruption which no longer allows the identification of individual structures, the medium (frozen water) and the cell surface have a relatively smooth appearance in replicas. Sometimes satisfactory structural preservation is obtained only in a small percentage of samples and variations in fixation and freezing procedures may be necessary.

Attempts to visualize individual Con A molecules on cells with more complex surface structures than erythrocytes may prove difficult. Where native Con A molecules cannot be unequivocally identified because of the substructure of the surface, ferritin-labelled Con A (see Chapter 8) may be used. Ferritin molecules, appearing with a diameter of approximately 17 nm on freeze-etching replicas allow easy detection of the Con A molecules (3,6). Quantitative studies however, are not possible since the Con A—ferritin conjugate no longer binds to all of the receptors (see also Chapter 13).

5. REFERENCES

1. Tillack, T. W. and Marchesi, V. T. (1970) 'Demonstration of the outer surface of freeze-etched red blood cell membranes', *J. Cell Biol.*, **45**, 649—53.
2. Pinto da Silva, P. and Branton, D. (1970) 'Membrane splitting in freeze-etching', *J. Cell Biol.*, **45**, 598—605.
3. Bächi, T. and Schnebli, H. P. (1975) 'Reaction of lectins with human erythrocytes. II. Mapping of Con A receptors by freeze-etching electron microscopy', *Exp. Cell Res.*, **91**, 285—95.
4. Howe, C. and Bächi, T. (1973) 'Localization of erythrocyte membrane antigens by immune electron microscopy', *Exp. Cell Res.*, **76**, 321—32.
5. Tillack, T. W. and Marchesi, V. T. (1972) 'The structure of erythrocyte membranes studied by freeze-etching. II. Localization of receptors for phytohemagglutinin and influenza virus to the intramembranous particles', *J. Exp. Med.*, **135**, 1209—27.
6. Pinto da Silva, P. and Nicolson, G. L. (1974) 'Freeze-etch localization of concanavalin A

receptors to the membrane intercalated particles of human erythrocyte ghost membranes',
Biochim. Biophys. Acta, **363**, 311—9.
7. Schnebli, H. P. and Bächi, T. (1975) 'Reaction of lectins with human erythrocytes. I.
Factors govering the agglutination reaction', *Exp. Cell Res.,* **91**, 175—83.

CHAPTER 15

Synopsis: When to Use Which Method?

HELMUT BITTIGER

In the preceding chapters a series of methods for the visualization of Con A receptors has been presented. Although each technique is designed for a particular application, a brief overall comparison appeared worthwhile.

1. LIGHT MICROSCOPY

Two methods for the visualization of Con A receptors at the light microscopical level are described: The fluorescence method (Chapters 5 and 6) and the peroxidase method (Chapter 7). Whereas the fluorescence technique is obviously restricted to light microscopy, peroxidase staining can also be applied in electron microscopy. The fluorescence method has the advantage of being applicable to viable cells at high optical resolution (0.1–0.2 μm). It is the only method allowing the direct investigation of the dynamics of membrane Con A receptors. Fluorescent ligands covalently bound to Con A can be used as direct markers for Con A receptors or when bound to anti-Con A or anti-IgG antibodies respectively as indirect markers in immunofluorescence techniques. The direct method being the simplest gives the best optical resolution but the lowest sensitivity, whereas the indirect fluorescence methods allow the detection of Con A receptors with high sensitivity due to the fluorescence amplification by multiple aggregates of fluorescein antibodies. The obvious price to be paid for this is a loss in resolution and a possible antibody-induced redistribution of Con A receptors in the cell membranes. Visualization of Con A receptors by the peroxidase method is achieved with the reaction product of its enzymatic activity. Simultaneously, if desired, conventional histochemical staining methods for cells can be employed that do not impair the observation of the brownish peroxidase reaction product.

2. ELECTRON MICROSCOPY

2.1 Thin Sectioning

The Con A—peroxidase technique, as already mentioned, is applicable to both light and electron microscopy, which is the unique advantage of this method. In addition, all reagents needed are easily available. A drawback of the method, however, lies in the relatively low electron microscopical resolution, as the reaction product of peroxidase diffuses and forms a thick coat on the membranes; post-staining of thin sections with uranyl acetate and lead citrate are only possible to a limited extent resulting in low contrast of the electron micrographs. The method is quite sensitive and is recommended for exploratory studies on Con A binding sites. The best resolution at the electron microscopical level is achieved with Con A—ferritin conjugates (100—200 Å) (Chapters 8—10). Not only has the iron core of the ferritin a very defined form and high electron density, but in addition all electron microscopical staining procedures are permitted, by which high resolution and contrast are obtained. A certain disadvantage of the method lies in the incomplete staining of the Con A receptors and in the necessity to prepare the conjugates in a time-consuming procedure; however, recently Con A—ferritin conjugates became available commercially (see appendix).

When the density and number of Con A receptors on cell membranes is low, an amplification (as in fluorescence microscopy) can be obtained by an indirect immunological sandwich technique using anti-Con A antibodies and ferritin-labelled anti-IgG antibodies (Chapter 13). The electron microscopical resolution is necessarily lower than with pure Con A—ferritin conjugates.

In an attractively simple way relatively high electron microscopical resolution can be obtained with the iron dextran technique (Chapter 12) where as in the peroxidase method free binding sites of membrane-bound Con A are occupied by iron dextran. As the method has not yet been widely applied, it is recommended that its suitability is assessed in each case.

2.2 Surface Replicas

One of the major disadvantages of the thin-sectioning techniques resides in the strictly two-dimensional representation of the structures to be studied. Three-dimensional evaluation of the distribution of membrane surface receptors with this method is possible to a limited extent by the tedious and time-consuming serial sectioning reconstruction technique. Surfaces can be studied more conveniently by replication techniques such as freeze etching and the haemocyanin method (Chapters 14 and 11). By freeze etching even single Con A molecules can be made visible on membrane surfaces. The etching of the suspension liquid is absolutely necessary to expose intact membranes as by freeze fracturing alone membranes are split in the middle of the lipid bilayer. Only pure water, not containing salts or cryoprotectants, sublimates in reasonable time and must therefore necessarily be used as suspension liquid. This may induce a certain damage to the cells during the preparation. Even more pronounced changes in cell structure seem likely in the substitution drying process in the haemocyanin method (Chapter 11). These structural changes may not be of importance in many investigations on the gross distribution of membrane surface Con A receptors. As equipment in the haemo-cyanin method a conventional platinum—carbon evaporation apparatus is needed, whereas the freeze-etching technique requires the more elaborate and expensive freeze-etching apparatus. Both methods can be used routinely and they are less time consuming than thin-sectioning techniques.

2.3 How Representative are Con A Receptor Staining Methods for Con A Binding Sites?

For electron microscopical quantitation of Con A binding sites the most reliable method is the freeze-etching method where single Con A molecules are visualized. All other methods give rather qualitative than quantitative results. The accessibility of Con A—ferritin conjugates to Con A binding sites is restricted mainly due to the large size of the conjugates (Chapters 8 to 10). Additional effects involve a possible decrease in interaction energy of Con A with its receptors as a consequence of the chemical coupling procedure. In comparison to the direct visualization of Con A molecules by freeze etching (Chapter 14) only about 20% of binding sites are stained by Con A—ferritin.

The problems of visualization with indirect methods as with peroxidase or iron dextran staining are discussed and demonstrated at length in Chapter 7. The main problems are the possible complete occupation of binding sites of the Con A molecule by membrane carbohydrate groups, thus not allowing the binding of peroxidase or iron dextran; in addition, the competitive binding of carbohydrate-containing markers to Con A may release Con A from its membrane surface receptor sites.

In conclusion, electron microscopical methods are of limited use for the quantitation of Con A receptors on membranes.

Section III:

QUANTITATIVE METHODS

CHAPTER 16

Quantitation of the Precipitin Reaction between Concanavalin A and Carbohydrate Polymers

I. J. GOLDSTEIN

The precipitation reaction between concanavalin A (Con A) and specific carbohydrate-containing macromolecules may be quantitated using procedures established for the precipitin reaction between antibody and antigen. Although turbidity measurements have been employed for studying Con A—polysaccharide interaction (1—8) this procedure is subject to numerous variables including time of incubation (8). The rate of turbidity development was also shown to differ with various Con A—polysaccharide ratios (8, cf. 9). For these reasons a determination of the protein or nitrogen in the Con A—carbohydrate complex is preferable.

The many variables which influence the Con A—polysaccharide interaction have been studied thoroughly (10). Maximum precipitation with dextran occurs in 24 h at 20—25° and pH 6.1—7.2, is unaffected by foreign proteins which do not contain determinant sugars for Con A and, provided the reaction is buffered, is virtually independent of ionic strength (up to 4 M NaCl).

The concentration of Con A used depends on the reactivity of the carbohydrate polymer with the lectin. Polysaccharides such as glycogen (11) or yeast mannan (12) require approximately 250—300 μg Con A per ml whereas less reactive polymers (e.g. the essentially linear dextran from *Streptococcus bovis*) required 1 mg protein per ml in our system (13). Good precipitin curves were established with a series of glycoprotein lectins and a myeloma IgM using 250 μg Con A (14).

1. PRECIPITIN REACTIONS IN SOLUTION

Essentially, increasing amounts of a carbohydrate polymer are added to a series of centrifuge tubes containing a constant quantity of Con A. The precipitates which form are centrifuged and washed. At this point protein may be determined directly, or the precipitates may be digested with 7 N sulphuric acid and nitrogen determined by a number of procedures. We (10) have used the method of Rosen (15) as modified by Schiffman and coworkers (16). More recently we have developed our own slight modification of Schiffman's procedure (17).

The fact that Con A is not a glycoprotein (18) makes it possible to analyse directly for carbohydrate in the Con A—carbohydrate precipitate. This is done most conveniently by the phenol—sulphuric acid method (19,20).

1.1 Precipitin Reaction (10,17,21)

To a series of 3 ml centrifuge tubes (e.g. Bellco Glass, Inc., Vineland, New Jersey) is added Con A (300 μg). Increasing amounts of a carbohydrate polymer (polysaccharide or glycoprotein) are added in the range of 10 μg to 1 mg. Some experimentation may be necessary to establish the proper range of carbohydrate polymer. The volume is adjusted to 1.0 ml such that the final incubation mixture is 1.0 M in NaCl and 0.018 M in phosphate buffer, pH 7.2. Although we have routinely used 1 M NaCl solution, physiological saline (0.15 M NaCl) may be employed with the same results.

After an incubation period of 24—36 h at 20—25° the tubes are centrifuged in a table model International Clinical centrifuge for 15 min at 3000 r.p.m. The supernatant solutions are carefully removed with a drawn out Pasteur pipette and the tubes placed in an inverted position over an absorbent paper towel in a test-tube rack and allowed to drain. At this point the precipitates may be washed (once or twice) by suspending in the same medium (0.5 ml) in which precipitation was conducted, briefly agitating the suspension on a Vortex mixer, centrifuging and draining as before. If purified Con A is used and the carbohydrate-containing polymer does not contain nitrogen, the washing step may be eliminated; this avoids loss of material due to solubility of the complex (see below) and additional manipulations.

1.2 Protein Determination in Con A—Polymer Precipitates

Protein determination may be conducted directly on the drained precipitates by dissolving in methyl α-D-mannopyranoside solution (10 mg/ml) or 0.1 N NaOH and measuring absorbance at 280 nm (22). A correction for turbidity due to polysaccharide or glycoprotein may be necessary. This can be achieved by constructing a standard curve relating the carbohydrate polymer concentration to absorbance (scattering) at 280 nm. Alternately, protein may be determined by dissolving the precipitate in 0.05 N NaOH (0.3—0.9 ml) and using a semimicro-Lowry procedure developed by Mage and Dray (23). This procedure has been adapted for use in the lima bean lectin blood group type A substance system (24).

We have routinely digested the precipitates in 0.05 ml 7 N H_2SO_4 (1 ml 95% H_2SO_4 and 4 ml H_2O) by heating in an electrically heated sand bath at 110° for

1 h. The temperature is gradually raised to 180° and maintained at this temperature for 1 h to complete the digestion. The samples are cooled and 30% H_2O_2 is added dropwise while maintaining a temperature of 110° and allowing 10—15 min between additions until the samples are water clear. The temperature is again slowly raised to 180° and maintained for 1 hr. Most of the water must be evaporated at approximately 130° or fluid will be lost from the tubes by spattering at 180°.

Nitrogen in the digested precipitates is determined by a modification (17) of the procedure of Rosen (15). Two ml of 1 M acetate buffer, pH 5.5 is added to each tube followed by 0.05 ml of N NaOH solution. The samples are thoroughly mixed. To 0.5 ml aliquot of each sample is added 0.5 ml of freshly prepared ninhydrin reagent prepared by mixing 1 ml of 4 M acetate buffer, pH 5.5, 3 ml of 3% ninhydrin in peroxide-free methyl cellosolve and 0.05 ml of 0.01 M KCN. The samples are mixed and incubated at 100° for 15 min. Ten ml of 50% aqueous ethanol is added immediately and the cooled samples read at 570 nm against a reagent blank.

A standard curve is constructed by performing ninhydrin analyses on a series of samples containing 0 to 40 μg nitrogen employing dried $(NH_4)_2SO_4$. The μg nitrogen in the original 2.0 ml volume is plotted against the absorbance at 570 nm. The μg nitrogen in the total digest can then be read directly off the standard curve. A precipitin curve is generated when μg nitrogen in the precipitate is plotted against μg carbohydrate polymer added (Figure 1).

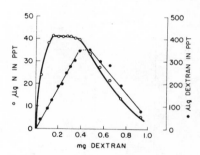

Figure 1. Quantitative precipitation curve of dextran B-1355-S with concanavalin A. The total amount of dextran in the precipitate is also illustrated. Concanavalin A, 42 μg of nitrogen. Published by permission of the Journal of Biological Chemistry. Ref. 10

1.3 Carbohydrate Content of Con A—Polymer Precipitates (10)

Washed precipitates are dissolved in 2 ml of 0.1 M potassium chloride—hydrochloric acid buffer, pH 1.8. An appropriate aliquot is removed for estimation of carbohydrate by the phenol—sulphuric acid method as follows: to 1.0 ml of the solution containing between 10 and 70 μg of carbohydrate is added 1.0 ml of 5% phenol solution [redistilled reagent grade phenol (5 g) dissolved in 100 ml water] and the solution mixed. Five ml of reagent grade 96% sulphuric acid is delivered rapidly from a fast flowing pipette directly to the liquid surface of each tube to produce good mixing. The tubes are agitated, cooled to room temperature, agitated again, and the orange—red colour read at 490 nm against a reagent blank. The amount of carbohydrate present is read from a standard curve relating μg sugar and absorbance at 490 nm. If the component sugars of the carbohydrate-containing polymer are known, a synthetic mixture composed of the component sugars in their approximate molar ratios may be employed. Otherwise D-glucose or D-mannose may be employed. The phenol—sulphuric acid procedure can be employed for virtually all classes of sugars with the exception of amino sugars (20).

1.4 'Solubility' of Con A—Carbohydrate Precipitates

It is often useful to carry out a study of the 'solubility' of the Con A—carbohydrate precipitates. Such data can furnish information on the affinity or strength of binding of Con A to the carbohydrate-containing polymer (10). This in turn allows some tentative conclusions regarding structural features of the carbohydrate moiety (e.g. extent and nature of branching and number of glycosyl units in the exterior oligosaccharide side chains).

A point on the precipitin curve, generally in the initial region of 'equivalence' is selected and the precipitation reaction is conducted in a series of 5 ml centrifuge tubes varying the volume from 0.5 to 4.0 ml keeping the amount of concanavalin A and polysaccharide constant. The precipitates are centrifuged and processed as above. A graph of μg of nitrogen precipitated against total volume is constructed, the 'solubility' of the precipitate being determined from the slope of the curve (Figure 2).

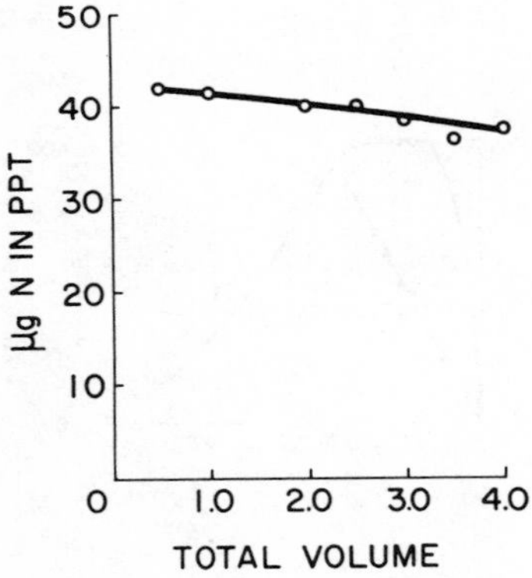

Figure 2. Effect of volume on concanavalin A-dextran precipitation. Concavalin A, 46 μg of nitrogen; dextran, 180 μg. Published by permission of the Journal of Biological Chemistry. Ref. 10

Carbohydrate polymers that are highly reactive with Con A, e.g. glycogen (11) and yeast mannan (12), exhibit low solubility (1—2 μg nitrogen/ml) whereas the precipitates formed between certain dextrans with few glucosyl end groups (13), plant and bacterial levans (25) or linear mannans with α-(1 $\rightarrow$ 2)-linked D-mannopyranosyl units (26) are more 'soluble' (3—30 μg nitrogen/ml).

2. CONCANAVALIN A—CARBOHYDRATE INTERACTIONS AS STUDIED BY TWO-DIMENSIONAL IMMUNODIFFUSION IN AGAR GEL

Two-dimensional immunodiffusion (Ouchterlony technique) provides a simple, rapid and useful approach to studies of Con A—carbohydrate polymer interaction. Preliminary information concerning reactivity, polydispersity and molecular weight of concanavalin A-reactive polymers may be obtained (27). No attempt will be made here to cover the fundamental aspects of immunodiffusion inasmuch as excellent texts are available [see for example the monographs by Crowle (28) and Clausen (29)].

Since the procedures involved in conducting immunodiffusion studies are almost too well known to comment on, only those aspects which are directly related to the use of Con A will be dealt with here.

2.1 Double Diffusion Technique

Almost any medium suitable for Ouchterlony plate studies may be employed provided that the pH is buffered close to neutrality. We regularly use a 1% solution of Noble agar (Difco Laboratories, Detroit, Michigan) in 0.15 M saline buffered with 0.1 M phosphate, pH 7.2 and containing 1% sodium azide as a preservative (27). The medium is heated to 100° and poured into 50 x 12 mm disposable Falcon plastic petri dishes (5 ml/dish).

Although any of a number of different hole patterns may be employed for two-dimensional immunodiffusion studies, we have generally used a pattern consisting of a central well surrounded by six peripheral wells. A solution of concanavalin A (1—10 mg/ml saline) is placed in the central well and solutions of the polymers to be tested for reaction with Con A are added to the peripheral wells. A saline blank and a positive control (e.g. glycogen, 1 mg/ml) should be included in all experiments. The tightly stoppered plates are incubated at 20—25° avoiding any draught or vibrational disturbance. Precipitin band formation generally occurs within 12—24 h (somewhat longer if plates are incubated at 4°).

The specificity of Con A—macromolecule interaction should always be examined (27). This is done most readily by flooding the plate with a 1% solution of methyl α-D-mannopyranoside, methyl α-D-glucopyranoside, D-glucose or D-man-nose, all inhibitors of Con A interaction (27). Band disappearance should occur within 30 min. For carbohydrate-containing polymers that are highly reactive with Con A (e.g. yeast mannans and some glycoproteins) it may require 20 h or more for the band to disappear.

Con A—biopolymer precipitin bands may be stained with protein dyes (amidoschwarz, coomassie blue, ponceau S), carbohydrate stains (p-phenyl-enediamine, or iodine in the case of amylopectin or glycogen) or treated with enzymes (dextranase to effect disappearance of dextran—Con A bands or α-amylase to dissipate Con A—glycogen precipitin bands).

2.2 Interpretation of Double Diffusion Assays

Some experimentation with the concentration of Con A and polysaccharide to be used is generally necessary. Con A at 1—3 mg/ml and carbohydrate polymer at 1 mg/ml are generally good concentrations at which to start. Polysaccharides or glycoproteins with few determinant sugars may require higher concentrations of Con A (e.g. 10 mg/ml) in order to form visible precipitation bands. Certain sparsely branched dextrans, for example, required 9 mg/ml Con A to yield a positive reaction (13). Negative reactions must be interpreted with caution, higher levels of Con A being used before accepting negative results. On the other hand, dilute solutions of carbohydrate polymers (≤0.1 mg/ml) run against a concentrated Con A solution could lead to precipitate formation in the polysaccharide well.

Con A-reactive, carbohydrate-containing macromolecules which are homogeneous usually give single, sharp bands whereas polydisperse preparations appear as diffuse bands of variable breadth (11,27). Mixtures of polymers may yield several discrete precipitation bands depending on the concentration and molecular weight of the individual components (27). The curvature of the bands may also give an indication of molecular weight. A straight band suggests that the biopolymer has a molecular weight similar to Con A (ca. 100,000 at pH 7); a band curving at its extremities toward the carbohydrate polymer well indicates a molecular weight greater than 100,000 whereas a band curving toward the protein well suggests a molecular weight <100,000 for the complex carbohydrate. In general, low molecular weight polysaccharides and glycoproteins have high diffusion coefficients and give bands closer to the Con A well. For comparison purposes it is, of course, necessary to examine polysaccharide or glycoprotein solutions of comparable concentrations.

3. REFERENCES

1. Cifonelli, J. A., Montgomery, R. and Smith, F. (1956) 'The reaction between concanavalin A and glycogen', *J. Am. Chem. Soc.*, **78**, 2485−8.
2. Cifonelli, J. A. and Smith, F. (1957) 'The interaction of bacterial polyglucosans with concanavalin A', *J. Am. Chem. Soc.*, **79**, 5055−7.
3. Cifonelli, J. A. and Smith, F. (1955) 'Turbidimetric method for the determination of yeast mannan and glycogen', *Anal. Chem.*, **27**, 1639−41.
4. Manners, D. J. and Wright, A. (1962) 'α-1,4-Glucosans. Part XIV. The interaction of concanavalin A with glycogens', *J. Chem. Soc.*, 4592−5.
5. Goldstein, I. J., Hollerman, C. E. and Merrick, J. M. (1965) 'Protein−carbohydrate interaction. I. The interaction of polysaccharides with concanavalin A', *Biochim. Biophys. Acta*, **97**, 68−76.
6. Smith, E. E., Smith, Z. H. G. and Goldstein, I. J. (1968) 'Protein−carbohydrate interaction. A turbidimetric study of the interaction of concanavalin A with amylopectin and glycogen and some of their enzymic and chemical degradation products', *Biochem. J.*, **107**, 715−24.
7. Doyle, R. J., Woodside, E. E. and Fishel, C. W. (1968) 'Protein−polyelectrolyte interactions. The concanavalin A precipitin reaction with polyelectrolytes and polysaccharide derivatives', *Biochem. J.*, **106**, 35−40.
8. Poretz, R. D. and Goldstein, I. J. (1968) 'Protein−carbohydrate interaction. XI. A study of turbidity as it relates to concanavalin A−glycan interaction', *Immunol.*, **14**, 165−74.
9. Hawkins, J. D. (1964) 'Some studies on the precipitin reaction using a turbidimetric method', *Immunol.*, **7**, 229−38.
10. So, L. L. and Goldstein, I. J. (1967) 'Protein−carbohydrate interaction. IV. Application of the quantitative precipitin method to polysaccharide−concanavalin A interaction', *J. Biol. Chem.*, **242**, 1617−22.
11. So, L. L. and Goldstein, I. J. (1969) 'Protein−carbohydrate interaction. XVII. The effect of polysaccharide molecular weight on the concanavalin A−polysaccharide precipitation reaction', *J. Immunol.*, **102**, 53−7.
12. So, L. L. and Goldstein, I. J. (1968) 'Protein−carbohydrate interaction. XIII. The interaction of concanavalin A with α-mannans from a variety of microorganisms', *J. Biol. Chem.*, **243**, 2003−7.
13. Goldstein, I. J., Poretz, R. D., So, L. L. and Yang, Y. (1968) 'Protein−carbohydrate interaction. XVI. The interaction of concanavalin A with dextrans from *L. mesenteroides* B-512-F, *L. mesenteroides* (Birmingham), *Streptococcus bovis*, and a synthetic α-(1 → 6)-D-glucan', *Arch. Biochem. Biophys.*, **127**, 787−94.

14. Goldstein, I. J., So, L. L., Yang, Y. and Callies, Q. C. (1969) 'Protein–carbohydrate interaction. XIX. The interaction of concanavalin A with IgM and the glycoprotein phytohemagglutinins of the waxbean and the soybean', *J. Immunol.*, 103, 695–8.
15. Rosen, H. (1957) 'A modified colorimetric analyses for amino acids', *Arch. Biochem. Biophys.*, 67, 10–5.
16. Schiffman, G., Howe, C. and Kabat, E. A. (1958) 'Immunochemical studies on blood groups. XXI. Chromatographic examination of constituents split from blood group A, B and O(H) substances and from type XIV pneumococcal polysaccharide by *Clostridium tertium* enzymes', *J. Am. Chem. Soc.*, 80, 6662–70.
17. Hayes, C. E. and Goldstein, I. J. (1974) 'An α-D-galactosyl-binding lectin from *Bandeiraea simplicifolia* seeds. Isolation by affinity chromatography and characterization', *J. Biol. Chem.*, 249, 1904–14.
18. Agrawal, B. B. L. and Goldstein, I. J. (1968) 'Protein–carbohydrate interaction. VII. Physical and chemical studies on concanavalin A the hemagglutinin of the jack bean', *Arch. Biochem. Biophys.*, 124, 218–29.
19. Dubois, M., Gilles, K. A., Hamilton, J. K., Rebers, P. A. and Smith, F. (1956) 'Colorimetric method for determination of sugars and related substances', *Anal. Chem.*, 28, 350–6.
20. Hodge, J. E. and Hofreiter, B. T. (1962) 'Determination of reducing sugars and carbohydrates', in *Methods in Carbohydrate Chemistry*, eds. Whistler, R. L. and Wolfrom, M. L. (Academic Press, New York and London). Vol. 1, 380–94.
21. Goldstein, I. J. (1972) 'Use of concanavalin A for structural studies', in *Methods in Carbohydrate Chemistry*, eds. Whistler, R. L. and BeMiller, J. N. (Academic Press, New York and London), Vol. 6, 106–19.
22. Young, N. M. and Leon, M. A. (1974) 'The affinity of concanavalin A and *Lens culinaris* hemagglutinin for glycopeptides', *Biochem. Biophys. Acta*, 365, 418–24.
23. Mage, R. and Dray, S. (1965) 'Persistent altered phenotype expression of allelic γ-G-immunoglobulin allotypes in heterozygous rabbits exposed to isoantibodies in fetal and neonatal life', *J. Immunol.*, 95, 525–35.
24. Galbraith, W. and Goldstein, I. J. (1972) 'Phytohemagglutinin of the lima bean (*Phaseolus lunatus*). Isolation, characterization, and interaction with type A blood-group substance', *Biochemistry*, 11, 3976–84.
25. So, L. L. and Goldstein, I. J. (1969) 'Protein–carbohydrate interactions. XXI. Interaction of concanavalin A with D-fructans'. *Carbohydr. Res.*. 10, 231–44.
26. Goldstein, I. J., Reichert, C. M., Misaki, A. and Gorin, P. A. J. (1973) 'An "extension" of the carbohydrate binding specificity of concanavalin A', *Biochim. Biophys. Acta*, 317 500–4.
27. Goldstein, I. J. and So, L. L. (1965) 'Protein–carbohydrate interaction. III. Agar gel-diffusion studies on the interaction of concanavalin A, a lectin isolated from jack bean with polysaccharides', *Arch. Biochem. Biophys.*, 111, 407–14.
28. Crowle, A. J. (1961), *Immunodiffusion* (Academic Press, New York and London).
29. Clausen, J. (1971), *Immunochemical Techniques for the Identification and Estimation of Macromolecules* (North American Publishing Company, Amsterdam and London).

CHAPTER 17

Titration of Anti-Con A Sera

H. P. Schnebli

Antibodies directed against Con A are easily obtained and often make this reagent even more useful (1). The titration of anti-Con A sera on the other hand requires a little trick: Haemagglutination inhibition assays analogous to the ones used to titrate antisera against viral agglutinins fail because Con A adsorbs to some serum glycoproteins resulting in non-specific inhibition. Ouchterlony microimmune diffusion tests are time consuming and relatively insensitive.

1. PRINCIPLE

Here we describe a simple 'indirect' haemagglutination assay which is based on the observation previously described (2) that human erythrocytes do not agglutinate with Con A if they are kept in suspension by shaking. After preloading the erythrocytes with Con A the excess of lectin is removed; subsequent addition of anti-Con A antibodies immediately results in massive agglutination.

2. SERA

Rabbit anti-Con A sera and IgG fractions therefrom were prepared as described earlier (1, see also Chapter 13 in this volume). As controls, rabbit sera of unrelated specificity (anti-Sendai virus) and the respective IgG fractions were used. Anti-Con A sera can also be obtained from Calbiochem (San Diego, Cal.).

3. METHOD

3.1 Preloading of Erythrocytes with Con A

Human erythrocytes from fresh, heparinized blood (group O, Rh +) are washed four times in phosphate-buffered saline (PBS) pH 7.4 and adjusted to 2×10^8 cells per ml. An equal volume of a Con A solution (1 mg/ml) is added. During a 10 min incubation at room temperature the cells are kept in suspension by occasional twirls on a Vortex mixer. The suspension is then diluted 20-fold with cold PBS and centrifuged for 3 min at 2000 r.p.m. in a clinical centrifuge. The supernatant fluid is discarded, the original volume of PBS replaced and the cells (slightly clumped) are quickly resuspended by vigorous shaking on the Vortex mixer for 10 to 20 seconds. This suspension, when kept agitated, is stable for at least 20 minutes.

The conditions of the incubation described above assure a reproducible preloading of approximately 90% of the available Con A receptor sites (1).

3.2 Reaction with Antisera

Twofold serial dilutions of antisera in PBS are prepared in advance in U-type microtitre plates (Cooke Engineering Co.) containing 25 μl per well. Now 25 μl of the Con A-preloaded erythrocytes (2×10^8 cells/ml) are added to each well. The plate is shaken for 30 to 60 seconds on a Tayo Bussan mixer and read within the next two to three minutes. As seen in Figure 1, specific antisera and IgG fractions cause massive agglutination (rows B and D, respectively). No agglutination is observed with sera and IgG of unrelated specificity (rows A and C, respectively) or when erythrocytes not preloaded with Con A are reacted with anti-Con A sera or

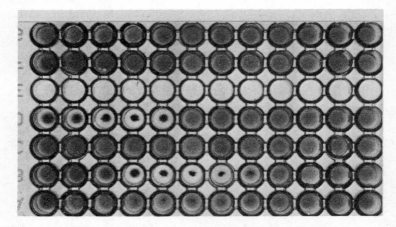

Figure 1. Titration of sera and IgG fractions against concanavalin A. A, Control serum (anti-Sendai). B, Anti-Con A serum. C, Control IgG (anti-Sendai) (1.05 mg/ml in the first well). D, Anti-Con A IgG (1.25 mg/ml in the first well). F and G same as B and D, respectively, except that the erythrocytes were not preloaded with Con A. Twofold serial dilutions in PBS from 1 to 11; controls without serum or IgG in 12. [Reproduced by permission of the copyright holder from *Experientia*, ref. 3.]

IgG (rows F and G, respectively). At high concentrations of serum and IgG agglutination is inhibited due to an excess of antibody (prozone effect).

4. TYPICAL RESULTS AND COMPARISON WITH DOUBLE DIFFUSION TEST

Results of antibody titration in sera prepared by us (1) or obtained from Calbiochem as well as IgG fractions are summarized in Table I. The assay is highly reproducible (same titre in 4 experiments) and more sensitive than the Ouchterlony microdiffusion test. We observed positive immunoprecipitin lines in Ouchterlony plates (3 mm distance between wells, 9 μl/well, 24 hours at room temperature) between Con A (100 μg/ml) and our rabbit anti-Con A serum diluted up to 1:40, and rabbit anti-Con A IgG diluted to 320 μg/ml, whereas no reactions were found with the unrelated IgG. A second weak precipitin line found between Con A and undiluted anti-Con A serum was identical with that obtained with undiluted anti–Sendai serum, indicating the presence of serum components different from IgG reacting with the lectin.

TABLE I.   Titration of anti-Con A sera and IgG fractions

Sera and IgG fractions	End titre
Serum: anti-Con A	1:256
anti-Con A (Calbiochem)	1:128
control (anti-Sendai)	>1:1
IgG: anti-Con A	66 μg/ml
control (anti-Sendai)	> 1000 μg/ml

5. PITFALLS

It is essential to use fresh human erythrocytes (blood group is not important) and the cells must not be exposed to proteolytic enzymes or sialidase.

The assay should be scored within a few minutes after mixing of Con A preloaded cells and sera: even with unrelated control sera we observed some slowly developing non-specific agglutination after about 10 minutes; this however could easily be separated from the rapid, specific agglutination shown in Figure 1.

6. REFERENCES

1. Bächi, Th. and Schnebli, H. P. (1975) 'Reaction of lectins with human erythrocytes. II. Mapping of Con A receptors by freeze-etching electon microscopy', *Exp. Cell Res.*, **91**, 285–95.
2. Schnebli, H. P. and Bächi, Th. (1975) 'Reaction of lectins with human erythrocytes. I. Factors governing the agglutination reaction', *Exp. Cell Res.*, **91**, 175–83.
3. Schnebli, H. P. and Bächi, Th. (1975) 'A simple titration assay for Anti-Concanavalin A Sera', *Experientia*, **31**, 1246–7.

CHAPTER 18

Survey and Evaluation of Labelling Methods and of Methods for Measuring Binding to Cells and Membrane Preparations

Kwen-Jen Chang
Pedro Cuatrecasas

Con A binds to a variety of cells through specific interactions with saccharide-containing cellular receptors and has been used as a molecular probe in studies of cell membrane function, structure and comparative studies of normal and transformed, agglutinated and non-agglutinated cells (see reviews, refs. 1 and 2). Fluorescein-, ferritin-, haemocyanin- and peroxidase-conjugated Con A have been widely used to study receptor site mobility and distribution as described elsewhere in this volume. Studies of the identification, characterization and purification of cellular receptors have been successfully accomplished in the past few years by the use of radioisotope-labelled biologically active ligands (e.g. neurotransmitter, drug) and peptides (e.g. hormones, antigens, lectins and toxins) (see reviews, refs. 3 and 4). The availability of radiolabelled Con A and special methodology for binding studies have permitted and encouraged intensive studies of Con A receptors.

1. LABELLING OF CON A

Con A was first labelled with [63]Ni by a replacement procedure in which [63]NiCl$_2$ was used to back-exchange the Mn^{2+} in Con A, and the resulting isotope-labelled Con A bound to cells with a background of 30 to 50% (5). Because of the high background and the inherent instability of [63]Ni-Con A, the originally reported binding studies with such [63]Ni-Con A have not been very reproducible. The results obtained contradicted in important ways more thorough studies using covalently labelled Con A (e.g. ^{3}H-acetylated (6) and ^{125}I-labelled Con A (7). [63]Ni-Con A has been used only rarely in recent years.

Covalent labelling of Con A has been achieved successfully by ^{3}H-acetylation (6, Chapter 21), ^{14}C-succinylation (8, Chapter 57) and ^{125}I-iodination (7, Chapters 19 and 20). ^{3}H-Acetylation and ^{14}C-succinylation yield stable isotopes of low specific activity. It has been shown in various kinds of studies that *mild* substitution does not alter the specificity and affinity of Con A for the hapten. Slightly modified affinity for certain polysaccharides has been reported for acetylated Con A (9). Extensive acetylation and succinylation of Con A can seriously affect the affinity and the physical—chemical properties of the molecules even though the specificity is not altered. Gunther and coworkers (8, Chapter 57) have shown that acetylated and succinylated Con A exist in the dimeric rather than the tetrameric state at pH 7.4, and the properties of mitogenicity, agglutination and inhibition of cap formation are significantly different from those of native Con A. Altered binding properties of succinylated Con A for 3T3 and SV3T3 cells have also been reported (10). Caution is needed in the interpretation of results of studies which use Con A derivatives having altered properties. It is not yet entirely clear that the altered biological responses of succinylated Con A are simply the result of the altered quarternary (dimeric) structure or of decreased affinity.

Among other disadvantages of [3]H- and ^{14}C-labelled Con A is the relatively low specific radioactivity which can be achieved (11). With such derivatives it may not be possible to detect special high affinity, low-capacity sites on cells, which may be biologically very important. For example, the mitogenic (8, 11) and insulin-like effects (12) of Con A occur at extremely low concentrations of the ligand, under conditions where only a small proportion of the total receptors are occupied. This alone suggests the existence of heterogenous and high-affinity sites which can only be detected with Con A of high specific activity.

[125]I-Labelled Con A has been prepared either by the lacto-peroxidase (7, 13, Chapter 19) or the chloramine T method (14, Chapter 20). Philips and coworkers (13) have labelled Con A with lactoperoxidase after adsorbing the molecule to 'Sephadex'; elution is achieved subsequently with α-methyl mannoside. However, this reaction requires large volumes of reaction mixture and ^{125}I-labelled Con A of high specific activity cannot be obtained since carrier I — must be added. ^{125}I-Con A of super-high specific activity (up to 2 atoms of carrier-free ^{125}I per 100,000 dalton Con A) can be prepared by the use of chloramine T (14, Chapter 20). The labelled lectin is purified by affinity chromatography on a 'Sephadex' column. The chloramine T method is a very simple and fast method and when properly performed the damage potentially caused by the oxidizing and reducing agents is not greater than that observed with the lactoperoxidase method. The fact that biologically fully active iodinated Con A can

be obtained with or without the presumed protection obtained by adding a specific hapten (i.e. α-methyl-D-mannopyranoside, 0.2 M) suggests that his method is well suited for innocuous labelling of this lectin. The main disadvantages of ^{125}I-labelling are the relatively short half-life (i.e. 60 days) of ^{125}I and the self-radiation damage occurring during storage. However, ^{125}I-Con A of high specific activity can be stored at $-20°$C for months without adversely affecting its biological activity if it is stored in dilute form and in the presence of high concentrations (protective) of albumin (e.g. 1%) (14).

The primary criterion of the labelled Con A product is that it retain its biological activity and specificity. It is thus advisable to purify the labelled Con A by affinity chromatography on a 'Sephadex' column. The purified, labelled Con A should therefore be checked for its ability to bind to cells and for the ability of α-methyl-D-mannopyranoside to inhibit or reverse the binding. At least 95% of the labelled Con A added to a cell or membrane suspension should bind, and at least 95% of bound Con A should be inhibited by adding the simple sugar. Similarly, addition of high, saturating concentrations of native Con A should inhibit 95% of the binding.

2. BINDING ASSAY

2.1 Separation of Bound from Free Con A

Binding assays for the Con A—receptor interaction are based on the successful and complete separation of receptor-bound from free Con A as well as on the estimation of the non-specific component of binding (i.e. that obtained in the presence of excess unlabelled Con A or the specific hapten, α-methyl-D-mannopyranoside).

One of the simplest techniques for measuring ligand binding to insoluble supports (e.g. cells or membranes) employs centrifugation as a means of separating bound from free ligand. The operation is performed at $4°$ to retard or prohibit dissociation of the complex, and the cell or membrane pellet is rapidly washed. With the use of suitable water-immiscible oils (15, 16) of properly selected density (e.g. with silicone fluid, dinonyl and dibutyl phthalate), or with a serum albumin cushion (13, Chapter 22), it is possible to separate the components rapidly and efficiently. The bound ligand can be obtained free from contaminating incubation medium, membrane or cell components which either float (e.g. fat cells) or sink (e.g. RBC, lymphocytes, membranes) under the influence of a centrifugal field. With a small, high-velocity centrifuge (e.g. Beckman 152 microfuge) the separation of free from bound ligand by bovine serum albumin cushion is achieved rapidly and yields very low non-specific background.

Con A can bind to glass and plastic surfaces, and this binding may be 'saturable' and inhibited by specific haptens. Thus, serious complications can occur when the binding assay is performed by simply centrifuging the tube containing incubation medium, washing the pellet and counting the total radioactivity in the tube; significant binding to the tube may occur (13). Similar problems may also occur if the binding assay is performed directly on monolayers of tissue culture cells. Transferring the reaction medium into a microfuge tube containing a layer of immiscible oil or albumin cushion and centrifuging to separate the cell or membrane components from the reaction medium which contains free, labelled Con A probably yields the best results (Chapters 22 and 23).

Filtration on microporous glass or synthetic polymer filters under vacuum provides another method whereby membrane- or cell-bound ligand is retained on the filter and free ligand is rapidly removed by washing through the filter. For such studies it is often a problem to select a filter which itself does not interfere with the binding measurement. Studies of the binding of Con A require the use of 'Teflon' filters to reduce the background binding to an acceptable level (14, Chapter 23).

2.2 Binding Properties of Con A

The binding of Con A to receptor sites is a temperature-dependent process. The rates of association and dissociation of the Con A—receptor complex are both more rapid at higher temperatures. The rate of association at room temperature is very rapid under most conditions of cell concentration and reaches equilibrium within 30 min. However, the time required to achieve equilibrium or steady state is a complex function of both the concentration of ligand and of the receptor binding site concentration (4, 17). The rate of dissociation of the Con A—receptor complex is extraordinarily slow, especially at low temperatures, but very rapid dissociation (at all temperatures) occurs upon the addition of a specific simple sugar (14).

2.3 Internalization and Non-specific Binding

Internalization of Con A may be a serious problem when binding studies are conducted at temperatures higher than $30°$, especially when intact cells with endocytotic or pinocytotic activity are used. It has recently been found that at least 95% of bound labelled Con A can be inhibited by initially adding α-methyl mannopyranoside to the Con A-containing binding assay, but only 60 to 80% of the cell-bound Con A can be dissociated if the sugar is added after Con A is bound to the cells (13). However, this residual, non-dissociable portion (20—40%) of the cell-bound Con A is still associated with the plasma membrane of the cell (13). Nearly all of the Con A bound to fat cells can be dissociated by sufficiently high concentrations of α-MM, even if temperatures of $30°$ are used to incubate the Con A—cell complex (14). Although many cell types do not appear to exhibit significant internalization of surface-bound Con A (even after prolonged incubations at $20°$ to $30°$), it is nevertheless advisable to determine this in all specific cases, and to consider conducting binding studies at temperatures below $24°$, even at $4°$, especially with cells of high endocytotic or pinocytotic activity. It must be remembered, however, that under such conditions (i.e. $4°$) the dimer—tetramer equilibrium of Con A may change such that the binding of a different molecular species is being studied.

The binding of labelled Con A in the presence of excess native Con A or α-methyl-D-mannopyranoside is usually considered to represent the non-specific component of binding. Under binding conditions such that most of the labelled Con A in the medium is bound to the cells or membranes, a condition which reflects the use of high receptor concentration relative to the dissociation constant of the complex (18), the inhibitory effectiveness of specific simple sugars will be greatly reduced because of the competition between receptor sites and sugar for the carbohydrate binding site of Con A. Under such conditions an artifactually low quantity of specific binding, and relatively high non-specific binding, may be observed. When α-MM is employed to estimate non-specific binding, at least 0.1 M α-MM must be used and the concentations of membranes or cells must be decreased so that the total cell-bound Con A is not more than 5% of the total Con A present. Caution must be paid to the non-specific binding of Con A to glass and plastic tubes as described above.

3. QUANTITATIVE ANALYSIS OF CON A–RECEPTOR INTERACTIONS

3.1 Considerations of Concentrations of Receptor and Labelled Con A

Binding data are usually interpreted according to Langmuir-type binding isotherms, competitive inhibition curves, Scatchard plots or double reciprocal plots to estimate the affinity and number of receptors of a given class (4). A serious problem relates to the aberrant data that may be obtained for such dissociable systems if the concentration of receptor used in the assay is greater than the dissociation constant of the interaction under study (18). Under such circumstances the affinity constant will be underestimated, and bizarre properties (e.g. cooperativity) may be unjustifiably invoked for the system (4,18). These conditions will often prevail in the study of biological receptors because the affinity of such systems is often very high while the specific radioactivity of the ligands used is not sufficiently high (^{3}H- or ^{14}C- and even ^{125}I-labelled Con A) to permit the detection of binding sites if these are present at concentrations well below the dissociation constant. Thus, in order to *measure* the receptors with accuracy by direct binding it is often necessary to use very high concentrations (exceeding K_D) of tissue material (18).

The equilibrium dissociation constant can be estimated from half-maximal binding of Langmuir binding isotherms or double reciprocal plots if the free, active labelled ligand concentration can be measured with accuracy and used in the abscissa when the bound, labelled ligand represents more than 5% of the total labelled ligand present in the medium. However, it is often difficult to estimate the free, active labelled ligand with precision because of possible degradation, deiodination, transpeptidation or other changes which may lead to heterogeneity of the labelled ligand. When the total ligand concentration is used in the abscissa to plot the binding saturation curve, the apparent dissociation constant (half-maximal binding concentration) is a linear function of the receptor or binding site concentration with a slope of 0.5 and an intercept equal to the dissociation constant (K_D) (18). Estimations of the affinity of complex formation from studies of the concentration of native ligand required to achieve half-maximal displacement of labelled Con A binding are also likely to underestimate the true affinity if the receptor and labelled ligand concentrations used are in excess of the K_D (19).

3.2 Scatchard Analysis

Scatchard analysis is most commonly used in the receptor field to estimate the affinity, number and possible heterogeneity of receptors (4). Scatchard analysis was originally derived for soluble macromolecules with independent, low-affinity binding sites for small ligands. Such analyses may not hold for the binding site of a large peptide to an insoluble membrane binding site. In the study of high-affinity systems frequently encountered in membrane—ligand interactions, especially when examined under conditions where a small proportion of the total available sites (present in large excess) are occupied, more than 10% of the ligand is usually bound in a manner which demonstrates a direct, linear increase in binding with increasing ligand concentration. Such data may result in a horizontal line in a Scatchard plot (20), superficially or falsely indicating the existence of an infinite number of such receptor sites (4).

If the affinity of the labelled ligand differs from that of unlabelled species, non-linear Scatchard plots can result when analysis is made from the competition of labelled ligand binding by varying the concentration of unlabelled ligand (23). This is particularly important for the use of ^{3}H-acetylated and ^{14}C-succinylated Con A, which may behave differently from native Con A (as discussed above).

Since Scatchard plots usually describe both extremities of the binding curve, even minor non-specific interactions (whether saturable or not) will grossly distort the curves, giving the impression of a 'second class' of binding sites of low affinity and high capacity. Perhaps the best information is obtained from the lowest saturation points, which describe the highest affinity processes; these occur as far as possible from the low-affinity and possible non-specific sites. Other possible causes for such non-linear Scatchard plots include positive or negative cooperativity between receptors and heterogeneity of receptor sites.

3.3 Polymerization of Con A

Polymerization of Con A may be especially important in the interpretation of binding data. The *apparent* behaviour of cooperativity, either negative or positive depending upon the affinity of the various polymer forms, may be observed in the Con A—receptor interaction (22, 23). It is well known that Con A can exist as a dimer, tetramer even high polymer depending on the temperature, pH and chemical modification. Furthermore, these association states are always in equilibrium and will depend on the Con A concentration under all conditions. Enhanced and reduced receptor binding of the self-associated ligand state would of course result in 'positive' and 'negative' cooperative behaviour, respectively. Such behaviour will be observed when the Con A concentration is varied and will be especially evident if the binding of tracer quantities of labelled lectin is examined in the presence of changing concentrations of the native protein.

Although the interpretation of Con A—receptor interactions is thus complicated by the above-mentioned difficulties, the use of simple Langmuir-type adsorption isotherms may yield accurate estimates of the *maximum* number of receptor binding sites per cell (or per mg of membrane protein).

3.4 Heterogeneity of Con A Receptors

In addition to the above-described methodological problems and special complicating properties (e.g. self-aggregation, states of altered affinity) of Con A, it is particularly difficult to estimate affinity constants for Con A binding to cells because of the very high probability that such 'receptors' represent very heterogeneous and varied classes of cell surface glycoproteins (11, 14). Since binding of Con A is presumably based on the specificity for sugar residues, since a continuum of relative affinities for simple as well as complex sugars almost certainly exists and since 'receptors' for the plant lectin have no true biological relevance (e.g. it is a foreign substance and there has been no opportunity for natural selection) absolute specificity for the binding cannot be expected to exist. Thus, it is probably improper to speak of 'receptors' for Con A in a strict sense. Since the lectin can potentially bind to a great many different cell surface glycoproteins, all the binding data will reflect a complex mixture of reactions which cannot be dissected (even if the data *appear* to reflect a homogenous, simple site). Care must therefore be exercised in interpreting affinity constants and numbers of sites from binding data. For these reasons, it is exceedingly hazardous to make firm conclusions from comparative binding studies (e.g. of normal *vs.* transformed cells) with respect to specific classes, numbers or affinities of sites. For example, a major change in a high-affinity, low-capacity site which is biologically very relevant may be quite readily missed.

4. MICROFLUOROMETRIC METHOD

Fluorescent labelled Con A has been used to study the mobility and distribution of binding sites as well as to analyze quantitatively the Con A receptors on the surface of normal and transformed, and agglutinated and non-agglutinated cells. A method which enables rapid qualitative measurements of the relative fluorescence intensity of single cells labelled with fluorescein isothiocyanate—Con A by microfluorometry has been developed by Neri and coworkers (Chapter 25). The major advantage of this microfluorometric method is that binding to individual cells can be studied and quantitative and qualitative comparisons can be done simultaneously in one preparation. In general, the problems and criteria discussed in the labelling and binding assay of radioisotope-labelled Con A should also be considered in the use of fluorescent labelled Con A.

5. REFERENCES

1. Lis, H. and Sharon, N. (1973) 'The biochemistry of plant lectins', *Ann. Rev. Biochem.*, **43** 541—74.
2. Nicolson, G. L. (1974) 'The interactions of lectins with animal cell surfaces', *Int. Rev. Cytol.*, **39**, 89—190.
3. Cuatrecasas, P. (1974) 'Membrane receptors', *Ann. Rev. Biochem.*, **43**, 169—214.
4. Cuatrecasas, P. and Hollenberg, M. D. (1976) 'Membrane receptors and hormone action', *Adv. Prot. Chem.*, **30**, in press.
5. Inbar, M. and Sachs, L. (1969) 'Structural difference in sites on the surface membrane of normal and transformed cells', *Nature*, **233**, 710—2.
6. Cline, M. J. and Livingston, D. C. (1971) 'Binding of ^{3}H-concanavalin A by normal and transformed cells', *Nature (New Biol.)*, **232**, 155—6.
7. Ozanne, B. and Sambrook, J. (1971) 'Binding of radioactively labelled concanavalin A and wheat germ agglutinin to normal and transformed cells', *Nature (New Biol.)*, **232**, 156—60.
8. Gunther, G. R., Wang, J. L., Yahara, Z., Cunningham, B. A. and Edelman, G. M. (1973) 'Concanavalin A derivatives with altered biological activities', *Proc. Nat. Acad. Sci. U.S.A.*, **70**, 1012—6.

9. Agrawal, B. B. L., Goldstein, I. J., Hassing, G. S. and So, L. L. (1968) 'Protein—carbohydrate interaction. XVIII. The preparation and properties of acetylated concanavalin A the hemagglutinin of the jack bean', *Biochemistry*, **7**, 4211—8.
10. Trowbridge, I. S. and Hilborn, D. A. (1974) 'Effects of succinyl-Con A on the growth of normal and transformed cells', *Nature*, **250**, 304—7.
11. Krug, U., Hollenberg, M. D. and Cuatrecasas, P. (1973) 'Changes in the binding of concanavalin A and wheat germ agglutinin to human lymphocytes during *in vitro* transformation', *Biochem. Biophys. Res. Commun.*, **52**, 305—12.
12. Cuatrecasas, P. and Tell, G. P. E. (1973) 'Insulin-like activity of concanavalin A and wheat germ agglutinin — direct interactions with insulin receptors', *Proc. Nat. Acad. Sci. U.S.A.*, **70**, 485—9.
13. Phillips, P. G., Furmanski, P. and Lubin, M. (1974) 'Cell surface interactions with concanavalin A', *Exp. Cell Res.*, **86**, 301—8.
14. Cuatrecasas, P. (1973) 'Interaction of wheat germ agglutinin and concanavalin A with isolated fat cells', *Biochemistry*, **12**, 1312—23.
15. Glieman, J., Østerlind, K., Vinten J. and Gammeltoft, S. (1972) 'A procedure for measurement of distribution spaces in isolated fat cells', *Biochim. Biophys. Acta*, **286**, 1—9.
16. Chang, K.-J. and Cuatrecasas, P. (1974) 'Adenosine triphosphate-dependent inhibition of insulin-stimulated glucose transport in fat cells', *J. Biol. Chem.*, **249**, 3170—80.
17. McPherson, R. A. and Zettner, A. (1975) 'A mathematical analysis of the incubation time in competitive binding systems', *Anal. Biochem.*, **64**, 501—8.
18. Chang, K.-J., Jacobs, S. and Cuatrecasas, P. (1975) 'Quantitative aspects of hormone—receptor interactions of high affinity: Effect of receptor concentration and measurement of dissociation constants of labelled and unlabelled hormones', *Biochim. Biophys. Acta*, **406**, 293—303.
19. Jacobs, S., Chang, K.-J. and Cuatrecasas, P. (1975) 'Estimation of hormone receptor affinity by competitive displacement of labeled ligand: Effect of concentration of receptor and of labeled ligand', *Biochem. Biophys. Res. Commun.*, **66**, 687—92.
20. Cuatrecasas, P., Hollenberg, M. D., Chang, K.-J. and Bennett, V. 'Hormone receptor complexes and their modulation of membrane function', in *Recent Progress in Hormone Research*, Vol. 31 (Academic Press, N.Y.), 37—94.
21. Taylor, S. I. (1975) 'Binding of hormones to receptors. An alternative explanation of nonlinear Scatchard plots', *Biochemistry*, **14**, 2357—61.
22. Nichol, L. W., Smith, M. D. and Ogston, A. G. (1969) 'The effects of isomerization and polymerization on the binding of ligand to acceptor molecules', *Biochim. Biophys. Acta* **184**, 1—10.
23. Cuatrecasas, P. and Hollenberg, M. D. (1974) 'Binding of insulin and other hormones to non-receptor materials: saturability, specificity and apparent 'negative cooperativity', *Biochem. Biophys. Res. Commun.*, **62**, 31—41.

CHAPTER 19

Preparation of ^{125}I-Concanavalin A by a Lactoperoxidase Affinity Labelling Method

PHILIP G. PHILLIPS
PHILIP FURMANSKI

1. PRINCIPLE OF THE METHOD

The effective radioiodination of a protein requires conditions for achieving a high specific activity with minimum damage. For proteins in which the retention of an activity is a consideration, for example, enzymes, antibodies or lectins, protection of the active site from modification, and removal of inactivated molecules are highly desirable. The method described here for the iodination of Con A accomplishes this, and, in addition, provides a rapid mechanism for the separation of iodinated Con A from unbound iodine and the other components of the reaction mixture. The basis of the method is the fact that Con A will reversibly bind to 'Sephadex' beads, and that the bound Con A can be iodinated using lactoperoxidase (E.C. 1.11.1.7).

2. METHOD

2.1 Adsorption of Con A to 'Sephadex'

15 ml of a 50% suspension of 'Sephadex' G-75 (Pharmacia Fine Chemicals) in phosphate-buffered saline (PBS) is added to 50 mg of Con A, dissolved in about

10 ml of PBS. This mixture is stirred gently for 30 min at room temperature, during which time Con A molecules with active sugar binding sites are attached to the 'Sephadex' beads. After 30 min, the stirring is discontinued and the 'Sephadex' is allowed to settle. The supernatant is then removed, and the beads are washed three times with fresh PBS, and finally resuspended in a total volume of 15 ml.

2.2 Iodination Reaction

The suspension of 'Sephadex' with bound Con A is stirred slowly, and the following are added: 600 μg of NaI, 5 mCi of carrier-free iodine-125 (Amersham Searle), 800 μg of lactoperoxidase (Sigma Chemical Company or Calbiochem), each dissolved in a small volume of PBS. 100 μl of 30% H_2O_2 are then slowly added and the stirring is continued for 30 min at room temperature.

2.3 Purification of the ^{125}I-Con A Product

At the completion of the reaction, the stirring is discontinued, and the gel permitted to settle out. The supernatant, containing free iodine-125 and other reactants, is removed and fresh PBS is added, with stirring, to resuspend the gel. This procedure is repeated six times, which eliminates most of the unbound label. The gel is then poured into a disposable column, fashioned in the following manner: a disposable 10 ml plastic pipette (Falcon Plastics) is cut off at the top and a glass-wool plug is forced down the pipette to form a retainer for the beads. An 18 gauge plastic tipped hypodermic needle, fitted with a piece of tygon tubing, is then glued onto the bottom of the pipette. A small clamp placed around the tygon tubing is used to control the flow rate through the column.

After settling in the column, the beads are washed with PBS until the radioactivity in the elute plateaus (usually between 5×10^4 and 1.0×10^6 c.p.m./ml). At this point 0.03 M glycine-HCl buffer, pH 2.0 (1) is added to the column, and fractions of about 4 ml are collected (see Figure 1). The fractions are collected into tubes containing 1 ml of 0.075 M Tris-HCl, pH 9.0, so that the final pH of the fractions is about 7. The tubes containing the ^{125}I-Con A are pooled and $MgCl_2$ and $CaCl_2$ (in that order) are added to a final concentration of 1.0×10^{-3} M. A precipitate which frequently forms in the tubes containing the highest concentrations of Con A is eliminated by centrifugation. This precipitate does not contain appreciable quantities of protein or radioactivity.

3. ANALYSIS AND STORAGE

We routinely determine the protein concentration of a sample from each fraction (2) and measure the level of radioactivity in the same sample. We find that between 75 and 90% of the original added Con A is recovered, and that the labelled Con A has a specific activity of between 4.0 and 6.0×10^4 c.p.m./μg. We have determined that an average 2.5 molecules of iodine-125 are coupled to each Con A tetramer, assuming a tetrameric molecular weight of 110,000 (3). Less than 1% of the label is present as free iodone-125 (Figure 2). If this level of contamination presents a significant problem, passage of the ^{125}I-Con A product through a column of 'Bio-Gel' P-10 (Biorad Laboratories), using PBS as an eluant, will

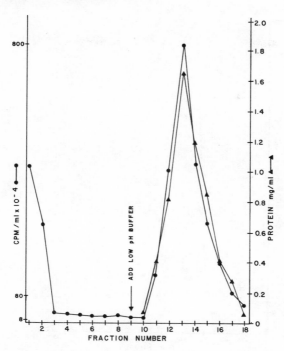

Figure 1. Elution of ^{125}I-Con A from 'Sephadex' G-75: At time zero, prior to fraction 1, a disposable column was loaded as described in the text. Fractions of 4 ml each were collected and an aliquot of each (after pH adjustment) was assayed for radioactivity and protein concentration

eliminate the remaining free iodine-125. Finally we have determined that the ^{125}I-Con A migrates as a single peak in disc gel electrophoresis (Figure 3). This peak has a relative migration which is identical to that of the original Con A. After analysis, the labelled Con A is stored frozen at -20°C.

4. COMPARISON WITH OTHER METHODS

There are several other methods for the iodination of proteins which have been, or could be used to label Con A (4—7 and Chapter 18). In addition, there are published methods for the acetylation (using [3]H- or ^{14}C-acetic anhydride, 3, 8—11 and Chapter 21), or for the succinylation (using ^{14}C-succinic anhydride, [3]) of Con A. These other methods have yielded Con A with varying specific activities, ranging from a few thousand c.p.m./μg to levels greater than those reported using this method. To separate the labelled Con A from the other reactants in these procedures, dialysis or chromatography have been used. We have found that dialysis

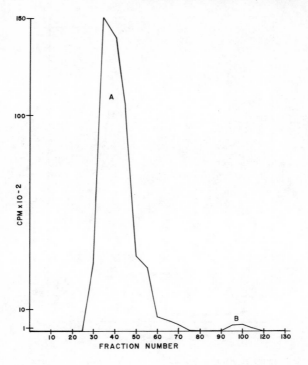

Figure 2. Determination of unbound ^{125}I contamination: A small amount of ^{125}I-Con A was loaded onto a column containing 85 ml of packed 'Bio-Gel' P-10. Fractions (1 ml each) were collected and assayed for radioactivity. Peak A corresponds to the void volume of the column and peak B is in the position of free iodine-125, as judged by independent chromatography on a similar column

inactivates a significant proportion of the labelled Con A. The method outlined here permits the iodination, isolation and purification of Con A in essentially a single step. Further, the reaction is carried out under conditions of limited iodination and protection of the lectin binding site.

5. ACKNOWLEDGEMENTS

This work was supported by grants from the NIH (5 R01 A109288, I R01 CA 15085), from the ACS (E557) and from the Milheim Foundation (72-17) to M. Lubin, and by fellowships to P.G.P. (Damon Runyon DRF-664 and NIH F02, CA50855) and to P.F. (Damon Runyon DRF-588 and NIH CA 41163). We thank Dr. M. Lubin for his advice and support during the course of these studies.

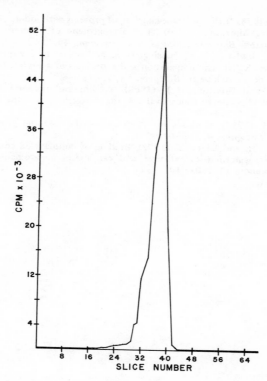

Figure 3. Gel electrophoresis of ^{125}I-Con A: A small sample of ^{125}I-Con A was run on 7.5% polyacrylamide gels in a β-alanine-glacial acetic acid buffer with a running pH of 4.3. After 2 h at 5 mA/gel, the gels were removed, fixed in 10% trichloroacetic acid, and then sliced for counting. In gels electrophoresed in the above manner and stained with coomassie blue, the label comigrated with unlabelled Con A

6. REFERENCES

1. Olson, M. O. J. and Liener, I. E. (1967) 'Some physical and chemical properties of concanavalin A, the phytohemagglutinin of the jack bean', *Biochemistry,* 6, 105–11.
2. Lowry, O. H., Rosenbrough, N. J., Farr, A. L. and Randall, R. J. (1951) 'Protein measurement with the Folin phenol reagent', *J. Biol. Chem.,* 193, 265–75.
3. Gunther, G. F., Wang, J. L., Yahara, I., Cunningham, B. A. and Edelman, G. M. (1973) 'Concanavalin A derivatives with altered biological activity', *Proc. Nat. Acad. Sci. U.S.A.,* 70, 1012–6.
4. Marchalonis, J. J. (1969) 'An enzymatic method for trace iodination of immunoglobulins and other proteins', 113, 299–305.
5. Helmkamp, R. W., Goodland, R. L., Bale, W. F., Spar, I. L. and Mutschler, L. E. (1960) 'High specific activity iodination of γ-globulin with iodine-131 monochloride', *Cancer Res.,* 20, 1495–500.

6. McFarlane, A. S. (1958) 'Efficient trace-labelling of proteins with iodine', *Nature*, **182**, 53.
7. Phillips, D. R. and Morrison, (1970) 'The arrangement of proteins in the human erythrocyte membrane', *Biochem. Biophys. Res. Commun.*, **40**, 284−9.
8. Agrawal, B. B. L., Goldstein, I. J., Hassing, G. S. and So, L. L. (1968) 'Protein−carbohydrate interaction. XVIII. The preparation and properties of acetylated concanavalin A, the hemagglutinin of the jack bean', *Biochemistry*, **7**, 4211−8.
9. Agrawal, B. B. L. and Goldstein, I. J. (1968) 'Protein−carbohydrate interaction. VII. Physical and chemical studies of concanavalin A, the hemagglutinin of the jack bean', *Arch. Biochem. Biophys.*, **124**, 218−29.
10. Frankel-Conrat, H. (1957) 'Methods for investigating the essential groups for enzyme activity', *Methods Enzymol.*, **4**, 247−69.
11. Kaneko, I., Satoh, H. and Ukita, T. (1972) 'Binding of radiolabeled concanavalin A and *Ricinus communis* agglutinin to rat liver and rat ascites hepatoma-nuclei', *Biochem. Biophys. Res. Commun.*, **48**, 1504−10.

CHAPTER 20

^{125}I-Labelled Concanavalin A of High Specific Activity

KWEN-JEN CHANG
PEDRO CUATRECASAS

1. PRINCIPLE

^{125}I-Iodinated concanavalin A (Con A) can be conveniently prepared by procedures employing the chloramine T—sodium metabisulphite method, as originally described by Greenwood and Hunter (1), with minor modifications (2). Specific activities as high as 2 Ci per μmole (20 μCi/μg) of ^{125}I-Con A (100,000 daltons) can easily be achieved without adversely affecting functional properties.

2. METHOD

2.1 Reagents

1. Sodium phosphate buffer, 0.25 M, pH 7.4.
2. Con A, 4 mg/ml in 0.1 M sodium phosphate buffer, pH 7.4.
3. Chloramine T, 5 mg/ml.
4. Na^{125}I (Union Carbide), carrier free, 1 mCi/μl.
5. Sodium metabisulphite, 10 mg/ml.

2.2 Procedure (1)

Twenty five μl of 0.1 M sodium phosphate buffer, pH 7.4, containing 50 to 100 μg of concanavalin A, are added to 100 μl of 0.25 M sodium phosphate buffer, pH 7.4, containing 1 to 2   mCi of carrier-free Na^{125}I. Twenty microlitres of freshly prepared chloramine-T (5 mg per ml) is added with gentle agitation and, after 30 sec at $24°$, 20 μl of sodium metabisulphite (10 mg per ml) is added to stop the reaction. After 10 sec, 0.2 ml of 0.1 M sodium phosphate, pH 7.4, containing 0.1% albumin is added. An aliquot of this reaction mixture is quickly withdrawn and diluted in 0.1 M sodium phosphate, pH 7.4, containing 0.1% albumin, to determine the total radioactivity and the percent of total radioactivity which has been incorporated into the protein. The remaining solution is then applied to a 'Sephadex' G-75 affinity column.

2.3 Purification

The iodinated Con A is purified by affinity chromatography on a 'Sephadex' G-75 column by immediately adding the remainder of the reaction mixture to a 3 ml column of 'Sephadex' G-75 which has been equilibrated with 50 mM Tris HCl, pH 7.5, containing 1 mM $MgCl_2$ and 1 mM $CaCl_2$. The flow of eluate is stopped for 30 min to allow the ^{125}I-Con A to fully interact with the column. The column is subsequently washed with about 50 ml of the same buffer until there is little or no radioactivity eluted from the column. The ^{125}I-Con A is selectively eluted with 0.3 M α-methyl mannopyranoside in 0.1 M Tris HCl, pH 8.5, containing 0.3% (w/v) bovine serum albumin. The flow of the elute is stopped temporarily (about 30 min) after adding the sugar solution. The major radioactive fractions (monitored by a hand-counting gauge) are pooled and dialyzed for 2 days at $4°$ against several large volumes of 50 mM Tris·HCl, pH 7.5 to remove α-methyl mannopyranoside. Alternatively, the ^{125}I-Con A bound to the 'Sephadex' G-75 column can be eluted with 0.2 N acetic acid, followed by neutralization with a strong phosphate buffer and NaOH. The final ^{125}I-Con A solution is stored in a $- 20°$ freezer.

3. RECOVERY AND SPECIFIC ACTIVITY

The percent incorporation of radioactivity is determined as follows: Aliquots (5 μl) of the reaction mixture are diluted to 1 ml with 0.1 M sodium phosphate buffer, pH 7.5 containing 0.7% bovine serum albumin. The radioactivity in 50 μl of the diluted reaction mixture is determined. An equal volume of cold 10% trichloroacetic acid is added. After 10 to 20 min at $4°$, the suspension is centrifuged. An aliquot (100 μl) of the supernatant is withdrawn for measurement of radioactivity. It is assumed that the radioactivity remaining in the supernatant represents non-incorporated [125]I. The fraction of ^{125}I incorporated into Con A is given by the equation

$$I = \frac{\text{R.A. before TCA} - \text{R.A. after TCA}}{\text{R.A. before TCA}}$$

where 'R.A. before TCA' and 'R.A. after TCA' refer to the radioactivity in aliquots before (50 μl) and after (100 μl) the addition of trichloroacetic acid, respectively. If the initial reaction mixture contains x μg of Con A and y μCi of ^{125}I, the specific activity (S.A.) of ^{125}I-Con A is given by the equation

$$S.A. = \frac{Iy}{x} \mu Ci/\mu g$$

The percent incorporation of radioactivity varies between 40 and 60% and the specific activity ranges from 3 to 20 μCi per μg, depending on the quantities of Con A and Na^{125}I used.

In a procedure identical to the one described above, the fraction of ^{125}I-Con A precipitated by trichloroacetic acid is determined in an aliquot of the final purified ^{125}I-Con A. About 90% of purified ^{125}I-Con A is TCA precipitable. The recovery of ^{125}I-Con A is determined by the equation

$$\text{Recovery} = \frac{\text{Total R.A. final}}{I(y)}$$

which varies between 50 and 80%.

Ninety percent of the purified ^{125}I-Con A can adsorb to 'Sephadex' G-75 using the conditions described under 'purification'. About 90% of the radioactivity can bind to fat cells, erythrocytes or liver membranes, and binding can be inhibited by 95 to 98% by adding 0.1 M α-methyl-mannopyranoside. At least 75 to 90% of the cell- or membrane-bound radioactivity can be dissociated upon subsequent addition of 0.1 M α-methyl mannopyranoside (2).

4. REFERENCES

1. Greenwood, F. C. and Hunter, W. M. (1963) 'The preparation of ^{131}I-labelled human growth hormone of high specific radioactivity', *Biochem. J.*, **89**, 114–23.
2. Cuatrecasas, P. (1973) 'Interaction of wheat germ agglutinin and concanavalin A with isolated fat cells', *Biochemistry*, **12**, 1312–23.

CHAPTER 21

Preparation of ³H-Concanavalin A

K. D. NOONAN

Concanavalin A is first prepared from jack bean or defatted jack bean meal according to the techniques described elsewhere in this book. This native Con A is then acetylated according to the following modification (1,2) of the procedure described by Fraenkel-Conrat (3).

1. METHOD

1.1 Acetylation of Con A

Native Con A (10 mg), dissolved in 1.0 ml phosphate-buffered saline (PBS, pH 7.2), is incubated at 0° for 30 min with 10 mM α-methyl mannopyranoside. This effectively blocks the active sites of the lectin, preventing acetylation of this portion of the molecule.

The Con A–PBS solution is then brought to pH 8.5 with solid Na_2CO_3. Care must be taken that the pH does not exceed pH 9.0 since above pH 9.0 the tetrameric Con A molecule will dissociate.

10 μCi ³H-acetic anhydride (sp. ac. 10–20 Ci/mMole) dissolved in dry acetone, benzene or dioxane is added, rapidly mixed with the Con A and the solution allowed to incubate at 0° for 90 min. It should be recalled that ³H-acetic anhydride in the presence of water will rapidly hydrolyse to acetic acid. Therefore care must be taken to dissolve the ³H-acetic anhydride in 'dry' organic solvents. This is particularly important if the acetic anhydride is to be stored between uses.

1.2 Separation from Unreacted Label

The radiolabelled Con A is placed on a 'Sephadex' G-25 column (90 cm x 1.5 cm, $V_0 = 75$ ml) and eluted with PBS (pH 7.2) at $4°$C. 2 ml fractions are collected and their absorbance monitored at 280 nm. Aliquots of each fraction are also counted in a liquid scintillation spectrometer. Most of the labelled Con A will be eluted from such a column in approximately 10 ml.

The ^{3}H-Con A is dialysed overnight against 4 litres of PBS (pH 7.2) to remove the α-methyl mannopyranoside from the active sites.

1.3 Concentration and Storage

The ^{3}H-Con A, which is generally at or near a concentration of 1 mg/ml, is then divided into 1 ml aliquots and frozen at $-10°$C. We have found that lyophilization of the labelled material frequently results in a loss of biological activity but Con A concentrated in polyethylene glycol retains its activity.

2. PROPERTIES OF ^{3}H-CON A

The ^{3}H-Con A which is prepared by the technique described has a specific activity between 1×10^6 c.p.m.'s and 1×10^7 c.p.m.'s/mg protein. Such specific activities compare favourably with those which can be obtained by iodination with ^{125}I.

^{3}H-Con A retains its ability to agglutinate transformed cells, runs with native Con A on 'Sephadex' and DEAE—cellulose chromatography and will compete with native Con A for binding sites on the cell surface, thus suggesting that acetylation does not markedly affect the biological properties of the Con A molecule.

3. EXAMPLE OF APPLICATION: BINDING TO FIBROBLASTS

The Con A binding assay which we have developed has been devised to limit the endocytosis of the Con A molecule by the cell. We have taken advantage of the fact that endocytosis is significantly reduced at $0°$ to limit what we believe to be non-specific uptake of the Con A molecule at temperatures above $15°$C.

3.1. Binding Assay

Fibroblasts are grown to approximately 80% confluency in plastic tissue culture dishes (3.5 cm diameter).

The plates are then incubated at $0°$ (on ice) for 5 min prior to the initiation of the assay.

The cells are washed twice with ice-cold phosphate-buffered saline (PBS, pH 7.2) and then incubated at $0°$ in 1.0 ml PBS containing the desired concentration of ^{3}H-Con A.

Incubation proceeds for 5 min after which the cells are washed 5 times with PBS and finally precipitated in ice-cold 10% trichloracetic acid (TCA).

The cells are scraped from the plate in TCA and the material solubilized in a commercial solubilizer (e.g. PCS or Aquasol) and counted.

3.2 Important Variables

If the binding assay is performed at 0° as described approximately 95% of the labelled Con A can be prevented from binding to the cells by preincubation of the lectin with 10 mM α-methyl mannopyrannoside (α-MM). However, it has been our experience with fibroblasts that only 50—60% of the labelled Con A can be prevented from associating with the cells if the binding is performed at 22° or above. Similarly binding assays performed at 22° suggest that two to threefold more Con A is associated with the cells at 22° than at 0°. We believe that this enhanced association of Con A with the cells at 22° is the result of endocytosis.

In most of the cell lines we have investigated the number of molecules of Con A bound to the cells remains the same whether the cells are attached to the plate or suspended in PBS. However, in a few cases (most notably CHO cells) the number of Con A molecules bound to the cell surface have increased following removal from the plate. Thus care must be taken to insure agglutinin binding remains the same whether the cells are on the plate or in suspension prior to attempting to use the assay described here for evaluating the role of lectin binding in agglutination.

3.3 Relationship between Binding and Agglutination

We have previously reported that transformed cells bind 2—3-fold more Con A than their normal counterparts. However a more extensive survey of cell types suggests that enhanced lectin binding is not a prerequisite of the transformed state. More importantly enhanced lectin binding does not assure the investigator that the cell which binds more lectin will be more agglutinable than its counterpart which binds fewer lectin molecules. In a number of cell lines which we have investigated the cell which binds 2—3 times more lectin may be much less agglutinable than the cell line which binds fewer Con A molecules. Thus great care must be taken in equating enhanced lectin binding with any biological parameter associated with cell growth.

4. COMPARISON WITH OTHER LABELLING METHODS

The primary advantage of ^{3}H-Con A, as compared to ^{125}I-Con A, is the stability of the isotope. ^{3}H-Con A can be prepared in large batches and frozen for up to 1 year with little or no loss in radioactivity or biological activity. ^{125}I-Con A, on the other hand, will show a progressive reduction in specific activity due to the decay of the ^{125}I and therefore cannot be prepared in large batches or stored for extended periods.

5. REFERENCES

1. Noonan, K. D. and Burger, M. M. (1973) 'Binding of [^{3}H] concanavalin A to normal and transformed cells' *J. Biol. Chem.*, **248**, 4286—92.
2. Noonan, K. D. and Burger, M. M. (1973) 'The relationship of concanavalin A binding to lectin-initiated cell agglutination', *J. Cell Biol.*, **59**, 134—42.
3. Fraenkel-Conrat, H. and Colloms, M. (1967), *Biochemistry*, **6**, 2740.

CHAPTER 22

The Microfuge Method of Measuring the Binding of Radiolabelled Concanavalin A to Cells

PHILIP G. PHILLIPS
PHILIP FURMANSKI

1. PRINCIPLE OF THE METHOD

Studies of the binding of lectins to cell surfaces are complicated by the fact that some of the lectin is loosely bound and removed by washing, and, at least for Con A, radiolabelled lectin binds to glass and plastic (1—3). This non-specific binding is saturable and inhibited by α-methyl-D-mannopyranoside, and is therefore a serious complication in cell binding studies.

The microfuge method described here effects a rapid and complete separation of bound from unbound lectin, with a minimum of manipulation. Further, there is no interference in the measurement by non-specific binding of Con A to glass or plastic.

2. ASSAY PROCEDURE

2.1 Preparation of Bovine Serum Albumin Cushions

Bovine serum albumin (BSA) (Sigma Chemical Company) is dissolved to a final concentration of 5% (w/v) in phosphate-buffered saline (PBS). It is not necessary to use freshly prepared BSA if a stock solution is sterilized (by filtration) after preparation and then stored under refrigeration. Microfuge tubes (Beckman Instruments, 450 lambda size) are loaded with approximately 300 lambda of the 5% BSA. The tubes are then spun in a Beckman microfuge for 30 seconds, to eliminate air locks which form when the BSA is placed into the tube.

2.2 Preparation of Cells

The binding assay can be done with tissue culture cells, ascites cells, blood cells or others. The preparation of cells for the assay should be by appropriate standard procedures, depending on the growth conditions of the cells. After harvesting (with EDTA), the cells should be washed several times with PBS.

2.3 Con A Binding Reaction

The test cells (generally $1-10 \times 10^6$ cells/assay, in 0.5 ml of PBS) are added to 0.5 ml of the appropriate concentration of labelled Con A, or other lectin in a 10 x 75 mm glass test-tube. The cells and the Con A are then mixed and allowed to react for 10 min at $4°C$.

2.4 Separation of Cells and Bound Con A from the Free Lectin

Following incubation, the tubes are pelleted in a table top centrifuge. The supernatant is removed and the cells are taken up in a small volume (about 50 lambda) of PBS. After gently resuspending the cells, the mixture is carefully layered onto the prespun cushions of BSA. The microfuge tubes are then closed and centrifuged at top speed for 1 min in the microfuge. During this time the cells migrate through the cushion to form a pellet at the bottom of the tube. Free lectin remains at the top of the cushion during this procedure and does not contaminate the lower level of the tube (by diffusion) for at least 30 min.

2.5 Preparation of Labelled Cells for Counting

Following centrifugation, the microfuge tubes are removed and the tube bottom, containing the cell pellet, is cut off with a razor blade. Using ^{125}I-Con A or ^{131}I-Con A, direct counting of the cell pellet in the microfuge tube tip is accomplished without removal of the cells, simply by placing the cut piece in a gamma counting tube. If beta-emmitting radiolabelled lectin is used, the cells should be lysed out of the tube with a solubilizer such as hydroxide of hyamine, dissolved in a scintillation fluid and then counted in a scintillation counter.

2.6 Specificity Controls

For each concentration of Con A used, controls are included consisting of cells and Con A plus 50 mM α-methyl-D-mannopyranoside (α-MM). In addition, a control consisting of labelled Con A but no cells, is included to test for sedimentation of labelled material (as in aggregates) in the microfuge procedure.

3. COMPARISON WITH OTHER METHODS

Most published methods for measuring lectin binding involve repeated washing of the cells following the binding reaction. We have found that repeated washing of cells tends to remove some of the bound lectin (1). In addition we, and others, have found that Con A will bind to glass (1) and to plastic (2,3). Therefore, using conventional methods, it is necessary to remove cells with bound lectin from the initial reaction vessel, before counting, to eliminate errors due to lectin bound to vessels. Table I shows the results of an experiment where three methods of measuring binding to cells were compared. BALB/3T3 cells transformed by Simian virus 40 (SV3T3) were removed from culture dishes with the aid of calcium magnesium-free PBS (CMF-PBS), which contained 0.02% ethylenediamine tetra-acetic acid (EDTA), and then washed three times with PBS. The experiment involved three different methods, of preparing the cells for counting as described in the legend to the table. It can be seen (see series 2) that a substantial amount of labelled Con A is bound to the initial reaction vessel. It can also be seen (compare series 2 to 3) that washing the cells several times has removed a portion of the bound Con A from the cells.

TABLE I.   Determination of Con A binding to cells by the microfuge method. Cells were removed from culture dishes with 0.02% EDTA in CMF-PBS and added to glass tubes containing 25 μg of ^{125}I-Con A, with or without α-MM (10^{-2} M). After 10 min at 4°C, the cells were processed as follows: series 1 cells were centrifuged, washed 3 times with PBS, and then counted in the same tubes in which the experiment was run; series 2 cells were also centrifuged and washed 3 times with PBS, but after washing, the cells were transferred to another tube. Both the original tube, empty of cells, and the new tube containing cells, were counted; series 3 cells were processed by the microfuge method. (Reprinted with the permission of Academic Press.)

Tube No.	Additions to Con A		Original tube (c.p.m.)	Microfuge or new test-tube (c.p.m.)
	α-MM	SV3T3 cells		
1A	0	+	15,611	
1B	+	+	3554	
1C	0	0	3314	
2A	0	+	4270	7083
2B	+	+	1563	568
2C	0	0	3376	246
3A	0	+		10,644
3B	+	+		754
3C	0	0		65

4. EXAMPLES OF APPLICATION

4.1 Con A Binding to Human Red Blood Cells

Freshly drawn, type 0+ human red blood cells (RBC's) were obtained from a blood bank. After washing three times with PBS, to remove plasma and the buffy coat, the cells were incubated for 10 min at 4°C with various concentrations of ^{125}I-Con A, either in the presence or absence of α-MM. Following the reaction, the cells were prepared for counting by the microfuge method and the results are shown in Figure 1. From the specific activity of the Con A used and the cell number, we have determined that the human RBC has about 2×10^6 specific Con A binding sites per cell (1).

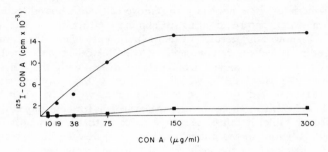

Figure 1. Binding of ^{125}I-Con A to human RBC's. RBC's were added to tubes containing various amounts of labelled Con A. After 10 min at 4°C the cells were prepared for counting by the microfuge method. ● no α-MM; ■ plus α-MM. [Reproduced with the permission of Academic Press, from *Exp. Cell Res.*, 86, 301–8 (1974).]

4.2 Con A binding to Tissue Culture Cells

BALB/3T3 (3T3) and SV3T3 cells were removed from growth surfaces with CMF-PBS plus EDTA and washed three times with PBS. Various concentrations of ^{125}I-Con A, in the presence or absence of α-MM, were mixed with the cells and allowed to react for 10 min at 4°C. The cells were then prepared for counting by the microfuge method, and the results are shown in Figure 2. We have calculated that there are 1.3×10^7 and 1.5×10^7 Con A binding sites per cell for 3T3 and SV3T3, respectively (1).

5. ACKNOWLEDGEMENTS

This work was supported by grants from the NIH (5 R01 A109288, I R01 CA 15085), from the ACS (E557) and from the Milheim Foundation (72-17) to M. Lubin, and by fellowships to P.G.P. (Damon Runyon DRF-664 and NIH F02, CA50855) and to P.F. (Damon Runyon DRF-588 and NIH CA 41163). We thank Dr. M. Lubin for his advice and support during the course of these studies.

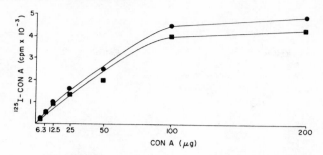

Figure 2. Binding of ^{125}I-Con A to 3T3 and SV3T3 cells. Either 3T3 or SV3T3 cells were mixed with various amounts of labelled Con A and treated as in Figure 1. ■ 3T3; ● SV3T3. [Reproduced with the permission of Academic Press, from *Exp. Cell Res.*, **86**, 301—8 (1974).]

6. REFERENCES

1. Phillips, P. G., Furmanski, P. and Lubin, M. (1974) 'Cell surface interactions with concanavalin A: location of bound radiolabeled lectin', *Exp. Cell Res.*, **86**, 301—8.
2. Phillips, P. G. and Lubin, M. 'Concanavalin A binding to surfaces, contribution of serum factors', *Exptl. Cell Res.*, submitted for review.
3. Berlin, R. D. and Ukena, T. E. (1972) 'Effect of colchicine and vinblastine on the agglutination of polymorphonuclear leucocytes by concanavalin A', *Nature (New Biol.)*, **238**, 120—2.

CHAPTER 23

Binding of ^{125}I-Labelled Concanavalin A to Cells

KWEN-JEN CHANG
PEDRO CUATRECASAS

The plant lectins, concanavalin A (Con A) and wheat germ agglutinin (WGA), have been very useful markers for exploring the surface topography of a variety of normal and neoplastic mammalian cells (1,2), as described elsewhere in the book. In few cases, however, have detailed binding studies been performed, and little information is available concerning the precise kinetic and quantitative properties of lectin—cell interactions. Iodinated plant lectin derivatives of very high specific radioactivity, prepared by the chloramine-T method as described in Chapter 20, have been useful in examining the detailed nature of the interaction of WGA and Con A with isolated fat cells, lymphocytes and erythrocytes, and for labelling the plasma membrane as markers in membrane isolation procedures.

1. FILTRATION ASSAY

1.1 Choice of Filter Material

A variety of filters of different chemical composition was tested to select those which adsorb the least amount of radioactivity in the absence of cells. The ones found to be most suitable, which were used in the present studies, were of nylon

(NRWP, Millipore Corp.) for WGA and of 'Teflon' (LSWP, Millipore Corp.) for Con A. Less than 1% of the total ^{125}I-WGA in the incubation medium adsorbs to the nylon filters and about 1.5% of the ^{125}I-Con A adsorbs to the 'Teflon' filters under our conditions in the absence of cells. The adsorption of radioactivity to the filters is unaffected by the presence of high concentrations of the specific simple sugars (N-acetyl-D-glucosamine or α-methyl-D-mannopyranoside) or by the presence of high concentrations (50 μg/ml) of the native plant lectins. Virtually all of the filters composed of cellulose or of cellulose esters which were examined bound prohibitively large amounts of the plant lectins, and this binding was modified by the presence in the buffer of the simple sugars or of the native plant lectin.

1.2 Procedure

The assay procedures used to measure the specific binding of ^{125}I-Con A to fat cells are similar to those described earlier for measuring the binding of ^{125}I-insulin (4,5). Isolated fat cells (about 10^4 cells/ml), prepared by the method of Rodbell (3), are incubated in disposable, polystyrene tubes (12 x 75 mm) for 30 to 40 minutes at 24° in 0.2 ml of Krebs—Ringer bicarbonate buffer containing 0.1% (w/v) albumin and the iodinated lectin (2×10^4 to 2×10^5 c.p.m. as indicated in legend of Figure 1). Three millilitres of the same buffer (ice-cold) are added to the tubes, the contents are poured on appropriate 'Millipore' filters (25 mm) positioned with vacuum in a multiple filtration manifold, and the filters are washed under vacuum with 10 ml of ice-cold Krebs—Ringer bicarbonate buffer containing 0.1% (w/v) albumin. The steps of dilution, filtration and washing consume 10 to 15 seconds. The filters are counted in a well-type gamma counter using disposable plastic tubes. The same assay procedure has also been used to measure ^{125}I-lectin binding to untransformed and transformed lymphocytes (6) and erythrocytes.

1.3 Specificity Control

All of the binding data is expressed in terms of 'specific' binding, as described in studies of the binding of ^{125}I-insulin to cells (4,5). Specific binding refers to the amount of ^{125}I-protein bound to the cells which can be specifically inhibited by the presence of an excess of native, unlabelled protein. Alternatively, this can be determined by measuring the difference in the amount of radioactivity bound to the cells in the absence and presence of high concentrations of the specific simple sugar which binds to the active sites of the plant lectin. The use of the native protein produces more profound and more reproducible displacement of binding than is observed with the simple sugars. The use of the native plant lectin has therefore been adopted in two sets, each consisting of duplicate or triplicate samples; the two sets differ only in the presence or absence of WGA (200 μg/ml) or native Con A (500 μg/ml). When the concentration of the iodinated WGA is less than 5 μg/ml virtually all of the bound radioactivity is replaced by the native protein.

The non-specific binding of iodo-Con A is generally quantitatively significant, and it is therefore essential in all cases to perform controls carefully and to make appropriate corrections for non-specific binding.

With the assay procedures described above it is possible to detect specific binding of the ^{125}I-plant lectins to fat cells with concentrations of iodoprotein as low as 10^{-11} M. For calculations of molar concentrations of the plant lectins the molecular weight of WGA is assumed to be 25,000 (7) and that of Con A 100,000 (8).

2. BINDING TO FAT CELLS

2.1 Characteristics of Binding

The binding of ^{125}I-Con A to fat cells is a saturable process with respect to the lectin (Figure 1). The apparent sigmoidicity might suggest some cooperative property of the binding sites or of the lectin. At low concentrations, the native protein paradoxically enhances the binding of the iodinated Con A to the cells (Figure 2). Although this could also be interpreted as a type of cooperative behaviour, it need not represent receptor—receptor interaction but may be an indication of receptor heterogeneity or lectin self-aggregation (9,10). Con A is

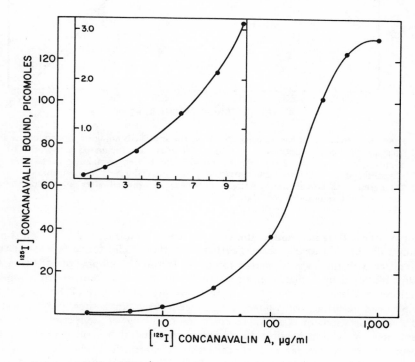

Figure 1. Specific binding of ^{125}I-concanavalin A to isolated fat cells as a function of the concentration of the plant agglutinin. Fat cells (2.2×10^5 cells) were incubated for 80 minutes at 24° in 0.2 ml of Krebs—Ringer bicarbonate buffer containing 0.1% (w/v) albumin and varying concentrations of ^{125}I-concanavalin A (log scale in main figure, not in insert). Specific binding was determined as described in the text. The binding observed with low concentrations of concanavalin A is depicted in the insert. Data from reference 13.

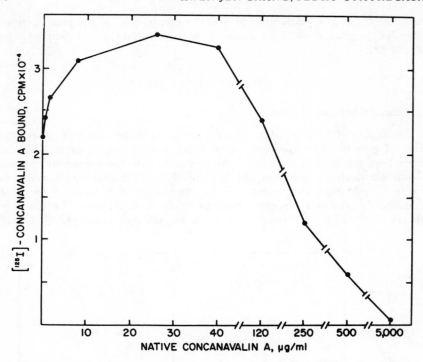

Figure 2. Effect of native concanavalin A on the specific binding of low concentrations of ^{125}I-concanavalin A to fat cells. Fat cells were incubated at 24° for 10 minutes in 0.2 ml of Krebs–Ringer bicarbonate buffer containing 0.1% (w/v) albumin and the indicated concentration of concanavalin A. ^{125}I-Concanavalin A (0.2 μg/ml) was added and specific binding was determined after incubating the samples for 70 minutes at 24°. Data from reference 13.

known to be a tetramer, and at the very low concentrations used ^{125}I-Con A may not be fully in a tetrameric state. Furthermore, the affinity for surface saccharides may differ with the state of aggregation of the lectin. The binding of ^{125}I-Con A is readily inhibited by a high concentration of the native protein; nearly complete inhibition is achieved by 5 mg/ml of the native protein (Figure 2). Because of the complexities resulting from possible heterogeneous binding sites, polymerization and self-association of Con A, receptor cross-linking or cooperative site–site interaction, it is not possible to designate meaningful kinetic expressions and interpretations to the fat cell–concanavalin A interaction. The data shown in Figure 1, however, indicate that a fat cell can bind a maximum of about 4×10^8 molecules of Con A (100,000 daltons).

2.2 Heterogeneity of Binding Sites

The known biological effects of WGA and Con A on isolated fat cells, such as stimulation of glucose transport, inhibition of lipolysis and inhibition of adenylate cyclase activity, are maximal with concentrations of these plant lectins which are very far below those required to saturate the binding sites for lectins (11). This points to the existence of functionally heterogeneous receptors for these lectins. This is supported further by the observation that whereas Con A inhibits adenylate cyclase activity and is mitogenic at concentrations below 50 μg/ml, higher concentrations cause stimulation of enzyme activity and are toxic to lymphocytes in tissue culture (6,11,12).

Vigorous digestion of fat cells with trypsin results in a modest loss of the binding capacity for Con A (13). Pronase digestion of cells is more effective than trypsin in causing a fall in the binding of the plant lectin to cells, although even under the most drastic conditions only 50% of the binding sites are lost. Neuraminidase digestion of cells does not alter the binding. Phospholipase C digestion causes a fall in the binding (13), possibly because of cell lysis. It is not possible to exclude the possibility that a large fraction of the total binding is to cell surface glycolipids rather than glycoproteins, especially in view of the considerable resistance of the binding to proteolytic digestion.

2.3 Dissociation of Con A—Cell Complex

Con A binds very rapidly to fat cells at 24° (Figure 3). The rate of spontaneous dissociation of the concanavalin A—cell complex is extremely slow. Even at 37° only slight dissociation occurs during 100 minutes, and at 4° the half-life is several days. The Con A—cell complex dissociates very rapidly after addition of the simple sugar, α-methyl-D-mannopyranoside, which has specificity for this plant lectin. The rate of dissociation in the presence of this sugar is temperature and sugar-concentration dependent (Figure 3). It is also demonstrated for WGA that the equilibrium of binding is achieved very rapidly, and the rate of dissociation of the WGA—cell complex is extremely slow whether it is measured by adding an excess of native, unlabelled protein or by removing the free, unbound labelled protein in the medium along with dilution of the cells. However, rapid and profound dissociation occurs after addition of the specific sugar, N-acetyl-D-glucosamine, or the glycoprotein ovomucoid (Figure 4). The very tight binding of lectins and the rapid dissociation of the lectin—cell complex induced by simple sugars have provided the basis for the use of plant lectins as useful markers for plasma membranes, as recently described (14).

2.4 Change in Surface Receptor Density during Change in Cell Size

Fat cells from small (about 140 g) rats can bind a maximum of about 3×10^8 molecules of WGA and 4×10^8 molecules of Con A per fat cell. Since these cells have an average diameter of about 35 μm, it can be calculated that the cell surface density of binding sites for wheat germ agglutinin is about 8.5×10^4 sites per μm^2 and for Con A about 11×10^4 sites per μm^2. Fat cells obtained from very large (about 550 g) rats can maximally bind very nearly the same number of WGA and Con A molecules per cell (Table I) as the small cells obtained from younger animals, despite a nearly 3-fold difference in the diameter of these cells. Starvation of the large rats for 3 to 10 days leads to very substantial reductions in the size of the fat cells obtained from the epidydimal fat pads (Table I). However, the

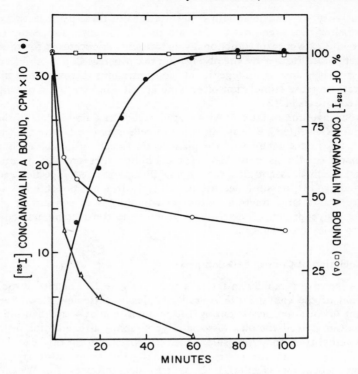

Figure 3. Rates of association and of dissociation of fat cell–concanavalin A complex. Fat cells were incubated at 24° in Krebs–Ringer bicarbonate buffer containing 0.1% (w/v) albumin and ^{125}I-concanavalin A (0.3 μg/ml), and specific binding to cells was determined at various times (●) by filtration on LSWP 'Millipore' membranes as described in the text. The dissociation was studied by incubating the cells with iodoconcanavalin A for 60 minutes as described above, followed by the addition of 5 mg/ml of native concanavalin A (□) or 50 mM α-methyl-D-mannopyranoside (○, △). The suspensions were then incubated for various times at 24° (○) or 37° (□, △). No discernable dissociation occurred in samples incubated with native concanavalin A at 24° (not shown in figure). Data from reference 13

maximum number of molecules of WGA, Con A or insulin which can bind per cell does not change despite these rapid and large changes in cell size. The cell surface densities of the receptors for these three proteins change in unison and the magnitude of the change can be more than 2- (3-day starvation) and 7- (10-day starvation) fold. It thus appears that at least, rapid changes in membrane surface area such as occur in starvation result from changes in the phospholipid rather than glycoprotein content of the membrane.

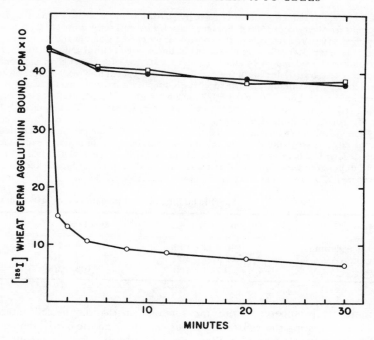

Figure 4. Dissociation of the fat cell—wheat germ agglutinin complex by native wheat germ agglutinin (●), α-methyl-D-mannopyranoside (□) and N-acetyl-D-glucosamine (○). After incubating fat cells for 50 minutes at 24° with 50 ng/ml of ^{125}I-wheat germ agglutinin, 250 μg/ml of the native protein (●, □) and 50 mM of the sugar (□, ○) were added and specific binding was determined after varying periods of continuing incubation at 24°. Data from Ref. 13.

3. GENERAL CONSIDERATION OF BINDING, MULTIVALENCY, CROSS-LINKING AND CELL AGGREGATION

Under the conditions of binding used here, plant lectins are binding almost exclusively to the surface of the cell. Equilibrium of binding is achieved quickly and a *plateau* of binding is maintained at that level for considerably long periods of time. Simple sugars or specific glycoproteins in solution (e.g. ovomucoid) cause immediate and profound dissociation of the lectin—cell complex, regardless of the time and conditions of exposure of the cell to the lectin. Further, ovomucoid—Sepharose derivatives are nearly as effective in dissociating the lectin. If phagocytosis or intracellular translocation of the lectins occurs under these conditions, it must represent an insignificant proportion of that which is specifically bound to the surface of the cell.

Although the binding of Con A to cells is clearly associated with carbohydrate components of the cell membrane, it is possible that a very substantial proportion of the binding could be occurring with glycolipids instead of glycoproteins.

TABLE I.   Effect of reduction of fat cell size by starvation on the density of various cell surface receptors. A very large (580 g) Sprague–Dawley rat was starved for 10 days, at which time it weighed 410 g. The fat pads of the control rat weighed 5 g and of the starved rat 0.5 g. The average diameter (300 cells were measured) of the isolated control cells was 90 μm and of the cells from the starved animal 34 μm. Fat cells (5 x 10^4 cells for wheat germ agglutinin and concanavalin A and 2 x 10^5 for insulin) were suspended in 0.2 ml of Krebs–Ringer bicarbonate buffer containing 0.1% albumin and the specific binding of saturating concentrations of ^{125}I-wheat germ agglutinin (130 μg per ml, 25 μCi per mg), ^{125}I-insulin (1.2 x 10^{-9} M, 2000 mCi/μmole) was determined as described in the text after incubating for 40 minutes at $24°$. The existence of virtual saturation with the concentrations indicated above was confirmed in these studies by the inclusion of samples in which the concentration of the proteins was varied (not shown in the table). Essentially the same results were obtained in four separate experiments performed as described here. In a similar experiment, cells (diameter, 100 μm) from control rats were found to bind the same amount of concanavalin A, wheat germ agglutinin and insulin as the cells (diameter, 65 μm) from 3-day starved rats. Data from ref. 13

	Specific binding of ^{125}I-protein (c.p.m./10^5 cells)	
[125]I-Protein	big cells	small cells
Wheat germ agglutinin	40,400 ± 1400	37,200 ± 1900
Concanavalin A	12,100 ± 300	11,700 ± 500
Insulin	8700 ± 400	9300 ± 300

Furthermore, it is apparent from the kinetics of the lectin–cell interaction, the effects of digesting the cells with various enzymes and the partial 'competition' of binding between lectins that there are multiple and heterogeneous binding 'receptors' for Con A (13). This heterogeneity, which probably reflects the lack of absolute specificity of the lectins for various complex carbohydrate determinants, is not surprising in view of the ability of even simple saccharides to bind to these plant lectins with varying degrees of affinity (1,2,15–17). It is at present not possible to determine whether changes in lectin–lectin interactions (e.g. such as self-aggregation) occur during the binding process which might account, at least in part, for the apparent complexities in the binding curves. The observation that a certain proportion of Con A molecules precipitate with time during incubation in NH_4HCO_3 at $37°$ (18) suggests that changes in the state of aggregation of this lectin could possibly occur during incubation with cells. More precise or detailed analysis of the available binding data is not yet possible because of the ambiguities resulting from the apparent multiplicity of binding sites, the large proportion of medium lectin which is bound to the cells when low concentrations of the lectin are used (this could result from a very large number of sites, or from a small group of sites of very high affinity) and the likelihood of important interactions between binding sites or between lectin molecules.

The well-known ability of Con A to *agglutinate* a variety of cells (reviewed 1,2), especially neoplastic or transformed cells, is generally believed to be based on the ability of these lectins to cross-link receptor sites by virtue of their multivalent structure. The ability of simple sugars to cause specifically rapid and marked dissociation of the lectin–cell complex must reflect the ability of these simple sugars to bind to unoccupied sites on lectin molecules which are already attached to the cell (13). The dissociation caused by the sugars present in the medium cannot result from the interaction of the sugar with the lectin free in solution, thus acting to prevent reassociation of the lectin with the cell, since the spontaneous rate of dissociation of the lectin—cell complex is negligible under the conditions utilized in these studies. The rates of spontaneous dissociation determined in the present studies are measured under experimental conditions which exclude reassociation of those lectin molecules (labelled) which dissociate from the cell—lectin complex. The lectin molecules attached to the cell must either have totally free and exposed sugar binding sites, or the free simple sugar in solution must be capable of binding to or interacting with free subsites which are contained in the same overall site which is forming the principal lectin—cell bond. The latter possibility is very unlikely on steric grounds in view of the ability of macromolecules such as ovomucoid (or ovomucoid—agarose) to dissociate the WGA—cell complex as effectively as the simple sugar. The simplest rationalization of these observations is that the plant lectins possess more than one binding site and that virtually all of the cell-bound lectins have at least one unoccupied carbohydrate binding site which is potentially capable of a 'special' kind of interaction with carbohydrate determinants. If a lectin molecule cannot *simultaneously* bind more than one saccharide with very high affinity, it is difficult to see how cross-linking alone can maintain cell aggregation unless other factors are involved. It is possible that entirely different kinds of bonds, independent of the continued presence of the lectin, are initially induced by the lectin and subsequently serve to maintain the agglutinated (cell—cell interaction) state of the cells. The fact that loss of lectin with simple sugar does not disaggregate cells, other considerations described in the earlier section on erythrocyte agglutination and the difficulties of maintaining multivalent binding of high affinity (to be described), suggest this possibility.

The existence of free saccharide binding sites on the cell-bound lectin molecules alone is not sufficient to explain the rapid dissociation which occurs upon addition of the specific simple sugar to the medium (13). The interaction of the sugar with one or more of these free binding sites must in addition induce a major change in the binding properties, conformation or quaternary structure of the lectin which results in the rapid release of the cell-bound lectin molecules. This could occur if binding of the sugar to the free lectin site results in a conformational change in the molecule which would in turn result in a major decrease in the affinity, or in an altered specificity, of another site which is already involved in the interaction of the lectin with the membrane. Alternatively, a protomeric lectin complex could possibly dissociate by the binding of the simple sugar to those components of the complex whose carbohydrate binding sites are free. In this case release of the lectin into the medium would not represent a true dissociation of the cell—lectin complex since the monomeric portion of the complex which was originally involved in the interaction with the membrane determinant would still be intact. This explanation is perhaps somewhat less likely than the former since the loss of cell-bound lectin which occurs upon addition of the sugar generally exceeds 75% of the cell-bound radioactivity after a reasonably short period of incubation.

It is known that concanavalin A in solution exists predominantly as stable dimers below pH 6 and as tetramers above pH 7 (see Chapter 2 in this book). Recent x-ray crystallographic studies which elucidate the structure of concanavalin A at 2 Å resolution reveal a tetrameric structure consisting of four identical saccharide binding sites (see Chapter 3). The orientation of a simple mono-saccharide (iodophenyl-D-glucopyranoside) in the binding site of Con A may be very different from that expected to occur with the non-reducing terminal residues of di- and oligosaccharides which presumably would be the natural receptors on the cell surface. The present studies support the view that certain simple sugars may bind differently than the cell surface saccharide receptors to Con A. The results would be readily explained within this framework if the special type of active-site occupation which exists with certain monosaccharides, and also possibly other analogous types of oligosaccharides (perhaps present in glycoproteins) give rise to a major change in the molecular properties of Con A such that the nature of the binding properties of the other sites on the molecule would be grossly altered.

The changes in the circular dichroic spectrum of concanavalin A in the near-u.v. region which occur upon binding of α-methyl-D-mannoside (19), and the changes in the precipitability of the lectin caused by this sugar (18) suggest that sugar binding under physiological conditions is accompanied by a conformational change in the protein. The binding of such a sugar to a given site may on this basis be associated with major changes in the binding properties of the other sites of the protomer, or possibly may change the quaternary structure of the protein, thus perhaps explaining the ability of simple saccharides to dissociate the cell-bound lectin, to prevent or reverse agglutinated cells and to prevent the biological effects of the lectin.

4. REFERENCES

1. Sharon, N. S. and Lis, H. (1972) 'Lectins: cell-agglutinating and sugar-specific proteins', *Science*, 177, 949—59.
2. Nicolson, G. L. (1975) 'The interactions of lectins with animal cell surfaces', *Int. Rev. Cytol.*, 39, 89—190.
3. Rodbell, M. (1964) 'Metabolism of isolated fat cells 1. Effects of hormones on glucose metabolism and lipolysis', *J. Biol. Chem.*, 239, 345—80.
4. Cuatrecasas, P. (1971a) 'Insulin—receptor interactions in adipose tissue cells: Direct measurement and properties', *Proc. Nat. Acad. Sci. U.S.A.*, 68, 1264—8.
5. Cuatrecasas, P. (1971b) 'Unmasking of insulin receptors in fat cells and fat cell membranes', *J. Biol. Chem.*, 246, 6532—42.
6. Krug, U., Hollenberg, M. D. and Cuatrecasas, P. (1973) 'Changes in the binding of concanavalin A and wheat germ agglutinin to human lymphocytes during *in vitro* transformation', *Biochem. Biophys. Res. Commun.*, 52, 305—12.
7. Nagata, Y. and Burger, M. (1972) 'Wheat germ agglutinin: Isolation and crystallization', *J. Biol. Chem.*, 247, 2248—50.
8. Edmundson, A. B., Ely, K. R., Sly, D. A., Westholm, P. A., Powers, D. A. and Liener, I. E. (1971) 'Isolation and characterization of concanavalin A polypeptide chains', *Biochemistry*, 10, 3554—9.
9. Cuatrecasas, P. and Hollenberg, M. D. (1974) 'Binding of insulin and other hormones to non-receptor materials: Saturability, specificity and apparent "negative cooperativity" ', *Biochem. Biophys. Res. Commun.*, 62, 31—41.
10. Nichol, L. W., Smith, M. D. and Ogston, A. G. (1969) 'The effects of isomerization and polymerization on the binding of ligands to acceptor molecules', *Biochim. Biophys. Acta.*, 184, 1—10.
11. Cuatrecasas, P. and Tell, G. P. E. (1973) 'Insulin-like activity of concanavalin A and wheat germ agglutinin — direct interactions with insulin receptors', *Proc. Nat. Acad. Sci. U.S.A.*, 70, 485—9.
12. Robbins, J. H. (1964) 'Tissue culture studies of the human lymphocytes', *Science*, 146, 1648.

13. Cuatrecasas, P. (1973) 'Interaction of wheat germ agglutinin and concanavalin A with isolated fat cells', *Biochemistry*, 12, 1312–23.
14. Chang, K.-J., Bennett, V. and Cuatrecasas, P. (1975) 'Membrane receptors as general markers for plasma membrane isolation procedures', *J. Biol. Chem.*, 250, 488–500.
15. Burger, M. M. and Goldberg, A. R. (1967) 'Identification of a tumor-specific determinant on neoplastic cell surfaces', *Proc. Nat. Acad. Sci. U.S.A.*, 57, 359–66.
16. So, L. L. and Goldstein, I. J. (1967) 'Protein–carbohydrate interaction. IX. Application of the quantitative hapten inhibition technique to polysaccharide–concanavalin A interaction', *J. Immunol.*, 99, 158–63.
17. Yariv, J., Kalb, A. J. and Levitzki, A. (1968) 'The interaction of concanavalin A with methyl α-D-glucopyranoside', *Biochim. Biophys. Acta*, 165, 303–5.
18. Cunningham, B. A., Wang, J. L., Pflumm, M. N. and Edelman, G. M. (1972) 'Isolation and proteolytic cleavage of the intact subunit of concanavalin A', *Biochemistry*, 11, 3233–9.
19. Pflumm, M. N., Wang, J. L. and Edelman, G. M. (1971) 'Conformational changes in concanavalin A', *J. Biol. Chem.*, 246, 4369–70.

CHAPTER 24

Measurement of Concanavalin A Binding to Isolated Membrane Fractions

T. W. KEENAN
I. H. MATHER
J. STADLER
W. W. FRANKE

1. PRINCIPLE

In this section we describe a radiochemical assay for determining the amount of concanavalin A that will bind to isolated plasma membrane and intracellular endomembrane fractions. An excess of radiolabelled Con A is incubated with the membrane fractions for a predetermined time interval. Unbound Con A is then removed by washing the membrane suspensions onto glass fibre filters and, after exhaustive washing, the radioactivity remaining on the filters is measured by scintillation counting. Provided that certain precautions are taken, the method is reproducible.

The assay provides a convenient method for determining the relative amounts of mannose-like residues exposed on the surfaces of isolated membrane fractions (1, 2). Further, as both intracellular endomembrane and plasma membrane fractions consist of closed vesicles of specific orientation (3–6), Con A can be used to advantage in studies of the asymmetry of distribution of carbohydrates on membrane surfaces.

2. METHODS

2.1 Membrane Fractions

The following membrane fractions from rat liver can be readily prepared by the referenced methods: nuclei and nuclear membranes (7), rough endoplasmic reticulum (8), smooth microsomes (8), Golgi apparatus (8, 9) and plasma membranes (8, 10, 11). Similar fractions from tissues such as mammary gland can be obtained by modifications of these methods (12—14). Milk fat globule membrane, which is derived from the apical plasma membrane of mammary secretory cells (13, 15, 16), can be isolated in large quantities (13, 17) and is useful as a source of plasma membrane for such studies. Since sucrose inhibits Con A binding to membranes (6, 18), fractions must be washed free of sucrose by repeated centrifugation in or exhaustive dialysis against phosphate-buffered saline (PBS) (0.15 M NaCl, 1 mM KH_2PO_4, 10 mM Na_2HPO_4, 1 mM $MgCl_2$, pH 7.2) (19). Washed fractions are stored at or below $0°C$ in PBS until required for assay.

2.2 Labelling of Concanavalin A

Radioactivity can be incorporated into the Con A molecule by any of the procedures outlined in this volume. For the studies described below, Con A was acetylated with ^{3}H-acetic anhydride as described by Fraenkel-Conrat (20) to yield a product with a specific activity of about 3.7×10^7 c.p.m./mg. Con A labelled by this method retains full binding activity (21).

2.3 Binding Assay

Assays were performed by incubating dispersed membrane fractions with the desired amount of Con A in PBS. In our studies the final volume was 1.0 ml and at least 0.33 μg Con A/μg membrane protein (as measured with the Folin reagent [22]) was employed. Samples were incubated in a reciprocating water bath at the desired temperature. Maximum binding of Con A is achieved at $19°C$ or $37°C$ after about 20 min, but only after about 60 min at $0°C$. After incubation for the desired time, the reaction mixture was transferred to a glass-fibre filter (Whatman, type GF/C, 2.4 mm) placed onto the six-place filtration mantle shown in Figure 1 (23); commercially available filter holders can also be used. The filter is washed under vacuum with four 5 ml aliquots of PBS. Further washing causes no additional decrease in radioactivity with filter loads of up to 300 μg of membrane protein. After washing, the filters are transferred to scintillation vials containing 1 ml of Nuclear Chicago solubilizer and incubated at $37°C$ for 1 h. Toluene-based scintillation fluid (10 ml) is then added and the radioactivity can be determined by standard methods. Control mixtures containing Con A, membrane and α-methyl mannoside (at least 2 mM) should always be incubated in parallel with the experimental samples. Specific binding is measured as the difference between the values obtained in the absence and presence of α-methyl mannoside. This corrects for adsorption, aggregation, entrainment or non-specific binding of Con A to either the membrane fraction or the glass-fibre filter. Typical background levels obtained in the presence of α-methyl mannoside were about 15% of the radioactivity measured in experimental samples.

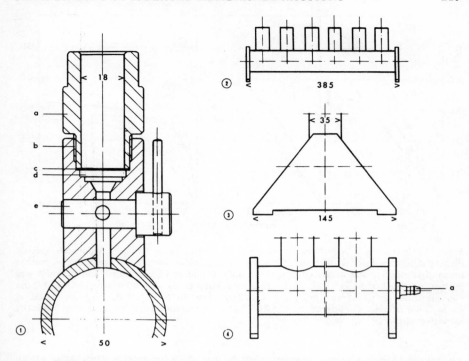

Figure 1. Filtration manifold for 25 mm 'Millipore' filters. 1. Individual filter holder, cross-section. Abbreviations: a, filter top with grip; b, screw threads; c, paper filter; d, sintered-glass filter; e, stopcock. 2. Side view of complete apparatus with six filter holders. 3. End view of complete apparatus. 4. Detail of the ends, side view. a, nozzle. Dimensions are given in mm. The apparatus is constructed from polyvinyl chloride, the stopcock from 'Teflon' and stainless steel

3. EXPERIMENTAL RESULTS AND DISCUSSION

3.1 Binding of Concanavalin A to Surface Membranes

Con A is readily bound by plasma membranes as judged by assaying bovine milk fat globule membrane or plasma membrane from rat liver (Figures 2 and 3). With milk fat globule membrane binding was linear with increasing membrane protein concentrations up to a ratio of membrane protein to Con A of about 4 to 1 at the three incubation temperatures employed (Figure 2). Within the linear range of the assay, binding was slightly greater at 19°C and slightly lower at 0°C than it was at 37°C. Con A binding increased in a nearly linear fashion as the Con A to membrane protein concentration was increased from a ratio of 1:8 to about 1:3. Further increases in the level of Con A did not appreciably increase the extent of Con A binding (Figure 2). Binding sites on the milk fat globule membrane became saturated within the first 20 min of incubation at either 19°C or 37°C. However, at 0°C saturation binding was not attained before about 60 min (Figure 2). At

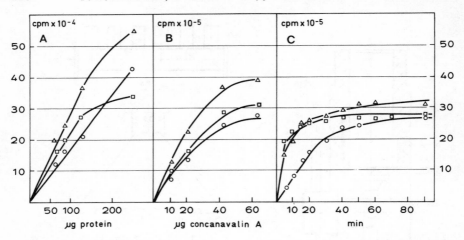

Figure 2. Concanavalin A binding to bovine milk fat globule membranes. Values are specific binding determined as the difference between binding in the absence and presence of α-methyl mannoside. Binding was measured at 0°C (○), 19°C (△) or at 37°C (□). In A and C, Con A was present at a level of 84 μg/assay, in A and B incubation times were 40 min and in B and C the concentration of membrane protein was 170 μg/assay. In B and C data are expressed as c.p.m./2 mg protein. (Reprinted from FEBS Letters (reference [6]) by permission from the managing editor)

saturation, the concentrations of Con A bound were nearly the same at all temperatures, with only slightly higher amounts being bound at 19°C than at 0°C or 37°C. With intact cells it has been observed that nearly double the amount of Con A can be bound at temperatures above about 12°C to 15°C (21). This increased binding could be due at least partially to endocytotic accumulation of Con A within cells, and thus many investigators have performed binding assays for brief periods at temperatures near 0°C (19, 21, 24, 25). That saturation of Con A binding sites on the milk fat globule membrane was not achieved before about 60 min at 0°C suggests that such short-term incubations with intact cells may not be adequate to determine the actual levels of Con A binding sites.

Freezing and thawing or prolonged incubation below 0°C did not appreciably affect the ability of milk fat globule membranes to bind Con A. Membrane preparations stored at −20°C for as long as four months and membrane preparations which had been frozen and thawed up to eight times bound about as much lectin as did freshly prepared milk fat globule membranes. In distinct contrast, freezing and thawing greatly increased the ability of endoplasmic reticulum or Golgi apparatus fractions to bind Con A. For example, freshly prepared rat liver Golgi apparatus bound only small amounts of Con A. Binding of Con A increased progressively through three cycles of freezing and thawing; further cycles yielded no additional increase in binding ability. This effect of freezing and thawing could be duplicated by subjecting Golgi apparatus fractions to ultrasound and is most probably due to an increased accessibility of the membrane-attached binding sites to the lectin. Using a variety of methods, it has been repeatedly observed that sugar residues are distributed along the inner face of intracellular membrane vesicles and along the outer face of plasma membrane vesicles (e.g. 3—6). The outer face of isolated plasma membrane vesicles is equivalent to the outer of environmental face of the cell surface membrane.

3.2 Binding of Concanavalin A to Intracellular Membranes

Plasma membranes from rat liver and bovine milk fat globule membranes bound Con A to nearly the same extent (Figure 3). These were by far the most active of the cellular membrane fractions evaluated for binding ability. Golgi apparatus fractions from rat liver and bovine mammary gland bound about half as much Con A as did plasma membranes or milk fat globule membranes (Figure 3). Much lower amounts of Con A were bound by rough endoplasmic reticulum or smooth microsomal fractions. Binding to total cell membrane fractions (total particulate fractions prepared by centrifugation of homogenates at 120,000 g for 1 h) on a protein basis occurred to about the same extent as the amount of Con A bound to isolated endoplasmic reticulum fractions (Figure 3).

Nuclear membranes are very similar to endoplasmic reticulum in morphology and composition (26, 27) and one would expect as a first approximation that nuclear membranes and endoplasmic reticulum would bind Con A to nearly the same extent. However, our results with rat liver nuclear membranes have been quite variable. With some preparations, nuclear and endoplasmic reticulum membranes were observed to bind nearly the same amounts of Con A (6). However, other

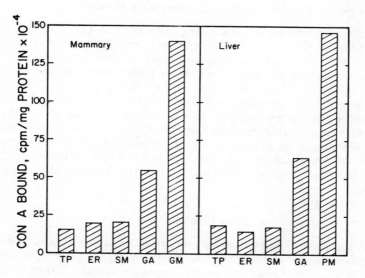

Figure 3. Concanavalin A binding by membrane fractions isolated from bovine mammary gland and rat liver. Specific binding was determined as described in the text. Abbreviations: TP, total particulate; ER, rough endoplasmic reticulum; SM, smooth microsomes; GA, Golgi apparatus; GM, milk fat globule membrane; PM, plasma membrane

nuclear membrane fractions were observed to bind levels of Con A approaching those bound by plasma membrane or milk fat globule membrane (unpublished observations). These extensive differences in the binding of Con A to different preparations of nuclear membranes cannot be accounted for by differences in the degree of contamination of the membranes with other fractions, for example with plasma membranes. In fact, the most highly purified nuclear membrane fractions examined were found to bind Con A most extensively (28). Similarly, Monneron and Segretain (29) have observed extensive binding of Con A to the nuclear membrane of calf thymocytes. It remains to be determined whether subtle differences in isolation methods or in the treatment of nuclear membranes after isolation can account for the above differences, and whether nucleic acids might interfere with the assay. Since our isolation of nuclear membranes involves sonication and a high salt extraction, the effect of sonifying and washing membranes with 1.5 M KCl on Con A binding was evaluated. Treatment of milk fat globule membranes with 1.5 M KCl increased specific binding of Con A on average to 120% of control levels. An even greater increase, to 180% of control levels, was noted when an endoplasmic reticulum fraction was treated in this manner. Thus, exposure of membranes to high concentrations of KCl can increase Con A binding ability on a protein basis. However, it remains to be determined whether this increase in specific binding activity is due either to an increased accessibility of membrane binding sites to Con A or merely reflects loss of protein that does not bind Con A.

3.3 The Effect of 'Triton' X-100 on Concanavalin A Binding

Detergents are reported to increase the accessibility of Con A to mitochondrial binding sites for this lectin (30, 31). However, at a final concentration of 1%, 'Triton' X-100 depressed the specific binding of Con A to milk fat globule membranes throughout the range of membrane protein concentrations tested (50 to 300 μg). Similarly, 'Triton' depressed specific binding to rat liver plasma membranes, nuclei and all intracellular endomembranes used in this study. The degree of depression of binding was variable but ranged between about 55 to 90% of the levels bound to untreated fractions. Depression was greatest at lower membrane protein concentrations, suggesting that the effect was due to dissolution of a fraction of the membrane protein. Whether detergents increase binding of Con A to intracellular endomembrane vesicles not disrupted by freezing and thawing was not evaluated.

4. CONCLUDING REMARKS

Con A specifically binds to isolated membrane fractions in a predictable manner. While various factors can influence the results, it can be used to determine the relative number or distribution of binding sites among different membrane fractions. Reacting Con A with isolated surface membranes overcomes some of the problems, particularly endocytosis and cell density effects, encountered in the measurement of Con A binding to the surface membrane of intact cells. Whether isolated plasma membranes bind Con A to the same extent as the plasma membrane of intact cells remains to be determined. It is entirely possible that the isolation method employed will affect the ability of surface or intracellular membranes to bind Con A.

5. ACKNOWLEDGEMENTS

Our work is supported by grants from the National Science Foundation (T.W.K.) and the Deutsche Forschungsgemeinschaft (W.W.F.). T.W.K. is supported by Research Career Development Award GM 70596 from the National Institute of General Medical Science.

6. REFERENCES

1. Inbar, M. and Sachs, L. (1969) 'Interaction of the carbohydrate-binding protein concanavalin A with normal and transformed cells', *Proc. Nat. Acad. Sci. U.S.A.*, 63, 1418—25.
2. Sharon, N. and Lis, H. (1972) 'Lectins: Cell agglutinating and sugar-specific proteins', *Science*, 177, 949—59.
3. Hirano, H., Parkhouse, B., Nicolson, G. L., Lennox, E. S. and Singer, S. J. (1972) 'Distribution of saccharide residues on membrane fragments from a myeloma-cell homogenate: Its implications for membrane biogenesis', *Proc. Nat. Acad. Sci. U.S.A.*, 69, 2945—9.
4. Nicolson, G. L. and Singer, S. J. (1974) 'The distribution and asymmetry of mammalian cell surface saccharides utilizing ferritin-conjugated plant agglutinins as specific saccharide stains', *J. Cell. Biol.*, 60, 236—48.
5. Kreibich, G., Hubbard, A. L. and Sabatini, D. D. (1974) 'On the spatial arrangement of proteins in microsomal membranes from rat liver', *J. Cell Biol.*, 60, 616—27.
6. Keenan, T. W., Franke, W. W. and Kartenbeck, J. (1974) 'Concanavalin A binding by isolated plasma membranes and endomembranes from liver and mammary gland', *FEBS Lett.*, 44, 274—8.
7. Kartenbeck, J., Jarasch, E. D. and Franke, W. W. (1973) 'Nuclear membranes from mammalian liver. VI. Glucose-6-phosphatase in rat liver, a cytochemical and biochemical study', *Exp. Cell Res.*, 81, 175—94.
8. Morré, D. J., Yunghans, W. N., Vigil, E. L. and Keenan, T. W. (1974) 'Isolation of organelles and endomembrane components from rat liver: Biochemical markers and quantitative morphometry', in *Methodological Developments in Biochemistry*, ed. Reid, E. (Longman, North-Holland), Vol. 4, pp. 195—236.
9. Morré, D. J. (1971) 'Isolation of Golgi apparatus', *Methods Enzymol.*, 22, 130—48.
10. Emmelot, P., Bos., C. J., Benedetti, E. L. and Rumke, P. (1964) 'Studies on plasma membranes. I. Chemical composition and enzyme content of plasma membranes isolated from rat liver', *Biochim. Biophys. Acta*, 90, 126—45.
11. Ray, T. K. (1970) 'A modified method for the isolation of the plasma membrane from rat liver', *Biochim. Biophys. Acta*, 196, 1—9.
12. Keenan, T. W., Huang, C. M. and Morré, D. J. (1972) 'Membranes of mammary gland. V. Isolation of Golgi apparatus and rough endoplasmic reticulum from bovine mammary gland', *J. Dairy Sci.*, 55, 1577—85.
13. Keenan, T. W., Morré, D. J., Olson, D. E., Yunghans, W. N. and Patton, S. (1970) 'Biochemical and morphological comparison of plasma membrane and milk fat globule membrane from bovine mammary gland', *J. Cell Biol.*, 44, 80—93.
14. Baumrucker, C. R. and Keenan, T. W. (1974) 'Membranes of mammary gland. VIII. Isolation and composition of nuclei and nuclear membranes from bovine mammary gland', *J. Dairy Sci.*, 57, 24—31.
15. Keenan, T. W., Morré D. J. and Huang, C. M. (1974) 'Membranes of mammary gland', in *Lactation: A Comprehensive Treatise*, eds. Larson, B. L. and Smith, V. R. (Academic Press, New York), Vol. 2, pp. 191—233.
16. Linzell, J. L. and Peaker, M. (1971) 'Mechanisms of milk secretion', *Physiol. Rev.*, 51, 564—97.

17. Mather, I. H. and Keenan, T. W. (1975) 'Studies on the structure of milk fat globule membrane', *J. Memb. Biol.*, **21**, 65—85.
18. Chang, K., Bennett, V. and Cuatrecasas, P. (1975) 'Membrane receptors as general markers for plasma membrane isolation procedures', *J. Biol. Chem.*, **250**, 488—500.
19. Noonan, K. D. and Burger, M. M. (1973) 'The relationship of concanavalin A binding to lectin-initiated cell agglutination', *J. Cell Biol.*, **59**, 134—42.
20. Fraenkel-Conrat, H. (1957) 'Methods for investigating the essential groups for enzyme activity', *Methods Enzymol.*, **4**, 247—69.
21. Noonan, K. D. and Burger, M. M. (1973) 'Binding of [^{3}H] concanavalin A to normal and transformed cells', *J. Biol. Chem.*, **248**, 4286—92.
22. Lowry, O. H., Rosebrough, N. J., Farr, A. L. and Randall, R. J. (1951) 'Protein measurement with the Folin phenol reagent', *J. Biol. Chem.*, **193**, 265—75.
23. Stadler, J. and Franke, W. W. (1974) 'Characterization of the colchicine binding of membrane fractions from rat and mouse liver', *J. Cell Biol.*, **60**, 297—303.
24. Parmley, R. T., Martin, B. J. and Spicer, S. S. (1973) 'Staining of blood cell surfaces with a lectin-horseradish peroxidase method', *J. Histochem. Cytochem.*, **21**, 912—22.
25. Barat, N. and Avrameas, S. (1973) 'Surface and intracellular localization of concanavalin A in human lymphocytes', *Exp. Cell Res.*, **76**, 451—5.
26. Franke, W. W. (1974) 'Nuclear envelopes. Structure and biochemistry of the nuclear envelope', *Phil. Trans. Roy. Soc. Lond. B.*, **268**, 67—93.
27. Franke, W. W. (1974) 'Structure, biochemistry and functions of the nuclear envelope', *Int. Rev. Cytol. Suppl.*, **4**, 71—236.
28. Franke, W. W.. Keenan, T. W., Stadler, J., Jarasch, E. D., Genz, R. and Kartenbeck, J. (1976) 'Nuclear membranes from mammalian liver. VII. Characteristics of highly purified nuclear membranes in comparison with other membranes', *Europ. J. Cell Biol.*, in press.
29. Monneron, A. and Segretain, D. (1974) 'Extensive binding of concanavalin A to the nuclear membrane', *FEBS Lett.*, **42**, 209—13.
30. Glew, R. H., Zatzkin, J. B. and Kayman, S. C. (1973) 'A comparative study of the interaction between concanavalin A and mitochondria from normal and malignant cells', *Cancer Res.*, **33**, 2135—41.
31. Glew, R. H., Kayman, S. C. and Kuhlenschmidt, M. S. (1973) 'Studies on the binding of concanavalin A to rat liver mitochondria', *J. Biol. Chem.*, **248**, 3137—45.

CHAPTER 25

A Quantitative Method for Measuring Fluorescence Intensity of Cells Labelled with Fluorescein Isothiocyanate - Concanavalin A

ANTHONY NERI
MARIE ROBERSON
STEVEN B. OPPENHEIMER

Methods involving the use of labelled concanavalin A (Con A) in an effort to better understand the cell surface have been numerous. Con A receptor site mobility and the distribution of these receptors on the surfaces of normal and transformed cells have been investigated by utilizing various labels conjugated to the Con A molecule. Labels such as fluorescein isothiocyanate, ferritin, haemocyanin and peroxidase coupled to Con A have been the most widely used. Although these labels when conjugated to Con A have provided information regarding receptor site mobility and distributional differences among cell types, only radioisotope-labelled Con A has been utilized to arrive at quantitative binding differences among different cell types (1—4).

1. PRINCIPLE

So far fluorochrome labels conjugated to Con A have been used mainly to discern distributional differences of Con A receptor sites on the surfaces of various cell types (5—8). We now describe a method which enables the investigator to quantitatively and rapidly measure relative fluorescence intensity of single cells labelled with fluorescein isothiocyanate—Con A (FITC—Con A) by microfluorometry.

2. EQUIPMENT

2.1 The Microscope

The Leitz Orthoplan microscope equipped with a Leitz fluorescence vertical illuminator is used to provide fluorescence excitation in incident light. In this system the exciting wavelength is directed onto the specimen from above through a series of excitation filters, suppression filters, mirrors and ultimately through the objective onto the specimen. The emitted light, along with non-specific light which reflects back, is captured by the objective and directed through a series of built-in suppression filters. The unwanted radiation is filtered out and light of a given wavelength is allowed to pass through. Incident illumination ultimately provides brighter and sharper images by simultaneously utilizing the objective as a light condensor. Therefore, as the numerical aperture of the objective is increased the specimen will receive more exciting radiation and hence will increase in fluorescence intensity. This type of illumination reduces light losses which can occur through scattering or dispersion, which is a problem of transmitted light excitation. The microscope also provides phase contrast illumination of the specimen with or without fluorescence. Therefore, the specimen can be brought into the plane of vision by merely utilizing the phase contrast system prior to photoexcitation and hence limit fluorescence fading to a minimum. The availability and use of this function can be of considerable importance when measuring fluorescence intensity.

Since the absorption wavelength of fluorescein—Con A is approximately 490 nm and the emitted wavelength is approximately 520 nm, the following light filters were used: excitation filter 5 mm BG12 and a 4 mm BG38 red suppression filter (both in lamp housing); suppression filters, K495 (built in the vertical illuminator) and K510 (exchangeable). The light source to be used in conjunction with the filters listed is an XBO 150 W, 20 V Xenon lamp (Osram, Berlin) powered by a Leitz power supply (#050230) rated at ±0.1% constant voltage supply. A Phaco 40/0.65 objective for phase contrast viewing of the specimen was used throughout this study. Other objectives such as Phaco 10/0.25 and Phaco öl 100/1.30 can also be used. When employing the Phaco öl 100/1.30 it is recommended that a non-fluorescent immersion oil with a refractive index of 1.515 is used. An appropriate source for the transmitted light for phase contrast observations can be supplied from a 6 V, 30 W low-voltage lamp powered by a Leitz variable transformer.

2.2 The Photometer

Microfluorometric measurements are performed with the Leitz-MPV photometer system with an adjustable measuring diaphragm. The photometer is attached to the Leitz Orthoplan fluorescence microscope by means of the Aristophot rail. The photomultiplier located in the photometer housing should be an EMI 9558 with photocathode type S20. This type of photomultiplier is recommended because of its spectral sensitivity for fluorescence measurements (9,10). The photomultiplier is connected to a Knott NSHM high tension supply unit and to an indicator (see *Figure 1). The high-tension supply unit set to deliver 1.1—1.3 kV to the photomultiplier for optimum response (11). Voltage settings below 0.7 kV and above 1.7 kV can cause improper and and noisy responses.

As indicator of fluorescence intensity we utilized a digital picoammeter manufactured by Kethly Instruments, model 445, with a display rate set at 1 per second. This type of indicator is used because of its sensitivity in reading small currents (10^{-12} amperes) which are generated by extremely faint fluorescent specimens.

3. MICROFLUOROMETRY

3.1 Preparation of the Specimen

The cell sample is treated with 50 μg/ml of FITC—Con A (obtained from Miles Yeda), usually for 10—30 minutes. It is important to remove as much unbound or free FITC—Con A as possible in order to limit background fluorescence to a minimum. This is usually accomplished by washing the treated cells twice in a suitable buffer. If aldehyde fixation is incorporated in the treatment of the specimen, the necessary control must be performed to determine non-specific autofluorescence induced by the fixative. We found that S180 tumour cells (12) treated only with phosphate-buffered saline containing 1% paraformaldehyde and 0.05% glutaraldehyde, pH 7.2, contributed as much as 6% fluorescence to the total fluorescence intensity. Therefore, these factors should be kept in mind during all measurements and appropriate corrections should be made. A small volume (25 μl) of the treated cell suspension is placed on a glass slide, covered over with a glass coverslip and placed on the microscope stage.

3.2 Focusing the Specimen

A brief diagrammatic representation of the microfluorometer is shown in Figure 1. By means of phase contrast illumination the individual cells are viewed through the binocular (J), focused, and brought into the plane of vision. By this

*Figure 1 was adapted and modified from Böhm and Sprenger, Histochemie *16*, pp. 103 (1968).

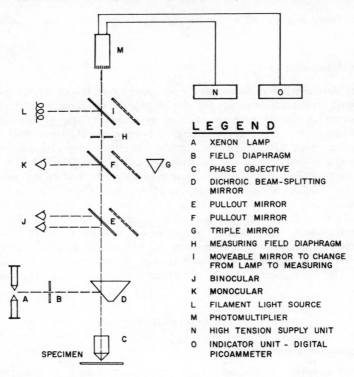

LEGEND

A XENON LAMP
B FIELD DIAPHRAGM
C PHASE OBJECTIVE
D DICHROIC BEAM-SPLITTING
   MIRROR
E PULLOUT MIRROR
F PULLOUT MIRROR
G TRIPLE MIRROR
H MEASURING FIELD DIAPHRAGM
I MOVEABLE MIRROR TO CHANGE
   FROM LAMP TO MEASURING
J BINOCULAR
K MONOCULAR
L FILAMENT LIGHT SOURCE
M PHOTOMULTIPLIER
N HIGH TENSION SUPPLY UNIT
O INDICATOR UNIT - DIGITAL
   PICOAMMETER

Figure 1.

method the cells have not yet been exposed to excitation radiation and hence fluorescence fading will be limited to a minimum prior to quantitation of fluorescence intensity. The binocular pull-out mirror (E) is moved out of place and light then reaches the monocular (K) by reflecting from the mirror (F) located in the photometer. The individual cells are refocused again according to the monocular located at the photometer housing. With the mirror (I) in place, light from the filament lamp (L) passes through the measuring diaphragm (H), reflects from both mirrors (F and G), and reaches the monocular. This light will then appear as a bright field in the centre of the darker viewing field as seen through the photometer monocular (K). By carefully adjusting the measuring diaphragm (H) it is possible to restrict all measurements of fluorescence intensity to individual cells. In order to maintain the diameter of the measuring field constant when measuring fluorescence intensity of many individual cells, it is advisable not to readjust the measuring diaphragm (H) between readings. Thereafter only cells which conform to the size of the measuring field are used for fluorescence intensity measurements. In this way, differences in fluorescence intensity readings contributed by variations in cell size can be kept at a minimum.

3.3 Measuring Fluorescence Intensity

With the cell in focus and in the measuring field, fluorescence intensity measurements can be made. The phase contrast illumination is turned off by means of a shutter located beneath the microscope stage (not shown in Figure 1). Exciting light from the Xenon lamp (A) is allowed to pass through an adjustable field diaphragm (B), deflected by a beam splitter (D) into the objective (C), and onto the specimen. By adjusting the field diaphragm (B) it is possible to limit excitation to an area that is approximately the same size as the measuring field. The emitted light reaches the monocular by passing through the beam splitter (D) and deflecting from the pull-out mirror (F). For fluorescence measurements, the pull-out mirror (F), and the mirror used to change from lamp to measuring (I) are moved out of place. When this is done the emitted light has a direct path to the photomultiplier (M). The fluorescence intensity is then displayed by the digital picoammeter indicator (O) in terms of amperes.

3.4 Important Variables

As an example, quantitative fluorescence measurements on mouse tumour cells (S180) treated with FITC—Con A are shown in Figure 2. These data indicate that S180 cells treated with 50 μg/ml of FITC—Con A at 4° or 37°C bind approximately the same amounts up to 20 minutes of incubation. But if the incubation time is extended to 30 minutes, the cells incubated at 37°C bind 30% more FITC—Con A. Although the 30% increase in mean fluorescence intensity might be a result of the cells internalizing the labelled lectin, it nevertheless remains clear that quantitative evaluations can be made. When comparing cells fixed prior to treatment with FITC—Con A (Figure 3) to cells fixed after treatment (Figure 2), the cells fixed prior to exposure with FITC—Con A had a mean fluorescence intensity of 24% less after 20 minutes, and 44% less after 30 minutes. It is clear that the fixative as used in this procedure (see legend, Figure 3) causes a reduction in the amount of FITC—Con A bound by S180 tumour cells.

4. APPLICATION: SINGLE CELL VARIATION OF CON A BINDING

Radioactivity measurements are made on whole populations of cells. Important differences in Con A binding of individual cells in the population cannot be made using radioactive Con A. The method employed here, however, enables easy measurement of fluorescence intensity of *individual cells*. For example, specific fluorescence intensity readings of ten different S180 cells exposed to FITC—Con A for 10 minutes, at 4°C, then fixed for 20 minutes, (see Figure 2) are as follows: 24.5×10^{-10}, 22.7×10^{-10}, 27.7×10^{-10}, 19.2×10^{-10}, 19.6×10^{-10}, 23.0×10^{-10}, 20.4×10^{-10}, 18.1×10^{-10}, 16.3×10^{-10} and 24.0×10^{-10} amperes. Non-specific fluorescence intensity due to fixation and background have been subtracted from these values.

5. RELIABILITY OF METHOD

The reliability of microfluorometric measurements as compared to radioisotope methods has been explored by Killander and coworkers (13). By using FITC-labelled antibody and ^{131}I-labelled antigen, they were able to correlate fluorescence

Figure 2. Sarcoma 180 (S180) cells were grown in male Swiss white mice for 7–14 days. The animals were sacrificed by cervical dislocation, and the peritoneal contents rapidly transferred to Dulbecco's phosphate-buffered saline pH 7.2 (DPBS). Cells were subjected to 6 washes by gentle centrifugation in DPBS at 1/5 speed in an International clinical centrifuge. Cells were then resuspended in 10 ml of cold DPBS, and a viability cell count using 0.1% trypan blue in DPBS was performed. Results from the dye exclusion test showed 99% viability. The experiment was performed using 5×10^6 cells per ml for all time points indicated. Fifty µg per ml (diluted in DPBS) of FITC–Con A (Miles-Yeda Laboratories) was then added to all individual cell suspensions and incubated at the appropriate temperatures. After incubations, cells were washed twice in 5 ml of cold DPBS and fixed with DPBS containing 1% paraformaldehyde and 0.05% glutaraldehyde for 20 minutes at 4°C. Fixed cells were again washed twice in 1 ml of cold DPBS and 25 µl of the cell suspensions were transferred onto glass slides and covered with glass coverslips. Each point in the graph represents the mean fluorescence intensity of ten randomly chosen, single cells, which have been corrected for autofluorescence (see text). Corrections for fluorescence contributed by the background have also been made. For measurement of background fluorescence, three empty areas adjacent to the cell which had been previously quantitated were measured, and the mean subtracted from the total fluorescence intensity. The data represented in this figure are: 50 µg per ml of FITC–Con A at 4°C (△–△), and at 37°C (○–○).

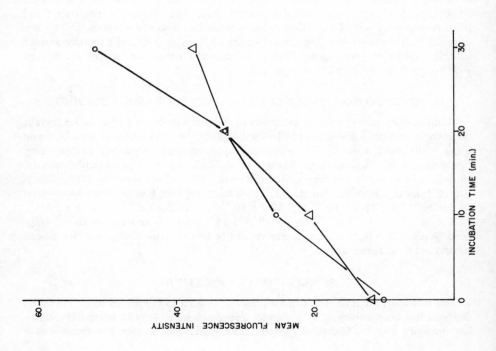

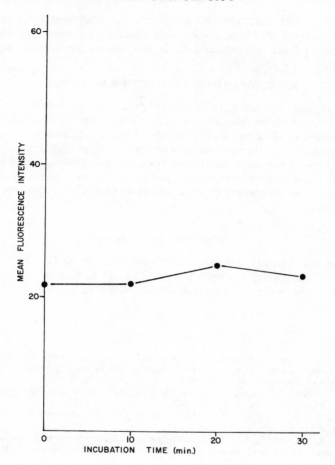

Figure 3. The effect of prior fixation with DPBS containing 1% paraformaldehyde and 0.05% glutaraldehyde on the mean fluorescence intensity of individual S180 cells. Prior to treatment with 50 μg per ml of FITC—Con A, the cells were exposed to the fixative for 20 minutes at 4°C. Cells were then washed twice in 10 ml of cold DPBS, incubated for various times with FITC-Con A at 37°C, washed twice and prepared for quantitation of fluorescence intensity, as described in Figure 2

intensity to radioactivity. They used FITC-anti-human IgM directed against formalinized human erythrocytes previously coated with ^{131}I-human IgM (antigen). The amount of FITC-anti-human IgM bound by the erythrocytes was correlated to the amount of ^{131}I-human IgM bound. Their results showed that fluorescence intensity (FITC-anti-human IgM) of single cells decreased as the amount of protein antigen (^{131}I-human IgM) coated on the surface of the cells is decreased. A strong correlation was found between the amount of antigen bound to the cell, when measured by radioactivity, and the fluorescence intensity. These experiments suggest that quantitative data obtained by measuring fluorescence intensity is as reliable as that which is obtained by measuring radioactivity.

6. COMPARISON WITH OTHER METHODS

The advantages that the microfluorometry methods described here offer over radioactivity methods in measuring Con A binding are: (i) simultaneously both quantitative and qualitative information about single cells in one preparation can easily be obtained, (ii) the method is more rapid and less costly and (iii) only by this method can quantitative information about Con A binding on a *specific* single cell be made. Radioactivity measurements must utilize large numbers of cells. The latter methods cannot provide information about Con A binding to specific single cells under study.

7. ACKNOWLEDGEMENTS

This research was supported by USPHS Research Grant CA12920 (to S.B.O.) from the National Cancer Institute. We thank Drs. D. L. Morton and W. D. Winters for the use of the fluorescence microscope and facilities, and Ralph Ridgeway for his illustrations.

8. REFERENCES

1. Kaneko, I., Satoh, H. and Ukita, T. (1972) 'Binding of radioactively labelled concanavalin A and *Ricinus communis* agglutinin to rat liver- and rat ascites hepatoma-nuclei', *Biochem. Biophys. Res. Commun.*, 48, 1504–10.
2. Inoue, M. (1974) 'Cell agglutination mediated by concanavalin A and the dynamic state of the cell surface', *J. Cell Sci.*, 14, 197–202.
3. Cline, M. J. and Livingston, D. C. (1971) 'Binding of ^{3}H-concanavalin A by normal and transformed cells', *Nature (New Biol.)*, 232, 155–6.
4. Ozanne, B. and Sambrook, J. (1971) 'Binding of radioactively labelled concanavalin A and wheat germ agglutinin to normal and virus-transformed cells', *Nature (New Biol.)*, 232, 156–60.
5. de Petris, S. (1974) 'Inhibition and reversal of capping by cytochalasin B, vinblastine and colchicine', *Nature*, 250, 54–5.
6. Loor, F. (1973) 'Lectin-induced lymphocyte agglutination. An active cellular process?', *Exp. Cell Res.*, 82, 415–25.
7. Inbar, M., Huet, C., Oseroff, A. R., Ben-Bassat, H. and Sachs, L. (1973) 'Inhibition of lectin agglutinability by fixation of the cell surface membrane', *Biochim. Biophys. Acta.*, 311, 594–9.
8. Nicolson, G. L. (1973) 'Temperature-dependent mobility of concanavalin A sites on tumour cell surfaces', *Nature (New Biol.)*, 243, 218–20.
9. Böhm, N. and Sprenger, E. (1968) 'Fluorescence cytophotometry: A valuable method for the quantitative determination of nuclear Feulgen-DNA', *Histochemie*, 16, 100–18.
10. Leitz Wetzlar (1974) *Microscope Photometer with Variable Measuring Diaphragms*, p. 15.
11. Leitz Wetzlar (1974) *MPV Microscope Photometer Instructions for Assembly and Operation*, p. 51.
12. Oppenheimer, S. B. (1973) 'Utilization of L-glutamine in intercellular adhesion: Ascites tumor and embryonic cells', *Exp. Cell Res.*, 77, 175–82.
13. Killander, D., Levin, A., Inoue, M. and Klein, E. (1970) 'Quantification of immunofluorescence on individual erythrocytes coated with varying amounts of antigen', *Immunol.*, 19, 151–5.

Section IV:
AGGLUTINATION

CHAPTER 26

Mechanisms of Cell Agglutination by Concanavalin A

BERNT T. WALTHER

1. INTRODUCTION AND DEFINITIONS

For some time it has been known that low concentrations of several plant agglutinins cause single transformed cells to agglutinate, while agglutination of normal cells can only be obtained after treatment with proteolytic enzymes (1,2) or at higher lectin concentrations. More recent work has tended to blur the sharp distinction between transformed and normal cell types based on their agglutinability (3). Nevertheless, plant lectin-mediated agglutination has been extensively investigated, since it seemed to reflect a fundamental alteration of the cell surface associated with (malignant) transformation (4,5). At first, agglutination was envisaged to proceed by simple agglutinin cross-links between receptors on two cells, and such receptors were thought to be either missing or unavailable (cryptic) on normal, untrypsinized cells (6,7). Later work on the agglutination reaction has revealed the involvement of more complex processes (8—11), and has contributed to the concept of membrane receptor mobility (12—16) and to the formulation of the 'fluid membrane' model (17).

The object of this paper is to discuss possible mechanisms through which concanavalin A mediates intercellular agglutination reactions. As other agglutinins appear to have agglutinating properties somewhat different from those of Con A (3—5), the mechanisms proposed for Con A may not be relevant for agglutinins in general. Unless otherwise specified, therefore, the term lectin is used in this paper to refer only to Con A.

A number of investigators currently favour models of agglutinability based on enhanced receptor mobility (paragraph 2.1). In several experimental situations, however, this model alone is insufficient to explain the occurrence of agglutination,* and alternative models must be sought. Some workers have proposed (18,19) that models of agglutination developed from analysis of chemically fixed cells also illustrate the mode of agglutination of live cells (paragraph 2.4). In contrast, we propose (20) that agglutination of live cells is a composite reaction involving several pathways and steps, and that only some of these reactions remain possible with chemically fixed cells (paragraph 2.5). In either case, it remains to be documented that observations made on chemically fixed cells are applicable to live cell agglutination reactions. It therefore is recommended that the available models of Con A agglutination are examined with caution, as further experimental work is clearly needed to understand fully the reaction mechanisms involved.

2. PROPOSED MECHANISMS

In this section several modes of Con A-mediated agglutination will be discussed. The first and most widely publicised mechanism involves enhanced lectin receptor mobility and agglutination through '(micro-) patches' of lectin receptors (8—11,21,27) (paragraph 2.1). Several surface topography features as well as surface metabolic and enzymatic factors have also been connected to cell agglutinability by Con A (paragraphs 2.2 and 2.3, respectively). Two recent models of Con A agglutination have been proposed based on experiments in which one of the two cells in an agglutination reaction has been treated with chemical fixatives such as glutaraldehyde. These models are the 'alignment' model (18,19) (paragraph 2.4) and the 'composite reaction' model (20) (paragraph 2.5).

2.1 Membrane Receptor Mobility

Originally lectins were thought to bind specifically to transformed cells and to cross-link such cells simply by simultaneous binding to the glycocalyxes of two

---

*'Agglutination' is defined in this paper as lectin-mediated intercellular aggregations which are blocked by the presence of specific carbohydrate-haptens, regardless of the method used to observe the agglutination phenomenon.

AGGLUTINATION   REACTIONS

(HYPOTHESIS)

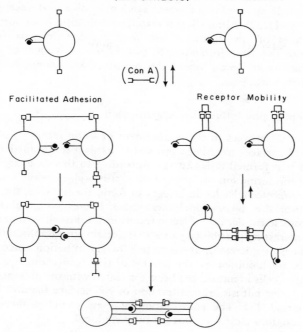

Figure 1. Hypothetical agglutination mechanisms (see text). The model shows how agglutination of cells may proceed simultaneously by two independent pathways, leading to a final aggregate with cells held together by *both* lectin bridges and by adhesion bonds. The hypothesis permits the agglutination reaction to proceed predominantly by one or the other mechanism depending on the particular experimental conditions. It is suggested that initial lectin bridges as defined in this text (and depicted in the 'facilitated adhesion' pathway), are not of sufficient strength to be detected as 'agglutination' if not stabilized by later adhesive bonds between the cells, except under conditions of minimal shear forces (see for example ref. 37). Adhesion reactions are depicted according to current models (see ref. 42, 54–60, 67 for further details). —□, lectin receptors; —●)—, two adhesive components

cells. This view had to be modified when it was shown that both agglutinable and non-agglutinable cells contain the same number of lectin receptors (22–27). Much current emphasis is placed on the phenomenon of lectin receptor mobility (8–11) as the basis for cell agglutinability (see Figure 1). In general, transformed cells unlike normal ones, display high membrane receptor mobility, allowing receptors to converge into patches or to 'clusters' during interaction with multivalent lectins or antibodies (12–16). (In the absence of lectin, receptors for Con A are randomly dispersed on all cell types studied.) It has been postulated that cross-linking of cells by lectins between 'patched' or 'clustered' receptor sites is much more effective in holding two (transformed) cells together (i.e. agglutinating them) than are cross-links formed by lectins between randomly dispersed receptors on two (normal) cells (8—11,27). Although this reaction mechanism is supported by several independent lines of investigation, and may account in part for enhanced agglutinability of live (transformed) cells (see Figure 1), it must be noted that at high concentrations of lectin where the receptors are 'immobilized' (28), cells always agglutinate best.

2.2 Surface Topography Influencing Agglutination

Architectural differences in the arrangement of lectin receptors on normal and transformed cell surfaces have been invoked to explain differential agglutinability of normal and transformed cells (29,31). According to this view, agglutinable cells may allow for the formation of *inter*cellular lectin bridges, while non-agglutinable cells only allow *intra*cellular lectin bridges to form, for instance, through shielding of lectin receptor sites by mucopolysaccharides. Some nonagglutinable cells may also have an insufficient density of lectin receptors on the cell surface (18).

A further interesting possibility exists that certain topographical features of the cell surface, such as microvilli, are essential for agglutination reactions, and that such structures are modified by the process of transformation. Agglutinability of normal untreated cells in mitosis has been correlated with the presence of microvilli on such cells durint mitosis (32). Alteration of cell surface features by lectins has also been reported (33). The contribution of these complex phenomena to cell agglutination reactions clearly are not yet fully explored.

2.3 Metabolic and Enzymatic Factors Affecting Cell Agglutination

An intriguing speculation that lectins possess (unknown) enzymatic activities is often discussed, but has yet to be documented by experimental data. The idea is that while lectins possess the ability to bridge cells by simultaneous binding to the glycocalyxes of two cells, this action is different from the agglutination itself, which is thought to occur by lectin-catalysed linkages between (unknown) substrates on the cell surfaces. Non-agglutinable cells are proposed to possess lectin receptors but to lack the putative substrate sites for the lectin's enzymatic action.

Aside from any enzymatic activity of Con A itself, it is possible that Con A could influence cellular enzymatic activities on the surface membranes through topographical rearrangement of carbohydrate moieties (21). Correlative evidence has been presented (34) that cyclic AMP levels govern the agglutinability of certain cells, and that cyclic AMP levels thus could be the central alteration in transformed cells. Cellular ATP levels have also been correlated with receptor mobility and agglutinability (35).

2.4 Alignment Model of Cell Agglutination

Recently a modification of the 'clustered' membrane receptor theory of cell agglutination (8-11,27) has been proposed based on observed agglutination properties of glutaraldehyde-fixed cells (18,19). It is suggested that mobile receptors are necessary for lectin agglutination, but that the required receptor movements in the membrane only involve small lateral displacements to align receptors on two different cells during an agglutination reaction. In contrast to the original model, extensive clustering of receptors according to the alignment model, inhibits rather than promotes cell agglutination. Transformed cells are thought to be agglutinable since these cells (but not untrypsinized, normal cells) allow for rapid lateral movements which are necessary to form multiple intercellular lectin bridges* when colliding cells are transiently bonded by lectin bridges between randomly dispersed receptors. The alignment model further postulates that irreversible cell–cell binding occurs subsequent to the formation of multiple lectin bridges, similar to an earlier prediction of formation of adhesive bonds upon Con A agglutination of cells (36). The experiments leading to the proposal of both the 'receptor alignment' model and 'composite reaction' model (next paragraph) of cell agglutination reactions are in principle quite similar, and the two models will be compared in paragraph 5.

2.5 Composite Reaction Model

The composite reaction model incorporates both the phenomenon of receptor mobility and the cell's adhesive properties† in accounting for cell agglutination by Con A (Figure 1). The 'composite reaction' model is predicated on the assumption that even if 'receptor alignment' or 'facilitated adhesion' mechanisms (paragraph 3.4) are necessary to account for agglutination reactions taking place under some experimental conditions, this does not rule out receptor clustering as a stabilizing force during Con A agglutination involving *live* cells. In this respect the model differs significantly from the alignment model. In essence the 'composite reaction' model suggests that two live cells may be stably bonded in one of two ways: by adhesive bonds after initial lectin bridges, or by lectin bonds after receptor clustering. In this respect also it differs from the receptor alignment model. Furthermore, the model suggests that the presence of lectin bridges enhances the rate of cellular adhesive events through a mechanism which we have termed 'facilitated adhesion' (paragraph 3.4). This facilitation is also proposed to work in reverse so that adhered cells can more easily be cross-linked by lectins. In the former respect the alignment model concurs with the composite reaction model, as the alignment model also postulates lectin-independent, irreversible cell–cell bindings *after* agglutination through aligned receptors. While both the alignment model and the composite reaction model serve to explain observations made on nucleated cells, they cannot account for haemagglutination reactions: in erythrocytes, where adhesion does not occur, receptor mobility has been ruled out as a factor in agglutination: lowering of the temperature to $0°$, where the receptors are

---

*'*Lectin bridges*' is used in this text as a concept distinct from agglutination. By lectin bridges is merely meant that two cells are held together by lectins in a way which is too *transient* or too *weak* to maintain a stable bonding between the two cells. The formation of irreversible aggregates (by whatever mechanism) in the presence of Con A is considered 'agglutination' (see footnote p. 232).

†'*Adhesion*' is defined as the inherent ability of cells to form aggregates in the absence of lectins, in a process which at least at this point in time has been shown to be insensitive to the common lectin-specific haptens.

immobilized (9,10), reduces the rate but not the extent of the agglutination reaction (37). However, as discussed in paragraph 6, Chapter 35, it is possible that assay methods involving different amounts of mechanical shear select for qualitatively different agglutination reactions. The agglutination reactions for which the alignment model and the composite model are tailored, are concerned with interactions with a binding strength comparable to those found in stable intercellular adhesions. Haemagglutination may represent a weaker type of agglutination reaction which is observed only because of experimental avoidance of shear forces. In such a case one may speculate that lectin bridges between random receptors would be of a stability sufficient for observation, while not being observed under conditions of greater mechanical shear forces.

3. RELATIONSHIP OF ADHESION TO AGGLUTINATION

Both of the more recent models for Con A-mediated cell agglutination have in common a proposal that adhesive reactions enter into the agglutination reactions in some manner. This section will discuss the interrelationships between adhesive and agglutinative events. Three alternative interpretations can be entertained regarding the relationship between lectin receptors and intercellular adhesive factors.

3.1 Agglutination and Adhesion: Identical Receptor Sites?

The evidence would tend to discount the possibility that lectin receptors and cell adhesion sites may be identical. First, univalent fragments of concanavalin A were found to block agglutination of embryonic cells while they did *not* block intercellular adhesion of such cells (38). Secondly, carbohydrate haptens which are effective in blocking agglutination of cells by Con A and by other lectins did *not* block adhesion of mouse fibroblast cell lines 3T3, 3T12, and SV3T3 cells (39). Furthermore, several agents, e.g. trypsin and EDTA, have different and opposite effects on cell adhesion and on agglutination of cells by Con A (36). However, while the possibility of identity between lectin receptors and adhesive sites is not supported by existing data, the evidence is not extensive enough to rule out the speculation that such an identity could be possible and even widespread (40).

3.2 Agglutination and Adhesion: Unrelated Receptor Sites and Independent Processes?

In this case it is proposed that cell adhesion and agglutination are entirely separate and unrelated. This assumption forms the basis for most current interpretations of the agglutination phenomenon. This mechanism derives support from the well-documented observation that under the conditions of the agglutination assay, adhesive reactions are severely limited (1,2). Even the altered adhesiveness of transformed cells (that is observed in the absence of lectins) results in relatively insignificant aggregation under the conditions of the visual agglutination assays (1,2); accordingly, changes in the agglutinability of transformed cells are unrelated to changes in adhesion (8–11,27).

3.3 Agglutination and Adhesion: Unrelated Receptor Sites but Interdependent Processes?

This concept suggests that although adhesion and agglutination could be completely *separate phenomena*, they may mutually *influence* each other. For example, once cells adhere to each other, the cross-linking of such cells by lectins is facilitated. Conversely, once cells are held together by lectin bridges, the establishment of adhesive bonds between such cells is facilitated. This is the essence of the 'composite reaction' mechanism of cell agglutination, depicted in Figure 1. The receptor alignment model, on the other hand, allows for adhesive reactions only *after* lectin agglutination has occurred. The effect of such adhesive reactions (induced cell—cell binding) according to both the receptor aligment model and the composite reaction model, is to render the overall agglutination of cells irreversible even to Con A-specific haptens. This argument is supported by the fact that agglutination of 'Sephadex' beads by Con A is reversible (18), while cell adhesive reactions have been shown to be irreversible in short but comparable time periods (41). With regard to the relative absence of intercellular adhesion of cells in visual agglutination assays, this does not preclude adhesive reactions from participating in *later steps* in the agglutination process in the presence of Con A (19,20,36). Rather, this observation signifies that *early* steps are rate limiting for adhesion under the experimental conditions (see Figure 2). It would seem unreasonable to assume that lectin bridges would *inactivate* cellular adhesive processes. Rather, it appears likely that once cells are brought into contact (by lectins), interactions which do not occur at a significant rate between separate cells in the absence of lectins, now become possible. We have termed such a phenomenon 'facilitated adhesion' and in the next paragraph we will discuss it further.

3.4 Facilitated Adhesion

This paragraph will present the concept of 'facilitated adhesion' in cell agglutination. Some of the experimental evidence will be briefly reviewed in paragraph 4.

Cells have an inherent capacity to aggregate to each other in the absence of lectins (see Figure 2). In Con A-mediated agglutination it is rather difficult to determine which two cells are held together only by lectins, and which cells have adhered to each other. The rate-limiting step in adhesive reactions varies with the exact experimental conditions, depending on a number of factors such as the collision frequency, collision geometry, prevailing shear forces, the amount of (interfering) glycocalyx, van der Waals forces between membranes, or adhesive site numbers (42). According to the 'facilitated adhesion' model, Con A bridges between glycocalyxes may overcome (early) rate-limiting steps in cell adhesion (Figure 2). Therefore it is proposed that Con A simply speeds up reactions which normally happen in the absence of lectins, without necessarily participating directly in the stable intercellular bonding. Under these conditions the rate-limiting step in the formation of intercellular bonds will reflect intrinsic adhesive properties of the cell. Rate-limiting steps in intercellular bonding are illustrated by the observation that two cells which are brought into close contact by micromanipulation, are able

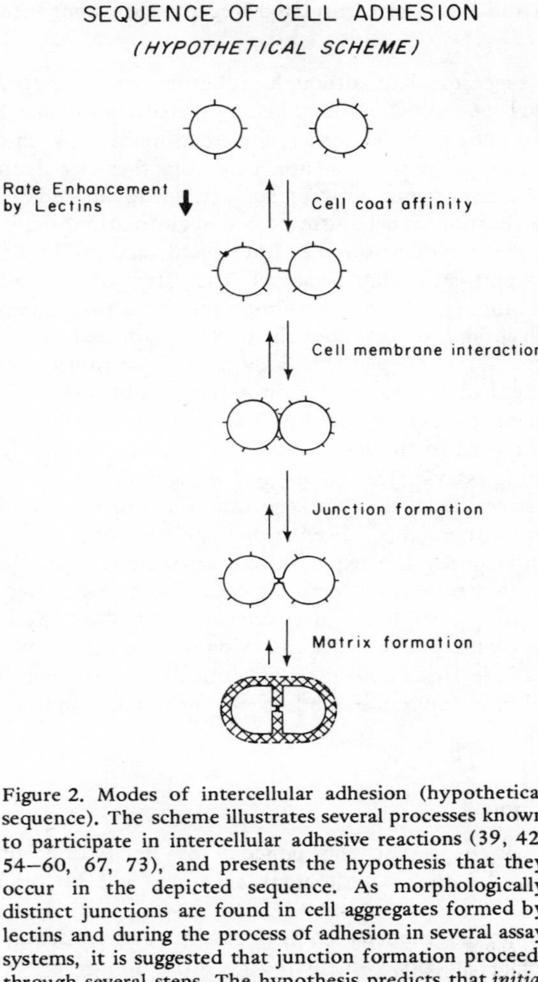

SEQUENCE OF CELL ADHESION
(HYPOTHETICAL SCHEME)

Figure 2. Modes of intercellular adhesion (hypothetical sequence). The scheme illustrates several processes known to participate in intercellular adhesive reactions (39, 42, 54–60, 67, 73), and presents the hypothesis that they occur in the depicted sequence. As morphologically distinct junctions are found in cell aggregates formed by lectins and during the process of adhesion in several assay systems, it is suggested that junction formation proceeds through several steps. The hypothesis predicts that *initial* interactions are rate limiting in intercellular adhesion under many experimental conditions. Lectins are known to bind to carbohydrates of the cell coat, and thus are suggested to enhance the rates of early adhesive steps ('facilitated adhesion' hypothesis; see text for details)

to form apparent cell junctions within a few seconds (43). Con A is envisaged as bringing about similar events by maintaining cells in transient, but somewhat rigid proximity. Essentially lectins enhance adhesion by increasing the effective cell concentration during the adhesion reactions.

Cell adhesiveness is generally thought to be affected by the process of transformation (41,42,44–49), but controversy exists as to the nature of such changes. The controversy is possibly due to the fact that cell adhesion encompasses many steps, and that different methods used to measure adhesion probably measure different adhesive steps. For instance, polyoma-transformed BHK cells are less adhesive than normal BHK cells as measured by the Coulter counter assay (46), but more adhesive as assessed by the cell layer assay (39,41). If transformation involved changes in both glycocalyx (50) and formation of cell contacts (51), as now appears likely, it is apparent how a method which measures 'early' adhesive events could show decreased adhesiveness in transformed cells, while a method measuring the formation of cell–cell contacts could show enhanced adhesiveness in transformed cells. Con A-mediated intercellular binding, if dependent on later adhesive steps, would show transformed cells to be more 'agglutinable' (see Figure 2). Enhanced Con A-agglutinability, while often found after transformation is by no means general* (3). The composite model of Con A-mediated agglutination is not dependent on a uniform change after transformation. The model simply proposes an agglutination reaction mechanism which involves cell adhesion. In a particular cell type, adhesive properties *may* or *may not* be decisive as to its agglutinability by Con A.

4. EVIDENCE FOR FACILITATED ADHESION

Originally we suggested that cell adhesion may occur after Con A agglutination of cells, thus rendering the Con A agglutination of trypsinized 3T3 cells irreversible (36). Similar conclusions have been reached by Rutishauser and Sachs (18,19). In the following, three types of more recent experiments (20) are discussed which suggest that adhesion properties may enter into the agglutination reaction in a more complicated manner which we have termed 'facilitated adhesion' (see Figure 1).

4.1 Agglutination in the Presence of Inhibitors of Adhesion

We have recently found that cytochalasin B inhibits intercellular adhesion of several cell types while having no effect on attachment of cells to non-cellular substrates (52). We therefore attempted to use this drug to study cell agglutination by the cell layer assay (36,53), under conditions where cell adhesion is blocked. Cytochalasin B severely limits the apparent agglutination of cells which are highly adhesive (Balb/c 3T12 cells) while having little effect on the apparent agglutination of cells with lower adhesiveness (20) and erythrocytes (37). This is the predicted result if adhesive processes participate in agglutination reactions. However, even in the absence of intercellular adhesion, the transformed cells (3T12) and normal, trypsinized cells (3T3) are more agglutinable than normal, EDTA-treated cells (3T3). Although this experiment provides suggestive evidence for the involvement of *more than one* process in the agglutination reaction, the results are inconclusive due to the ambiguities in the inhibition caused by cytochalasin B. Ideally cell adhesion should be blocked by *specific* inhibitors of cell adhesion. This approach is impossible as the molecular mechanisms of intercellular adhesion are not known at this time (42,54–60).

*B. Glimelius, B. Westermark, and J. Ponten, personal communication.

Considerable evidence has been accumulated on the effects of cytochalasins on lectin-mediated cell agglutination and on distribution of lectin receptors on the cell surface (61—65). Anchoring of receptors to the cellular cytoskeleton of micro-filaments and microtubules is probably responsible for the degree (or lack) of mobility of receptors on the cell surface (61—63). It is therefore possible that cytochalasin B in the above experiments has its effects on the cytoskeleton and thereby affects the two processes (adhesion and agglutination) not because they are interdependent processes, but simply because of their common dependency on cytoskeletal factors, or indirectly through cytoskeletal effects on the cell surface morphology.

4.2 Agglutination under Conditions of Limited Membrane Receptor Mobility

Recent evidence for facilitated adhesion comes from studies on agglutination of cells in which one or both of the participating cells have been treated with glutaraldehyde. The adhesive properties of such cells have recently been described (66,67). Figure 3 illustrates that glutaraldehyde fixation severely limits cell agglutination as measured by the cell layer assay (36,53) in accordance with other published reports (8,18). The agglutination phenomenon persists only in the case where the single cell and not the cell layer has been glutaraldehyde fixed (Figure 3D) (asymmetrical sensitivity). The ability of a fixed, single cell to be agglutinated is independent of whether the layer to which it is presented is composed of normal (limited membrane receptors mobility), or transformed (high receptor mobility) cells (see Figure 4). This asymmetrical sensitivity of cell agglutinability to the glutaraldehyde fixation process is unexpected, since it does *not* correlate with the presence or absence of membrane receptor mobility properties of the cells participating in the agglutination reaction. However, agglutinability is closely correlated with the persistence of adhesive properties of glutaraldehyde-fixed cells (66,67). In the experiments shown in Figure 3D where *neither* the fixed single cells nor the normal (untrypsinized) 3T3 cell layer should possess membranes where 'clustering' of receptors is possible (10), agglutinability *still* exists (20). While the correlation with persistent adhesion properties of the interacting cells is striking, the observed rates of agglutination are clearly much faster than the (background) adhesive rates (Figure 3D). However, measured adhesive rates do not accurately reflect cell 'adhesiveness', since rates are limited by early adhesive steps (paragraph 3.4). Our interpretation of these kinetics is that cross-linkages of cells to the layer by lectins *facilitate adhesion* reactions, and it is this process of 'facilitated adhesion' and not the lectin bridges which causes the cells to *stay attached* to the cell layer. Thus, in situations where both lectin cross-linkages between layers and single cells are possible (Figure 4B) and where membrane receptor mobility (10) exists on one of the interacting cells (Figure 4C), single cells are *not* stably bound to the layers in the presence of Con A as adhesion is blocked in these cases (66,67). The observations on agglutinability between fixed and live cells need, of course, to be extended to many more cell types and agglutinins. Additional work is needed to identify the agglutination systems in which adhesive processes are participating in the overall reaction, in order to rule out the possibility that the involvement of adhesive processes in agglutination reactions is a unique aspect of the cell layer assay system.

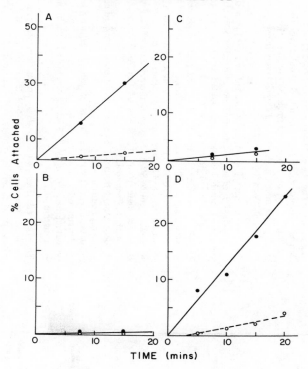

AGGLUTINATION PROPERTIES OF CELLS FIXED
WITH GLUTARALDEHYDE

Figure 3. The effect of glutaraldehyde fixation on agglutination of 3T3 cells. Agglutination was measured by the cell
layer assay as described (20, 36, 53). Single cell suspension
containing 5×10^4 cells/ml were obtained by trypsin dissociation for all experiments. Single cells or cell layers were
treated with glutaraldehyde (3%, 12 hours, followed by
extensive washes) prior to assay as follows: A. No glutaraldehyde pretreatment; note special scale on ordinate in part A.
B. Both cell layer and single cells pretreated with glutaraldehyde. C. Cell layers pretreated with glutaraldehyde; untreated
single cells. D. Single cells pretreated with glutaraldehyde;
untreated cell layers. ●, Con A added to single cell suspension
(80 μg/ml); ○, no Con A added. Background adhesive rates
are low due to early rate-limiting steps in the absence of
Con A (20), and thus do not accurately reflect inherent cell
adhesiveness in the various situations. The asymmetrical
sensitivity of adhesion to the process of glutaraldehyde
fixation will be discussed elsewhere (67). At this point, the
data provide qualitative if not quantitative support for
'facilitated adhesion'. The low rates are, however, internally
consistent with the hypothesis of 'facilitated adhesion'

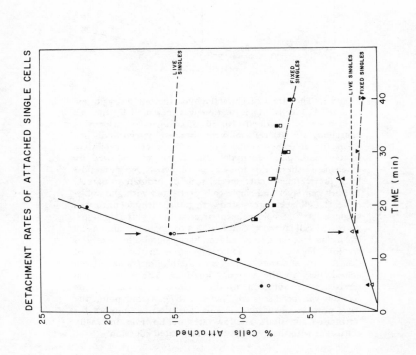

Figure 4. Detachment of previously attached single cells from a cell layer by methyl α-mannoside. Cell agglutination was measured by the cell layer assay (20,36, 53) with single cell concentrations of 5 x 10⁴ cells/ml. Glutaraldehyde-fixed (3%, 12 hours), trypsin-dissociated single Balb/c 3T3 cells were used in all experiments with SV 40-3T3 cell layers (closed symbols) or 3T3 cell layers (open symbols).

(— · —), methyl α-mannoside added at $t = 15$ min to pre-attached *fixed* single cells, final concentration 50 mM;
(——), methyl α-mannoside added at $t = 15$ min to pre-attached *live* single cells;
(——), *no* methyl α-mannoside added, *fixed* single cells.
△   attachemt to 3T3 layers in the absence of Con A;
▲   attachment to SV 3T3 layers in the absence of Con A;
○   attachment to 3T3 layers in the presence of Con A;
●   attachment to SV 3T3 layers in the presence of Con A;
□   detachment from 3T3 layers of Con A-attached cells by methyl α-mannoside;
■   detachment from SV 3T3 layers of Con A-attached cells by methyl α-mannoside;
▽   detachment from 3T3 layers of adhered cells by methyl α-mannoside;
▼   detachment from SV 3T3 layers of adhered cells by methyl α-mannoside.

The dashed lines without data points indicate previously reported data for (bottom dashed line) stability of live cell adhesion to a cell layer (41) and (top dashed line) live cell agglutination to a cell layer (36)

4.3 Reversibility of Agglutination

Further support for interdependence of adhesion and agglutination reactions comes from experiments in which the reversibility of the agglutination reaction was studied by the cell layer assay. Figure 4 shows that with live trypsinized 3T3 cells, no reversibility of the agglutination reaction is observed (36). We reasoned that if this observation was the result of the formation of irreversible adhesive bonds, one might, in the case of fixed single cells, be able to observe reversibility, because such cells are 2- to 3-fold less adhesive than live cells (67), and live to fixed cell adhesive bonds are in addition slightly more reversible than are adhesive bonds between live cells (67). The data in Figure 4 show that if the reversibility of the agglutination reaction is tested with (trypsin-dissociated) fixed 3T3 cells attached to a live (untreated) 3T3 layer in the presence of Con A, some of the attached cells are released by the appropriate hapten. The remaining cells are detached at a much slower rate, comparable to the rate at which *adhered* fixed cells are detached with or without the addition of hapten. It is tempting to speculate that the detached cells represent the *newly attached cells* which are bound to the layer by *lectins*, but which have not yet (or only partially) adhered to the cell layer (see Figure 2). While such interpretations are consistent with the available data, other interpretations are also possible. Nevertheless, the hypothesis that partial retention of adhesiveness after glutaraldehyde fixation (67) is responsible for fixed cell agglutinability is attractive, since it is otherwise difficult to explain why, on one hand, there is not even a minimal agglutination between *two fixed* (transformed) cells, and why, on the other hand, fixed cells can be agglutinated to *normal* cell layers at all. Membrane receptor mobility should be lacking in *both* cases and agglutination should be *impossible in both* cases. Yet, agglutination is observed, but only in the case where (facilitated) adhesion is possible (Figure 3D), and *not* in the cases where adhesion is blocked (66) (Figure 3B and C).

5. SYNOPSIS

In the preceding sections we have discussed agglutination reactions between live cells in one case, and between live and fixed cells in the other case. Different reaction mechanisms have been proposed for the agglutination pathways in these two situations. For live cell agglutination, except for erythrocyte agglutination (37), the mobile receptor model plausibly accounts for agglutination while it is unfit to account for agglutination properties of fixed cells (18,20). Agglutination between live and fixed cells can be accounted for by two alternative models. While previous workers have employed fixed to live cell agglutination to explain live cell agglutination (18,19), it is proposed in this paper that agglutination between live cells is a composite reaction involving two interdependent mechanisms, only one of which remains partially intact after treatment of cells with chemical fixatives, leaving fixed cells able to agglutinate to live cells under certain conditions.

It is interesting that novel reaction mechanisms for Con A-mediated agglutination have been proposed based on experimentation with new methods allowing separate pretreatments of the cells to participate in an agglutination reaction, namely the nylon fibre assay (68) and the cell layer assay (36,53). What remains to be documented is that such pretreatments (e.g. fixation) do not result in agglutination reactions which are not normally taking place with untreated cells.

Such novel reactions could be caused by the fixatives themselves or by the effect of the fixatives on the cells.

A common feature of both models for fixed to live cell agglutination is that they for the first time interrelate cell adhesive properties to the agglutination phenomenon. However, there are several differences in the way adhesive reactions are perceived to be involved. First, the alignment model holds that since agglutination does not happen when both cells have been pretreated with Con A, adhesive reactions cannot initiate agglutination reactions. In this case, the composite model holds that since no lectin bridging is possible, there is consequently no agglutination. Secondly, the alignment model holds that formation of clustered receptor sites prior to agglutination is inhibitory to agglutination, while the composite reaction mechanism holds that the formation of clusters before or during agglutination would stabilize or initiate agglutinative bonding, but that such reactions alone are not sufficient to give irreversible aggregation in the absence of adhesive interactions. While mechanistic differences between the two models are apparent, this should perhaps not overshadow the fact that the two models are based on similar experimental data obtained in two different systems, perhaps indicating that mechanisms such as 'receptor alignment' and 'facilitated adhesion' may describe more general phenomena not limited to one experimental system. A decision on the relative merits of the 'alignment' model and the 'composite reaction' model seems premature at this juncture. Furthermore, at least parts of the two models of fixed to live cell agglutination are not necessarily mutually exclusive.

Recently possible connexions between cell surface morphology and cyclic AMP levels have been suggested (69,70). Cyclic AMP has been shown to decrease the number of microvilli on fibroblast surfaces (70). If microvilli are necessary for Con A agglutination (32), cyclic AMP regulation of cell agglutination may be envisaged (34,70). Microvilli have also been implicated in intercellular adhesion (42,53,71,72), suggesting a possible morphological basis for 'facilitated adhesion' during Con A agglutination of fibroblasts.

In conclusion, it seems abundantly clear from much recent work that the apparently simple process of agglutination of cells by multivalent plant lectins such as concanavalin A has turned out to be a complex reaction which in nucleated cells probably also involves intercellular adhesive reactions. The current data are consistent with several models of agglutination proposed from analysis of agglutination reactions under different experimental conditions. These models have yet to be reconciled. Further work is also required to establish whether in fact agglutination reactions with fixed and live cells represent some of several agglutination reactions possible with live cells. Our present understanding of the Con A-mediated agglutination mechanisms is still incomplete, and the final clarification of these problems may still hold many surprises for future research.

6. ACKNOWLEDGEMENTS

The author is grateful to Drs. Pamela Talalay and William J. Snell for many helpful discussions and to Mrs. Dorothy Regula for expert help in preparing this manuscript. Supported by a Special Fellowship from the Leukemia Society of America and a Fellowship from NTNF, Norway. Publication No. 837 from the McCollum-Pratt Institute, Johns Hopkins University.

7. REFERENCES

1. Burger, M. M. (1967) 'Identification of a tumor-specific determinant on neoplastic cell surfaces', *Proc. Nat. Acad. Sci. U.S.A.*, **57**, 359—66.
2. Inbar, M. and Sachs, L. (1969) 'Interaction of the carbohydrate-binding protein concanavalin A with normal and transformed cells', *Proc. Nat. Acad. Sci. U.S.A.*, **63**, 1418—25.
3. Sharon, N. and Lis, H. (1972) 'Lectins: Cell-agglutinating and sugar-specific proteins', *Science*, **177**, 949—59.
4. Sachs, L. (1974) 'Regulation of membrane changes, differentiation and malignancy in carcinogenesis', *Harvey Lect.*, **68**, 1—35.
5. Lis, H. and Sharon, N. (1973) 'The biochemistry of plant lectins (phytohemagglutinins)', *Ann. Rev. Biochem.*, **42**, 541—74.
6. Burger, M. M. (1969) 'A difference in the architecture of the surface membranes of normal and virally transformed cells', *Proc. Nat. Acad. Sci. U.S.A.*, **62**, 994—1001.
7. Inbar, M. and Sachs, L. (1969) 'Structural difference in the sites on the surface membranes of normal and transformed cells', *Nature*, **223**, 710—12.
8. Inbar, M., Huet, C., Oseroff, A. R., Ben-Bassat, H. and Sachs, L. (1973) 'Inhibition of lectin-agglutinability by fixation of the cell surface membrane', *Biochim. Biophys. Acta.*, **311**, 594—9.
9. Nicolson, G. L. (1973) 'Temperature-dependent mobility of concanavalin A sites on tumor cell surfaces', *Nature (New Biol.)*, **243**, 218—20.
10. Rosenblith, J. Z., Ukena, T. E., Yin, H. H., Berlin, R. D. and Karnovsky, M. J. (1973) 'A comparative evaluation of the distribution of concanavalin A-binding sites on the surfaces of normal, virally-transformed, and protease-treated fibroblasts', *Proc. Nat. Acad. Sci. U.S.A.*, **70**, 1625—9.
11. Noonan, K. D. and Burger, M. M. (1973) 'The relationship of concanavalin A binding to lectin-initiated cell agglutination', *J. Cell Biol.*, **59**, 134—42.
12. Taylor, R. B., Duffus, W. P. H., Raff, M. C. and DePetris, S. (1971) 'Redistribution and pinocytosis of lymphocyte surface immunoglobulin molecules induced by anti-immunoglobulin antibody', *Nature (New Biol.)*, **233**, 225—9.
13. DePetris, S. and Raff, M. C. (1972) 'Distribution of immunoglobulin on the surface of mouse lymphoid cells as determined by immunoferritin electron microscopy. Antibody-induced, temperature dependent redistribution and its implication for membrane structure', *Eur. J. Immunol.*, **2**, 523—35.
14. Unanue, E. R., Perkins, W. D. and Karnovsky, M. J. (1972) 'Ligand-induced movement of lymphocyte membrane macromolecules. I: Analysis by immunofluorescence and ultrastructural radiography', *J. Exp. Med.*, **136**, 885—906.
15. Kourilsky, F. M., Silvestre, C., Neauport-Sautes, C., Loosfelt, Y. and Dausset, J. (1972) 'Antibody-induced redistribution of HL-A antigens at the cell surface', *Eur. J. Immunol.*, **2**, 249—57.
16. Loor, F., Forni, L. and Pernis, B. (1972) 'The dynamic state of the lymphocyte membrane. Factors affecting the distribution and turnover of surface immunoglobulins', *Eur. J. Immunol.*, **2**, 203—12.
17. Singer, S. J. and Nicolson, G. L. (1972) 'The fluid mosaic model of the structure of the cell membranes', *Science*, **175**, 720—31.
18. Rutishauser, U. and Sachs, L. (1974) 'Receptor mobility and the mechanism of cell—cell binding induced by concanavalin A', *Proc. Nat. Acad. Sci. U.S.A.*, **71**, 2456—60.
19. Rutishauser, U. and Sachs, L. (1975) 'Cell:cell binding induced by different lectins', *J. Cell Biol.*, **65**, 247—57.
20. Walther, B. T. and Roseman, S. (1975) 'Concanavalin A-mediated agglutination of Balb/c 3T3 cells treated with glutaraldehyde', *J. Cell Biol.*, **67**, 445a.
21. Edelman, G. M., Cunningham, B. A., Reeke, G. N., Becker, J. W., Waxdal, M. J. and Wang, J. L. (1972) 'The covalent and three-dimensional structure of concanavalin A', *Proc. Nat. Acad. Sci. U.S.A.*, **69**, 2580—4.
22. Inbar, M., Ben-Bassat, H. and Sachs, L. (1971) 'A specific metabolic activity on the surface membrane in malignant cell-transformation', *Proc. Nat. Acad. Sci. U.S.A.*, **68**, 2748—51.
23. Malucci, L. (1971) 'Binding of concanavalin A to normal and transformed cells as detected by immunofluorescence', *Nature (New Biol.)*, **233**, 241—4.

24. Cline, M. J. and Livingstone, D. C. (1971) 'Binding of ^{3}H-concanavalin A by normal and transformed cells', *Nature (New Biol.)*, **232**, 155–6.

25. Ozanne, B. and Sambrook, J. (1971) 'Binding of radioactively-labeled concanavalin A and wheat germ agglutinin to normal and virus-transformed cells', *Nature (New Biol.)*, **232**, 156–60.

26. Arndt-Jovin, D. J. and Berg, P. (1971) 'Quantitative binding of ^{125}I-concanavalin A to normal and transformed cells', *J. Virol.*, **8**, 716–21.

27. Nicolson, G. L. (1972) 'Modification of cell membrane site topography by proteolytic enzyme', *Nature (New Biol.)*, **239**, 193–7.

28. Yahara, I. and Edelman, G. M. (1973) 'The effects of concanavalin A on the mobility of lymphocyte receptors', *Exp. Cell Res.*, **81**, 143–5.

29. Moore, E. G. and Temin, H. M. (1971) 'Lack of correlation between conversion by RNA tumour viruses and increased agglutinability of cells by concanavalin A and wheat germ agglutinin', *Nature*, **231**, 117–8.

30. Kapeller, M. and Doljanski, F. (1972) 'Agglutination of normal and RSV-transformed chick embryo cells by concanavalin A and wheat germ agglutinin', *Nature (New Biol.)*, **235**, 184–5.

31. Burger, M. M. and Martin, G. S. (1972) 'Agglutination of cells transformed by Rous sarcoma virus by wheat germ agglutinin and concanavalin A', *Nature (New Biol.)*, **237**, 9–12.

32. Porter, K., Prescott, D. and Frye, J. (1973) 'Changes in surface morphology of Chinese Hamster ovary cells during the cell cycle', *J. Cell Biol.*, **57**, 815–36.

33. Rossmando, E. F., Steffek, A. J., Mujurid, D. K. and Alexander, S. (1974) 'Scanning electron microscopy observations on cell surface changes during aggregation of Dictyostelium Discoideum', *Exp. Cell Res.*, **84**, 73–8.

34. Willingham, M. C. and Pastan, I. (1974) 'Cyclic AMP mediates the concanavalin A agglutinability of mouse fibroblasts', *J. Cell Biol.*, **63**, 288–94.

35. Vlodavsky, I., Inbar, M. and Sachs, L. (1973) 'Membrane changes and adenosine triphosphate content in normal and malignant transformation', *Proc. Nat. Acad. Sci. U.S.A.*, **70**, 1780–4.

36. Rottmann, W. L., Walther, B. T., Hellerqvist, C. G., Umbreit, J. and Roseman, S. (1974) 'A quantitative assay for concanavalin A-mediated cell agglutination', *J. Biol. Chem.*, **249**, 373–80.

37. Schnebli, H. P. and Bächi, Th. (1975) 'Reaction of lectins with human erythrocytes. I. Factors governing the agglutination reaction', *Exp. Cell Res.*, **91**, 175–183.

38. Steinberg, M. S. and Gepner, L. A. (1973) 'Are concanavalin A receptor sites mediators of cell–cell adhesion', *Nature (New Biol.)*, **241**, 249–51.

39. Walther, B. T. and Roseman, S. (1975) 'Intercellular adhesion. Properties and experimental distinction between two early steps in adhesive reactions', manuscript in preparation.

40. Evans, P. M. and Jones, B. M. (1974) 'Studies of cellular adhesion-aggregation', *Exp. Cell Res.*, **88**, 56–62.

41. Walther, B. T., Ohman, R. and Roseman, S. (1973) 'A quantitative assay for intercellular adhesion', *Proc. Nat. Acad. Sci. U.S.A.*, **70**, 1569–73.

42. Curtis, A. S. G. (1974) 'Cell adhesion', *Progr. Biophys. Molec. Biol.*, **27**, 317–86.

43. Löwenstein, W. R. (1973) 'Membrane junctions in growth and differentiation', *Fed. Proc.*, **32**, 60–4.

44. Coman, D. R. (1961) 'Adhesiveness and stickiness; two independent properties of the cell surface', *Cancer Res.*, **21**, 1436–8.

45. Weston, J. and Roth, S. (1969) 'Contact inhibition: Behavioral manifestations of cellular adhesive properties *in vitro*', in *Cellular Recognition*, eds. Smith, R. T. and Good, R. A. (Applecraft, Century, Crofts, N.Y.), pp. 29–38.

46. Edwards, J. G., Campbell, J. A. and Williams, J. F. (1971) 'Transformation of polyoma virus affects adhesion of fibroblasts', *Nature (New Biol.)*, **231**, 147–8.

47. Halpern, B., Pejsachowicz, B., Febvre, H. L. and Barski, G. (1966) 'Differences in patterns of aggregations of malignant and non-malignant mammalian cells', *Nature*, **209**, 157–9.

48. Johnson, G. S. and Pastan, I. (1972) 'Cyclic AMP increases the adhesion of fibroblasts to substratum', *Nature (New Biol.)*, **236**, 247–9.

49. Dorsey, J. K. and Roth, S. (1973) 'Adhesive specificity between normal and transformed fibroblasts', *Devel. Biol.*, 33, 249—56.
50. Oseroff, A. R., Robbins, P. W. and Burger, M. M. (1973) 'The cell surface membrane: Biochemical aspects and biophysical probes', *Ann. Rev. Biochem.*, 42, 647—82.
51. Abercrombie, M. (1970) 'Contact inhibition in tissue culture', *In vitro*, 6, 128—47.
52. Hellerqvist, C. G., Rottmann, W. L., Walther, B. T. and Roseman, S. (1975) 'Cell adhesion' in *Mammalian Cells, Probes and Problems*, eds. Richmond, C. R., Peterson, D. J., Mulaney, P. J. and Anderson, E. C. E.R.D.A. Symposium Series CONF.-731007, Oak Ridge, USA, pp. 284—96.
53. Walther, B. T. and Rottmann, W. L. (1975) 'Quantitation of fibroblast agglutination by the cell-layer assay' this volume, Chapter 35.
54. Weiss, L. (1960) 'The adhesion of cells', *Int. Rev. Cytol.*, 9, 187—225.
55. Steinberg, M. S. (1970) 'Does differential adhesion govern self-assembly in histogenesis? Equilibrium configurations and the emergence of a hierarchy among populations of embryonic cells', *J. Exp. Zool.*, 173, 395—434.
56. Roseman, S. (1971) 'The synthesis of complex carbohydrates by multiglycosyltransferase systems and their potential function in intercellular adhesion', *Chem. Phys. Lipids.*, 5, 270—97.
57. Parsegian, V. A. and Gingell, D. (1972) 'Some features of physical forces between biological membranes', *J. Adhesion*, 4, 283—306.
58. Hausman, R. E. and Moscona, A. A. (1975) 'Purification and characterization of the retina-specific cell-aggregating factor', *Proc. Nat. Acad. Sci. U.S.A.*, 72, 916—20.
59. Gerisch, G. (1972) 'Cell aggregation and differentiation in Dictyostelium', *Current Topics in Dev. Biol.*, 3, 157—97.
60. Pessac, B. and Defendi, V. (1972) 'Cell aggregation: Role of acid mucopolysaccharides', *Science*, 175, 898—900.
61. Ukena, T. E., Borysenko, J. Z., Karnovsky, M. J. and Berlin, R. D. (1974) 'Effects of colchicine, cytochalasin B, and 2-deoxyglucose on the topographical organization of surface-bound concanavalin A in normal and transformed fibroblasts', *J. Cell Biol.*, 61, 70—82.
62. DePetris, S. (1975) 'Concanavalin A receptors, immunoglobulins, and θ antigen of the lymphocyte surface', *J. Cell Biol.*, 65, 123—46.
63. Edelman, G. M., Yahara, ?. and Wang, J. L. (1973) 'Receptor mobility and receptor—cyto-plasma interactions in lymphocytes', *Proc. Nat. Acad. Sci. U.S.A.*, 70, 1442—6.
64. Wessels, N. K., Spooner, B. S., Ash, J. F., Bradley, M. O., Luduena, M. A., Taylor, M. A., Wrenn, J. T. and Yamada, K. M. (1971) 'Microfilaments in cellular and developmental processes', *Science*, 171, 135—43.
65. Godman, G. C., Miranda, A. F., Deitch, A. D. and Tannenbaum, S. W. (1975) 'Action of cytochalasin D on cells of established lines. III. Zeiosis and movements at the cell surface', *J. Cell Biol.*, 64, 644—67.
66. Walther, B. T. (1975) 'Adhesive properties of Balb/c 3T3 cells fixed with glutaraldehyde', *Fed. Proc.*, 34, 498.
67. Walther, B. T. and Roseman, S. (1975) 'Adhesive properties of mouse normal and transformed fibroblasts after treatment with glutaraldehyde', manuscript in preparation.
68. Edelman, G. M., Rutishauser, U. and Millete, C. F. (1971) 'Cell fractionation and arrangement on fibers, beads, and surfaces', *Proc. Nat. Acad. Sci. U.S.A.*, 68, 2153—7.
69. Hsie, A. W., Jones, C. and Puck, T. T. (1971) 'Further changes in differentiation state accompanying the conversion of Chinese hamster ovary cells to fibroblastic form by dibutyryl adenosine 3,'5'-monophosphate and hormones', *Proc. Nat. Acad. Sci. U.S.A.*, 68, 1648—52.
70. Willingham, M. C. and Pastan, I. (1975) 'Cyclic AMP modulates microvillus formation and agglutinability in transformed and normal mouse fibroblasts', *Proc. Nat. Acad. Sci. U.S.A.*, 72, 1263—7.
71. Shimada, Y., Moscona, A. A. and Fischman, D. A. (1974) 'Scanning electron microscopy studies of cell aggregation: Cardiac and mixed retina—cardiac cell suspensions', *Devel. Biol.*, 36, 428—46.
72. Howard, P. F., Wetzel, B., Walther, B. T. and Roseman, S. (1975) 'Scanning electron

microscopy studies of intercellular contacts in the cell layer assay' *J. Cell Biol.,* **67**, 179a.
73. Umbreit, J. and Roseman, S. (1976) 'Two stages in intercellular adhesion', *J. Biol. Chem.,* **250**, 9360−8.

CHAPTER 27

Survey of Agglutination Techniques: How they Affect the Result

HANS PETER SCHNEBLI

Section IV of this book is a representative collection of techniques for assessing lectin-mediated cell agglutination. The number and in particular the diversity of the methods are a tribute to the difficulties involved in quantitating a complex phenomenon like cell agglutination. Not surprisingly, results obtained with different methods often deviate significantly and in some cases appear to be contradictory. The aim of this chapter is thus to discuss some of the experimental parameters that may critically influence the results obtained. It is hoped that this, together with the detailed analysis of the agglutination mechanism(s) of the preceding chapter, may help to reconcile some of the conflicting results reported in the literature.

1. GENERAL CONSIDERATIONS

Lectin-mediated cell agglutination in many ways resembles agglutination by antibodies. As in serology, twofold serial dilutions of lectins are often used in agglutination experiments and the titration curves obtained (Figure 1) are in every respect similar to the well-known serological titration curves. Note that the curves shift to lower concentrations and (again like serological titration curves) 'sharpen' with time (Figure 1). The essential point is that within the time of a reasonable experiment, agglutination develops progressively and does not even approach an

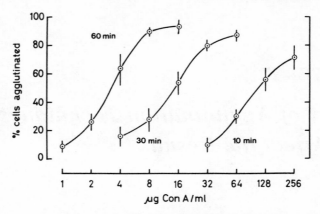

Figure 1. Time and concentration dependence of agglutination by Con A. Neuraminidase-treated human erythrocytes were incubated with the indicated concentrations of Con A in a microtitre plate. The plate was shaken one minute at a time on a Tayo Bussan mixer (42 Hz) with one minute intervals of rest (see Chapter 31 for details). After 10, 30 and 60 minutes portions of the cell suspensions were removed from the wells and the percentage of agglutinated cells determined by counting under the microscope

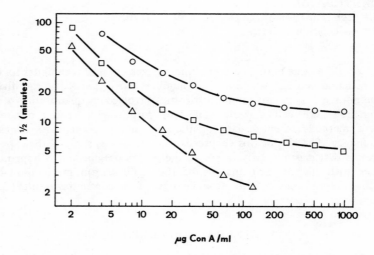

Figure 2. Effect of temperature and Con A concentration on rate of agglutination. The assays were done according to the 'microagglutination method' described in Chapter 31. Each point represents the average of 4 to 10 experiments and represents the time required to agglutinate approximately 50% of the cells under the given conditions. Cells: trypsinized human erythrocytes. ○, 0°; □, 24°; △, 37°. Reproduced by permission of Academic Press, from *Exp. Cell Res.*, 91, 175–83 (1975).]

end-point equilibrium (Figure 2). It follows that comparisons of agglutination experiments are only valid if rates or some reasonable approximation of rates are measured.

2. IMPORTANT EXPERIMENTAL VARIABLES

2.1 Handling of Cells

Suspensions of blood and lymphoid cells suitable for agglutination experiments can easily be prepared, and the agglutinating properties of these cells are very reproducible and stable during limited storage (a few hours at $0°$) which is an important point. Suspensions of cells from solid tissues can only be prepared, if at all, by using digestive enzymes and are therefore not widely studied.

Cells from monolayer cultures are usually suspended by an EDTA—saline treatment which is considered to be innocuous. These cells, however, tend to become more agglutinable upon standing in the unfavourable suspension medium, perhaps due to leakage of enzymes leading to autolysis of the cell surface. Tissue culture media with serum cannot be used, as serum glycoproteins (and glucose in the case of Con A) may interfere with the agglutination reaction.

Cultured 'normal' cells suspended with the aid of proteases show increased agglutinability, while most transformed cells are not affected. On the other hand, hyaluronidase treatment may in certain cases be used to amplify a small difference in agglutinability between normal and transformed cells. Where enzymes (particularly proteases) are used, it must be realized that these not only alter the agglutinability of the cells drastically but also adsorb tenaciously to cell surfaces.

The agglutinating properties of tissue culture cells also appear to vary with culture conditions, such as time after subculturing and cell density. To avoid these possible density-dependent effects, most investigators use cells from subconfluent cultures for agglutination studies.

Perhaps the most serious problem with nucleated cells is the background clumping, and often the conditions of an agglutination experiment have to be compromised because of it. Spontaneous aggregation is more pronounced in epithelial than in fibroblastic cells, and occurs at a rate characteristic for each cell type. Background clumping is often also caused by DNA leakage from damaged cells and this may be reduced by the addition of protease-free DNAase (1 μg/ml) to the cell suspension. In the case of lymphocytes dead cells can be removed by simply filtering through a 5 mm layer of cotton wool.

2.2 Cell Density and Lectin Concentration

As in any other reaction, the rate of agglutination depends on the concentration of the reactants, i.e. the cell density and the lectin concentration. With nucleated cells the agglutination rate is roughly proportional to the cell density, while erythrocyte agglutination (as measured in the microtitre assay, Chapter 31), for reasons that are not understood, appears to be much less sensitive to cell density. In practice one chooses a cell density (usually 10^6 to 10^8 cells/ml) that is convenient for the measurement of agglutination or for scoring. Standardization appears to be more difficult if the cell density is changed during the assay, e.g. by centrifugation and resuspension (Chapter 29). Interesting variations include the microhaemadsorption assay (Chapter 34) and in the cell layer assay (Chapter 35) where a relatively dilute cell suspension is confronted with the high cell density of a (standardized) confluent monolayer.

Since cells may differ in their agglutinability by several orders of magnitude, it often becomes necessary to vary the lectin concentration in order to obtain a measurable rate. The rate of agglutination is roughly proportional to the lectin concentrations in most cases, but this relationship is not always straightforward: as shown in Figure 2 for erythrocytes, the rate of agglutination becomes concentration independent at high lectin concentration, while at low concentrations the proportionality exists. For the interpretation of 'titration curves' where the lectin concentration is varied while all other conditions are kept constant, it is important that at low concentration the rate is proportional to the lectin concentration and it will be shown below that one can determine agglutination rates quite accurately from such titration curves. Agglutination experiments with a single lectin concentration are much more difficult to interpret, expecially if the time is also fixed, as unfortunately it often is (see paragraph 3).

2.3 Time of Reaction

It goes without saying that, when measuring reaction rates (as is usually the case with agglutinations, see paragraph 1), the time of reaction must be controlled carefully, and it is obvious that the time required for measuring the degree of agglutination (or scoring) should be short compared with the reaction time. This is a serious consideration in those assays that involve time-consuming microscopic evaluation (Chapter 29) and the hanging drop method (Chapter 28). The problem can be avoided by choosing long reaction times or by stopping the reaction either by dilution (particle counter methods), removal of non-agglutinated cells (cell layer and microhaemadsorption method) or by allowing the cells to settle out (microtitre and microhaemagglutination method).

Long reaction times not only improve the reproducibility and sensitivity of the assays but are necessary if there is a lag in the agglutination reaction (see e.g. Figure 2). On the other hand, long reaction times are often troublesome, because of increased background clumping.

In practice, experiments with a single fixed time point are often satisfactory. However, it is essential to demonstrate that true rates are measured by checking the assay at different time points.

Parenthetically, it may be stated that non-agglutinable cells probably do not exist; it is much more likely that any cell (provided it binds lectin) can be agglutinated if treated with high enough concentrations of lectin and for extended periods of time.

2.4. Temperature

While everybody agrees that temperature affects the rate of agglutination reactions, the claim that Con A-mediated cell agglutination is blocked at $0°$ is debatable. As will be shown below in the paragraph on 'Scoring', the experiments where no agglutination was observed at $0°$, either did not allow quantitative interpretations or the method for quantitation did not allow the calculation of rates.

We have carefully studied the effect of temperature on erythrocyte agglutination with the microagglutination technique (Chapter 30). From data shown in Figure 2 it is evident that lowering of the temperature affects the rate of the reaction, but even at $0°$ agglutination is not blocked. The temperature coefficients for Con A binding and agglutination (Table I) calculated from Figure 2 and from independent binding studies are of the same magnitude as those of most chemical reactions and are incompatible with a temperature transition point. This has now been confirmed and extended with other methods and other cells: as shown in Table II, agglutination measured by the conventional microtitre technique was only slightly reduced by lowering of the temperature to $0°$. In another study, Oppenheimer was able to follow agglutination of teratoma cells at $0°$ with a particle counter method (Chapter 33).

TABLE I.   Temperature dependence of the erythrocyte agglutination rate. [Reproduced by permission of Academic Press from *Exp. Cell Res.*, 91, 175 (1975).]

	Q_{10}
Con A binding	1.4
Agglutination at low concentrations	1.35
Agglutination at high concentrations	1.65

TABLE II.   Effect of temperature on agglutination

Cell type	Lowest agglutinating Con A concentration (μg/ml)	
	$0°$C	$24°$C
Erythrocytes (human) trypsinized	0.8	0.7
Erythrocytes neuraminidase treated	1.0	1.0
T−Lymphocytes (mouse)	13.0	4.0
L 1210 (mouse leukaemia)	2.4	1.1
TIMI (mouse leukaemia)	1.4	1.1
EAC (mouse carcinoma)	2.0	1.0
BHK (hamster fibroblast) trypsinized	0.4	0.4

Whether or not cells agglutinate at $0°$ has implications for mechanism(s) of agglutination which require receptor mobility, and since Con A dissociates into dimers at $0°$ concerning the valency of the lectin. In practice, however, routine agglutinations are most easily done at room temperature, but it must be kept in mind that results will vary with large changes of ambient temperature.

2.5 Mixing

Agglutinations depend a great deal on the kind and amount of mixing during the reaction. A slight to moderate amount of agitation increases agglutination while vigorous shaking may in some cases prevent cells from agglutinating. It is important to note that not all cells and not even all lectins are affected in the same way or to the same degree. The explanation for this is not at all clear, but whatever the reason, it is necessary to standardize agitation during an agglutination reaction.

The simplest and presumably most reproducible methods are those that do not involve any agitation (after the initial mixing), e.g. the conventional microtitre method, the microhaemadsorption and the microhaemagglutination methods. Another group of methods empirically introduces a moderate amount of shearing forces (different in each case!) by a variety of shakers and rocking platforms operating at 20 to 70 cycles per min; these include the hanging drop method, the particle counter methods and the cell layer assay. In one case (microagglutination method) high shearing forces are produced by a vibrator (42 cycles per second).

2.6 pH, Ionic Environment

Most lectins are optimally active in the neutral to slightly acidic range. The rule thus is to use a medium tolerated by the cells — the lectins are a lot less fussy, except for the fact that serum and glucose must be avoided. Although Con A has a requirement for divalent cations, agglutination proceeds normally even in Ca^{2+}- and Mg^{2+}- free salt solutions, presumably because the short time of the experiments is not sufficient for the removal of the ions non-covalently bound to the lectin, or because of leakage of Ca^{2+} from the cells.

3. QUANTITATION

3.1 Determination of the Degree of Agglutination

A number of methods involve quantitation of the *degree* of agglutination under a given set of conditions, while in another group of methods (titration assays) the conditions are varied in order to obtain a certain proportion (usually 50%) of agglutinated cells.

Quantitation of the degree of agglutination is fairly straightforward and may involve: microscopical counting of free and agglutinated cells (Chapters 28, 29), the use of a particle counter (Chapters 32, 33), counting of erythrocytes adhering to a lectin-coated cell layer (Chapter 34) or the determination of radiolabelled cells bound to a cell layer (Chapter 35). The titration methods may involve determination of the *time* required to agglutinate approximately 50% of the cells at a given lectin concentration (Chapter 31) or determination of the *lectin concentration* required to agglutinate approximately 50% of the cells at a given time (Chapters 30, 31); in this last type of method the length of the experiment is determined by the time the free cells need to settle to the bottom of the well and to become unavailable for the reaction. In principle, any degree of agglutination (provided it can be observed accurately) can be used in the titration assays, but it is obviously best to work in the steep portion of the titration curve (around 50% agglutination).

Clump size is probably not a valid measure, although the clump size may roughly correlate with the degree of agglutination. The size of a clump furthermore depends on a number of extraneous factors like salts, especially divalent cations.

3.2 Calculation of Rates

At first sight it would appear that the methods that actually measure the degree of agglutination would be more accurate than the titration assays, and that agglutination rates calculated from these methods would be more precise, but this is actually not the case. The titration assays have the great advantage of providing a direct measure of the agglutination rate. At low concentrations the rate of agglutination is directly proportional to the lectin concentration (slope of 1 in Figure 2; see also Figure 3 in Chapter 31). The rate of agglutination is therefore *inversely proportional* to the concentration of lectin that gives 50% agglutination in a given time (in practice: the lowest concentration of lectin that gives a positive reaction under a given set of conditions). It is *not* necessary to extrapolate to initial velocities (nor to use short times in the experiments), since the reaction rates do not decrease significantly with the time (Figure 2); note also that agglutination reactions sometimes have a lag phase.

With methods that measure the degree of agglutination, the kinetics of the reaction can be calculated from time course experiments. On the other hand, when all such measurements are made at a single time point, calculation of rates becomes very difficult. The quantitation of the degree of agglutination, especially with the particle counter methods, is highly reproducible ($\pm$ 5% between experiments, see Chapter 33). Still, accurate *rates* will only be obtained from intermediate degrees of agglutination (steep portion of titration curves) since for example 80% to 90% and 90% to 95% agglutinated cells (measured under the same conditions and time), both reflect twofold increments in agglutination rates.

CHAPTER 28

A Standard Assay for Agglutination with Lectins (Hanging Drop Method)

MAX M. BURGER

Agglutination with plant or invertebrate lectins has several features, which differ from classical antibody—antigen precipitations. Thus the stoichiometry is different in as much as most lectins seem to be at least tetravalent in their native state, while antibodies are divalent. Lectins have peculiar temperature dependencies. Thus, while concanavalin A can disaggregate at low temperatures, wheat germ agglutinin does not show such a pronounced temperature dependence of subunit interactions. Furthermore, temperature seems to influence also the cell surface as evidenced by the shifts in temperature dependence of the agglutination process induced by altering the surface composition with fatty acids with high- and low-phase transition temperatures (1).

When measuring agglutination, one should be aware that this process consists of at least two components (2): the binding of the lectin to the cell surface and subsequently the stable agglutination of two and more cells by ways and means, which may be mediated by a series of steps other than the simple bridging via the lectin molecule.

Until the molecular mechanism(s) for lectin agglutination are better understood, it is particularly important to standardize the conditions for the agglutination procedure — or if new methods are used to report them in sufficient detail to permit comparisons between methods and results.

1. PREPARATION OF CELLS

1.1 Cells Grown in Suspension

Cells grown in suspension are usually grown in various culture media often including different amounts and different types of sera. Since serum glycoproteins may — and in many cases do — bind lectins, it is important to be meticulous about the standardization of the rinsing procedures when handling cells grown in suspension.

Cells are suspended from the original source (ascites fluid, suspension culture *in vitro*, etc.) at concentrations of $1-2 \times 10^6$/ml. All dilutions or rinses are carried out with calcium- and magnesium-free phosphate-buffered saline solution (CMF-PBS*) prewarmed to $37°$. Cells are spun out of a 3 ml suspension in a 15 ml capacity tube of 1 cm diameter at minimal speed (about $600 \, g$) for 2.5 minutes. Longer centrifugation or too high speeds, that may be required for larger volumes, should be avoided since many cells get damaged, particularly during centrifugations, as we noticed from the loss of material absorbing at 260 nm, loss of plating efficiency and sometimes even from trypan blue uptake. In order to minimize contact with the centrifuge tube, no angle rotors, but only swinging bucket rotors should be used. In general, speed and time of centrifugation should be determined for every cell type and sometimes even for different growth conditions, since the fragility of a cell was often found to be influenced by the growth stage, serum content, ionic composition of the medium and other factors.

Cells are resuspended in CMF-PBS to $1-2 \times 10^6$/ml by very gentle pipetting with a wide-mouth (2 mm diameter) pipette and respun under the same conditions. The cell pellet is resuspended and recentrifuged once more, and after this second rinse finally suspended at $1-2 \times 10^6$ cells/ml in CMF—PBS, if no divalent ions are required for the particular lectin used. Even if the particular lectin required divalent cations, they are preferably added to the lectin solution directly to avoid long contact between the cells and divalent cations, since these ions increase the spontaneous background agglutination of the control. If cations have to be added with the lectin, similar amounts of cations have, of course, to be added to a control suspension without lectin.

1.2 Cells Grown on Culture Dish Surfaces

Cultured cells cannot be removed by enzymes (proteases) since such enzymes may alter the agglutinability, particularly agglutination of untransformed

---

*CMF-PBS: per litre: NaCl, 8 g; KCl 0.2 g; Na_2HPO_4, 1.15 g; KH_2PO_4, 0.2 g.

cells (4, 5). Although EDTA-containing buffers, which remove cells from the culture dishes, may damage the cell surface, it seems that EDTA does not alter the agglutinability with the lectins tested so far.

After aspirating the medium from cells grown on petri dishes or culture flasks, they are rinsed 5 times with CMF-PBS for about five minutes. The incubation is carried out in a 37° bath and the petri dishes are resting on a temperature-controlled platform (Gerhardt type H52, Carl Bittmann, Basel) at 37° in between operations. Not much time is allowed between addition and removal of the buffer since some cell lines tend to come off more easily (transformed cells) and since some cells within one cell line may be lost (mitotic and early G_1 cells), which would lead to a distortion of the overall results, due to selection of certain cell populations. An additional five rinses with CMF-PBS containing EDTA* will round off almost all cells. Some cell types, that are particularly well attached, can be removed by repeated incubations with the CMF-PBS—EDTA solution for five to ten minutes. Once 70 to 90% of the cells are rounded off, they can be brought into suspension by directing a gentle stream of medium at an angle of 45° towards them. The pipette used for this operation should have a wide mouth (2 mm diameter) to avoid damage to the cells that are sucked back and forth through the opening. Bubbles and frothing lead to damage and rupture of cells. If cells were ripped off too early they often preserve their pseudopodes and shapes other than that of a sphere. Such cells are usually severaly damaged as evidenced by their impregnation with trypan blue.

1.3 Cells from Intact Tissue Sources

Tissue pieces are gently teased apart with dissecting needles and incubated in CMF-PBS containing EDTA and 30% calf serum for 10 to 30 minutes, depending on size and dissociability. After further mechanical dissociation in a tissue press, or by passing the pieces through cheese cloth, the resulting suspension is centrifuged under the conditions described above (600 g for 2.5 min) and resuspended. To remove large clumps, the suspension is passed through the cheese cloth and permitted to sediment in the centrifuge tube at 1 g for 3 minutes through 3—4 cm depth. The supernatant is now treated as described for suspension cultures above.

2. AGGLUTINATION PROCEDURE

Of the cell suspension, 0.09 ml (2×10^6 cells/ml) is added to a porcelain plate with 12 wells, each 5 mm deep and 20 mm in diameter; 0.01 ml of various lectin concentrations in PBS is added and mixed with the cell suspension by gently aspirating the suspension 5 times into a Pasteur pipette. After 1 minute, 20 μl are transferred into one of the wells of a microscope slide containing two or more wells, while the other well receives a control suspension of cells without lectins.

In order to avoid evaporation of the small sample volumes during the agglutination process, which may not only be irregular from sample to sample, but

---

*CMF-PBS—EDTA: the same as CMF-PBS except that 0.2 g EDTA (= versene) is added per litre.

which may amount to 40% in small samples over a 20 minute aggregation period, we have chosen the hanging drop procedure, where the sample can be kept to less than 20 μl without appreciable loss of volume, if done properly. The slide is turned upside down and firmly attached to the temperature block adjusted to 22°. The temperature block is mounted on a rocking platform driven by 2 motors. One motor provides a circular motion of 20 r.p.m. with a circle diameter of 70 mm. The other motor goes simultaneously through 20 rocking motions per minute from an angle of +22.5° via − 22.5° to + 22.5°. The two motions have to be synchronous to guarantee a standardization of the shear forces that act on the agglutinating cells in the hanging drop preparation. The samples are scored after 4 minutes of mixing.

If no temperature block and no rocker is available, agglutination can be carried out by rocking the inverted drop slides by hand, or, as pointed out earlier (3), in plastic immunologic microtitre plates or in 3 ml samples in regular 3.5 cm diameter plastic culture dishes. Standardization of mixing and of the temperature is, however, poor and the results are less reliable.

Optimal incubation times, as well as temperature, buffers, mixing conditions, cell densities, etc. do differ between different lectins and should be determined for each individual lectin by careful kinetic analysis. For concanavalin A, e.g. where agglutination is a slower process than with wheat germ agglutinin, an optimal time lies between 10 and 15 minutes, and temperatures are very critical.

From different degrees of agglutination (scoring discussed below) at different lectin concentrations, the concentration at which half maximal agglutination is reached, can be determined for the chosen standard time.

From several time points initial velocities can be extrapolated. They are very hard to obtain due to poor accuracy at early time points, which require large numbers of determinations at later time points. On the other hand, such initial velocity values give, of course, a more accurate picture of the kinetics of the reaction.

3. SCORING

The degree of aggregation can be scored by several methods (3). We have examined the following procedures and considered them as unsuitable: (a) Examination by eye of the degree of granulation of the flocculated cell suspension may be clinically useful but is not sufficiently quantitative. Furthermore the size of the aggregates seems to influence the final result as seen by eye, whereas for a kinetic treatment one prefers to know the total number of cells that enter aggregates, be that two-cell or 50-cell aggregates. (b) Gravity is used to measure erythrocyte aggregation in whole blood from patients. Such sedimentation rates in capillaries at 1 g of lectin aggregated cells could not be determined accurately enough since the boundary between aggregated cells and single cells or cell-free medium turned out to be too poorly defined and inaccurate (3). (c) Nephelometric determination of the amount of light scattered measures again aggregate size rather than the number of cells that enter an aggregate. The same criticism holds for measurements of the light absorbed or transmitted, since the size of the cell clumps was found to depend on many extraneous factors like pH, salt concentration in general, specific ions like Ca^{2+} or Mg^{2+}, serum proteins, small organic components like spermin, spermidin, etc. In certain concentration ranges these components may not at all alter the total number of cells that have agglutinated, while influencing strongly the clump size that can withstand the shear forces during the agglutination assay.

We have chosen a method whereby the ratio of single cells to agglutinated cells is determined by counting those two types of cells with the microscope (3) or the Coulter counter.

3.1 Microscope Method

The advantage of this procedure over the Coulter counter procedure is the fact that dead cells (polygonal, no nucleus or even ghosts) and large fragments of cells are not counted since they can be visually distinguished from undamaged cells after a short training time. Furthermore both procedures eliminate the serious danger of measuring the size of the aggregate, which is, as pointed out in the introduction, often not dependent on the amount of lectin but on extraneous conditions like Ca^{2+} content, serum components, etc.

Since the periphery of the hanging drop may — and often does — have more cells in the aggregated state, 3 sample fields with a total of at least 100 cells in each are counted for single and aggregated cells, both in the periphery of the drop, and another 3 fields it the centre of the drop. A beginner preferably recounts the same field twice to check whether his differential counting can be repeated or whether his judgement still fluctuates somewhat.

The present scoring used in our laboratory is very slightly modified from our last description in *Methods in Enzymology* (3): 0—12.5% cells agglutinated (2 and more cells in clumps) = 0; 12.6—37.5% = (+); 37.6—56% = +; 56.1—69.0% = +(+); 69.1—79.0% = ++; 79.1—84.9% = ++(+); 85—89.9% = +++; 90.0—94.0% = +++(+); 95.0—100% = ++++. On a simplified scale 50% agglutinated cells = +; 75% = ++; 88% = +++; and > 95% = ++++. Such a scale, that is shifted towards the higher values, should avoid rating mediocre or poor agglutination as satisfactory.

3.2 Coulter Counter

The orifice of the counting chamber of earlier Coulter counter models were too small, and frequent clogging made the use of this convenient procedure prohibitive.

Before agglutination the number of single cells per ml are counted first. After agglutination the window discriminators are set such as to screen out all aggregated cells and count again single cells only. With a channel analyser attachment to the Coulter counter the number and size of the larger aggregates can be assessed in a reliable fashion. The total amount of agglutinated cells is, however, calculated from the decrease in single cells. The uncertainty therefore remains whether some cells might have been damaged or ruptured. This technical problem could introduce a serious error into the scoring.

For rapid kinetic determination of agglutination the Coulter counter has still a certain advantage over the microscope. The lag between removing the sample from the mixing platform, and the counting of all 6 fields under the microscope, may be too extensive and seriously influence the accuracy of the time points. Still, the drawbacks of the use of the Coulter pointed out above and in the first paragraph of this chapter on scoring should be kept in mind.

4. COMMENTS AND PITFALLS

4.1 Spontaneous Unspecific Agglutination

Transformed fibroblasts particularly have a tendency to agglutinate without lectins. Such background agglutinations should not simply be deducted from the measured agglutination with the lectin. They should rather be accepted as a warning, since the readiness for spontaneous agglutination can promote or potentiate agglutination with the lectin and create false positive results. Experiments, where more than one control cell suspension agglutinates to + or more, are discarded and cells are freshly prepared.

Yet another control for the specificity of the agglutination by the lectin can be carried out. The particular lectin dose, which was previously found to agglutinate to ++, can be incubated for 10 minutes with 0.02 M of the carbohydrate hapten. If after addition of the cell suspension to that preincubation, agglutination of the particular batch of cells is not completely inhibited, unspecific agglutination has to be considered, and those cells should be discarded.

4.2 False Positive Results with Cells that Normally Agglutinate Only Poorly

Untransformed fibroblasts agglutinate usually quite poorly. The most common problem for the inexperienced person is to find agglutinabilites for untransformed cells that are too high. This is in most of the cases due to rough handling of the untransformed cells during the preparation of a unicellular suspension. Thus removal of the untransformed cells, which have a tendency to stick better to the substratum, is sometimes and unfortunately done without keeping the cells warm, by prolonged incubation in CMF—PBS—EDTA, or by trying to dissociate cells that come off as cell sheets, with quick and careless aspirating with the pipette. A possible interpretation for false positive results under such circumstances may be the activation or the release of proteolytic enzymes in damaged cells.

4.3 Other Possible Problems

It is quite informative to set up a balance sheet for the total number of cells still present while removing and washing them. It is often observed, particularly by the beginner, that centrifugation leads to a loss of 10—40% of the number of cells originally in the culture dish. Such problems might easily lead to a selected population of cells, although it is tacitly assumed that a random population was assayed.

Many mixing procedures using automatic devices lead to an asymmetric distribution of the cells in the particular suspension. In the hanging drop procedure, for instance, many non-agglutinated cells can be found at the centre tip of the drop, mimicking a large clump. Gentle tapping of the microscope edge towards a hard object will disperse non-agglutinated, but not agglutinated cells.

If cells are gently removed from the culture vessel, and if agglutinations are done for a while, results become remarkably reproducible and reliable. Standard deviations for the determination of lectin concentrations necessary for half maximal agglutination can be between 20—25%.

5. AN EXAMPLE OF THE USE OF THIS AGGLUTINATION METHOD

Soon after the purification of the wheat germ agglutinin (6), and before it was found that concanavalin A had — as much as wheat germ agglutinin — a tendency to agglutinate transformed cells more than untransformed cells (7), we felt it important to correlate loss of certain growth control properties with the alterations in the cell surface monitored by agglutinability.

Saturation densities in culture under well-defined conditions (same amounts of serum, optimal pH, etc,) seemed to correlate with the degree of tumourogenicity of 3T3 mouse fibroblast cells and their transformed derivative lines (8). It was, therefore, important to compare the degree to which these various cell lines agglutinated with the loss of growth control (high saturation densities). Such a correlation was indeed found with wheat germ agglutinin and is shown in Figure 1. A similar correlation was observed for the agglutinability with concanavalin A. While these correlations need not be absolute and leave open a series of critical questions, they encouraged us at the time to amplify the evidence for the relation of loss of growth control and surface architecture.

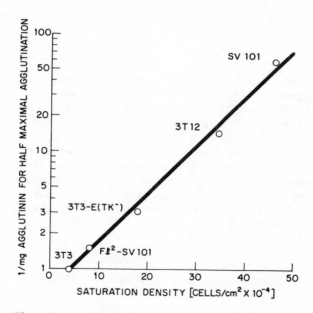

Figure 1. Correlation between agglutinability (ordinate) with wheat germ agglutinin and loss of growth control in culture (saturation density on abscissa). Agglutinability changes are similar for concanavalin A. SV101 is a SV40 virus transformed 3T3 line, 3T12 is a spontaneously transformed 3T3 line; 3T3-E (TK⁻) is a thymidine kinase-less 3T3 mutant and Fl²-SV101 is a variant of the SV101 transformed line which has regained its property to grow to low saturation and simultaneously the surface has returned to poor agglutinability. [Reproduced with permission of the copyright holder, from reference (9).]

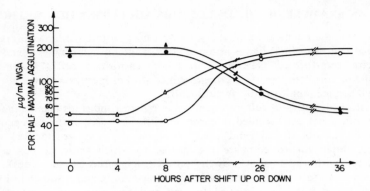

Figure 2. Appearance and disappearance of surface alteration monitored with wheat germ lectin (ordinate) after shift in temperature of
ts-3 transformed BHK cells. ▲ — — — ▲ clone 7-C and ● — — — ● clone 1
shifted from the non-permissive temperature (39°) to the permissive
temperature (32°) at time 0. △ — — — △ clone 7-C shifted from the
permissive temperature (32°) to the non-permissive temperature (39°)
at time 0. (From Eckhart et al. (10))

Among the oncogenic viruses some temperature-sensitive mutants could be
found which permit the expression of the transformed state at the low permissive
temperature. On the other hand, cells carrying this virus behaved at higher
temperature, where the viral gene product could not be expressed any more, as
normal cells that are subject to normal growth control and also normal
morphology. If the transformed properties of a cell line could be shifted from the
normal to the transformed state, the obvious question arose, whether the surface
properties monitored by lectin agglutinability could also be shifted simultaneously.
Figure 2 demonstrates that this was the case for a cell transformed with an
oncogenic DNA virus (10), while similar data were found for cells transformed with
an oncogenic RNA virus (11).

6. ACKNOWLEDGEMENT

This work has been supported by grant No. 3.1330.73SR of the Swiss National
Foundation.

7. REFERENCES

1. Horwitz, A. B., Hatten, M. E. and Burger, M. M. (1974) 'Membrane fatty acid replacements
   and their effect on growth and lectin induced agglutinability', *Proc. Nat. Acad. Sci. U.S.A.*,
   71, 3115—9.
2. Rapin, A. and Burger, M. M. (1974) 'Tumor cell surfaces: General alterations detected by
   agglutinins', *Advan. Cancer Res.*, **20**, 1—91.
3. Burger, M. M. (1974) 'Assays for agglutination with lectins', in *Methods in Enzymology*,
   Vol. XXXII, Biomembranes, Part B eds. in chief Colowick, S. P. and Kaplan, N. O.
   (Academic Press), pp. 615—21.

4. Pollack, R. E. and Burger, M. M. (1969) 'Surface specific characteristics of contact inhibited cell line containing the SV 40 viral genome', *Proc. Nat. Acad. Sci. U.S.A.*, **62**, 1074—6.

5. Inbar, M. and Sachs, L. (1969) 'Structural differences in sites on the surface membrane of normal and transformed cells', *Nature*, **223**, 710.

6. Burger, M. M. and Goldberg, A. R. (1967) 'Identification of a tumor specific determinant on neoplastic cell surfaces', *Proc. Nat. Acad. Sci. U.S.A.*, **57**, 359—66.

7. Inbar, M. and Sachs, L. (1969) 'Interaction of the carbohydrate binding protein concanavalin A with normal and transformed cells', *Proc. Nat. Acad. Sci. U.S.A.*, **57**, 1418—25.

8. Aaronson, S. A. and Todaro, G. T. (1968) 'Basis for the acquisition of malignant potential by mouse cells activated *in vitro*', *Science*, **162**, 1024—6.

9. Burger, M. M. (1970) 'Changes in the chemical architecture of transformed cell surfaces' in *Permeability and Function of Biological Membranes*, eds. Bolis. Katchalsky, Keynes, Loewenstein and Pethica (North Holland Publishing Co.), II/5, pp. 107—19.

10. Eckhart, W., Dulbecco, R. and Burger, M. M. (1971) 'Temperature-dependent surface changes in cells infected or transformed by a thermosensitive mutant of Polyoma virus', *Proc. Nat. Acad. Sci. U.S.A.*, **68**, 283—6.

11. Burger, M. M. and Martin, G. S. (1972) 'Agglutination of cells transformed by Rous sarcoma virus by wheat germ agglutinin and concanavalin A', *Nature (New Biol.)*, **237**, 9—12.

CHAPTER 29

Quantitation of Concanavalin A-induced Lymphocyte Agglutination by Optical Microscopy

FRANCIS LOOR

The degree of agglutinability of lymphocytes by lectins is usually evaluated in a conventional way, i.e. by leaving the cells to sediment in tubes or wells after their treatment with a series of twofold dilutions of the lectins and by inferring the end point of agglutination from a macroscopic examination of the cell sediment (1). However, the limit at which all cells are still agglutinated is not always clear cut and a series of intermediate stages between full agglutination and no agglutination might be found extending over a few tubes.

1. PRINCIPLE

Since it is at the limiting lectin dose that one can expect a maximal effect of any other factor influencing the agglutination process, it was necessary to develop a more precise quantitation of the degree of cell agglutination. This was realized by microscopic examination of the lectin-treated cells either in a semi-quantitative way or in a fully quantitative way. Here this simple method is described briefly for concanavalin A-induced lymphocyte agglutination, though it works with other lectins as well (2) and should be valid for any type of cell.

2. METHOD

2.1 Preparation of Cell Suspensions

Cell suspensions are prepared from various mouse lymphoid organs (spleen, lymph nodes or thymus). The lymphoid tissues are either teased with needles and

267

forceps or pressed through a screen mesh to release cells in cold medium kept on ice. A suitable medium for cells that do not have to be put in culture afterwards is Dulbecco's PBS (Ca^{2+} and Mg^{2+} free) supplemented with 0.5% bovine serum albumin (Sigma, St. Louis, Missouri) and adjusted to pH 7.2.

The cells are filtered through double-layered cotton gauze to eliminate cell clumps. They are washed by 3 to 4 cycles of centrifugation ($200-300\,g$, 10 min) in the refrigerated centrifuge and resuspension of the pellet in cold fresh medium.

Finally they are resuspended at a concentration of 25×10^6 cells/ml in fresh medium. Trypan blue exclusion test always give values higher than 90—95% viability at this stage.

2.2 Agglutination Reaction Procedure

The concanavalin A is dissolved in physiological saline (0.9%) containing 1 mM $MnCl_2$ and 1 mM $CaCl_2$. Aliquots of 0.1 ml of the cell suspension are then mixed with 0.1 ml of Con A at the desired concentration in plastic culture tubes (Falcon Plastics No. 2054, 12 mm diameter, round-bottom shape). The tubes are incubated for 30 min at $37°$ in a water bath, during which time they are occasionally gently shaken or rotated by hand, just enough to keep the cells in suspension and not allow them to settle down. The cells are then centrifuged ($200-300\,g$, 10 min), resuspended in 1.0 ml medium, pelleted again and finally resuspended (by hand shaking) in $50-100\,\mu l$ of medium (2 drops from a Pasteur pipette). One droplet of the cell suspension is then put on a microscope slide, covered with a coverslip and the small chamber with the cell suspension is sealed with nail polish and the cells are immediately observed under the phase contrast microscope.

2.3 Quantitation

The cell suspension can be examined at rather low magnification (100—500 times) to get a semi-quantitative evaluation of the degree of agglutination. This can be represented by the usual 1 + to 4 + designation and is sufficient for many experiments. An example is shown in Figure 1.

A more quantitative estimate of the agglutination reaction can be obtained by counting at higher magnification (500—1250 times) the relative numbers of free and clumped cells, a cell clump being an aggregate of at least 4 cells. A few thousand cells per sample should be counted to get a good quantitation of the degree of agglutination. The semi-quantitative estimate of cell agglutination is of course much faster but less accurate and much more subjective than when cells are counted individually. A 4 + agglutination usually corresponds to more than 80% of the cells being clumped; a 3 + agglutination to a 50—80% cell clumping; a 2 + agglutination to a 30—60% cell clumping and a 1 + agglutination to 20—40% of the cells being clumped. By the semi-quantitative method there appears to be no background agglutination in the absence of lectins. However, by quantitative microscopic examination, one finds some background cell aggregation which is variable from organ to organ and from mouse to mouse. This background clumping usually is not more than a few percent, but it may reach 15% in some cases. It should be pointed out here that the background of lymphocyte clumping is very

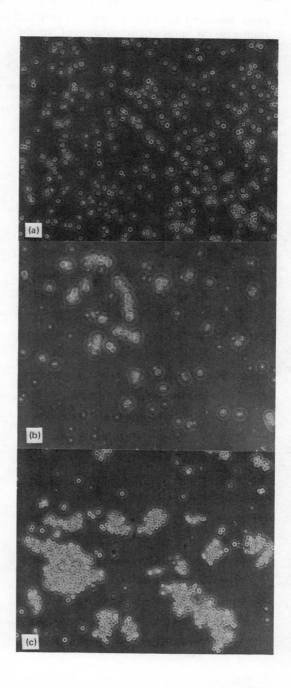

Figure 1. Microscopic evaluation of the degree of agglutination by concanavalin A. (a) No agglutination; (b) ++; (c) ++++. Cells were activated mouse thoracic duct lymphocytes. The Con A dose was 5 μg/ml. In sample (a) there was 10^{-4} M cytochalasin B in the medium and in sample (b) there was 10^{-2} M NaN$_3$. Both cytochalasin B and NaN$_3$ were found to be inhibitors of the Con A-induced lymphocyte agglutination (2)

much increased by some batches of foetal calf serum, that are commonly used in studies on the mitogenic stimulation of lymphocytes by concanavalin A.

3. REFERENCES

1. Kabat, E. A. and Mayer, M. M. (1967), *Experimental Immunochemistry*, 2nd edn. (G. C. Thomas, Springfield, USA).
2. Loor, F. (1973) 'Lectin-induced lymphocyte agglutination. An active cellular process?', *Exp. Cell Res.*, **82**, 415–25.

CHAPTER 30

A Microhaemagglutination Inhibition Assay for Concanavalin A Receptor Activity

DAVID F. SMITH
EARL F. WALBORG, JR.

1. PRINCIPLE OF THE METHOD

Agglutinins are multivalent agents which bind to specific cell surface determinants and induce cytoagglutination by cross-linking receptors on different cells. Several different classes of agglutinins have been identified: e.g. the haemagglutinating myxoviruses, antibodies against MN blood group antigen and lectins, plant proteins which interact with specific saccharide determinants. Cell surface macromolecules which possess receptor activity for these agglutinins have been isolated in soluble form (1–5). These soluble macromolecules have been assayed for agglutinin receptor activity by their ability to inhibit agglutinin-induced

271

cytoagglutination. Inhibition of cytoagglutination depends on competition between the cell-bound and soluble receptors for the available agglutinin. This principle provided the basis for a method to assay concanavalin A (Con A) receptor activity of saccharides, glycopeptides and glycoproteins (6).

2. MATERIALS

2.1 Concanavalin A

This lectin was purchased as a crystalline (2x) product from Miles Laboratories, Inc., Kankakee, Ill. or Miles Yeda, Rehovot, Israel. The Con A was supplied and stored in saturated NaCl. Alternately the Con A was purified by the method of Agrawal and Goldstein (7). A stock solution (0.4 mg Con A/ml) was prepared in calcium- and magnesium-free phosphate-buffered saline, pH 7.4 (CMF-PBS). This buffer (8) had the following composition: 8.0 g NaCl, 0.2 g KCl, 1.65 g Na_2HPO_4 (anhydrous), 0.2 g KH_2PO_4, diluted to 1 litre with water. The concentration of Con A was determined spectrophotometrically using an extinction coefficient of 13 for a 1% solution at 280 nm in a 1 cm path length. The stock solution (0.4 mg Con A/ml) was routinely prepared fresh 12 to 20 hours prior to use in the assay and stored at $4-8°$. On the day of the assay the stock solution was removed from the cold and allowed to warm to $22°$. Dilutions of the Con A stock solution (1:2, 1:3, 1:4, 1:6 . . . 1:192) were prepared in CMF-PBS.

2.2 Guinea Pig Erythrocytes

Guinea pig erythrocytes, which are highly agglutinable by Con A, were utilized in the assay. Guinea pigs of the American Slick or Shorthair strains have been utilized in these studies. No significant variations between these strains or between individuals within a strain have been observed. Presumably other guinea pig strains would also be suitable for use in this assay. Approximately 2 ml of blood was obtained by heart puncture and immediately diluted with 0.5 ml of Alsever's solution (20.5 g D-glucose, 8.0 g Na citrate $·2H_2O$, 0.55 g citric acid and 4.20 g NaCl diluted to 1 litre with water). The diluted blood, stored at $4-8°$, could be utilized for up to 3 days after collection. On the day of the assay approximately 1 ml of the blood was diluted to 15 ml with CMF-PBS. The erythrocytes were collected by centrifugation (250g, 10 min) and washed three times in a graduated centrifuge tube with 15 ml of CMF-PBS. The volume of the packed (250g, 10 min) erythrocytes was measured and diluted with CMF-PBS to obtain a suspension containing 1% (v/v) erythrocytes.

Rabbit erythrocytes which are also highly agglutinable by Con A may be utilized, however sufficient data or experience with these cells have not been accumulated to warrant inclusion herein. Using the method described here, erythrocytes from rat (Sprague Dawley strain) or human were only agglutinable by high concentrations (130 μg/ml) of Con A (9)*.

---

*Reference 9 contains an error in Table II on p. 863. The Con A concentrations should be >130, >70, >70, >70 and 0.5 μg/ml, reading from top to bottom.

2.3 Preparation of Test Solutions of Saccharides, Glycopeptides and Glycoproteins

Saccharides were prepared as 25 mM solutions in CMF-PBS. Glycopeptides and glycoproteins were prepared at concentrations of 3 to 6 mg/ml in CMF-PBS. Dilutions (1:2, 1:3, 1:4, 1:6 . . .) of these stock solutions were prepared in CMF-PBS. Ovalbumin glycopeptide, prepared according to Montgomery and coworkers (10), was included in each assay to serve as an internal standard for Con A receptor activity and to check the reproducibility of the assay.

2.4 Microtest Tissue Culture Plates

The assays were performed in Microtest tissue culture plates (No. 3034 Microtest Plates, Falcon Plastics, 1950 Williams Dr., Oxnard, Calif.). These plates were utilized as supplied and discarded after use. Cleaning and reuse of the plates was not found to be satisfactory. The plates have 60 separate wells arranged in 6 rows. Each well has a volume of approximately 15 μl.

2.5 Inverted Microscope

An inverted (tissue culture) microscope was utilized. A Zeiss UPL or similar type microscope is suitable. A total magnification of 100 x was utilized to visualize the contents of the wells of the Microtest plates.

3. ASSAY OF SPECIFIC HAEMAGGLUTINATION ACTIVITY OF CON A

To each well of a Microtest plate was added 5 μl of CMF-PBS. Following this 5 μl aliquots of each of the Con A serial dilutions were placed in separate wells of the Microtest plate. The Con A dilutions were added in the order of increasing concentration to avoid significant cross-contamination of the Con A dilutions. Several millilitres of the erythrocyte suspension (well mixed) were poured into a small beaker and 5 μl aliquots of the suspension added to each well of the Microtest plate. Care should be taken to obtain a uniform suspension of erythrocytes when taking aliquots. Erythrocytes were added to the wells in the order of increasing Con A concentration. The Microtest plates were covered and allowed to stand for 30 min at 22°. All pipetting was performed using an Eppendorf 5 μl micropipette with disposable plastic tip. After 30 min the end point of the assay was determined by microscopic examination.

The wells of the Microtest plates are in the shape of truncated cones, the smallest cross-sectional area being at the bottom of the plate. Non-agglutinated erythrocytes settle to the bottom of the well and concentrate at the junction of the side and bottom of the well. When viewed microscopically, one observes a red ring of erythrocytes enclosing a less dense, evenly distributed layer of erythrocytes. Agglutinated erythrocytes settle as clumps of erythrocytes with incomplete ring formation or none at all. The end point of the haemagglutination assay was defined as the lowest concentration of Con A which demonstrated haemagglutination activity, i.e. the lowest concentration of Con A in which complete ring formation was not observed. Haemagglutination titres were routinely performed in triplicate and the average titre utilized to calculate haemagglutination activity. End points in triplicate assays rarely differed by more than one dilution, consequently maximum variability was approximately ±33%.

The specific haemagglutination activity of Con A preparations was determined by defining one unit of haemagglutination activity as the amount of Con A per well at the end point. The specific activity was expressed as haemagglutination units (HAU) per mg Con A. The specific haemagglutination activity of Con A, obtained commercially or purified by the method of Agrawal and Goldstein (7), was 22,000 ± 2000 HAU/mg.

4. ASSAY OF CON A HAEMAGGLUTINATION INHIBITORY ACTIVITY

Aliquots (5 µl) of the test inhibitor dilutions were added to the Microtest plates. Immediately thereafter, aliquots (5 µl) of a Con A solution (600 HAU/ml), i.e. 3 HAU, were added to each well of the Microtest plate. This Con A solution (600 HAU/ml) was prepared by appropriate dilution of the Con A stock (0.4 mg/ml) solution. The dilution factor was calculated on the basis of the haemagglutination assay. The Con A aliquots were added in the order of increasing inhibitor concentrations to avoid contamination of the Con A solution with inhibitors. The plates were covered and allowed to stand at 22° for 1 h, at which time 5 µl aliquots of the erythrocyte suspension were added to each well. After another 30 min the inhibition of haemagglutination was evaluated microscopically. The end point of the haemagglutination inhibition assay was defined as the lowest concentration of inhibitor which demonstrated haemagglutination inhibition activity, i.e. the lowest concentration of inhibitor in which complete ring formation was observed. Haemagglutination inhibition titres were routinely performed in triplicate and the average titre utilized to calculate haemagglutination inhibition activity. End points in triplicate assay rarely differed by more than one dilution, consequently maximum variability was approximately ±33%. Specific haemagglutination inhibition activities were calculated by defining one unit of haemagglutination inhibition activity as the amount of inhibitor per well at the end point. The specific activity was expressed as haemagglutination inhibitor units (HAIU) per mg inhibitor.

The use of small volumes (5 µl) of test inhibitor dilutions permitted the determination of specific, Con A inhibitory activities in triplicate at inhibitor concentrations as high as 1 mg/ml with as little as 100 µg of test inhibitor. This sensitivity is extremely valuable when working with cell surface glycopeptides which are generally obtained in less than milligram quantities.

For direct comparison of the haemagglutination inhibition activities of several different inhibitors, it is best to perform as many of the assays as possible using the same Con A solution (600 HAU/ml) and erythrocyte suspension. In addition the effect of the inhibitor on the agglutinability of erythrocytes, in the absence of Con A, should also be investigated. In certain cases test substances have exhibited haemagglutination activity at high concentrations, but haemagglutination inhibition activity at lower concentrations. In the case of certain sialoglycopeptides this may be related to their high charge density, a property which permits the formation of ionic bridges between cells.

The Con A haemagglutination inhibition activities of several known inhibitors of Con A are reported in Table I. The specific inhibitory activities of the monosac-

TABLE I. Specific activity of inhibitors of Con A-induced haemagglutination (6)

Inhibitor	Specific inhibition activity[a]	
	HAIU/mg	HAIU/μmole
Methyl α-D-mannopyranoside		300
p-Nitrophenyl α-D-mannopyranoside		300
Methyl α-D-glucopyranoside		60
Ovalbumin	130	6000[b]
Ovalbumin glycopeptide[c]	2800	6200[d]

[a]Specific activities are expressed as haemagglutination inhibition units (HAIU). One HAIU is defined as the minimum concentration of inhibitor necessary to inhibit completely 3 haemagglutination units of lectin.
[b]Assuming a molecular weight of 45,000 daltons (11).
[c]Prepared by the method of Montgomery and coworkers (10).
[d]Calculated using a molecular weight of 2210 daltons, determined by gel filtration (12) and aspartic acid composition.

charides correlated well with the reported specificity of Con A (13). Ovalbumin, a glycoprotein containing only one oligosaccharide moiety showed Con A receptor activity. Ovalbumin glycopeptide which contains the entire oligosaccharide moiety exhibited a 20-fold increase in specific activity (HAIU/mg) over the parent glycoprotein, indicating that the oligosaccharide moiety bears the Con A receptor activity. As indicated by the specific activities expressed on a molar basis (HAIU/μmole), the Con A receptor activity of ovalbumin was quantitatively recovered in the glycopeptide.

5. EXAMPLE OF APPLICATION: PURIFICATION OF HEPATOMA CELL SURFACE GLYCOPEPTIDES WITH CON A RECEPTOR ACTIVITY

The purification of Con A receptor sites from the surface of AS-30D rat ascites hepatoma cells was monitored using this haemagglutination inhibition assay (6). Sialoglycopeptides were released from the cell surface by incubation of intact cells with papain. The sialoglycopeptides were resolved into three molecular weight classes (C-SGP-A, C-SGP-B, and C-SGP-C) by gel filtration on 'Sephadex' G-50. The lowest molecular weight component (C-SGP-C) was digested with pronase to minimize variation in peptide chain length and to obtain the limit glycopeptides. The product of pronase digestion (P-SGP-C) was resolved by ion-exchange chromatography on diethylaminoethyl (DEAE)-cellulose into five carbo-hydrate-containing fractions (DC1 . . . DC5). Each of these glycopeptide fractions was assayed for Con A receptor activity and the results summarized in Table II. Con A receptor activity was present in C-SGP-A and C-SGP-C. Pronase digestion of C-SGP-C resulted in a 22% loss of Con A receptor activity. Loss of lectin receptor activity has also been observed after pronase digestion of human erythrocyte glycopeptides (4). Ion exchange chromatography of P-SGP-C resolved two glyco-peptide fractions, DC1 and DC2, which exhibited specific activities higher than the

TABLE II.   Purification of Con A receptor activity from cell surface sialoglycopeptides of rat ascites hepatoma, AS-30D (6)

		Con A receptor activity[a]	
Fraction	Yield of C-SGP[b] (mg/100 mg)	Specific activity (HAIU/mg)	Total activity (HAIU/fraction x 10^{-3})
C−SGP−A	40	450	18
C−SGP−B	15	< 70	
C−SGP−C	45	400	18
P−SGP−C	35	400	14
DC1	5	1000	5
DC2	5	1100	6
DC3	8	200	2
DC4	7	100	1
DC5	9	<70	

[a] Receptor activities are expressed as haemagglutination inhibition units (HAIU). One HAIU is defined as the minimum concentration of inhibitor necessary to inhibit completely 3 haemagglutination units of lectin.
[b] C−SGP represents the glycopeptide fraction cleaved from the surface of intact AS−30D tumour cells by papain.

parent fraction, P-SGP-C. The recovery of the Con A receptor activity from the ion exchange column was quantitative.

This method for assaying lectin receptor activity has been utilized to monitor the isolation of cell surface glycopeptides possessing potent and/or specific lectin receptor activities (6,14). These glycopeptides will be utilized in structural investigations designed to elucidate the molecular requirements for binding of a variety of lectins to the surface of normal and neoplastic cells.

6. ACKNOWLEDGEMENTS

This study has been supported by research grants from the National Cancer Institute (No. CA 11710) and the Paul and Mary Haas Foundation.

D.F.S. was a recipient of a Rosalie B. Hite Predoctoral Fellowship in Cancer Research.

7. REFERENCES

1. Kathan, R. H., Winzler, R. J. and Johnson, C. A. (1961) 'Preparation of an inhibitor of viral hemagglutination from human erythrocytes', J. Exp. Med., 113, 37−45.
2. Ohkuma, S. and Shinohara, T. (1967) 'MN blood-group active sialoglycopeptides released from human erythrocytes by pronase treatment', Biochim. Biophys. Acta., 147, 169−71.
3. Gardas, A. and Koscielak, J. (1971) 'A, B. and H blood group specificities of glycoprotein and glycolipid fractions of human erythrocyte membrane. Absence of blood group active glycoproteins in the membrane of non-secretors', Vox Sang., 20, 137−49.
4. Kornfeld, R. and Kornfeld, S. (1970) 'The structure of phytohemagglutinin receptor sites from human erythrocytes', J. Biol. Chem., 245, 2536−45.
5. Akiyama, Y. and Osawa, T. (1971) 'Isolation of glycoproteins from human erythrocytes possessing inhibitory activity against various phytohemagglutinins', Proc. Japan Acad., 47, 104−9.

6. Smith, D. F., Neri, G. and Walborg, E. F., Jr. (1973) 'Isolation and partial characterization of cell-surface glycopeptides from AS-30D rat hepatoma which possess binding sites for wheat germ agglutinin and concanavalin A', *Biochemistry*, **12**, 2111—8.

7. Agrawal, B. B. L. and Goldstein, I. J. (1965) 'Specific binding of cancanavalin A to cross-linked dextran gels', *Biochem. J.*, **96**, 23c—25c.

8. Cronin, A. P., Biddle, F. and Sanders, F. K. (1970) 'Wheat germ agglutinin conjugated with fluorescein isothiocyanate: Staining of normal and transformed Chinese hamster cells', *Cytobios*, **2**, 225—31.

9. Leseney, A. M., Thomas, M. W., O'Neill, P. A. and Walborg, E. F., Jr. (1974) 'Role of papain-labile, cell-surface glycopeptides in agglutination of rat erythrocytes by concanavalin A and wheat germ agglutinin', *Colloques Internationaux du Centre National de la Recherche Scientifique No. 221, Methodologie de la Structure et du Métabolisme de Glycoconjugués*, Editions du C.N.R.S., Paris, pp. 859—68.

10. Montgomery, R., Lee, Y. C. and Wu, Y. (1965) 'Glycopeptides from ovalbumin. Preparation, properties and partial hydrolysis of the asparaginyl carbohydrate', *Biochemistry*, **4**, 566—77.

11. Warner, R. C. (1954) In *The Proteins*, eds. Neurath, H. and Bailey, K. (Academic Press, New York, N.Y.), Vol. 2A, p. 443.

12. Bhatti, T. and Clamp, J. R. (1968) 'Determination of molecular weight of glycopeptides by exclusion chromatography', *Biochim. Biophys. Acta.*, **170**, 206—8.

13. So, L. L. and Goldstein, I. J. (1967) 'Protein—carbohydrate interaction. IX. Application of the quantitative hapten inhibition technique to polysaccharide—concanavalin A interaction', *J. Immunol.*, **99**, 158—63.

14. Neri, G., Smith, D. F., Gilliam, E. B. and Walborg, E. F., Jr. (1974) 'Concanavalin A and wheat germ agglutinin receptor activity of sialoglycopeptides isolated from the surface of Novikoff hepatoma cells', *Arch. Biochem. Biophys.*, **165**, 323—30.

CHAPTER 31

A Microagglutination Method Suitable for Kinetic Studies

H. P. SCHNEBLI

This method was developed to observe the progression of the agglutination reaction as a function of time and to determine the effect of mechanical shearing on agglutination (1).

1. PRINCIPLE

The method is a variation of the microagglutination assay originated by Severs (2). Cells are incubated with a large range of Con A concentrations in U-type microtitre plates. In contrast to the conventional microtitre assays, here the cells are kept in suspension by shaking throughout the experiment. At intervals the plates are scored for macroscopically observable cell aggregation.

2. MATERIALS AND EQUIPMENT

2.1 Equipment

The microtitre system (Cooke Engineering Co., Alexandria, Va.) is used. While it is convenient to have the complete set of microtitre items, only the Tayo Bussan mixer, an enlarging mirror for rapid visual scoring and U plates are essential.

2.2 Cell Suspensions

After harvesting and/or enzyme treatment, the cells are washed in Ca^{2+}, Mg^{2+} and glucose-free phosphate-buffered saline (PBS) and suspended in the same at twice the optimal cell density. Optimal densities of some cells are given in Table I; visual scoring is more difficult in assays with lower or higher densities. Erythrocyte suspensions are used the same day, all other suspensions within 2—3 hours (stored cold).

TABLE I.   Optimal cell densities for agglutination assays in microtitre plates (50 μl)

	Cell density (cells/ml)
Human erythrocytes (trypsinized, neuraminidase-treated)	10^8
Mouse thymocytes	5×10^7
TIMI (mouse leukaemia, *in vitro*)	2×10^7
L1210 (mouse leukaemia, ascites)	2×10^7
Ehrlich ascites carcinoma (mouse)	10^7
BHK 21 (hamster fibroblasts, *in vitro*)	10^7

3. METHOD

3.1 Setting Up of the Assay

Twenty-five μl each of a twofold serial dilution of Con A in PBS (3) are placed in U-type plates (from 1 to 12). Every second or third row in addition receives α-methyl-D-glucoside (5 μl of a 0.55 M solution in PBS) as a control for non-specific aggregation.

The reaction is initiated by the addition of 25 μl of the prepared cell suspensions to each well of the plate. With the calibrated droppers provided by Cooke this can be accomplished in less than 30 seconds. The plates are then covered with adhesive tape and shaken on the Tayo Bussan mixer for one minute at a time with one-minute intervals of rest, during which the plates are observed on the magnifying mirror and scored for macroscopically observable aggregation.

3.2 Scoring

Figure 1 shows a comparison of this method with the conventional microtitre techniques and illustrates the scoring with neuraminidase-treated human erythrocytes (1) as an example.

In the conventional technique the microtitre plates are left undisturbed for 60 minutes after the initial mixing. This allows the non-agglutinated cells to settle in form of a small knob at the bottom of the well while agglutinated cells do not settle (Figure 1, panel b).

If the cell suspension is agitated throughout the reaction, however, non-agglutinated cells remain in suspension while agglutinated cells appear at first as a granular

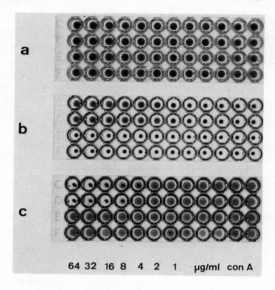

64 32 16 8 4 2 1 µg/ml con A

Figure 1. Comparison of conventional microtitre technique and the microagglutination method suitable for kinetic studies. Human erythrocytes treated with neuraminidase (10^8 cells with 50 U of *Vibrio cholerae* neuraminidase, Calbiochem, in one ml, $37°$, 60 min) are incubated with twofold serial dilutions of Con A as indicated. Rows E and F are controls and contain in addition 0.05 M α-methyl-D-glucoside. Panel a: beginning of the experiments. Panel b: the plate was left undisturbed for 60 minutes (conventional method). Agglutination occurs at Con A concentrations above 2 µg/ml (rows C and D, 1—6); non-agglutinated cells settle and form small knobs. Panel c: the plate was shaken for one minute at a time alternating with one-minute intervals of rest (see text) and photographed at 60 minutes. Agglutination occurs at Con A concentrations above 4 µg/ml (rows C and D, 1—5). Non-agglutinated cells appear as a homogeneous suspension

aggregation and later form large clumps (Figure 1, panel c). Agglutination is scored positive when the suspension appears distinctly more granular than the controls immediately adjacent (Figure 1, panel c, 4 µg Con A/ml represents the limit).

3.3 Variation: Measurement of Rapid Agglutination Reactions

Since the method described so far obviously is not applicable for agglutination reactions that occur within the first minute or two, the following variation was introduced: Twenty-five µl of *one* concentration are placed into each well of two rows of a U plate. One of the rows receives α-methyl-D-glucoside as above and the plate is now agitated on the Tayo Bussan mixer. Cells are added to one experimental and one control well (while shaking!) at regular intervals of 5 or 10 seconds. After all wells have received cells, the plate is placed onto the magnifying mirror and scored ten seconds after the last addition of cells. Scoring can be done with 'one look' so that the time required for agglutination to occur can be calculated easily from the number of intervals and the time delay of scoring.

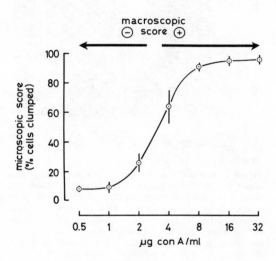

Figure 2. Comparison of macroscopic and microscopic scoring of agglutination. Neuraminidase-treated human erythrocytes were incubated with serial dilutions of Con A under the same conditions as in Figure 1, panel c. The proportion of agglutinated cells was determined microscopically after 60 minutes

3.4 Quantitative Interpretations

The validity of the macroscopic scoring was checked in several experiments by microscopical determination of the proportion of agglutinated cells. In Figure 2 such a comparison is shown for an experiment equivalent to the one shown in Figure 1. This and similar experiments not shown here establish that the visual scoring method determines quite accurately the midpoint of the agglutination titration curve. In other words, the macroscopically positive score, when first observed, indicates half maximal agglutination. Hence we assume that the time required for macroscopically observable agglutination to occur at any one concentration of Con A, roughly corresponds to the half time ($T\frac{1}{2}$) of the reaction at this particular Con A concentration.

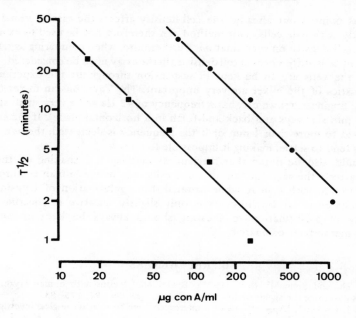

Figure 3. Time dependence of agglutination by Con A. The assays were done as described in the text. Each point is the average of at least 4 experiments and represents the time when macroscopic agglutination was first observed at the lectin concentration indicated. ○, L1210 cells; □, L1210 cells, neuraminidase treated

4. EXAMPLES

This method has originally been used to investigate the kinetics of erythrocyte agglutination (1), but more recently we have extended the studies to other cells. Figure 3 shows as an example the kinetics of agglutination of L1210 cells. It can be seen that the neuraminidase-treated L1210 cells react about 6.5 times faster than the untreated cells over the whole range of Con A concentrations tested. The slope of lg $T\frac{1}{2}$ against lg of the lectin concentration in both cases is nearly one, indicating a second-order reaction.

5. COMMENTS

The most serious disadvantage of this method is the need for visual scoring, which requires some practice. Differences in scoring among individual workers can be avoided if everyone checks his scoring microscopically, as illustrated in Figure 2.

Because of their colour, erythrocytes are easier to work with than other cells. Nucleated cells can be dyed (e.g. by addition of 0.005% of crystal violet to a cell suspension of $10^7 - 10^8$ cells/ml), yet dyed cells are not very suitable, as they are slightly more agglutinable than untreated cells.

As was pointed out already, the cell density affects the ease of visual scoring, particularly with pale cells. This method can therefore not be used to examine the effect of cell density on agglutination. Furthermore, when comparing agglutination of different cells, differences in cell densities in the assay must be considered.

Since the cells are to be kept in suspension throughout the experiment, the characteristics of the mixer are very important. The Tayo Bussan mixer (supplied by Cooke Engineering) we use, has a frequency of 42 Hz and a very small amplitude (0.3–0.4 mm sideways and back and forth in a horizontal plane). If the amplitude is increased to more than 1 mm or if the frequency is decreased, the non-agglutinated cells tend to settle, making it impossible to score.

It should also be noted that mechanical shearing (i.e. shaking on the mixer) severely affects the agglutination of some cells: e.g. native human erythrocytes do not agglutinate with Con A while agitated, but agglutination of trypsinized and neuraminidase-treated erythrocytes is only slightly sensitive to shearing (1). The sensitivity of agglutination to shearing should always be kept in mind when different methods are compared.

6. REFERENCES

1. Bächi, Th. and Schnebli, H. P. (1975) 'Reaction of lectins with human erythrocytes. I. Factors governing the agglutination reaction', *Exp. Cell Res.*, **91**, 175–83.
2. Severs, J. L. (1962) 'Applications of a microtechnique to viral serological investigations', *J. Immunol.*, **88**, 320–34.
3. Merchant, D. J., Kahn, R. H. and Murphy, W. H., Jr. (1964) in *Handbook of Cell and Organ Culture*, (Burgess Publishing Co., Minneapolis, Minn.), p. 217.

CHAPTER 32

Quantitation of Agglutination with an Automatic Particle Counter

MARK C. WILLINGHAM

1. PRINCIPLE

Although direct microscopic observation of agglutinated cells can give a rough estimate of the degree of cell clumping, small changes in agglutination are difficult to detect. We have therefore developed a method to allow quantitation of these agglutination processes. This method which depends on electronic discrimination of single and clumped cells, enables more sensitive measurements, removes some of the subjectivity of visually judging these results, and allows comparisons of results between different individuals. Any particle counter that allows sizing of cells can be used in this method. Another method utilizing similar instrumentation has been developed independently (see S. B. Oppenheimer, Chapter 33), but differs from our procedure in that disappearance of single cells is used as the criterion of agglutination. In our procedure the increase in clumped cells is added as an additional measurement.

2. METHOD

2.1 Cell Suspensions

Our studies have mainly dealt with cultured rodent cell lines, both normal (3T3-4, NRK) and transformed (MSV-3T3, MSV-NRK, L929, SV40-3T3) and mutants of these lines (3T3cAMP^{tcs}) (1, 3). These cells are routinely grown in Dulbecco—Vogt's modified Eagle's medium supplemented with 10% calf serum (Colorado Serum Co.) and penicillin—streptomycin (50 U/ml each) (Flow Laboratories Inc., Rockville, Md.) in a humidified atmosphere of 95% air—5% CO_2 at 37°.

Flasks (75 cm^l) containing subconfluent cells ($\sim$1—2 x 10^6 cells) were removed from the incubator. The medium was poured off and the flask was rinsed twice with 10—20 ml of phosphate-buffered saline (PBS) (Ca^{2+}-, Mg^{2+}-free) at room temperature.

Some cells (e.g. L929) are removable by this procedure alone or by gentle agitation by pipetting the medium in the flask over the cells with an automatic (piggyback) pipetter. Other cells (e.g. 3T3-4) require the use of EDTA to aid cell removal from the substratum. For consistency with all types of cells, we added 10 ml of PBS with 0.1 mM EDTA to each flask for 5 min at 23°. The cells were then removed by gentle shaking.

The cell suspensions were then spun in 40 ml concial centrifuge tubes (800 g, 5 min), the cell pellets resuspended in PBS (without EDTA) and aliquots counted with a Particle Data Inc. (Elmhurst, Ill.) (model no. 112TA) cell counter. This instrument was used for all the procedures described. Cells (1.5—2.0 x 10^5) were added in 3.0 ml PBS to 20 cm^2 (diameter 5 cm) plastic culture dishes (Falcon Plastics, Div. of BioQuest, Oxnard, Calif.) for the agglutination assay.

2.2 Agglutination Procedure

In these studies we mainly used concanavalin A (Con A) as the agglutinating agent, but this method applies to any agglutinin. Con A (Miles Laboratories, Inc., Kankakee, Ill.) was dissolved in saturated NaCl and stored at 25 mg/ml at 4°. Fresh dilutions of this stock solution were added to each dish of suspended cells to give final concentrations of 0—100 μg/ml, the added volume being less than 0.3 ml per dish.

Cells removed by EDTA treatment will stick to plastic dishes, whether Falconized (tissue culture grade) or not (bacterial grade). This stickiness seemed to be independent of Con A. If the cells were allowed to settle onto the plastic and then gently resuspended, they subsequently showed less adherence. Therefore, after the addition of Con A to the dishes, each was allowed to sit undisturbed for 3 min. Then the cells were resuspended by gentle pipetting with a Pasteur pipette. The dishes were placed on a rack in a shaker bath (Fermentation Design, model no. W3325-1) at 23° and swirled at 25 r.p.m. for 10 min. The contents of each dish were then poured into an Accuvette counting vial containing 17 ml of 'Isoton' solution (Coulter Electronics, Inc., Hialeah, Fla.) and assayed as described below.

2.3 Counting Procedure

The automatic particle counter we used (Particle Data Inc. cell counter, model 112TA) allows the discrimination of particle size on the basis of analysis of pulses produced as particles pass through an orifice. Large particles produce larger pulses than small particles. By adjusting discrimination windows (threshold settings) one can choose the particle size that is counted. Since our counter was not equipped with automatic monitoring of multiple channels, we chose to measure selected windows of varying size manually.

First we adjusted lower threshold and current-gain values to produce a half maximal cell number count for single cells (standard mean count, SMC), at approximately ¼ of the maximal pulse height (all counts normalized manually for each sample to adjust for electrolyte differences). Cultured cells (L929, 3T3) have a cell volume in the range of 3500—7000 μm^3. The settings used to produce this range at an SMC of ¼ the maximal pulse height were a current of 1/2 and gain of 2—1/8.

Agglutinated cells produce pulse heights higher than single cells, roughly proportional to the size of the clump. Therefore, two cells clumped together produce a pulse height approximately twice the size of a single cell, and thus are counted in a window that includes a threshold setting of 2 x SMC for single cells. The SMC for single cells is, for the purpose of our studies, assumed to be roughly the same as the SMC of assays in the absence of agglutinin, since the assay conditions were selected to minimize spontaneous agglutination. That this assumption is reasonable has been confirmed by direct microscopic observation of the cell suspensions.

The procedure for quantitation, then, involved counting the number of particles above a fixed lower threshold setting, and then varying the lower threshold to higher and higher values. Thus, counts were made with a window of 50 →, 100 →, 200 →, and so forth. These counts were expressed graphically as a percentage of the total number of particles in the sample, assumed to be the number of real particles in the broadest window (50 →). An example of the curves obtained in the presence or absence of an agglutinin is shown in Figure 1.

The curve counts taken in the absence of agglutinin (presumed to represent single cells) was used to determine an SMC, and thus twice the SMC could be determined (see Figure 1). As an example, an SMC was determined from the curve in Figure 1 as a lower threshold setting of 240, so that twice the SMC would be 480. The counts at this 2SMC point could then be determined. The curve produced in the presence of an agglutinin are plotted similarly.

2.4 Calculation of Agglutination Index

Since the values at 2SMC should give some indication of clumps composed of two cells, this point was arbitrarily chosen to represent agglutination. The agglutination index is derived from these values as follows:

$$A.I. = \frac{(A) - (B)}{(B)} \times 100$$

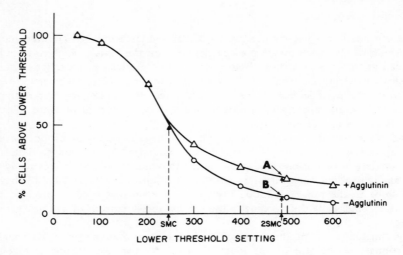

Figure 1. Particle size distribution curves in the presence and absence of agglutinin. Counts were taken with lower threshold window settings of 50 →, 100 →, and so forth. These values were then expressed as a percentage of the total number of particles in the sample count (per 0.5 ml counting volume, assumed to be the same as the 50 → count). These were plotted against the lower threshold setting of each window and used to create the curve shown in the absence of agglutinin. The half maximal (50%) point was determined from this curve, and the lower threshold value at this point (SMC) was found. This would represent the volume of a single cell if computed with the appropriate constant ($K = 24.5$ in this case, standardized with mulberry pollen). The 2SMC threshold value was then found and a second curve in the presence of agglutinin after a 10 min assay was constructed, each point being expressed as a percentage of the total particle number in that particular sample (in the presence of agglutinin). The curve derived in this fashion is shown and is considerably different from the control curve (in the absence of agglutinin), particularly at the 2SMC point. From these two curves the percentage values at the 2SMC threshold setting are read. In this case the value in the absence of agglutinin (B) was 10.0 and the value in the presence of agglutinin (A) was 21.0. The agglutination index (A.I.) for this reaction is thus 110 (for calculations see text)

where A represents the percentage of cells at 2SMC in the presence, and B in the absence, of an agglutinin. The agglutination index represents the percentage increase (at 2SMC) above the control (without agglutinin) curve due to the presence of the agglutinin. Thus, the higher the degree of agglutination, the higher the A.I. Multiple counts at each threshold value can be taken to give some indication of the range of values, and thus the range of the A.I.

3. EXAMPLES OF AGGLUTINATION MEASUREMENTS

Figure 2 shows the agglutination indices measured for highly agglutinable transformed cells, and for 'normal' cells with low agglutinability, in the presence of concanavalin A. The amount of agglutination seen visually with transformed cells is

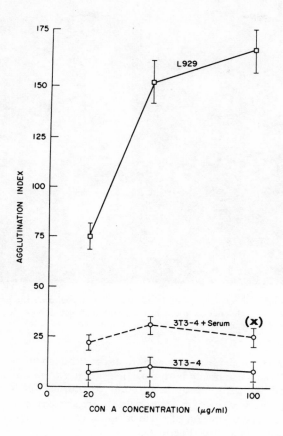

Figure 2. Examples of agglutination indices in normal
and transformed cultured cells. The agglutination
indices of 3T3-4 (normal) and L929 (transformed)
cells were determined as described, in assays containing 20, 50 or 100 μg/ml Con A. The range of values of
the individual counts allowed the calculation of the
range of values for the agglutination indices as shown
by the bars at each point. L929 cells are highly
agglutinable. Treatment of the L cells with 1 mM
Bt_2cAMP decreases their agglutinability (shown on
this figure at 100 μg/ml Con A as (x)). 3T3-4 cells, on
the other hand, have very low agglutinability under
the conditions of this assay. Addition of new serum to
these cells (15 min before the assay) causes a fall in
cAMP levels and a small but definite increase in
agglutinability

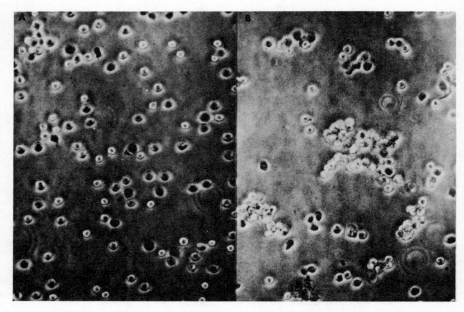

Figure 3. Appearance of suspensions of L929 cells after an agglutination procedure. L929 cells were assayed by the agglutination procedure described in the absence (A) or presence (B) of 100 μg/ml con A. This degree of agglutination corresponded to an A.I. of 160

shown in Figure 3. Small changes in agglutination, although not evident microscopically, are detected quantitatively by the particle counter. An example of such a small change induced by addition of new serum to 'normal' cells is also shown in Figure 2. Other subtle changes in agglutinability are also easily determined with this method: Cyclic AMP inhibition of agglutinability (3) is readily evident as a decline in the agglutination index (see Figure 2).

4. COMMENTS

4.1 Cell Cultures

The conditions of cell culture are an important consideration in the reproducibility of agglutination measurements.

Medium changing in culture vessels should be frequent, at least every two days, particularly in the days just preceding assays. Some cells (1) are sensitive to recent medium addition, so the last medium change should be 12—24 hours before an assay. Trypsinization can alter the properties of cells, not only immediately after exposure (1—2 hours), but for days after exposure. Trypsin subculturing results in altered morphology (2) for the first cell cycle after planting, at least until the next mitotic event. In confluent (G_1-arrested) normal cells, this period can extend for 24—48 hours. Agglutination assays should therefore be postponed until the cells regain normal morphology and motility. Cell density can also affect agglutination, both lectin mediated and spontaneous (3). Therefore, we routinely measured agglutination at comparable subconfluent densities.

4.2 Important Variables

The conditions for the agglutination assays described here were selected to maximize the difference in Con A agglutinability between transformed and normal cells. More vigorous shaking or longer incubation times result in increased agglutination of all cell types, including normal cells, and increase the spontaneous aggregation in the absence of lectin. Spontaneous aggregation was also high if some cells were allowed to grow to confluency. Hence, the most reproducible assays were at subconfluent cell densities.

Among the factors that determine the conditions for seeing the largest differences in agglutination between normal and transformed cells, two main considerations seem particularly important. First, the number of collisions determine the extent of agglutination. This is determined mainly by a) the length of time of the agglutination assay, longer times allowing more collisions, and b) the number of cells in a particular volume of suspension, more cells allowing more collisions. Secondly, the velocity of the collisions are important. Low-velocity agitation seems essential to show the difference between normal and transformed cells. Too high an agitation rate results in higher-velocity collisions which produce agglutination even in normal cells. Thus, high-velocity agitation with large numbers of cells for long periods produces extensive agglutination in *both* normal and transformed cells. It is only when low-velocity, short-interval, low cell number conditions exist that transformed cells maximally show their much higher Con A-mediated agglutinability.

4.3 Patterns of Agglutination

Generally, agglutination in suspension shows two patterns of cell clumping. One involves the formation of clumps composed of two to five cells/clump. The other involves the formation of large aggregates of 50 or more cells. The reasons for these patterns are unclear. Some cell lines show a preference for a particular pattern while others show a mixture of the two. This quantitation procedure, however, detects both types equally well.

Large cell clumps result in a more significant decrease in the *total* number of particles. The increase in counts at 2SMC would thus not have to represent a very large number of particles to produce a large increase in the A.I., since the values are expressed as the *percentage* of the total number of particles. Small cell clumps may produce little or no decrease in the total particle number, since apparently the counter may recognize a two-cell clump as two separate cells at a low threshold setting. These small clumps produce a large increase at the 2SMC setting, however. Thus, both large and small cell clumps are detected by this method.

5. Abbreviations

EDTA = ethylenediamine tetraacetic acid
MSV = Murine sarcoma virus

NRK = normal rat kidney cells
3T3—4 = NIH Swiss 3T3 mouse embryo fibroblasts
L929 = methylcholanthrene-induced murine tumour cells
SV40 = Simian virus 40

6. REFERENCES

1. Willingham, M. C., Carchman, R. A. and Pastan, I. (1973) 'A mutant of 3T3 cells with cyclic AMP metabolism sensitive to temperature change', *Proc. Nat. Acad. Sci., U.S.A.,* **70**, 2906—10.
2. Willingham, M. C. and Pastan, I. (1975) 'Cyclic AMP and cell morphology in cultured fibroblasts: Effects on cell shape, microfilament and microtubular distribution and orientation to substratum', *J. Cell Biol.,* **67**, 146—59.
3. Willingham, M. C. and Pastan, I. (1974) 'Cyclic AMP mediates the concanavalin A agglutinability of mouse fibroblasts', *J. Cell Biol.,* **63**, 288—94.

CHAPTER 33

Quantitative Assay for Measuring Cell Agglutination in Small Volumes

S. B. OPPENHEIMER

Much early work in the area of lectin-mediated agglutination has been accomplished by estimating the amount of agglutination under the microscope and recording these estimates using a serological scale of 0 to ++++. It became clear that such an assay could not be used to measure rates of agglutination carefully or to measure small, and often highly significant, differences in agglutinability between test suspensions.

1. PRINCIPLE

In this chapter, a rapid, convenient and highly reproducible assay for measuring the kinetics of lectin-mediated agglutination will be described. This assay (1) utilizes an electronic particle counter (2) coupled with a gyratory shaker with a large radius of gyration (3). The shaker facilitates agglutination by allowing intercellular collisions to occur in the experimental suspensions. A large radius of gyration provides adequate wave motion in the suspension facilitating the use of tiny sample volumes. In this way the system is miniaturized allowing 10 experiments for every one achieved with a standard shaker.

The assay is very convenient. Cell suspensions (0.2 ml samples in 1 dram screw-cap vials) are rotated at 70 r.p.m., removed at various time points, diluted with 10 ml of buffer such as 'Isoton' (Coulter Electronics, Hialeah, Fla.) and counted with an electronic particle counter. Agglutination is measured as the disappearance of the defined single-cell population into aggregates over time. In this way, initial agglutination kinetics are rapidly obtained.

2. PARTICLE COUNTER OPERATION

The electronic particle counter measures the change in conductivity of the buffer solution in the aperture when part of the solution is displaced by a particle impermeable to the buffer. The counter discriminates between particles of different volumes and can determine the number of particles in the suspension as the suspension is aspirated through the aperture. The assay method described here depends upon determining the number of single cells remaining in a rotated suspension as a function of time, when the disappearance of single cells results from agglutination or adhesion of the cells to each other. A multichannel analyser must be used to adequately determine size distributions of any aggregates formed as a result of the disappearance of single cells. In practice, for convenience and speed, such an analysis need not be done. Excellent initial agglutination kinetics can be obtained, as will be seen, by measuring single-cell disappearance only, provided that appropriate controls are carried out.

2.1 Single-cell Size Distribution

A size distribution of the single-cell population is obtained before agglutination assays are performed. Figure 1 shows a size distribution of single mouse ascites teratoma cells obtained with a model B Coulter counter. A haemocytometer is used to confirm that the peak contained in the window (in this case a window of 10 to 100) represents the single cells. The number of cells remaining in the 10—100 window as a function of time is a direct measure of the degree of agglutination, if single-cell disappearance results from adherence to each other and not cell lysis or leakage.

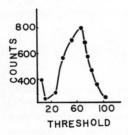

Figure 1. Size distribution of teratoma cells. Cells were counted using model B Coulter counter settings of: 1/amplitude = 8 and 1/aperture current = 0.707. Increasing threshold settings represent increasing particle volume. The cells in the 10—100 window were sized at 12 ± 3 μm

2.2 Aperture Size and Coincidence Errors

Best results are obtained when the aperture size is 3.6 times larger than the largest particle to be measured (5). Since the method used for measuring agglutination depends upon determining the number of single cells remaining in a shaken suspension as a function of time, and does not depend upon counting aggregates which have formed over time, it is of little concern whether or not these aggregates approach or even exceed the diameter (usually 100 μm) of the aperture, though such aggregates could conceivably block the aperture. In practice, aperture blockage seldom occurs.

Ideally, perfect counting would be obtained if single particles traversed the critical capillary sensing zone of the particle counter before entry of another particle. In dense suspensions, coincidence losses in the count result because of multiple passages (6). Coincidence errors may be obviated by applying appropriate correction factors. A more satisfactory procedure is to dilute the suspensions to be counted to the point where coincidence effects are negligible (about $10^4 - 10^5$ cells/ml) (5).

2.3 Validity of the Assay

The validity of the assay, i.e. that the disappearance of single cells represents agglutination, and not, for example, lysis or leakage has been established (1,2,4). Leaky cells, being of lower electrical impedance than intact cells would measure as smaller than their true volume (the actual measured size would depend upon the degree of leakiness). Hence, such leaky cells and cellular debris can be counted in a window representative of particles smaller than intact cells. Such a 'debris' window (3—10, in the case of teratoma cells) detects cell damage. Figure 2 shows that when cells are exposed to a 0.1% 'Alconox' detergent solution, counts in the debris window increase nearly tenfold, concomitant with a decrease of approximately 50% in single-cell counts. Such a debris window, representing particles of smaller size than intact single cells, is therefore a useful indicator of cell damage.

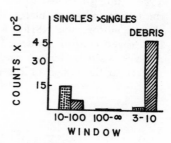

Figure 2. Detection of cellular damage with the Coulter counter. Equal aliquots of teratoma cells were exposed to a 0.1% 'Alconox' detergent solution (lined figures), or Hank's balanced salts control solution (stippled figures) and counted at the indicated window settings

Microscopic and macroscopic examination of vial contents and dye exclusion tests, in addition to a debris window are routinely used to confirm that single-cell disappearance does indeed represent agglutination.

2.4 Reproducibility of the Method

The electronic particle counter provides a statistically reproducible and reliable means for measuring single-cell disappearance in agglutination studies (1,2,4,7—11). The use of small volumes of cell suspension and a gyratory shaker having a large radius of gyration was also shown to be reliable (1,2,7—11). Replicate determinations using 2 or 3 vials per time point indicate that maximum variation in single-cell number in a given experiment is less than 10% and average variation of replicate vials is 3% (1). This error includes errors and variations introduced by differences in the geometry of the vials, pipetting and in the particle counter. In most experiments, mean values for time points in repeated experiments differ by an average of 2%. Maximum variation for mean values obtained for a given time point in a repeated experiment is less than 10% (1). For example, at 35 min, 1 mg/ml Con A, an initial cell density of 7×10^5 teratoma cells/ml, 68 r.p.m. and 17°, mean values obtained for the percentage agglutination (percentage of single cells which have formed aggregates) in four separate experiments were 64%, 67%, 67% and 66% (1).

3. SHAKING TECHNIQUE AND SHAKER CONSTRUCTION

A gyratory shaker is used to induce collisions between cells in the agglutination experiments. It is desirable to reduce the volume of cell suspension used in experiments to conserve cells, media and lectins. In order to accomplish this, any shaker used for such experiments must propagate a good circular wave motion in small volumes of cell suspension. Henkart and Humphreys (3) have modified an original gyratory shaker technique developed by Moscona (12) by increasing the radius of gyration of the shaker from ⅜ inch to 2 inches. The 2 inch radius easily permits a circular wave in 0.2 ml of cell suspensions in 1 dram vials at moderate speeds (70 r.p.m.), while a small radius does not permit adequate swirling of small volumes.

A shaker with a large radius of gyration is easily constructed by connecting two 12 inch square aluminium plates with 4 lengths of ¼ inch by ³⁄₁₆ inch heavy walled Tygon tubing fastened to the ends of four ¼ inch by 1 inch bolts fastened through the upper and lower plates. Rotary motion is provided by a Bodine shunt-wound electric motor, type NSH-12R with a 18:1 reduction gear, equipped with a Minarik D.C. speed control, model SH-14 (both available from Minarik Electric Co., Los Angeles, Calif.). Henkart and Humphreys (3) note that there are a few critical aspects to the design of the shaker. It is essential that the motor drive shaft be

Figure 3. Gyratory shaker with large radius of gyration. Construction of this shaker is described in the text (after Henkart and Humphreys [3])

above the exact centre of the bottom plate, that a ball bearing be positioned in the centre of the top plate to engage the drive shaft complex and that the Tygon legs be of equal length and tight, so that tilting of the top plate does not occur during rotation. A photograph of the shaker is shown in Figure 3.

Hundreds or vials can rapidly be filled with an automatic micropipette, placed on the shaker, removed at time intervals and assayed with the electronic counting method.

4. APPLICATION OF THE METHOD

This assay has been used to measure small changes in the initial kinetics of cell agglutination and aggregation (1,7—11). Figure 4 shows that the initial kinetics of teratoma cell agglutination, mediated by concanavalin A, vary with respect to parameters such as temperature. The data describe the reproducibility obtained

Figure 4. Effect of temperature on the kinetics of cell agglutination by concanavalin A. Teratoma cells were incubated at 17° (●) or 0° (▲) with 1 mg/ml Con A, or at 17° (○) or 0° (△) without Con A, allowed to agglutinate, and the agglutination was measured as the percentage of single cells which have agglutinated under the standatd conditions of pH 7.5 HEPES-buffered glucose-free Hank's balanced salts solution, 1×10^6 cells/ml and 68 r.p.m., using the assay described in the text

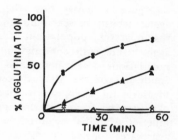

Figure 5. Age-dependent agglutination with lectins. (1) Con A, 20 μg/ml; (2) *Ricinus communis* agglutinin, 20 μg/ml; (3) wheat germ agglutinin, 20 μg/ml. Embryos were dissociated in calcium—magnesium-free sea water, rotated in this solution with the lectins indicated above at pH 7.8, 68 r.p.m. and 17°. The percent agglutination was determined after 60 min. Each point gives the range of values obtained in 3 repeated experiments. In each experiment, duplicate vials were used for each time point. Values have been corrected for background aggregation occurring in calcium—magnesium-free sea water. In all experiments described the appropriate hapten sugars inhibited lectin-mediated agglutination

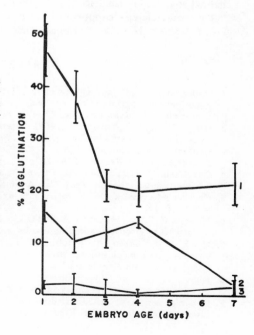

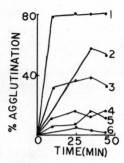

Figure 6. Kinetics of Con A agglutinability of suspensions of different sea-urchin cell types. (1) Micromeres; (2) 50% micromeres, 50% mesomeres and macromeres; (3) mesomeres and macromeres; (1, 2, 3) Con A 1 mg/ml; (4) mesomeres and macromeres; (5) 50% micromeres, 50% mesomeres and macromeres; (6) micromeres; (4, 5, 6) without Con A. Embryos were dissociated, separated using a Ficoll gradient and rotated and the percentage agglutination was determined as the percentage of single cells which have agglutinated under the standard conditions described by Roberson and Oppenheimer (8)

using replicate vials for each time point and illustrate the degree of quantification obtainable for initial agglutination rate measurements (1).

Figure 5 shows how the assay was used to describe the dependency of agglutinability of sea-urchin embryo cells on embryonic age (8). Figure 6 indicates that migratory populations of sea-urchin embryo cells (micromeres) are 65% more agglutinable with concanavalin A than other less active sea-urchin embryo cell populations (mesomeres and macromeres) after 45 min, under the standard conditions of the assay procedure (8). It should be noted that since this assay measures the disappearance of single cells into aggregates and does not measure size of aggregates formed during agglutination experiments, it should be utilized where kinetics of initial agglutination are required. In order to determine the sizes of aggregates formed during later stages of the agglutination process, either multi-channel analysis or visual scoring should be performed (13).

This research was supported by United States Public Health Service Research grant CA 12920 from the National Cancer Institute.

5. REFERENCES

1. Oppenheimer, S. B. and Odencrantz, J. (1972) 'A quantitative assay for measuring cell agglutination: Agglutination of sea urchin embryo and mouse teratoma cells by concanavalin A', *Expt. Cell Res.*, 73, 475—80.
2. Orr, C. W. and Roseman, S. (1969) 'Intercellular adhesion. I. A quantitative assay for measuring the rate of adhesion', *J. Membr. Biol.*, 1, 109—25.
3. Henkart, P. and Humphreys, T. (1970) 'Cell aggregation in small volumes on a gyratory shaker', *Expt. Cell Res.*, 63, 224—7.
4. Oppenheimer, S. B., Edidin, M., Orr, C. W. and Roseman, S. 'An L-glutamine requirement for intercellular adhesion', *Proc. Nat. Acad. Sci., U.S.A.*, 63, 1395—402.
5. Counter Electronics (1966) *Instruction and Service Manual for the Model 'B' Coulter counter*, p. 15.
6. Mattern, C. F. T., Brackett, F. S. and Olson, B. J. (1957) 'Determination of number and size of particles by electrical gating:blood cells', *J. Appl. Physiol.*, 10, 56—70.
7. Krach, S. W., Green, A., Nicolson, G. L. and Oppenheimer, S. B. (1974) 'Cell surface changes occurring during sea urchin embryonic development monitored by quantitative agglutination with plant lectins', *Expt. Cell Res.*, 84, 191—98.
8. Roberson, M. and Oppenheimer, S. B. (1975) 'Quantitative agglutination of specific populations of sea urchin embryo cells with concanavalin A', *Expt. Cell Res.*, 91, 263—8.
9. Oppenheimer, S. B. (1973) 'Utilization of L-glutamine in intercellular adhesion: ascites tumor and embryonic cells', *Expt. Cell Res.*, 77, 175—82.

10. Oppenheimer, S. B. and Humphreys, T. (1971) 'Isolation of specific macromolecules required for adhesion of mouse tumour cells', *Nature*, **232**, 125—7.
11. Oppenheimer, S. B., Potter, R. L. and Barber, M. L. (1973) 'Alteration of sea urchin embryo cell surface properties by mycostatin, a sterol binding antibiotic', *Develop. Biol.*, **33**, 218—23.
12. Moscona, A. A. (1961) 'Rotation-mediated histogenic aggregation of dissociated cells', *Expt. Cell Res.*, **22**, 455—75.
13. Edwards, J. (1973) 'Intercellular adhesion', in *New Techniques in Biophysics and Cell Biology*, Vol. 1, eds. Pain, R. H. and Smith, B. J. (Wiley, London), pp. 1—27.

CHAPTER 34

A Microhaemadsorption Method for Assessment of Concanavalin A-induced Cell Agglutination

P. FURMANSKI
P. G. PHILLIPS

1. PRINCIPLE OF THE METHOD

The conventional test for the interaction of cells with Con A is agglutination. Single-cell suspensions are incubated with the lectin and visually, or electronically monitored for the formation of large cell aggregates. This method has several disadvantages, including (a) the need for relatively large numbers of cells in suspension, with few or no cell clumps initially; (b) the need for dissociating agents or mechanical scraping to remove the cells from the substratum on which they are grown; and, (c) the inability to score reactivity on an individual cell basis, or to assess reactivity in heterogeneous cell populations.

We have described a method for determining the interaction of Con A with cells, based on the Con A-mediated adsorption of red blood cells to the test cells (1). This method overcomes the disadvantages of the agglutination assay.

301

2. PREPARATION OF CELLS FOR THE ASSAY

2.1 Tissue Culture Test Cells

The test cells, Balb/3T3 and SV3T3 used in our studies, or other cell types, are grown using standard tissue culture methodology. The cells are removed from their growth vessels using trypsin — EDTA, and 200—500 cells are seeded into the individual wells of Microtest plates (Falcon Plastics). For most purposes the wells can be filled using a Pasteur pipette and a rubber bulb; for more critical applications, a Hamilton syringe with a repeating dispenser is used.

The cells are incubated in the Microtest plates for 24—48 h, washed several times by flooding the entire plate with phosphate-buffered saline (PBS), and used in the assay.

2.2 Red Blood Cells

Freshly drawn human (type O) blood or outdated units of human (type O) blood obtained from the local blood bank are used as a source of indicator red blood cells (RBC's).

Prior to use the blood is centrifuged, and the plasma and buffy coat discarded. The RBC's are washed three times with PBS and resuspended to a haematocrit of 2% in PBS.

Other types of RBC's (including sheep and rabbit) can be used as indicators. However, we have found that the most consistent results, with minimum interference by background adsorption, are achieved using human type O RBC's.

3. ASSAY METHOD

When the microhaemadsorption method is used to compare the interaction of Con A with Balb/3T3 (normal) and SV3T3 (transformed) cells, different results are obtained depending on how the assay is carried out (1). If the tissue culture test cells are treated with Con A, and then adsorption of uncoated RBC's is determined, the results are analogous to those obtained in the agglutination assay: transformed cells react strongly, while normal cells react only at very high lectin concentrations, if at all. However, if untreated test cells are incubated with Con A-coated RBC's, transformed cells exhibit only a 2- to 4-fold greater reactivity than normal cells.

We have suggested (1) that these two modes of assay reflect a difference between simple binding of lectin to the surface of the tissue culture test cell, and the subsequent availability of the bound lectin to participate in agglutination (or haemadsorption). This is in accord with data from our own and other laboratories on binding and agglutination (1—5).

3.1 Assay by Treatment of Tissue Culture Cells

Seeded washed microtest plates are prepared as described above (2.1). Solutions of Con A in PBS are added to the individual wells, and the plate is incubated at 37° for 10 min. The cells are then washed by flooding the plate with PBS, and uncoated indicator RBS's are added to fill each well. After further incubation (37°, 10 min), the plates are rinsed with PBS and scored (see paragraph 3.3).

3.2 Assay by Treatment of Red Blood Cells

Washed suspensions of human type O RBC's (see paragraph 2.2) are incubated with Con A at 37° for 10 min. Unbounded Con A is removed by centrifugation, the RBC's are washed three times with PBS, and resuspended in PBS to a haematocrit of 2%. The individual wells of seeded, washed microtest plates (see paragraph 2.1) are filled with the RBC suspensions, incubated at 37° for an additional 10 min, washed to remove unadsorbed RBC's and scored (see paragraph 3.3).

3.3 Quantitation of Haemadsorption

Con A-mediated adsorption of the RBC's to the test cells is scored by estimating the proportion of the test cells which have RBC's attached. A convenient scale is 0 to ++++ reactive; with 0 = 0—1%; ± = 1—10%; + = 10—25%; ++ = 25—50%; +++ = 50—75%; and ++++ = 75—100%.

3.4 Specificity Controls

Microhaemadsorption is inhibited by α-methyl-D-mannopyranoside (50 mM) or α-methyl-D-glucopyranoside (100 mM). Inhibition is observed when the sugars are added to the microtest plate wells together with the indicator RBC's, or when the sugars are present during the treatment of either the test cells or RBC's with Con A.

Pretreatment of both the RBC's and the tissue culture test cells with Con A inhibits microhaemadsorption, suggesting that free Con A binding sites are required for agglutination or adsorption (1).

3.5 Useful Modifications

Removal of unbound RBC's from microtest plates is facilitated by (1) including BSA (0.1%) in the RBC suspension; (2) tapping the microtest plate against a hard object; or (3) completely filling the plate with PBS, covering the plate tightly with the lid and incubating the plate in an inverted position, rocking gently.

The assay described here is performed in microtest plates which are easily handled, use very small quantities of reagents, and are readily scored since the entire well can be seen in the field of a 10x phase objective. However, we have successfully applied the assay to cells grown in 16 mm wells (Linbro FB16-24TC dishes), or 35 or 60 mm standard tissue culture petri dishes.

Van Nest and Grimes (6) have reported that removal of unbound RBC's from haemadsorption assays performed in 35 mm dishes is facilitated by rocking the dishes during incubation with the RBC's.

Rittenhouse and Fox (7) have reported the quantitation of micro-haemadsorption by measuring the haemoglobin content in cell lysates from the reaction dishes. Chromium-51 labelled RBC's can similarly be used in quantitation.

Smets and De Ley (8) have described the separation of microhaemadsorption reactive and non-reactive cells by centrifugation in a Ficoll—Isopaque gradient.

4. TYPICAL APPLICATION OF MICROHAEMADSORPTION

The demonstration of Con A binding sites on SV3T3 cells using the micro-haemadsorption assay, is shown in Figure 1.

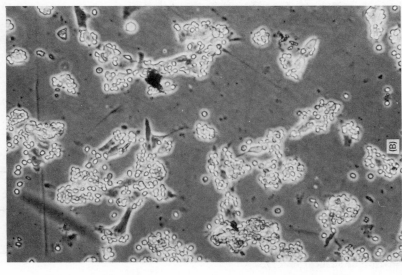

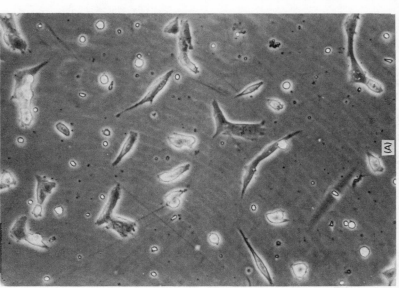

Figure 1. Adherence of Con A-coated indicator RBC's to SV3T3 test cells. (A) RBC's with no Con A, non-reactive; (B) RBC's treated with 20 μg/ml Con A, reactive. Scale represents 100 μ. [Reprinted with permission of the Society of Experimental Biology and Medicine from *Proc. Soc. Exp. Biol. Med.*, 140, 216–219 (1972).]

5. ACKNOWLEDGEMENTS

This work was supported by grants from the NIH (5 R01 A109288, I R01 CA 15085), from the ACS (E557) and from the Milheim Foundation (72-17) to M. Lubin, and by fellowships to P.G.P. (Damon Runyon DRF-664 and NIH F02, CA50855) and to P.F. (Damon Runyon DRF-588 and NIH CA 41163). We thank Dr. M. Lubin for his advice and support during the course of these studies.

6. REFERENCES

1. Furmanski, P., Phillips, P. G. and Lubin, M. (1972) 'Cell surface interactions with concanavalin A: Determination by microhemadsorption', *Proc. Soc. Expt. Biol. Med.*, **140**, 216—9.
2. Phillips, P. G., Furmanski, P. and Lubin, M. (1974) 'Cell surface interactions with concanavalin A: Location of bound radiolabeled lectin', *Expt. Cell Res.*, **86**, 301—8.
3. Burger, M. M. (1969) 'A difference in the architecture of the surface membrane of normal and virally transformed cells', *Proc. Nat. Acad. Sci. U.S.A.*, **62**, 994—1001.
4. Ozanne, B. and Sambrook, J. (1971) 'Binding of radioactively labelled concanavalin A and wheat germ agglutinin to normal and virus-transformed cells', *Nature (New Biol.)*, **232**, 156—60.
5. Arndt-Jovin, D. J. and Berg, P. (1971) 'Quantitative binding of ^{125}I-concanavalin A to normal and transformed cells', *J. Virol.*, **8**, 716—21.
6. Van Nest, G. A. and Grimes, W. J. (1974) 'Concanavalin A-induced agglutination and tumorigenicity in virally and spontaneously transformed cells derived from Balb/c mice', *Cancer Res.*, **34**, 1408—12.
7. Rittenhouse, H. G. and Fox, C. F. (1974) 'Concanavalin A mediated hemagglutination and binding properties of LM-cells', *Biochem. Biophys. Res. Commun.*, **57**, 323—31.
8. Smets, L. A. and De Ley, L. (1974) 'Cell cycle dependent modulations of the surface membrane of normal and SV40 virus transformed 3T3 cells', *J. Cell Physiol.*, **84**, 343—8.

CHAPTER 35

Quantitation of Fibroblast Agglutination by the Cell Layer Assay

B. T. WALTHER
W. L. ROTTMANN

The cell layer assay was originally developed to quantitate intercellular adhesion reactions *in vitro* (1,2). The adaptation of this assay to measure agglutination reactions (3,4) follows the pattern by which other adhesion assays have been adapted for use as agglutination assays with only small changes in design. The 'aggregate size' adhesion assay developed by Moscona's group (5), for example, formed the basis for the visual agglutination assay (6,7), and the electronic particle counter assay for adhesion (8) inspired a quantitative method for measuring cell agglutination (9). Thus, lectin agglutination has often been studied with methodology first developed to measure intercellular adhesion. Only recently have attempts been made to evaluate the possible role of adhesive processes in the overall agglutination reaction (4,10).

S.E.M. OF INTERCELLULAR JUNCTIONS

2A: 6,000 x (90 DEG.)

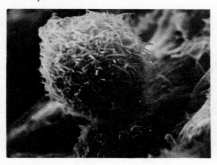

2B: 50,000 x (87 DEG.)

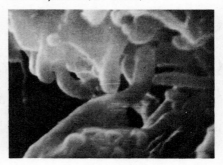

2C: 15,000 x (74 DEG.)

2D: 15,000 x (80 DEG.)

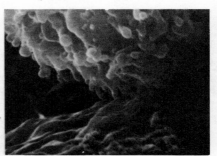

(3T12)$_{SC}$ (3T12)$_{ML}$
(2 x 10⁶ cells/ml)
Incubation 4 mins.

Figure 1. Morphology of attachment sites of single cells to a cell layer. BALB/c 3T12 cells were seeded at 2×10^5 cells/cm² on Falcon plastic dishes and allowed to grow overnight in Dulbecco's medium supplemented with 10% calf serum. Single 3T12 cells (2×10^6 cells/ml) were prepared as described in the text and incubated with the cell layer in medium B* for 4 minutes, in order to study only the earliest adhesive events (1). The incubation was stopped by adding equal volumes of 6% glutaraldehyde in medium B. After 2 hours at 37°, the cells were shifted to 0° for a further 10 hours of fixation in the same medium. Cells were dehydrated stepwise in ethanol and dried by the critical-point method. Specimens were coated with carbon and a further thin coat of gold-palladium (60:40). Cells were viewed at 20 kV with a Hitachi HiScan HS-2R scanning electron microscope (12). Magnifications are as indicated, and the numbers in parenthesis refer to tilt angles. Figures 2C and 2D together allow a stereo view of the junction area between the single cell and the cell layer shown in Figure 2A. Figure 2B shows a higher magnification of the area shown in Figures 2C and 2D. Positive identification of the newly adhered single cells has been achieved in more recent work (27)

---

*Medium B = HEPES buffered Hank's balanced salt solution (without glucose) pH 7.3.

The cell layer assay measures intercellular adhesion by determination of the *rate* of binding of single cells in suspension to a confluent cell layer. The morphology of the cellular processes involved in adhesion is depicted in Figure 1, in which single 3T12 cells had been allowed to adhere to a (homologous) layer of 3T12 cells. Non-adhered cells were washed off, and the adhered cells were examined by the scanning electron microscope. As can be seen, adhesive contacts are made at multiple sites by microvilli from both the single cells and from the cell layer. Evidently the adhesive process involves highly complex microvillar structures. The junction site is probably rather transient in appearance, since the actual sites and extent of microvillar contacts may be changing continuously. However, adhesive interactions do not require all of these structures intact, since single cells fixed with glutaraldehyde also adhere to a live cell layer (11) and such cells can also be agglutinated to a layer with concanavalin A (10).

1. PRINCIPLE

The principle of the cell layer assay method is illustrated in Figure 2. A uniformly labelled suspension of single cells is introduced over a confluent cell layer. After various times, the non-attached cells are washed off, and adhesion is measured directly by counting the radioactivity in the cells remaining on the cell layer. Each cell type attaches to its homologous cell layer at a characteristic rate which is, however, quite sensitive to the method used for preparing the single-cell suspension (12). Lectins such as concanavalin A were found to *enhance* the rate of attachment of single cells to a cell layer, and this stimulation forms the basis for quantitation of agglutination by the cell layer assay (3,4). The cell layer method determines the agglutination properties of cells in two states: *single cells*, obtained

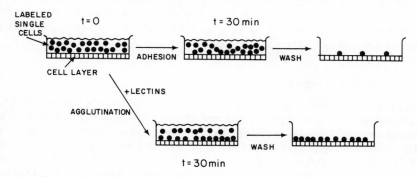

PRINCIPLE OF CELL-LAYER ASSAY

Figure 2. Principle of the cell layer assay. See text for description. Labelled single cells in suspension are symbolized by closed circles; cells in the cell layer by open squares. The enhanced number of labelled single cells which remain on the layer after the washing procedure upon addition of lectins to the single-cell suspension, forms the basis for quantitative measurements of cell agglutination by this assay procedure

by dissociation of cells in a layer, and *cell layers*, attached to various substrata. Since agglutination properties of cells are changed by dissociation procedures, the agglutinative properties of a cell type should be studied in both states.

2. PREPARATION OF CELLS

Two mouse fibroblast cell lines were used, Balb/c 3T3 and Swiss 3T3. Both cell lines were grown in Dulbecco's modified Eagle's medium supplemented with 10% calf serum at $37°$ in a water-saturated atmosphere of 95% air—5% CO_2 in Falcon tissue culture dishes. The cells were examined for mycoplasma by Microbiological Associates, Bethesda, Md., and were found free from contamination. Stocks of the two lines were maintained in sealed ampoules in liquid nitrogen, and cells were passed no more than 20 times for the experiments described below. The Balb/c 3T3 line was highly contact inhibited, reaching a saturation density of 2.6×10^4 cells/cm^2, while the Swiss 3T3 cells attained a density of 5×10^4 cells/cm^2. Cells were transferred before reaching confluency. For passage, the plates were first washed with phosphate-buffered saline, and then treated with 0.25% trypsin (GIBCO) for 10 to 15 min at $37°$.

2.1 Preparation of Cell Layers

Cell layers were prepared from confluent cultures of cells essentially as described (1,2). Each well of multiple-well tissue culture dishes (model FB16-24TC, Linbro Chemical Co., New Haven, Conn.) was inoculated with 2 mls of a suspension of trypsin-dispersed single cells (1×10^5 cells/ml of growth medium) 24 hours before use. The high cell density (1×10^5 cells/cm^2) ensured complete coverage of the plastic substratum. For trypsin treatment of cell layers, growth medium was removed by aspiration, the cell layers were washed twice with medium B (1,4), and treated with 0.5 ml of 0.025% trypsin (GIBCO 0.25% trypsin diluted 10-fold in medium B). After incubation for 4 min at $37°$, the trypsin solution was removed by gentle aspiration and the cell layers were washed twice with growth medium. The cell layers were incubated in growth medium for 15 min, and inspected for complete confluency by the phase microscope.

2.2 Preparation of Single Cells for Assay

Radioactively labelled single cells were prepared as described previously (1,2). Confluent cultures were trypsin dispersed as described above and inoculated at a concentration of 1.5×10^6 cells per Falcon tissue culture plate (78.5 cm^2). The freshly inoculated cells were maintained in growth medium containing 5—10 μCi/ml of L-^{3}H-leucine (5 Ci/mmole) for 12 to 18 hours, followed by a chase period in unlabelled growth medium for at least 2 hours prior to harvesting. The cells were removed from the culture dish by treatment with either 0.25% trypsin or 0.5 mM EDTA in phosphate-buffered saline (as indicated) for 10 to 15 min at $37°$. The dissociated cells were diluted in 20 volumes of growth medium, collected by centrifugation at 50g for 10 min, suspended in 20 ml of medium B, recentrifuged, and finally suspended in medium B at a concentration of 2.5×10^4 cells/ml. Cells usually incorporated 0.2 to 0.5 c.p.m. of L-^{3}H-leucine per cell; essentially all of the labelled material was precipitable with cold trichloroacetic acid.

3. ASSAY PROCEDURE

Immediately prior to the assay, cell layers (untreated or trypsin treated) were rinsed twice with medium B, and 1 ml of the labelled single-cell suspension was added to each well of the multiple-well dish (0.5 ml of suspension per cm^2 of cell layer). Concanavalin A in medium B was added with a micropipette (routinely 50 μl of 1 mg/ml stock solution) and mixed by swirling (paragraph 4.1).

Alternatively, preincubation of cell layers with Con A, in the absence of single cells, was followed by removal of unbound Con A before introduction of the labelled single-cell suspension over these layers (paragraph 4.2). Mixtures were incubated at 37° with gentle agitation (60 strokes per min, 8 cm per stroke) in a Warner—Chilcott reciprocal water bath shaker. After incubation for the required time, the unattached single cells were removed by gentle aspiration. The cell layers were washed three times with 1 ml aliquots of medium B, lysed with 0.5 N NH$_4$OH, and radioactivity was determined by standard liquid scintillation techniques as previously described (1,2).

Results are expressed as the percentage of the total cells in suspension which attach to the cell layer as a function of time. Under standard conditions the single-cell concentrations were 2.5–5 x 10^4 cells/ml (see Figure 3).

4. EXAMPLES OF APPLICATION

4.1 Agglutination of Single Cells to a Layer by Soluble Con A

In this assay a labelled single-cell suspension is introduced over a layer, with Con A being added to the suspension at time zero. Attachment proceeds without detectable lag periods (4).

It is important to use concentrations of single cells that are not aggregated in suspension by the addition of Con A to the suspension (see Figure 3). Generally this requirement is met by using single-cell concentrations of 5 x 10^4 cells/ml or less. Above these levels, high concentrations of Con A cause clumping of cells in the suspension, resulting in lower rates of attachment to the layer. However, below 10^5 cells/ml, the fraction of cells attached in an incubation period is independent of the exact single-cell concentration (Figure 3), allowing for ready comparison of results from different experiments.

As illustrated in Figure 4, the rate of attachment of cells to a layer in the presence of Con A is strongly dependent on the Con A concentration. Furthermore, the observed attachment is largely inhibited by methyl α-mannoside. Judged by this criterion, the attachment is truly an agglutination reaction, even though the concept of agglutination measured by binding of single cells to a layer may appear confusing at first glance. A series of other criteria for agglutination reactions, such as sensitivities towards EDTA and trypsin treatment of the two participating cells, hapten-inhibition specificities and sensitivity toward low temperatures have been tested (4) and all conform with the findings obtained using the visual agglutination

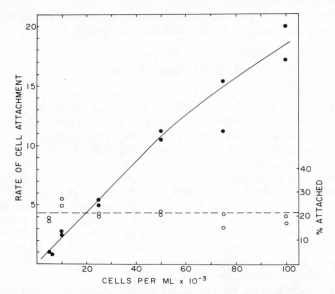

Figure 3. Effect of single-cell concentration on the rate of cell attachment. Aliquots (1.0 ml of suspensions containing the indicated concentrations of trypsin-dispersed Swiss 3T3 cells) were exposed to untreated Swiss 3T3 cell layers. Concanavalin A was added as a 50 μl aliquot of a 1 mg/ml solution and mixed by swirling (final concentration 48 μm/ml). Cell attachment was measured as described in the text. The rate of cell attachment (left ordinate) is expressed as the number of labelled cells attached (x 10^{-3}) to the cell layer per 10 min incubation (●). The percentage of attachment (right ordinate) represents the percentage of the initial number of cells attached during a 10 min incubation (○)

assay. Thus the cell layer assay allows quantitative kinetic measurements of cell agglutination phenomena.

4.2 Agglutination of Single Cells to Layers containing Prebound Con A

In this assay a labelled single-cell suspension is introduced in the *absence* of soluble Con A to a layer to which Con A has been bound by prior incubation. (Unbound Con A is removed by several washes). Since detachment of bound Con A upon incubation with a single-cell suspension is too low to account for agglutination of single cells by the mechanism described in paragraph 3 (4), this assay overcomes the limitations placed on the concentration range of single cells in suspension.

We have measured the amount of Con A bound to a cell layer during preincubation experiments by using varying amounts of labelled Con A for differing lengths of time (4). Thus we were able to prepare layers with varying and known amounts of prebound Con A. When the attachment of single cells to such layers was

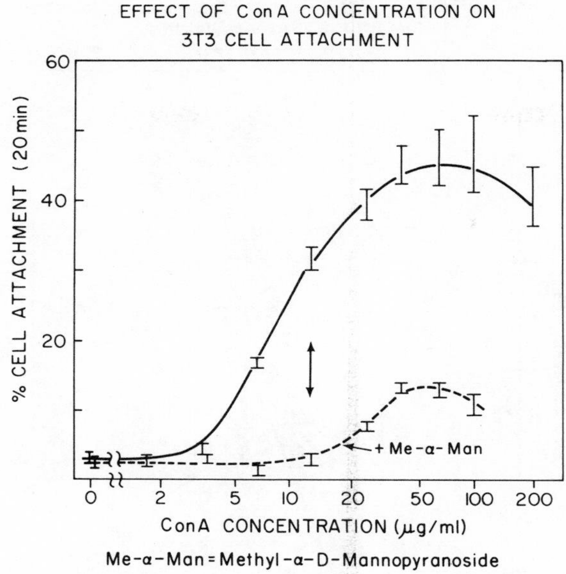

EFFECT OF Con A CONCENTRATION ON
3T3 CELL ATTACHMENT

Me−α−Man = Methyl−α−D−Mannopyranoside

Figure 4. Effect of Con A concentration on the rate of
BALB/c 3T3 cell attachment to a homologous cell layer.
Untreated 3T3 cell layers were prepared as described in
the text. Equal volumes of a suspension of trypsin-
dissociated single 3T3 cells (50,000 cells/ml) and a 2-fold
concentrated solution of concanavalin A in medium B
(plus or minus 2.6 mM methyl α-mannoside) were rapidly
mixed and a 1.0 ml aliquot of the mixture was added to
each cell layer. This resulted in a final concentration of
25,000 cells/ml, and the indicated concentrations of
concanavalin A, plus (− − −) or minus (——) 1.3 mM
methyl α-mannoside. The percentage of cells attached at
10 and 20 min was measured in order to determine the
rates of cell attachment. Rates are expressed as the
percentage of cells attached in 20 min. Each bar represents the range of three determinations. The double-
headed arrow denotes a Con A concentration where the
Con A-mediated agglutination appears to be almost completely inhibited by 1.3 mM methyl α-mannoside

compared to the amount of Con A bound during the preincubation experiment (4),
a striking correlation was found (Figure 5). Below a threshold level of 0.2 μg
Con A/cm^2 cell layer, no stimulated attachment was observed. Above this density
of Con A on the layer, attachment was approximately proportional to the Con A
density on the layer. This attachment is a true agglutination reaction based on its
inhibition by methyl α-mannoside (4,15).

In further experiments (15) we have found that with the assay described here,
the attachment of single cells to a layer containing prebound Con A, is proportional
to the single-cell number in suspension over a wide cell concentration range

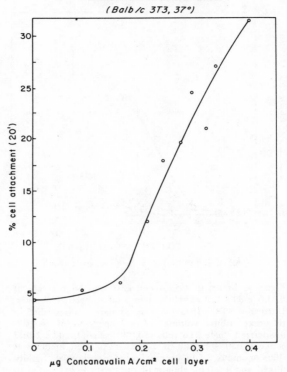

Figure 5. Relationship between cell attachment and the amount of concanavalin A prebound to the cell layers. The observed cell attachment (measured as described in the text) was plotted against the amount of concanavalin A bound during preincubation experiments. The binding of the unlabelled concanavalin A was estimated from, and assumed to be idenitical with, the binding of ^{14}C-concanavalin A (labelled by acetylation, ref. 13). BALB/c 3T3 cell layers were 2 cm[2] in area, and the single BALB/c 3T3 cells were obtained by trypsinization. Points are averages of triplicate measurements

(Figure 6). Even at high concentrations of single cells where the number of single, attached cells is greater than the number of cells in the layer, the Con A-mediated rate of attachment is unchanged. In this case it seems likely that cells are not attaching directly to the layer, but are attaching to 'single' cells already bound to the layer. One interpretation of this finding is that cells which are agglutinated by Con A to a cell layer, have higher affinity for other single cells in suspension, although there (probably) is no Con A to directly mediate the binding of additional single cells (see references 16,17 for a discussion of similar or related phenomena).

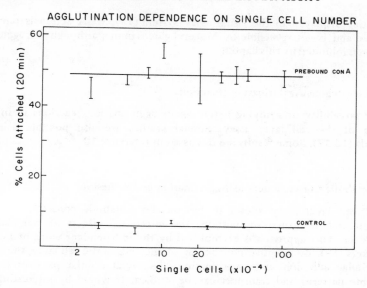

Figure 6. The effect of single-cell number on the attachment of cells to layers containing prebound Con A. Labelled single-cell suspensions of BALB/c 3T3 cells were obtained by trypsinization and incubated over 2 cm² of BALB/c 3T3 cell layers which had been preincubated for 30 min with medium B (control) or with medium B containing 50 μg/ml of Con A (prebound Con A), followed by washes with medium B to remove unbound Con A. In all cases the labelled cells were suspended in 1 ml of medium B. After 20 min, non-attached cells were removed by the standard washes and attachment of cells measured as described in the text. As reported elsewhere (1,2) the number of adhered cells is proportional to the single-cell number (lower curve), so that the fraction of adhered cells compared to the total number of cells in suspension is constant. Similar results are obtained with attachment of cells to a layer containing prebound Con A (top curve) except that the attached fraction constitutes a much larger percentage of the input cell number

Cell adhesion is also proportional to the cell number in suspension as can be seen in Figure 6. (1,12)

4.3 Agglutination Reactions with Multiple Lectins

The cell layer assay allows the simultaneous participation of multiple lectins during the agglutination reactions. Thus we have observed (15) that the action of Con A and wheat germ agglutinin (WGA) are approximately additive, suggesting that these lectins interact with different membrane receptors, in accordance with other evidence in the literature (18). The cell layer assay is thus useful in elucidating whether lectins of similar or different hapten sensitivities in fact bind to similar or different receptors on the cell surface. We have observed, however, that the properties of WGA in the cell layer assay are somewhat different from the properties displayed in the visual assay system (15); WGA action is temperature sensitive and is not reversible by *N*-acetyl-glucosamine, although the agglutination reaction *is* inhibited by this hapten.

4.4 Heterologous Agglutination Reactions

The possibility of studying heterologous agglutination reactions is an unusual feature of the cell layer assay. Similar studies are also possible using other methods (16,17). Some results are discussed in reference 10.

4.5 Metabolic Consequences of Agglutination and Adhesion

The cell layer assay allows studies on the metabolic consequences of both adhesion and agglutination, without the interfering effects of cell dissociation procedures. This approach is exemplified by the independent work of Levine and coworkers (19). By selective prelabelling of the cells in the cell layer, the effects of intercellular adhesion and agglutination on several cellular parameters such as polysome patterns and transport can be studied. It would be interesting by this approach to explore the possibility that cells communicate 'messages' to each other, either directly or indirectly (20,21).

Various forms of metabolic cooperation (24,25) are also amenable to study by the cell layer assay. By preloading metabolically deficient cells (the cell layer), and introducing the metabolically competent cells as single cells, it seems feasible to develop quantitative procedures for evaluation of metabolic cooperation to complement current autoradiographic procedures.

4.6 Lectin-resistant Cells and Cell Adhesion

The cell layer assay would seem to facilitate the analysis of the changes in cells selected for growth resistance to lectins (22). The adhesiveness and agglutinability of such cells could be tested with undissociated cells (in the cell layers) either in a homologous reaction or in a heterologous reaction with a reference cell type such as BALB/c 3T12 cells. A screening of cell adhesiveness in cells resistant to multiple lectins should contribute to our understanding of the process of intercellular adhesion (23).

4.7 Ultrastructural Aspects of Agglutination

The cell layer assay offers many clear advantages for studies on ultrastructure of interacting cell surfaces. Cells engaged in an adhesive reaction (see Figure 1) or in agglutinative reactions, can be identified both by transmission (26) and by scanning electron microscopy (12,27). Since these events are taking place simultaneously in tens of thousands of cells located in the same plane, their analysis is greatly facilitated. Furthermore, the interacting surfaces can be completely visualized at high tilt angles in the scanning microscope (see Figure 1).

5. GENERAL COMMENTS AND REMAINING PROBLEMS

It is essential for assay purposes that the cell population be uniformly labelled in terms of radioactivity per cell. This can be verified autoradiographically by exposing labelled cells to emulsions for different lengths of time. To ensure uniform labelling of cells, the time of exposure to the radioisotope should not be shorter than one cell cycle.

The choice of substratum for the cell layers is decided by the stability it confers on the layer during the assay procedure. Most cell types can be readily maintained on plastic, but for certain cells, glass or other modified substrata give more stable layers. Integrity of the layer is essential for assay purposes, since exposure of the substratum will allow cell—substratum adhesion in addition to the background cell—cell adhesion during the agglutination reaction.

The number of cells used to form cell layers should be sufficient to ensure complete confluency of the cell layer, which may result in some stacking or partial overlapping of the cells in the layer. Excessive stacking should be avoided, since superconfluent cell layers of many cell types detach from the substratum in serum-free medium.

The temperature of the cell layers should be maintained at $37°$ throughout the experiment since cell agglutination is extremely sensitive to temperature. Furthermore, the time interval between the replacement of growth medium by the incubation medium B and the introduction of the single-cell suspension should be rigorously defined, because the morphology of cell layers changes upon prolonged exposure to serum-free medium.

The proportion of cell layer surface area and single-cell suspension volume used in the assay must be chosen such that sedimentation-limited kinetics is avoided. For the cell types studied thus far, a volume to surface ratio $\leqslant 0.5$ ml/cm^2 has satisfied this requirement.

It is essential that the washing procedures employed at the end of an agglutination assay be rigorously standardized. Generally the layers were washed three times with 0.5 ml of medium B per cm^2 cell layer. The number of washes used to remove non-attached cells may be determined by the stability of a given cell layer. The observed Con A-mediated attachment decreases with increasing numbers of washes used at the end of the incubation. Presumably a higher number of washes removes more loosely attached cells and leaves behind on the layer only the more tightly bound cells. In the case of cell adhesion, one observes qualitative differences in the mechanism of cell binding to a layer, when one compares the cells that are removed by the first wash and those cells still attached after three washes (12). A similar study remains to be done with the agglutination reaction.

Some aspects of the assay require further study. Reproducibility was considered acceptable, although occasional variations in absolute rates of agglutination were observed with different cell preparations. The reasons for this variability are not understood at this time. The cell type under investigation must be capable of forming a layer which adheres to a fixed substrate. However, it is possible to use the assay with cells such as erythrocytes by employing a reference cell layer of a different type, such as 3T3 cells (28). Agglutination measurements using highly adhesive cells (1) are not possible without modifications of the assay procedure (4,15), since the high adhesive background rate otherwise impairs the quantitative measurement of Con A-mediated stimulation of attachment of single cells to the layers. However, we have recently measured agglutination of 3T12 cells (highly adhesive cells) by the cell layer assay in short (5—10 min) incubations of a 3T12 cell layer with a suspension of single 3T12 cells containing Con A (15). In this situation the attachment to the layer is largely Con A mediated, since cell adhesion only proceeds after a lag period of approximately 5 min (1,2), while cell agglutination by Con A proceeds without a detectable lag period (4).

6. COMPARISON WITH OTHER METHODS

The cell layer assay method has been adapted for quantitative measurements of concanavalin A-stimulated rates of single-cell attachment to a cell layer. The parameters of the assay have been established with both BALB/c and Swiss 3T3 cells, which gave similar results in a number of experiments. An assay similar in principle to the one described here has been applied to leucocytes which have low intercellular adhesive properties, but which are agglutinated by Con A (14).

The assay offers the following advantages over the previously published methods that measure agglutination by visual estimation of the extent of cell aggregation in the presence of lectin (6,7); (a) The method reported here is an objective, quantitative assay. (b) It permits measurement of lectin-stimulated agglutination above a background rate of intercellular adhesion. (c) The assay offers the unique advantage of studying agglutination properties of undissociated cells (in the cell layer). The effects of dissociation procedures on agglutination properties can in turn be carefully studied by independent manipulations of the two agglutinating cells. (d) The strength of the agglutinative interactions can readily be assessed by changing the procedure for removal of unbound cells from the layer. A large number of cells which are scored as unbound by the regular exhaustive washing procedure, have been found to have interacted with the layer through Con A with bonds of less strength, but sensitive to specific haptens (15). In the case of cell adhesion, similar procedures reveal two qualitatively different adhesive steps involving bonds of unlike sensitivity to mechanical shear (12). It is interesting to speculate that agglutinative interactions studied under different amount of mechanical shear, may also be different in nature. In fact methods designed to minimize shear forces tend to attribute different properties to the agglutination reaction than do methods which employ considerable shear forces (29). It remains to be documented whether under conditions of minimal shear forces, lectins can agglutinate cells loosely through randomly dispersed receptors, and whether agglutination involving stronger bonding of cells also involves micropatches of receptors as well as intercellular adhesive phenomena (10).

7. ACKNOWLEDGEMENTS

This work was performed while BTW was Fellow of the Natl. Cystic Fibrosis Research Foundation and Chr. Michelsen Fond, Norway. WLR was recipient of Postdoctoral Fellowship from Natl. Cancer Institute, US Natl. Institute of Health. Publication No. 836 from the McCollum-Pratt Institute, Johns Hopkins University.

8. REFERENCES

1. Walther, B. T., Ohman, R. and Roseman, S. (1973) 'A quantitative assay for intercellular adhesion', *Proc. Nat. Acad. Sci. U.S.A.*, **70**, 1569—73.
2. Walther, B. T. and Ohman, R. (1974) 'The cell layer assay', in *Methods in Enzymology*, Vol. XXXII, Biomembranes, part B, eds. Fleischer, S. and Packer, L. (Academic Press), pp. 603—9.
3. Rottmann, W. L., Hellerqvist, C. G., Umbreit, J. and Walther, B. T. (1973) 'A quantitative assay for intercellular adhesion and its application to cell agglutination by lectins', *Fed. Proc.*, **32**, 556.
4. Rottmann, W. L., Walther, B. T., Hellerqvist, C. G., Umbreit, J. and Roseman, S. (1974) 'A quantitative assay for concanavalin A-mediated cell agglutination', *J. Biol. Chem.*, **249**, 373—80.
5. Lilien, J. E. and Moscona, A. A. (1967) 'Cell aggregation: Its enhancement by a supernatant from cultures of homologous cells', *Science*, **157**, 70—2.
6. Burger, M. M. (1969) 'A difference in the architecture of the surface membranes of normal and virally transformed cells', *Proc. Nat. Acad. Sci. U.S.A.*, **62**, 994—1001.
7. Inbar, M. and Sachs, L. (1969) 'Interaction of the carbohydrate-binding protein concanavalin A with normal and transformed cells', *Proc. Nat. Acad. Sci. U.S.A.*, **63**, 1418—25.
8. Orr, C. W. and Roseman, S. (1969) 'Intercellular adhesion: A quantitative assay for measuring the rate of adhesion', *J. Membr. Biol.*, **1**, 109—24.
9. Oppenheimer, S. B. and Odencrantz, J. (1972) 'A quantitative assay for measuring cell agglutination', *Exp. Cell Res.*, **73**, 475—80.
10. Walther, B. T. (1976) 'Mechanisms of cell agglutination by concanavalin A', this volume, Chapter 26.
11. Walther, B. T. (1975) 'Adhesive properties of glutaraldehyde-fixed Balb/c 3T12 cells', *Fed. Proc.*, **34**, 498.
12. Walther, B. T. and Roseman, S. (1975) 'Intercellular adhesion. Properties and experimental distinction between two early steps in adhesive reactions', manuscript in preparation.
13. Agrawal, B. B. L., Goldstein, I. J., Hassing, G. S. and So, L. L. (1968) 'Protein—carbohydrate interaction. XVIII. The preparation and properties of acetylated concanavalin A, the hemagglutinin of jackbean', *Biochemistry*, **7**, 4211—8.
14. Berlin, R. D. and Ukena, T. E. (1972) 'The effect of colchicine and vinblastine on the agglutination of polymorphonuclear leucocytes by concanavalin A', *Nature (New Biol.)*, **238**, 120—2.
15. Walther, B. T. and Roseman, S. (1975) 'Agglutination of Balb/c 3T12 and 3T3 cells', *J. Cell Biol.*, **67**, 445a.
16. Rutishauser, U., Yahara, I. and Edelman, G. (1974) 'Morphology, mobility and surface behaviour of lymphocytes bound to nylon fibers', *Proc. Nat. Acad. Sci. U.S.A.*, **71**, 1149—53.
17. Rutishauser, U. and Sachs, L. (1974) 'Receptor mobility and the mechanism of cell—cell binding induced by concanavalin A', *Proc. Nat. Acad. Sci. U.S.A.*, **71**, 2456—60.
18. Janson, V. K., Sakamoto, C. K. and Burger, M. M. (1973) 'Isolation and characterization of agglutinin receptor sites', *Biochim. Biophys. Acta.*, **291**, 136—43.
19. Levine, E. M., Jeng, D. and Chang, Y. (1974) 'Contact inhibition, polysomes and cell surface membranes in cultured mammalian cells', *J. Cell Physiol.*, **84**, 349—61.
20. Kolodny, G. M. (1974) 'Transfer of macromolecules between cells in contact' in *Cell Communications*, ed. Cox, R. P. (John Wiley and Sons, Ltd.), pp. 97—111.
21. Roseman, S. (1974) 'Complex carbohydrates and intercellular adhesion', in *The Cell Surface in Development*, ed. Moscona, A. A. (John Wiley and Sons, Ltd.), pp. 255—71.
22. Ozanne, B. and Sambrook, J. (1971) 'Isolation of lines of cells resistant to agglutination by concanavalin A from 3T3 cells transformed by SV40', in *The Biology of Oncogenic Viruses*, ed. Silvestri, L. G. (North Holland Publish. Co.), pp. 248—57.
23. Roseman, S. (1970) 'The synthesis of complex carbohydrates by multiglycosyltransferase systems and their potential function in intercellular adhesion', *Chem. Phys. Lipids*, **5**, 270—97.
24. Subak-Sharpe, H., Bürk, R. R. and Pitts, J. D. (1969) 'Metabolic cooperation between biochemically marked mammalian cells in culture', *J. Cell Sci.*, **4**, 353—67.

25. Stoker, M. G. P. (1967) 'Transfer of growth inhibition between normal and virus-transformed cells', *J. Cell Sci.,* **2,** 293–304.
26. Walther, B. T., Rauch, B. and Roseman, S. (1975) 'Sequential reactions in intercellular adhesion', manuscript in preparation.
27. Howard, P. F., Wetzel, B., Walther, B. T. and Roseman, S. (1975) 'Scanning electron-microscope studies of intercellular contacts in the cell layer assay', *J. Cell Biol.,* **67,** 179a.
28. Rittenhouse, H. G. and Fox, C. F. (1974) 'Concanavalin A mediated hemagglutination and binding properties of LM cells', *Biochem. Biophys. Res. Commun.,* **57,** 323–31.
29. Schnebli, H. P. and Bächi, T. (1975) 'Reaction of lectins with human erythrocytes. I. Factors governing the agglutination reaction', *Exp. Cell Res.,* **91,** 175–83.

Section V:
SEPARATION METHODS

A

Preparation of Immobilized Concanavalin A

CHAPTER 36

Affinity Chromatography on Immobilized Concanavalin A

KENNETH O. LLOYD

1. PRINCIPLE OF THE METHOD

The rationale for the use of insolubilized lectins to isolate and study complex carbohydrates follows from the earlier use of immobilized antibodies to isolate antigens (1). Although the method is limited to the isolation of carbohydrates for which specific lectins exist, the ease of preparation of large amounts of these reagents makes it a very attractive procedure. The utility of the method is further enhanced by the ability to elute the adsorbed components with easily available, low molecular weight inhibitors at neutral pH, in contrast to the more vigorous conditions necessary to elute many antigens from their insolubilized antibodies.

This chapter describes the preparation of insoluble forms of concanavalin A and

323

some examples of its use to isolate and fractionate some glycoproteins and polysaccharides. The method utilizes the ability of the immobilized reagent to adsorb specific glycoproteins or polysaccharides from mixtures containing other, non-reactive, polymers. The insolubilized concanavalin A is then washed with a buffer to elute non-specifically adsorbed components and the desired component is then eluted with an inhibitor such as D-glucose, methyl α-D-glucopyranoside or methyl α-D-mannopyranoside. Similar separations can be performed by precipit- ation with soluble concanavalin A (2,3) but using the insolubilized lectin simplifies the separation of the lectin from the reactive component.

2. PREPARATION OF INSOLUBILIZED CONCANAVALIN A

2.1 Direct Coupling to Agarose

A convenient procedure utilizes the cyanogen bromide-activated agarose method developed by Porath and coworkers (4) as applied to concanavalin A by Lloyd (5) and by Aspberg and Porath (6). A suitable preparation of agarose beads (e.g. 'Sepharose' 4B, Pharmicia Co.) is activated by adding CNBr (500 mg). dissolved in water (10 ml), to a suspension of beads (50 ml settled volume) in water (50 ml). The mixture is stirred at room temperature and the pH adjusted to 11 with 2 N NaOH. The pH is kept between 11 and 11.2 for 7 min by the addition of more base. The mixture is then filtered on a sintered-glass Buchler funnel and washed with cold water (200 ml) and cold 0.7 M NaCHO$_3$ (250 ml). To the concanavalin A to be coupled (650 mg), in 40 ml of 1 M NaCl, is added 425 mg of NaCHO$_3$ and then the activated 'Sepharose'. After 2 days at 4° with occassional stirring, the product is made into a column. The column is washed at 4° with 0.07 M NaHCO$_3$ (200 ml), 0.1 M sodium borate pH 8.5 in 1 M NaCl (1 litre), 0.1 M sodium acetate pH 4.1 in 1 M NaCl (1 litre) and finally 1 M NaCl containing 5 mM MnCl$_2$ and 5 mM CaCl$_2$. The last solution restores the cations necessary for the activity of concanavalin A which may have been lost by eluting with the pH 4.1 buffer. The derivative may be stored at 4° in 1 M NaCl either with or without 0.01% sodium azide. The proportions described result in gels containing about 5 mg of concanavalin A per ml of gel. Other concentrations may be prepared by varying the amount of CNBr used to activate the agarose (7). The capacity of the gel may be determined using dextran (5) or a radiolabelled glycoprotein (8). However, since the capacity of the gels for different components varies considerably, it is recommended that the capacity for each polysaccharide or glycoprotein be determined separately. Other workers have used variations of this basic procedure: [i] the washing procedure has been simplified (9), [ii] the coupling has been carried out in the presence of methyl α-D-mannoside (10) and [iii] the unreactive groups on the agarose have been blocked with glycine (11).

Concanavalin A coupled to agarose is available commercially from Pharmacia Fine Chemical Co., Miles Laboratories, Inc. and Sigma Chemical Co.

Caution

Cyanogen bromide is a highly toxic compound and suitable precautions should be taken during the experiment. It is advisable to work in a fume hood and to avoid acidifying the filtrates before disposal.

2.2 Coupling to Agarose via Spacer Arms

The advantages of using spacer arms separating the gel and the coupled ligand have been emphasized by Cuatrecasas (12). Such a derivative of concanavalin A has been described by Choi and Jenson (8). They used activated N-hydroxysuccinimide esters of diaminodipropylaminosuccinyl-agarose to prepare a gel containing 25 mg concanavalin A/ml of gel.

2.3 Immobilization by Cross-linking and Related Methods

2.3.1 Polymerization of Concanavalin A with Glutaraldehyde

The procedure of Avrameas and Ternyck (13) can be used. To a stirred solution (50 ml) of concanavalin A (822 mg) in 0.1 M sodium phosphate buffer, pH 7.0 is added aqueous 25% (v/v) glutaraldehyde (0.4 ml) over a period of 1 min. The reaction mixture is stirred in an ice bath for 1 h. The solution becomes turbid in a few minutes and gelatinized after 15 min. The suspension is centrifuged and the supernatant solution poured off. Virtually all the protein becomes insolubilized. The cross-linked reagent is washed by resuspension in 1 M NaCl (14) or in two portions of 0.15 M NaCl followed by three portions of either veronal buffer, 0.1 M pH 7.5 or acetate buffer, 0.15 M pH 5.0, both containing $MgCl_2$, $MnCl_2$ and $CaCl_2$ at 0.001 M final concentration (15). The reagent can be used in batch procedures (15) or can be made into a column by mixing with 'Bio-Gel' P10 (2.5 g, BioRad Laboratories, Richmond, Ca. U.S.A.) previously swollen in 1 M NaCl (14).

2.3.2 Preparation of Poly-L-leucyl Concanavalin A

Concanavalin A can also be insolubilized by polymerizing poly-L-leucyl side chains on to the protein (5) by the procedure of Kaplan and Kabat (16). Concanavalin A (15 mg) in 1 M NaCl (1 ml) is diluted with 0.07 M $NaHCO_3$ (2.0 ml). The solution is cooled to 4°, stirred gently, and solid L-leucine-carboxyanhydride (30 mg, Cyclo Chemical Co.) is added. After a few minutes a flocculent white precipitate forms. After 15 h at 4° with occasional stirring, the precipitate is removed by centrifugation and washed five times with 4 ml of 0.15 M NaCl in 0.018 M Na phosphate buffer, pH 7.2. When a weight ratio of anhydride/protein of 2.0 is used, almost all the concanavalin A becomes insolubilized. The reagent is stored in 1 M NaCl at 4°. The material can be made into a column by mixing an amount containing 120 mg of protein with 12 g of moist 'Bio-Gel' P2 previously swollen in the above phosphate buffer.

2.4 Coupling to Polyacrylamide Beads by the Acyl Azide Method

A very good alternative method for coupling proteins to solid-phase supports has been devised by Inman and Dintzis (17,18) using hydrazide derivatives of polyacrylamide beads as carrier. For the coupling of proteins the hydrazide polymer is converted to the azide form by treatment with nitrous acid; primary amino groups of lysyl and N-terminal residues react and form stable amide bonds.

$$\overset{O}{\overset{\|}{C}}-NH_2 \xrightarrow{NH_2-NH_2} \overset{O}{\overset{\|}{C}}-NH-NH_2 \xrightarrow[0^\circ C]{HNO_2} \overset{O}{\overset{\|}{C}}-N_3$$

$$\xrightarrow[pH\ 8-10]{R-NH_2} \overset{O}{\overset{\|}{C}}-NH-R$$

Polyacrylamide supports offer some advantages, like high chemical stability and low non-specific adsorption. Polyacrylamide hydrazide beads are commercially available from Bio Rad Laboratories under the trade name 'Hydrazide Bio-Gel' P.

3. ADSORPTION AND DESORPTION PROCEDURES

3.1 Column Methods

Most workers have preferred to work with gels in the form of columns either alone, in the case of concanavalin A—agarose, or mixed with 'Bio-Gel' polyacrylamide beads, in the case of the cross-linked preparations. It appears that the concanavalin A derivatives are able to remove reactive components from a mixture when they are present in either high or low concentrations. Application of the sample can be carried out in any convenient buffer (pH 6.5—7.5); many workers use buffers containing Mn^{2+} and Ca^{2+} (ca. 5 mM). After applying the sample the column should be eluted with the starting buffer until the protein or carbohydrate content of the eluate reaches background levels. These procedures can be carried out at 2° to 4° or at room temperature (see, however, paragraph 3.4).

3.2 Batch Procedures

This procedure has not been used extensively, however, Avrameas and Guilbert (15) used it to purify peroxidase and glucose oxidase from crude preparations. In such an experiment, 100 mg of glutaraldehyde cross-linked concanavalin A was mixed with 18.5 mg of a partially purified preparation of peroxidase in veronal buffer, 0.1 M pH 7.5 containing 1 mM Ca^{2+}, Mg^{2+} and Mn^{2+}. The mixture was stirred gently for 15 min at room temperature and then centrifuged at 5000g for 15 min at 4°. The precipitate was washed with the same buffer until the washings had an optical density of less than 0.02 at 280 nm. The adsorbed enzyme was then eluted with the same buffer to which methyl α-D-mannoside had been added to a final concentration of 0.2 M. The eluate was dialysed overnight against distilled water and freeze dried.

3.3 Use in the Presence of Detergents

Concanavalin A has been used successfully in the presence of 'Triton' X-100 (10) NP-40 (19) and sodium deoxycholate (20).

3.4 Desorption Procedures

Porath (7) has discussed the theoretical and practical considerations in choosing gradient or batch-wise elutions. Most workers have used one-step elutions with D-glucose, sucrose, methyl α-D-glucopyranoside or methyl α-D-mannopyranoside. More complete desorption should be obtainable by using an inhibitor with a higher binding affinity for concanavalin A than the bound component. However, since α-D-mannopyranose appears to have the optimum configuration and many glycoproteins contain such residues, this is not always feasible. Gradient elutions have the advantage that preferential elution of components having different affinities for concanavalin A may be obtained. Such a separation was obtained for carcinoembryonic antigen (CEA) using a two-step elution procedure (21).

It is generally considered that immobilized concanavalin A preparations are effective over a reasonable range of temperatures (2° to room temperature). Norden and O'Brien (22), however, reported that although acid β-galactosidases are adsorbed by concanavalin A—agarose at 2°, they could not be eluted by 0.75 M methyl α-D-mannoside at this temperature. Full recovery was not obtained until the elution was carried out at 23°. The result is surprising in that concanavalin A precipitates better with dextrans at room temperature than at 4° (23).

Borate-containing buffers may be useful in eluting very strongly adsorbed components since it has been shown that sodium borate inhibits the precipitation of polysaccharides by soluble concanavalin A (24). Kennedy and Rosevear (25) have used sequential elutions with phosphate and borate (0.01 M) buffers to separate glycogen and other polysaccharides into two populations not detected by other methods. These fractionations are thought to depend on the different physical characteristics and on the length and distribution of branches in the molecules. It would seem likely that the interaction with concanavalin A is disrupted by complexing of the borate with the polysaccharides although interaction with the concanavalin A or agarose cannot be ruled out.

3.5 Re-use

Insolubilized concanavalin A prepared either as concanavalin A—agarose or by cross-linking can be re-used many times. After use the column should be 'stripped' with a concentrated solution of inhibitor, e.g. 1.0 M methyl α-D-mannoside and then washed with starting buffer or with 1 M NaCl containing 5 mM Mn^{2+} and Ca^{2+}. After being used five to ten times the preparations begin to lose some activity although Avrameas and Guilbert (15) reported that gluteraldehyde cross-linked concanavalin A lost as much as 50% of its activity after being used five times. The use of Mn^{2+}- and Ca^{2+}-containing buffers and lower temperatures may help to prevent inactivation.

4. PITFALLS

4.1 Irreversible and Non-specific Adsorptions

Non-specific adsorptions appear not to be a problem using most forms of insolubilized concanavalin A. Data on the recovery of reactive components and on the possibility that some components become lost because of irreversible adsorption is more variable. The recovery of polysaccharides such as glycogen and dextrans is generally good (5,14). This is also the case with most glycoproteins (5,14). Exceptions, however, do occur as in a study of Hunt and coworkers (26) on mouse L cell glycoproteins in which it was found that 50% of the radioactivity of an applied sample was not eluted from the column. Similarly, Choi and Jensen reported (8) that bound components not eluted with 0.2 M methyl α-D-mannoside could subsequently be eluted with a buffer containing 10 M urea and 1% SDS. Interferon appears to interact with concanavalin A—agarose by virtue of hydrophobic interactions as well as sugar recognition and cannot be eluted with sugars from some preparations of adsorbents (27). Ethylene or propylene glycols in the presence of methyl α-D-mannoside or tetraethylammonium chloride can, however, be used to desorb interferon from the columns.

A more common observation is on the inability of all the concanavalin A-reactive components to be bound to the column. This was first observed in a study of blood group glycoproteins (5) in which it was found that a non-adsorbed fraction still retained its ability to be precipitated by soluble concanavalin A. Similar observations have been made with a human erythrocyte glycoprotein (10). The phenomenon has been explained as being due to greater numbers of reactive sites per molecule being required for binding to immobilized concanavalin A than for precipitation in solution (5). Since heterogeneity in their carbohydrate chains is

Table I.   Polysaccharides and glycoproteins purified by using immobilized concanavalin A

Component purified	Form of concanavalin A	Reference
Dextrans, yeast mannan	Bound to 'Sepharose'	5
Gastric blood group glycoproteins	Insolubilized with L-leucyl-N-carboxyanhydride	5
Glycogen, dextrans, IgM	Glutaraldehyde cross-linked	14
Human serum components	Bound to agarose	6
Peroxidase, glucose oxidase, serum glycoproteins	Glutaraldehyde cross-linked	15
Chorionic gonadotropin, lutenizing and FS hormones	Bound to 'Sepharose'	29
Teichoic acid	Bound to 'Sepharose'	30
Porcine pancreatic lipase	Bound to 'Sepharose'	31
Lymphocyte membrane glycoproteins	Bound to 'Sepharose'	20
Hepatitis B antigen	Bound to 'Sepharose'	32
Bovine rhodopsin	Bound to 'Sepharose'	33
Antitrypsin	Bound to 'Sepharose'	34
Human glycogen synthetases	Bound to 'Sepharose'	35
γ-Glutamyl transferase	Bound to 'Sepharose'	36
Human erythrocyte glycoprotein	Bound to 'Sepharose'	10
Carcinoembryonic antigen	Bound to 'Sepharose'	21
α-Fetoprotein	Bound to 'Sepharose'	37
β-Galactosidases	Bound to 'Sepharose'	22
Rabbit galactosyl transferase	Bound to 'Sepharose'	38
Lymphocyte membrane glycoproteins	Bound to 'Sepharose'	8
Mouse L cell glycoprotein	Bound to 'Sepharose'	26
HL−A antigen	Bound to 'Sepharose'	39
Human platelet glycoprotein	Bound to 'Sepharose'	28
Glucomannan	Gluteraldehyde cross-linked	40

a common property of glycoproteins, most glycoproteins will contain a population of molecules not having a suffficient number of reactive sites to be bound to insolubilized concanavalin A.

4.2 Leaching of Concanavalin A

One of the drawbacks of affinity chromatography is a tendency for the bound ligand to slowly dissociate from the resin and to appear in the eluates. Concanavalin A–agarose columns have a slight tendency to leach concanavalin A, however, the property appears to be variable (28) and some preparations appear to be more stable than others. Products isolated by this method should be checked by polyacrylamide gel electrophoresis for trace contamination with concanavalin A. Cross-linking of the agarose before coupling (7) may help to reduce this problem. Attention should also be paid to the thoroughness of washing of the coupled gel before use. The use of polyfunctional spacers to separate the gel from the protein may give more stable substitution than direct coupling to agarose (12).

5. EXAMPLES OF SEPARATIONS

Immobilized concanavalin A has been used quite widely for the isolation and fractionation of polysaccharides and glycoproteins. A list of these applications is given in Table I. Specific examples of its use are discussed in more detail in the next sections.

6. REFERENCES

1. Silman, I. H. and Katchalski, E. (1966) 'Water-insoluble derivatives of enzymes, antigens and antibodies', *Ann. Rev. Biochem.*, **35**, 873–908.
2. Leon, M. A. (1967) 'Concanavalin A reaction with normal human immunoglobulin G and Myeloma immunoglobulin G', *Science*, **158**, 1325–6.
3. Lloyd, K. O. and Bitoon, M. A. (1971) 'Isolation and purification of a peptido-rhamomannan from the yeast form of *Sporothrix schenckii*', *J. Immunol.*, **107**, 663–71.
4. Porath, J., Axén, R. and Ernbach, S. (1967) 'Chemical coupling of proteins to agarose', *Nature*, **215**, 1491–2.
5. Lloyd, K. O. (1970) 'The preparation of two insoluble forms of the phytohemagglutinin concanavalin A, and their interactions with polysaccharides and glycoproteins', *Arch. Biochem. Biophys.*, **137**, 460–8.
6. Aspberg, K. and Porath, J. (1970) 'Group specific adsorption of glycoproteins', *Acta. Chem. Scand.*, **24**, 1839–41.
7. Porath, J. (1974) 'General methods and coupling procedures', in *Methods in Enzymology*, Vol. XXXIV, eds. Jackoby, W. B. and Wilchek, M. (Academic Press, New York), pp. 13–30.
8. Choi, Y. S and Jenson, J. C. (1975) 'Concanavalin A binding proteins of lymphoid cell surface', *J. Exp. Med.*, **140**, 597–602.
9. Edelman, G. M., Rutishauser, U. and Millette, C. F. (1971) 'Cell fractionation and arrangement of fibers, beads and surfaces', *Proc. Nat. Acad. Sci. U.S.A.*, **68**, 2153–7.
10. Findlay, J. B. C. (1974) 'The receptor proteins for concanavalin A and *Lens culinaris* phytohemagglutinin in the membrane of the human erythrocyte', *J. Biol. Chem.*, **249**, 4398–403.
11. Porath, J. and Fryklund, L. (1970) 'Chromatography of proteins on dipolar ion adsorbents', *Nature*, **226**, 1169–70.

12. Parikh, I. March, S. and Cuatrecasas, P. (1974) 'Topics in the methodology of substitution reactions with agarose', in *Methods in Enzymology*, Vol. XXXIV, eds. Jakoby, W. B. and Wilcheck, M. (Academic Press, New York), pp. 77–102.
13. Avrameas, S. and Ternyck, T. (1969) 'The cross-linking of proteins with glutaraldehyde and its use in the preparation of immunoadsorbents', *Immunochem.*, 6, 53–66.
14. Donnelly, E. H. and Goldstein, I. J. (1970) 'Glutaraldehyde–insolubilized concanavalin A: an adsorbent for the specific isolation of polysaccharides and glycoproteins', *Biochem. J.*, 118, 679–680.
15. Avrameas, S. and Guilbert, B. (1971) 'Biologically active water-insoluble protein polymers. Their use for the isolation of specificity interacting proteins', *Biochemie*, 53, 603–14.
16. Kaplan, M. E. and Kabat, E. A. (1966) 'Studies of human antibodies. IV. Purifications and properties of anti-A and anti-B obtained by absorption and elution from insoluble blood group substance', *J. Exp. Med.*, 123, 1061–81.
17. Inman, J. K. and Dintzis, H. M. (1969) 'The derivatisation of cross-linked polyacrylamide beads. Controlled introduction of functional groups for the preparation of special purpose, biochemical adsorbents', *Biochemistry*, 8, 4074–82.
18. Inman, J. K. (1974) 'Covalent linkage of functional groups, ligands and proteins to polyacrylamide beads', in *Methods in Enzymology*, Vol. XXXIV, eds. Jacoby, W. B. and Wilchek, M. (Academic Press. New York), pp. 30–58.
19. Lloyd, K. O. (1975), unpublished results.
20. Allan D., Auger, J. and Crumpton, M. J. (1972) 'Glycoprotein receptors for concanavalin A isolated from pig lymphocyte plasma membranes by affinity chromatography in sodium deoxycholate', *Nature (New Biol.)*, 236, 23–5.
21. Rogers, G. T., Searle, F. and Bagshawe, K. D. (1974) 'Heterogeneity of carcinoembryonic antigen and its fractionation by Con A affinity chromatography', *Nature*, 251, 519–21.
22. Norden, A. G. W. and O'Brien, J. S. (1974) 'Binding of human liver β-galactosidases to plant lectins insolubilized on agarose', *Biochem. Biophys. Res. Commun.*, 56, 193–9.
23. So, L. L. and Goldstein, I. J. (1967) 'Protein–carbohydrate interactions. IV. Application of the quantitative precipitin method to polysaccharide–concanavalin A interaction', *J. Biol. Chem.*, 242, 1617–22.
24. Svensson, S., Hammarström, S. and Kabat, E. A. (1970) 'The effect of borate on polysaccharide–protein and antigen–antibody reactions and its use for the purification and fractionation of crossreacting antibodies', *Immunochem.*, 7, 413–22.
25. Kennedy, J. F. and Rosevear, A. (1973) 'An assessment of the fractionation of carbohydrates on concanavalin A–"Sepharose" 4B by affinity chromatography', *J. Chem. Soc. Perkins Trans.*, 2041–6.
26. Hunt, R. C., Bullis, C. M. and Brown, J. C. (1975) 'Isolation of a concanavalin A receptor from Mouse L cells', *Biochemistry*, 14, 109–15.
27. Davey, M. W., Sulkowski, E. and Carter, W. A. (1975) 'Purification of human interferon on concanavalin A–agarose', *Fed. Proc.*, 34, 638.
28. Nachman, R. L., Hubbard, A. and Feris, B. (1973) 'Iodination of the human platelet membrane. Studies of the major surface glycoproteins', *J. Biol. Chem.*, 248, 2928–36.
29. Dufau, M. L., Tsuruhara, T. and Catt, K. J. (1972) 'Interaction of glycoprotein hormones with agarose–concanavalin A', *Biochem. Biophys. Acta.*, 278, 281–92.
30. Doyle, R. J., Birdsell, D. C. and Young, F. E. (1973) 'Isolation of the teichoic acid of *bacillus subtilis* 168 by affinity chromatography', *Prep. Biochem.*, 3, 13–8.
31. Garner, C. W. and Smith, L. C. (1972) 'Porcine pancreatic lipase. A glycoprotein', *J. Biol. Chem.*, 247, 561–5.
32. Neurath, A. R., Prince, A. M. and Lippin, A. (1973) 'Affinity chromatography of hepatitis B antigen on concanavalin A linked to agarose', *J. Gen. Virol.*, 19, 391–4.
33. Steinemann, A. and Stryer, L. (1973) 'Accessibility of the carbohydrate moiety of rhodopsin', *Biochemistry*, 12, 1499–502.
34. Liener, I. E., Garrison, O. R. and Pravda, Z. (1973) 'The purification of human serum antitrypsin by affinity chromatography on concanavalin A', *Biochem. Biophys. Res. Commun.*, 5, 436–43.
35. Sølling, H. and Wang, P. (1973) 'A rapid method for the purification of glycogen synthetases I and D utilizing concanavalin A bound to agarose', *Biochem. Biophys. Res. Commun.*, 53, 1234–9.

36. Takahashi, S., Pollack, J. and Seifter, S. (1974) 'Purification of γ-glutamyltransferase of rat kidney by affinity chromatography using concanavalin A conjugated with "Sepharose" 4B', *Biochem. Biophys. Acta.*, **371**, 71—5.

37. Pagé, M. (1973) 'α-Foetoprotein: Purification of "Sepharose"-linked concanavalin A', *Canad. J. Biochem.*, **51**, 1213—5.

38. Podolsky, D. K., Weiser, M. M. LaMont, J. T. and Isselbacher, K. J. (1974) 'Galacto-syltransferase and concanavalin A agglutination of cells', *Proc. Nat. Acad. Sci.*, **71**, 904—8.

39. Billing, R. J. and Terasaki, P. I. (1974) 'Purification of HL-A antigen from normal serum', *J. Immunol.*, **112**, 1124—30.

40. Koleva, M. I. and Achtardjieff, C. (1973) 'Isolation of an electrophoretically homogeneous glucomannan using glutaraldehyde-insolubilized concanavalin A', *Carbohyd. Res.*, **31**, 142—5.

B

Separation of Soluble Macromolecules

CHAPTER 37

Some Examples of Isolations and Separations using Immobilized Concanavalin A: Blood Group Glycoproteins, Glucosylated Teichoic Acid, Ovalbumin, Carcinoembryonic Antigen

KENNETH O. LLOYD

1. FRACTIONATION OF BLOOD GROUP GLYCOPROTEINS

Unlike the A, B and H active glycoproteins from human ovarian cyst fluids, the blood group preparations from stomach mucosa are able to precipitate with concanavalin A (1,2). This reactivity is due to the presence of terminal 2-acetamido-2-deoxy-α-D-glucosyl residues which are not found in the ovarian cyst preparations (1). Using precipitation methods it is possible to fractionate hog gastric mucin samples into two fractions. One of these has only blood group activity whereas the other has both blood group and concanavalin A determinants (1). Hog blood group substance has also been fractionated on an insolubilized concanavalin A column (3). Poly-L-leucyl concanavalin A (containing 120 mg of concanavalin A) was made into a column (8.0 x 1.8 cm) by mixing it with 12 g of moist 'Bio-Gel' P2. The column was equilibrated with 0.15 M NaCl in 0.18 M sodium phosphate buffer, pH 7.2 and the sample (11.8 mg) applied in the same buffer. Continued elution removed an unabsorbed fraction (Figure 1). The adsorbed component was then eluted with 0.02 M methyl α-D-mannoside in 0.15 M NaCl in 0.18 M sodium phosphate buffer. Samples were isolated by dialysis and lyophilization.

333

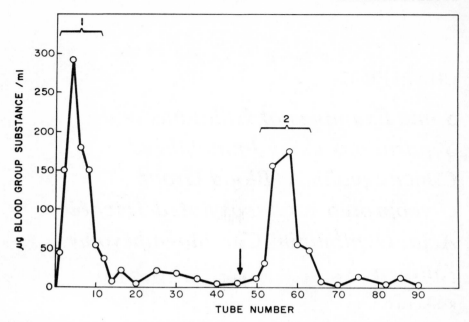

Figure 1. Separation of hog gastric mucin blood group glycoproteins on immobilized concanavalin A. The arrow indicates initiation of elution with 0.02 M methyl α-D-mannoside. (Reproduced by permission of Academic Press, Inc.)

Analytical data showed that the recovery was 92% and that the two fractions had essentially the same sugar composition as the original sample. When the ability of the fractions to precipitate concanavalin A in solution was examined it was found, unexpectedly, that the unadsorbed fraction (Figure 1) still retained some ability to precipitate concanavalin A (even though the column was not overloaded). Figure 2 compares these two fractions with the fractions obtained by a precipitation method (1) in their ability to precipitate with concanavalin A.

It was concluded that although affinity chromatography eliminates the problems involved in removal of the concanavalin A which are inherent in the precipitation method, the fractionation was not as efficient in providing a concanavalin A receptor-free sample.

2. ISOLATION OF GLUCOSYLATED TEICHOIC ACID

Doyle and coworkers (4) have described the rapid purification of α-glucosylated teichoic acid from *Bacillus subtilis* by affinity chromatography on concanavalin A—agarose. To a column (2.5 x 20 cm) of immobilized concanavalin A was added 45 mg of autolysate in 0.03 M Tris HCl buffer pH 7.3. The column was eluted with the same buffer and samples were collected and analysed (Figure 3). The glycan, peptide and peptidoglycan fragments emerged first as a heterogeneous fraction. The adsorbed teichoic acid was then eluted with 0.03 M methyl α-D-glucoside in the above buffer. The product was isolated by dialysis and lyophilization.

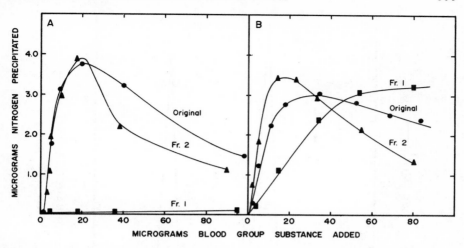

Figure 2A. Precipitation characteristics of hog gastric mucin blood group glycoproteins fractionated by precipitation with concanavalin A. Fractions 1 and 2 are the non-precipitated and precipitated fractions respectively. (Redrawn from reference 1 and reproduced by permission of the Williams and Wilkins Co.)

Figure 2B. Precipitation characteristics of hog gastric mucin blood group glycoproteins fractionated on immobilized concanavalin A. Fractions 1 and 2 are the unadsorbed and absorbed fractions respectively. (Redrawn from reference 3 and reproduced by permission of Academic Press Inc., New York.)

Some phosphorus-containing material always eluted in the first peak. The authors suggest that this may represent a wall contaminant or a non-glycosylated teichoic acid.

Apart from the ease of operation another advantage of the method is its mildness since it avoids the use of harsh extraction procedures such as hot or cold TCA, hydrazine or NaOH. The teichoic acid prepared by this method was found to have a higher sedimentation coefficient than preparations obtained by conventional methods. The composition of the teichoic acids prepared by the two methods was comparable.

3. FRACTIONATION OF OVALBUMIN

Although ovalbumin contains terminal α-D-mannosyl residues (5) it is not precipitated by concanavalin A in solution (6). Presumably this is because there is only one oligosaccharide chain on each ovalbumin molecule and therefore it is not able to form an insoluble lattice with the reagent. Despite its unprecipitability, however, ovalbumin is partly adsorbed by concanavalin A—agarose (6). In a typical experiment ovalbumin (5.7 mg) was applied to a column (18 x 1 cm) of con-canavalin A—'Sepharose' in 0.15 M NaCl in 0.018 M sodium phosphate buffer pH 6.5. Elution with this buffer was contained until the unadsorbed components (fraction 1) had been eluted. A gradient from 150 ml of buffer to 150 ml of 0.02 M methyl α-D-mannoside was then applied. A second component was then eluted

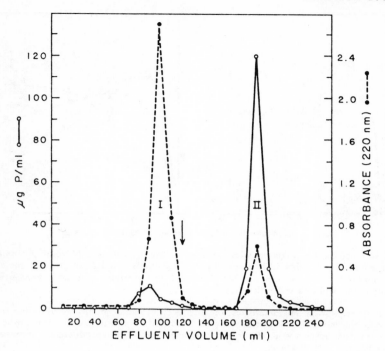

Figure 3. Affinity chromatography of a cell wall autolysate from *Bacillus subtilis* 168 on concanavalin A—agarose column. (Reproduced with permission of the authors and Marcel Dekker, Inc.)

(fraction 2). Both components were isolated by dialysis and freeze drying; the yields of fractions 1 and 2 were 3.0 mg (53%) and 2.3 mg (40%) respectively. Neither fraction was able to precipitate with soluble concanavalin A.

An explanation of the basis for the fractionation appeared when the carbo-hydrate composition of the fractions was determined. The original ovalbumin has a mannose:*N*-acetyglucosamine ratio of 2.2 whereas fraction 1 had a ratio of 2.0 and fraction 2 had a ratio of 2.8. It is clear that fraction 2 consisted of a population of molecules containing a sufficient number of terminal α-D-mannosyl residues for them to be bound by the immobilized concanavalin A. Huang and coworkers (5) have shown that such heterogeneities do exist in the carbohydrate chains of ovalbumin.

4. FRACTIONATION OF CARCINOEMBRYONIC ANTIGEN (CEA)

As discussed earlier, most workers have used a single-step elution to remove bound components from insoluble concanavalin A columns. The present example shows the advantage of using elutions with inhibitors of different concentrations. In this experiment, Rogers and coworkers (7) used concanavalin A—'Sepharose' to separate CEA into three distinct fractions having different antigenic properties. A

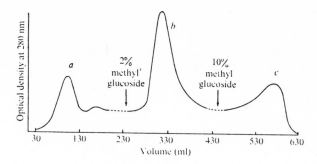

Figure 4. Affinity chromatography of partially purified carcinoembryonic antigen on concanavalin A—agarose. (Reproduced with permission of the authors and Nature.)

sample of CEA (98 mg) was dialysed against 0.1 M sodium acetate buffer (pH 6) containing 1 M NaCl and 10^{-3} M of $CaCl_2$, $MgCl_2$ and $MnCl_2$. The solution was applied to a column of concanavalin A—'Sepharose' (66 ml) and eluted with the buffer. After unbound fraction 1 was collected, a 0.11 M solution of methyl α-D-glucoside was applied and fraction 2 was collected. A third fraction was eluted with 0.55 M inhibitor (Figure 4). The fractions were pooled and isolated by membrane ultrafiltration. Immunological studies showed that fractions 1 and 2 were two forms of CEA with immunologically distinct determinants whereas fraction 3 was similar to a non-specific cross-reacting antigen (NCA) previously identified in CEA samples. These workers suggest that these fractions will enable them to improve the specificity of the CEA assay.

5. REFERENCES

1. Lloyd, K. O., Kabat, E. A. and Beychok, S. (1969) 'Immunochemical studies on blood groups. XLIII. The interaction of blood group substances from various sources with a plant lectin, concanavalin A', *J. Immunol.*, **102**, 1354—62.
2. Clarke, A. E. and Denborough, M. A. (1971) 'The interaction of concanavalin A with blood group substance glycoproteins from human secretions', *Biochem. J.*, **121**, 811—6.
3. Lloyd, K. O. (1970) 'The preparation of two insoluble forms of the phytohemagglutinin concanavalin A, and their interactions with polysaccharides and glycoproteins', *Arch, Biochem. Biophys.*, **137**, 460—8.
4. Doyle, R. J., Birdsell, D. C. and Young, F. E. (1973) 'Isolation of the teichoic acid of *Bacillus subtilis* 168 by affinity chromatography', *Prep. Biochem.*, **3**, 13—8.
5. Huang, C. C., Mayer, H. E. and Montgomery, R. (1970) 'Microheterogeneity and paucidispersity of glycoproteins. I. The carbohydrate of chicken ovalbumin', *Carbohydr. Res.*, **13**, 127—37.
6. Lloyd, K. O. (1974), unpublished data.
7. Rogers, G. T., Searle, F. and Bagshawe, K. D. (1974) 'Heterogeneity of carcinoembryonic antigen and its fractionation by Con A affinity chromatography', *Nature*, **251**, 519- 21.

CHAPTER 38

Separation of Glycoprotein Hormones with Agarose-Concanavalin A

Maria L. Dufau
Kevin J. Catt

The binding affinity of concanavalin A for carbohydrate moieties has been applied to chromatographic separation of a variety of glycoproteins present in plasma and tissue extracts (1). Such procedures were initially developed for fractionation of plasma glycoproteins (2), and have been more recently applied to chromatography of glycoprotein hormones of pituitary and placental origin (3). The presence of carbohydrate residues in the pituitary trophic hormones, thyroid-stimulating hormone (TSH), luteinizing hormone (LH) and follicle-stimulating hormone (FSH), and in the placental gonadotropins, human chorionic gonadotropin (hCG) and pregnant mare serum gonadotropin (PMSG), renders these molecules amenable to fractionation by group-specific affinity chromatography on agarose-coupled concanavalin A. Each of these trophic hormones is composed of two subunits, a relatively constant α subunit with similar structural features in all glycoprotein hormones, and a series of structurally distinct β subunits which confer

339

hormonal and target-cell specificity upon the native hormones formed by association of the α and β subunits. The glycoprotein nature of these several trophic hormones and their subunits has been shown to favour binding to agarose—concanavalin A, with subsequent displacement by sugars such as α-D-methyl glucopyranoside and mannopyranoside.

1. PRINCIPLE

Chromatography on agarose—concanavalin A has been applied to the purification of ^{125}I-labelled human chorionic gonadotropin, luteinizing hormone and follicle-stimulating hormone, yielding tracer preparations with improved binding affinity for specific antibodies and gonadal receptor sites. This procedure also permits the extraction of glycoprotein hormones from biological fluid (urine, plasma and cerebrospinal fluid) in high yield and with a moderate degree of purification after elution with methyl glucopyranoside.

The methodological applications of these procedures to purification of labelled gonadotropins are described in detail in this chapter. During initial experiments with agarose—concanavalin A, the affinity gel was prepared by coupling the purified lectin, obtained from Calbiochem, to agarose ('Sepharose' 6B, Pharmacia) which was activated by treatment with CNBr as previously described (3). More recently, the agarose—concanavalin A preparation has been obtained in prepared form from Pharmacia, and has been found to give results identical with those obtained by the use of the gel-lectin complex prepared in the laboratory.

2. GENERAL CHEMICAL CONSIDERATIONS

The most detailed information about carbohydrate residues of the glycoprotein hormones is available for human chorionic gonadotropin, a glycoprotein hormone of molecular weight 38,000 which is formed by the combination of α and β subunits of molecular weight 15,000 and 23,000, respectively. Human chorionic gonadotropin contains approximately 30% carbohydrate, distributed in seven saccharide units. The relevant loci for interaction with convanavalin A are four carbohydrate chains which are being linked to asparaginyl residues at positions 52 and 72 in the α subunit, and at positions 13 and 30 in the β subunit. The monosaccharide sequence in the branch carbohydrate units, determined by sequential removal of the monosaccharides with specific glycosidases, is N-acetyl-neuraminic acid (NANA) or fucose, followed by galactose, N-acetylglucosamine and mannose (4). Luteinizing hormone (LH), TSH and FSH probably contain basically similar saccharide chains, although ovine and bovine LH are devoid of sialic acid, and human LH contains less sialic acid than hCG. The α subunits of hCG, hLH, ovine LH and bovine LH contain significantly more mannose than the corresponding β subunits, and are generally richer in sugar residues (4—7). The binding of glycoprotein hormones to concanavalin A is of particular interest in relation to the carbohydrate content and composition of these molecules, and the variable presence of terminal sialic acid residues in the carbohydrate side chain. The more deeply situated mannose residues, with preferential binding affinity for the lectin, seem to be adequately accessible to permit interaction with the concanavalin A binding site. Also, N-acetylglucosamine has been shown to possess weak binding affinity for concanavalin A (8), and may participate in the binding of glycoprotein hormones to the gel—lectin complex. Conversely, N-acetylneuraminic acid does not bind to concanavalin A, and its presence may even reduce the affinity of glycoproteins for the complex, as indicated by the enhanced binding of desialylated hCG preparations described below. More detailed delineation of the carbohydrate chain structure of the glycoprotein hormones should permit a more precise definition of the role of the individual sugar residues in determining the overall affinities of these molecules for Concanavalin A.

3. FRACTIONATION OF HUMAN CHORIONIC GONADOTROPIN BY CHROMATOGRAPHY ON CONCANAVALIN A

3.1 Intact Human Chorionic Gonadotropin

When partially purified human chorionic gonadotropin with biological activity of 8000—10,000 IU/mg as determined by radioligand—receptor assay (9) was adsorbed to an 0.5 x 14 cm column of 'Sepharose'—concanavalin A, the hormonal activity could be eluted as a single peak with buffer containing 0.2 M methyl α-D-glucopyranose or mannopyranoside. The recovery from such columns, calculated by radioimmunoassay (10) and binding assay, was always over 85%. The biological activity estimated in pooled fractions with gonadotropic activity showed an increased potency of 12,000 to 14,000 IU/mg, representing a purification factor of about 1.5-fold (Figure 1). When crude human gonadotropin (Pregnyl, Organon;

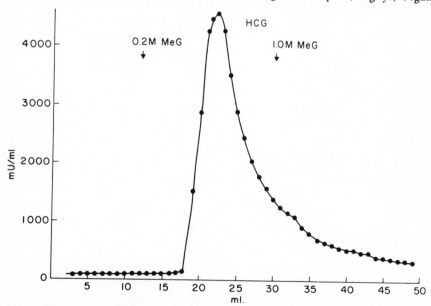

Figure 1. Elution pattern of human chorionic gonadotropin after adsorption to an agarose—concanavalin A column. The first twelve (1 ml) fractions were eluted by a solution of bovine α globulin (1 mg/ml) in phosphate-buffered soline (pH 7.4). Fractions 13—30 with 0.2 M methyl α-D-glucopyranoside (MeG) and fractions 35—50 with 1.0 M methyl α-D-glucopyranoside

1500 IU/mg) was fractionated by a similar procedure, the purification factor was 3—4-fold compared to the original material, with equally high recovery.

3.2 Chemically Modified Gonadotropins

Chromatography of desialylated human chorionic gonadotropin on agarose—concanavalin A usually gave elution profiles which were distinctly retarded in comparison to the pattern of intact unlabelled and ^{125}I-labelled hCG preparations, with relatively poor recovery (30%) of the asialo hormone (Figure 2). In addition, when preparations of asialo-agalacto human chorionic gonadotropin were adsorbed to 'Sepharose'—concanavalin A, no elution was observed with concentrations of methyl α-D-glucopyranoside up to 1 M, or even after treatment with 1.0 M disodium EDTA. When tracer ^{125}I-labelled intact human chorionic gonadotropin was applied simultaneously with the asialo-agalacto hCG preparation, the radioactive hormone was eluted with recovery of less than 5%. The majority of the radioactive hCG remained attached to the column, suggesting interaction with the asialo-agalacto chorionic gonadotropin preparation. A similar phenomenon was observed when mixtures of ^{125}I-labelled hCG and unlabelled asialo-agalacto hCG were subjected to electrofocusing in 4% acrylamide gel in a pH gradient from 3 to 10. In such experiments, the ^{125}I-hCG tracer did not focus in the usual position of pH 3—4, but instead was located as a broad peak at pH 6.3, where asialo-agalacto hCG was detected by radioimmunoassay. When such mixtures were subjected to gel chromatography on 'Sephadex' G-200, the radioactive material was similarly eluted with the asial-agaloacto hCG ($K_{a\,v} = 0.58$) instead of at the position characteristic of hCG ($K_{a\,v} = 0.41$).

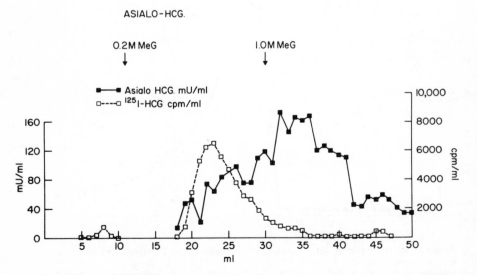

Figure 2. Chromatography of desialylated human chorionic gonadotropin (■———■, mIU/ml) and ^{125}I-labelled intact human chorionic gonadotropin (□———□, c.p.m./ml)

When ^{125}I-labelled α and β subunits of human chorionic gonadrotropin were adsorbed to agarose—concanavalin A columns, the elution patterns were essentially the same as those observed for the intact human gonadotropin preparations.

4. PURIFICATION OF ^{125}I-LABELLED GONADOTROPINS FOR RECEPTOR BINDING STUDIES AND RADIOIMMUNOASSAY

4.1 Purification of ^{125}I-hCG and ^{125}I-hLH

Chromatography of radioiodinated glycoprotein hormones on 'Sepharose'—concanavalin A columns provides a useful technique for preparation of tracers with improved binding properties for target cell receptors (3,11,12,14). For purification of ^{125}I-labelled hCG, LH or FSH, the iodination mixture from chloramine-T (14) or lactoperoxidase (13,15,16) labelling is transferred to a 5 × 140 mm column of 'Sepharose'—concanavalin A (Figure 3), followed by passage of 12 ml of phosphate-buffered saline containing 0.01% bovine γ-globulin. The labelled hormone is then eluted with the same solution containing 0.2 M methyl glucopyranoside (3). The best tracer is more highly retarded, and is recovered in fractions 25—33 from the affinity column. This tracer showed high binding to antibody-coated tubes (3,10)

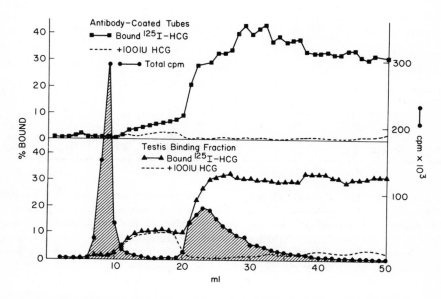

Figure 3. Purification of ^{125}I-labelled human chorionic gonadotropin tracer. The elution profile from an agarose—concanavalin A column is expressed in c.p.m. (●———●). Two radioactive peaks are shown at fractions 6—11, at fractions 20—30. Binding of tracer from each fraction to plastic tubes coated with anti-human chorionic gonadotropin serum (10) is shown in A (■———■) and binding to the rat testis receptor fraction in B (▲———▲). Displacement of binding to the solid-phase antibody or to the rat testis binding fraction (14) in the presence of excess human chorionic gonadotropin (100 IU) is shown in A and B (— — — —)

(up to 43% of 100,000 c.p.m. or 2 ng), to testis particulate receptor fractions (up to 60%), and to soluble testis and ovarian receptor (11,12) preparations (40%). In each case, complete displacement of binding to antibody or receptors was achieved by addition of excess hormone. A typical elution profile of tracer purification is shown in Figure 3, and the binding of hCG tracer prepared by cellulose adsorption (14) and concanavalin A purification to soluble testicular receptors are compared in Table I. Tracer prepared by a combination of the two procedures gives the highest specific uptake by testicular receptors and the lowest blank value, and such tracer preparations have been of considerable value for the use in binding studies using solubilized receptors of testis and ovary. However, for the majority of binding studies, tracer hormone purified by chromatography on agarose—concanavalin A has been satisfactory.

TABLE I.   Binding of tracer preparations (^{125}I-hCG) by soluble gonadotropin receptors. (Bound ^{125}I-hCG — percent of total c.p.m.)

Method of preparation	Cellulose purification				'Sepharose'—concanavalin A purification		Cellulose followed by 'Sepharose'—concanavalin A	
Column fraction no.	5	6	7	8	Fraction A (18—24)	Fraction B (25—33)	A	B
Content of labelled tracer (percent)	12	49	29	14	81	19	80	20
(a) Bound to soluble receptor	22	18	20	21	17	24	19	27
(b) Bound to soluble receptor +10 μg hCG	10	3	3	3	1	1	1	1
Specific binding	12	15	17	18	16	23	18	26

Chromatography of radioiodination reaction mixtures containing ^{125}I-labelled human luteinizing hormone or follicle-stimulating hormone on 'Sepharose'—concanavalin A was also found to be applicable to the purification of these glycoprotein hormones after labelling by the chloramine T (14) or lactoperoxidase procedures (15). In each case, the labelled hormone was adsorbed to the column and later eluted with 0.2 M methyl gluco- or mannopyranoside.

4.2 Purification of ^{125}I-hFSH

Human follicle-stimulating hormone tracer labelled by the lactoperoxidase procedure and purified on agarose—concanavalin A also displayed improved binding properties for hormone receptor sites, and is of value for binding studies with particulate and soluble receptor preparations of immature and adult rat testis (Figure 4).

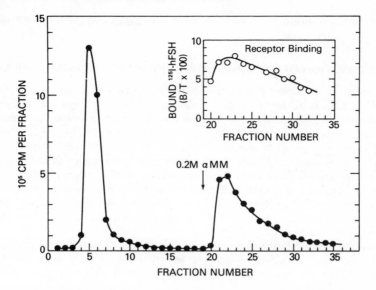

Figure 4. Elution of ^{125}I-hFSH from 'Sepharose'—concanavalin A by α-methyl-D-mannoside (α-MM). The insert shows the binding activity of the labelled hormone to rat testis homogenates in each fraction (13)

5. CHROMATOGRAPHY OF SOLUBLE LH/hCG RECEPTORS AND RECEPTOR—HORMONE COMPLEXES

The agarose—concanavalin A procedure can also be applied to the chromato-graphy of soluble hCG/hLH receptors extracted from testis and ovary by non-ionic detergents (11). When soluble free receptors or the ^{125}I-hCG—receptor complex were applied to the column, a peak of contaminating protein with no binding activity was eluted with the buffer, while binding activity or radioactive labelled receptor—hormone complex was recovered by elution of the column with competing sugar, indicating the presence of carbohydrate moieties on the receptor molecule. In every instance an increase of 5- to 10-fold in specific binding activity of the receptors was noted.

6. BINDING STUDIES WITH ANTIBODY-BOUND ^{125}I-hCG

The ability of labelled gonadotropins to interact with 'Sepharose'—concanavalin A is preserved after binding of the tracer hormone to specific antibody. Previous combination with antibody had no significant effect upon the binding of LH or hCG to concanavalin A. This observation is consistent with the lack of involvement of carbohydrate residues in the immunological activity of the hormone (17) and indicates that the relatively large complex formed by combination with antibody did not significantly reduce the accessibility of the carbohydrate components responsible for interaction with concanavalin A. On the other hand, glycoproteins extracted from biological fluids by adsorption to agarose—concanavalin A can then be assayed without elution, by subjecting the gel complex to radioimmunoassay or radioligand—receptor assay.

7. EXTRACTION OF GONADOTROPINS FROM BIOLOGICAL SAMPLES

Extraction of chorionic gonadotropin from human pregnancy plasma in high yield (84—95%) was achieved by this procedure as shown in Figure 5. The pattern of elution was similar to the one observed with human chorionic gonadotropin preparations with a main peak of elution in fractions 18—27 (Figure 5). Fractionation of monkey pregnancy plasma with a similar total recovery showed only a recovery of 43% associated with the fraction corresponding to intact hCG while 25% was associated with an early peak in fractions 4—11, indicating the probable existence of a species devoid of carbohydrate residues. The elution pattern of the major peak of monkey plasma chorionic gonadotropin also suggests that the circulating hormone does not include a significant proportion of desialylated gonadotropin.

Chromatography of human pregnancy urine on agarose—concanavalin A gave an elution profile that was clearly more retarded than that of pregnancy plasma. The hormone activity was eluted over a considerable wider area with an initial major

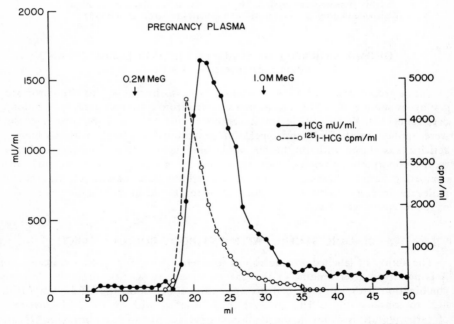

Figure 5. Fractionation of 16-week human pregnancy plasma and elution of ^{125}I-labelled human chorionic gonadotropin tracer from an agarose—concanavalin A column. The radioactive human chorionic gonadotropin elution pattern is expressed in c.p.m./ml (o — — — o), and the unlabelled human chorionic gonadotropin eluted from the column was detected by radioimmunoassay and expressed as mIU/ml (●——————●)

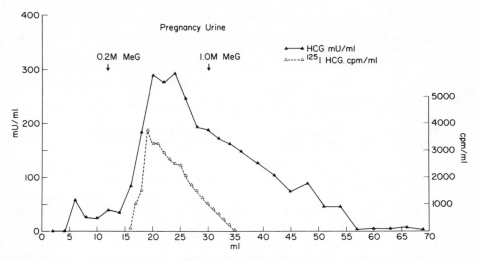

Figure 6. Chromatography of desialylated human chorionic gonadotropin (▲— — —▲, mIU/ml) and ¹²⁵ I-labelled intact human chorionic gonadotropin on agarose—concanavalin A (△— — —△, c.p.m./ml)

peak in fractions 20–30 and additional immunoreactive material also in fractions 30–50. The elution characteristics of the latter component were similar to those of desialylated human chorionic gonadotropin. In addition, gonadotropin present in plasma and urine or normal female, male and postmenopausal females can be extracted by this procedure and subjected to measurement by radioimmunoassay and or bioassays.

These examples of extraction and chromatography of glycoprotein hormones on agarose-coupled concanavalin A illustrate several of the uses of carbohydrate group-specific affinity fractionation procedures. The method does not provide a high-resolution method for isolation of the individual glycoprotein hormones, but is useful for batchwise extraction and partial purification of the glycoprotein hormone fraction from body fluids and tissue extracts. The most valuable aspects of the technique in our experience have been apparent during purification of radioactive tracer hormones after radioiodination, and during enrichment of partially purified hormone preparations. The specific examples cited in this chapter indicated the application of these procedures to LH, FSH and hCG, and their potential value for fractionation of other glycoprotein hormones such as TSH and PMSG. In particular, isotopic labelling for receptor studies has been most satisfactorily performed by lactoperoxidase iodination followed by purification on agarose—concanavalin A, and this combination can be recommended as a highly effective method for the preparation of radioiodinated glycoproteins.

8. REFERENCES

1. Leon, M. A. (1967) 'Concanavalin A reaction with normal immunoglobulin G and myeloma immunoglobulin G', *Science*, **158**, 1325–7.

2. Aspberg, K. and Porath, J. (1970) 'Group specific adsorption of glycoproteins', *Acta. Chem. Scand.,* **24**, 1839—41.
3. Dufau, M. L., Tsuruhara, T. and Catt, K. J. (1972) 'Interaction of glycoprotein hormones with agarose-concanavalin A', *Biochim. Biophys. Acta.,* **278**, 281—92.
4. Bahl, O. P. (1973) 'Chemistry of human chorionic gonadotropin', in *Hormonal Proteins and Peptides,* ed. Li, C. H. (Academic Press, Inc., N.Y.), pp. 171—99.
5. Closset, J. and Hennen, G. (1972) 'Chemical and immunological characteristics of the subunits of human pituitary, FSH and LH', in *Structure—Activity Relationships of Protein and Polypeptide Hormones,* eds. Margoulies, M. and Greenwood, F. C. (Excerpta Medica), pp. 325—6.
6. Papkoff, H. and Samy, T. S. (1967) 'Isolation and partial characterization of the polypeptide chain of ovine interstitial cell-stimulating hormone', *Biochim. Biophys. Acta.,* **147**, 175—7.
7. Reichert, L. E., Rasco, M. A., Ward, D. N., Niswender, G. D. and Midgley, A. R. (1969) 'Isolation and properties of subunits of bovine pituitary luteinizing hormone', *J. Biol. Chem.,* **244**, 5110—7.
8. Goldstein, I. J., Hollerman, C. E. and Smith, E. E. (1965) 'Protein—carbohydrate interaction. II. Inhibition studies on the interaction of concanavalin A with polysaccharides', *Biochemistry,* **4**, 875—83.
9. Catt, K. J., Dufau, M. L. and Tsuruhara, T. (1972) 'Radioligand receptor assay of luteinizing hormone and chorionic gonadotropin', *J. Clin. Endocrinol. Metab.,* **34**, 123—32.
10. Catt, K. J. (1970) 'Solid phase radioimmunoassay', *Acta. Endocrinol. (Copenhagen) Suppl.,* **142**, 222—42.
11. Dufau, M. L., Charreau, H. E. and Catt, K. J. (1973) 'Characteristics of a soluble gonadotropin receptor from the rat testis', *J. Biol. Chem.,* **248**, 6973—81.
12. Dufau, M. L., Charreau, H. E., Ryan, D. and Catt, K. J. (1974) 'Soluble gonadotropin receptors of the rat ovary', *FEBS Lett.,* **39**, 149—53.
13. Ketelslegers, J-M. and Catt, K. J. (1974) 'Receptor binding properties of ^{125}I-FSH prepared by enzymatic iodination', *J. Clin. Endocrinol. Metab.,* **39**, 1159—62.
14. Catt, K. J., Dufau, M. L. and Tsuruhara, T. (1971) 'Studies on radioligand-receptor assay system for luteinizing hormone and chorionic gonadotropin', *J. Clin. Endocrinol. Metab.,* **32**, 860—3.
15. Thorell, J. I. and Johansson, B. G. (1971) 'Enzymatic iodination of polypeptides with ^{125}I to high specific activity', *Biochim. Biophys. Acta.,* **251**, 363—6.
16. Dufau, M. L., Podesta, E. and Catt, K. J. (1975) 'Physical characteristics of the gonadotropin receptor—hormone complexes found *in vivo* and *in vitro*', *Proc. Nat. Acad. Sci. U.S.A.,* **23**, 359—68.
17. Schurrs, A. W. M., De Jager, E. and Homan, J. D. H. (1968) 'Studies on human chorionic gonadotropin', *Acta. Endocrinol.,* **59**, 120—38.

CHAPTER 39

Isolation of Dopamine-ß-hydroxylase

Dominique Aunis
Maria-Teresa Miras-Portugal

The isolation of dopamine-ß-hydroxylase (DBH), the final enzyme in the biosynthesis of noradrenaline from tyrosine, is of increasing importance because more and more studies in neurochemistry are being devoted to this enzyme. Until now, several purification methods have been developed but they appear rather fastidious due to the requirement of a number of steps. The use of concanavalin A has been successful in purifying the enzyme and the present chapter describes the method of isolating dopamine-ß-hydroxylase from bovine adrenal medulla.

1. PRINCIPLE OF THE METHOD

Bovine adrenal medullary dopamine-ß-hydroxylase (3,4-dihydroxy-phenylethylamine, ascorbate: O_2 oxidoreductase (ß-hydroxylating) E.C. 1.14.17.1) is a glycoprotein with a reported sugar content of between 4.6 g/100 g of protein (1) and 5.69 g/100 g of protein (2). The carbohydrate moiety is made up of the characteristic sugar components of glycoproteins including mannose, glucose, galactose, N-acetylglucosamine, N-acetylgalactosamine and N-acetylneuraminic acid (2). The presence of sugars has stimulated several groups to study the

349

interactions between DBH and lectins (3—5). Thus, DBH was found to interact strongly with concanavalin A (Con A) and the resulting insoluble complex was shown to be reversibly dissociated by α-methyl-D-mannoside (3—5). As the chromaffin granule proteins (chromogranins) were found to have no affinity for this lectin (3,5), DBH was fortunately the only protein to be retained on a column of Con A covalently bound to 'Sepharose' 4B, allowing one to easily purify the enzyme.

2. PREPARATION OF MATERIALS

2.1 Isolation of Chromaffin Granules

Chromaffin granules are routinely prepared by the improved procedure described by Smith and Winkler (6) as modified by Helle (7). Contamination accounts for less than 1% of the preparation (5,7). The washed pellet of chromaffin granules (5) is suspended in cold, deaerated water (1 ml of water for each gram of starting dissected adrenal medulla) and frozen and thawed twice. Centrifugation at 100,000g for 60 min in a R-40 rotor (Spinco centrifuge) allows the separation of the supernatant fluid, containing soluble proteins, from the pellet which is composed of membranes. The supernatant is collected and diluted with an equal volume of 0.2 M sodium acetate buffer (pH 6.5) containing 2 M NaCl and 2 mM each of $CaCl_2$ and $MgCl_2$. Protein concentration is about 1.5 mg/ml as measured by the Folin method of Lowry and coworkers (8).

2.2 Preparation of Insolubilized Con A

Con A covalently bound to 'Sepharose' 4B can be synthetized as described in the present book or purchased from Pharmacia Fine Chemicals (Uppsala, Sweden). When purchased, the gel sediment comes in 0.1 M sodium acetate buffer (pH 6.5) containing salts (1 M NaCl; 1 mM each of $MgCl_2$, $MnCl_2$ and $CaCl_2$) and 0.01% merthiolate. As merthiolate is a relatively strong inhibitor of DBH (Aunis and coworkers, unpublished observations), the gel has to be washed with the same buffer but free of merthiolate.

The gel is poured into a column (0.9 x 20 cm) to give a column bed 0.9 x 15 cm, when purification from 50 g of adrenal medulla is desired. The column is then washed with 100 ml of the acetate buffer at a flow rate of 20 ml/h. All subsequent operations are routinely carried out at $4°C$.

3. TECHNIQUE OF PURIFICATION OF DBH FROM CHROMAFFIN GRANULES

3.1 Absorption of DBH

The soluble lysate of the chromaffin granules, containing DBH, chromogranins, catecholamines, ATP and ions is passed through the column at a flow rate of 20 ml/h. After absorption of proteins, the column is washed extensively with the acetate buffer containing 1 M NaCl and 1 mM each of $CaCl_2$ and $MgCl_2$, until no protein can be detected in the effluent (a good criterion is the spectrophotometric measurement at 280 nm, when the absorbance is $< 0.01-0.005$). The volume of the washing buffer is about 250—300 ml.

3.2 Desorption of DBH

Elution of DBH is performed by passing a solution of 0.1 M α-methyl-D-mannoside in the acetate buffer free of salts. The enzyme immediately emerges from the column. However, as the interaction between DBH and Con A is a balanced binding, 200 ml were found to be necessary for complete recovery of the enzyme.

3.3 Concentration of DBH

The eluate containing DBH is collected and its concentration appears to be most satisfactory by ultrafiltration in an Amicon Diaflo cell through XM-100 A membranes under compressed N_2 gas, although slight inactivation can occur. The further removal of α-methyl-D-mannoside and salts is obtained by dialysis against water.

3.4 Re-use of the Con A—'Sepharose' 4B column

The Con A—'Sepharose' 4B column can be re-used by washing the column with 100 column volumes of deaerated, distilled water. The column is then allowed to equilibrate with 100 column volumes of 0.1 M sodium acetate buffer containing 1 M NaCl and 1 mM each of $CaCl_2$, $MgCl_2$ and $MnCl_2$.

When the column is not to be used during a prolonged period of time, merthiolate has to be added to prevent bacterial growth.

Successive washings, absorptions and elutions do not alter the binding capacity of the Con A, as we have observed in our laboratory where a column has been used for one year without a change in its performance.

4. PITFALLS

The major hazard is the elution of DBH with some chromogranin. This can be due to insufficient washing before elution with α-methyl-D-mannoside. Chromogranins are acidic proteins and may bind to Con A by protein—protein interactions. To circumvent contamination, extensive washing in buffer of high ionic strength is required but this procedure does not inactivate the enzyme since binding on Con A seems likely to protect DBH from denaturation.

5. COMPARISON WITH OTHER METHODS

The purification and recovery factors are summarized in Table I. From 50 g of original dissected bovine adrenal medulla, 4 mg of DBH can be readily obtained in two days, the enzyme being pure as judged by polyacrylamide gel electrophoresis (Figure 1), immunodiffusion and NH_2-terminal amino acid analyses (9).

The present method has the great advantage of providing an enzyme preparation of the highest specific activity [320 μmoles of octopamine/30 min/mg of protein or

TABLE I.   Isolation^a of bovine adrenal medullary dopamine-β-hydroxylase by the use of immobilized concanavalin A

Steps	Activity (μmoles/30 min)	Protein (mg)	Specific activity (μmoles/30 min/mg)
Washed chromaffin granules	4550	525	8.6
Soluble lysate	2200	130	17.0
Eluate from Con A column^b	1056	3.3	320

^{a}Isolation of DBH from 42 g of starting dissected adrenal medulla.
^{b}After concentration in an Amicon Diaflo cell (XM-100 A membranes).

600 μmoles/30 min/mg of protein, depending upon the conditions of measurement (3,5,10)]. When compared to the multistep methods originally described by Friedman and Kaufman (11) or Foldes and coworkers (12), the present method is characterized by its simplicity. Moreover, it appears preferable to a recent method using affinity chromatography on tyramine coupled to 'Sepharose' (13) because the latter method needs the synthesis of a somewhat sophisticated matrix.

A relatively high amount of DBH can be easily obtained. Such preparations are thus available for the study of the structure (see 9) which, of course, requires a lot of material, but other purposes may be envisaged. On the other hand, the soluble lysate of chromaffin granules can be obtained free from DBH, allowing for example the further purification of chromagranin A without contaminating DBH (which is known to be highly antigenic and to interfere in immunological studies using chromogranin A).

6. ATTEMPTS TO PURIFY DBH FROM OTHER ORGANS

Recently our group has purified DBH from human serum. The method of purification includes, among several steps, chromatography on immobilized Con A (14,15). The principle and details of the method are the same as described here, except that DBH is not pure at this stage but requires additional treatment due to the presence in serum of large amounts of glycoprotein.

A similar procedure has also been used for the purification of DBH from human pheochromocytoma; the enzyme has been obtained pure in one step (Miras-Portugal and Aunis, unpublished results).

7. REFERENCES

1. Wallace, E. F., Krantz, M. J. and Lovenberg, W. (1973) 'Dopamine-β-hydroxylase: a tetrameric glycoprotein', *Proc. Nat. Acad. Sci. U.S.A.*, **70**, 2253–5.
2. Aunis, D., Miras-Portugal, M. T. and Mandel, P. (1974) 'Bovine adrenal medullary dopamine-β-hydroxylase : studies on the structure', *Biochim. Biophys. Acta.*, **365**, 259–73.
3. Rush, R. A., Thomas, P. E., Kindler, S. H. and Udenfriend, S. (1974) 'The interaction of dopamine-β-hydroxylase with concanavalin A and its use in enzyme purification', *Biochem. Biophys. Res. Commun.*, **57**, 1301–5.
4. Wallace, E. F. and Lovenberg, W. (1974) 'Studies on the carbohydrate moiety of dopamine-β-hydroxylase : interaction of the enzyme with concanavalin A', *Proc. Nat. Acad. Sci. U.S.A.*, **71**, 3217–20.

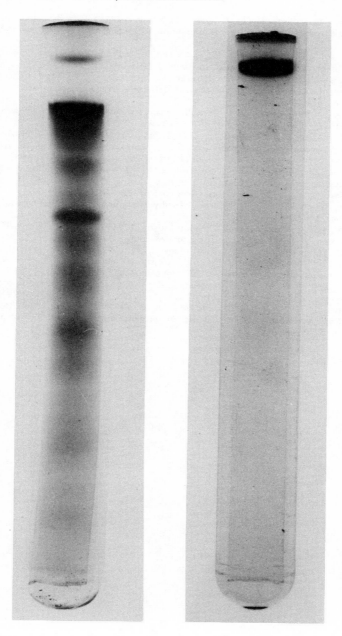

Figure 1. Polyacrylamide gel electrophoresis of total soluble proteins in chromaffin granule lysate (left) and of DBH eluted from Con A–'Sepharose' 4B column (right) [Reproduced by permission of copyright holder from reference 5.]

5. Aunis, D., Miras-Portugal, M. T. and Mandel, P. (1975) 'Bovine adrenal medullary dopamine-β-hydroxylase : studies on interaction with concanavalin A', *J. Neurochem.*, **24**, 425—31.

6. Smith, A. D. and Winkler, H. (1967) 'A simple method for the isolation of adrenal chromaffin granules on a large scale', *Biochem. J.*, **103**, 480—2.

7. Helle, K. B., Flatmarck, T., Serck-Hanssen, G. and Lönning, G. (1971) 'An improved method for the large-scale isolation of chromaffin granules from bovine adrenal medulla', *Biochim. Biophys. Acta.*, **226**, 1—8.

8. Lowry, O. H., Rosebrough, H. J., Farr, A. L. and Randall, R. J. (1951) 'Protein measurement with the Folin phenol reagent', *J. Biol. Chem.*, **193**, 265—75.

9. Aunis, D., Allard, D., Miras-Portugal, M. T. and Mandel, P. (1975) 'Bovine adrenal meduallary chromogranin A : studies on the structure and further evidence for identity with the dopamine-β-hydroxylase subunit', *Biochim. Biophys. Acta*, **393**, 284—95.

10. Rush, R. A., Kindler, S. H. and Udenfriend, S. (1974) 'Homospecific activity, an immunologic index of enzyme homogeneity : changes during the purification of dopamine-β-hydroxylase', *Biochem. Biophys. Res. Commun.*, **61**, 38—44.

11. Friedman, S. and Kaufman, S. (1965) '3,4-Dihydroxyphenylethylamine β-hydroxylase', *J. Biol. Chem.*, **240**, 4763—73.

12. Foldes, A., Jeffrey, P. L., Preston, B. N. and Austin, L. (1972) 'Dopamine-β-hydroxylase of bovine adrenal medulla. A rapid purification procedure', *Biochem. J.*, **126**, 1209—17.

13. Aunis, D., Miras-Portugal, M. T. and Mandel, P. (1973) 'Bovine adrenal medullary dopamine-β-hydroxylase : purification by affinity chromatography, kinetic studies and presence of essential hystidyl residues', *Biochim. Biophys. Acta.*, **327**, 313—27.

14. Miras-Portugal, M. T., Aunis, D. and Mandel, P. (1975) 'Etude structurale de la dopamine-β-hydroxylase purifiée du sérum humain', *C.R. Acad. Sci. (D) (Paris)*, **280**, 479—82.

15. Miras-Portugal, M. T., Aunis, D. and Mandel, P. (1975) 'Human serum dopamine-β-hydroxylase: purification, molecular weight, presence of sugars and kinetic properties', *Biochimie*, **57**, 669—75.

CHAPTER 40

Separation of Virus-induced Antigens by Affinity Chromatography on Concanavalin A-'Sepharose'

GARY H. COHEN
MANUEL PONCE DE LEON
HELENA HESSLE
ROSELYN J. EISENBERG

Concanavalin A (Con A), a protein isolated from the jack bean, forms specific complexes with certain polysaccharides and glycoproteins (1—3). The reaction between Con A and the carbohydrate-containing molecule is specific for α-D-glycosyl-type residues (4—6). However, it is important to note that Con A also binds to and may precipitate polyelectrolytes and polysaccharides which are devoid of terminal mannopyranoside (or related) residues (7—9). The fact the Con A is able to interact with two different classes of compounds, and by different mechanisms, increases the potential usefulness of the lectin in purification of viral antigens.

In applying Con A affinity chromatography to the separation of glycoprotein antigens from herpes simplex virus-infected (HSV) cells we found two distinct groups of viral antigens were isolated (10). Group I includes glycoprotein antigens which bind to Con A—'Sepharose' by a specific mechanism. These antigens are eluted from the column by α-methyl mannoside (α-MM) or α-methyl glucoside (α-MG). Group II contains antigens which bind to Con A—'Sepharose' by a non-specific or electrolytic mechanism (7—9). These antigens are eluted by high concentrations of salts such as NaCl. In this chapter, we will describe the method used for accomplishing this separation and shall present some experiments which well help clarify the chemical and biological nature of the two groups of viral antigens separated by Con A affinity chromatography.

1. MATERIALS AND PREPARATIONS

1.1 Source of Biological Material

Virus proteins may be divided into three groups based on their function:

1) structural proteins present in mature virions;
2) the viral-induced enzymes which are present in infected cells but may or may not be present in mature virions;
3) the so-called soluble antigens, other than enzymes, that are generally found in large quantity in virus-infected cells.

In order to characterize individual virus-coded proteins in terms of function as well as structure, two major lines of approach have been employed. The first and most direct approach involves mechanical or chemical disintegration of preparations of purified virus into component parts. The primary advantage of this procedure is that it avoids the problems of contamination of viral antigens with host-cell components. However, there are several serious disadvantages to this approach. It is necessary to obtain large quantities of the virus in purified form. In the case of enveloped viruses such as HSV, purification of infectious virus has only recently been achieved (11). Furthermore, the mature virion does not usually contain all of the proteins coded for by the virus. A second approach has involved fractionation of the soluble antigen components of the infected cell. One advantage of this approach is that large amounts of the antigen are readily obtained. Moreover, this method allows detection and isolation of non-virion antigens. However, a serious disadvantage of this approach is the heavy contamination of the soluble antigen fraction with host-cell components. Since detection of viral components in infected cells relies heavily on immunological procedures, non-immunogenic or weakly immunogenic viral components may be difficult to study. Because of the heavy contamination of extracts with host antigens, it is important that the antiserum preparations do not react with these proteins. Two procedures are utilized to minimize this problem: 1) antisera made against extracts of infected cells are thoroughly absorbed with normal cell extract to produce sera which react only with unique virus-induced antigens or 2) infected cells of one species are employed to make antisera which are then used to assay the virus-induced antigens produced by a cell of a different species.

In studying the antigens of herpes simplex virus-infected cells we utilized the second approach involving fractionation of the soluble antigen components of the infected cell. Further details on the uses and pitfalls of this technique have been described by Wilcox and Cohen (12).

2. AFFINITY CHROMATOGRAPHY WITH CON A–'SEPHAROSE'

2.1 Choice of Buffers and Eluants

The choice of starting buffer depends to a large extent on the stability of the substance to be adsorbed. To purify viral antigens from HSV-infected cells we routinely employed 0.01 M phosphate buffer, pH 7.2 (PB). Buffers employed in other systems are listed in Table I.

To remove glycoproteins bound to Con A–'Sepharose' by a specific mechanism, α-MM or α-MG at a concentration of 0.2 M is generally employed (Table I). To elute HSV glycoproteins we dissolved the saccharide in water rather than in PB (10). For an unexplained reason, phosphate buffer prevented the elution of glycoproteins from Con A–'Sepharose'. We routinely employed the α-MM supplied by Sigma Chemical Co. (St. Louis, Mo. 63178). This preparation, in contrast to the product sold by other companies, contained no ultraviolet absorbing material.

To remove substances bound to Con A–'Sepharose' by a non-specific (ionic) mechanism, a strong salt solution such as 2 M NaCl was usually employed. However, we have successively eluted material with 0.2 M NaCl.

A typical elution protocol would consist first of a PB wash to remove unbound material, elution of non-covalently bound material with 0.2 M α-MM or a gradient of α-MM from 0.1 M to 1 M followed by a second PB wash and finally elution with 2 M NaCl. The Con A-Sepharose column is then washed extensively with PB and stored as a suspension in this buffer at $4°C$. A preservative (1% toluene) is added for long-term storage.

2.2 Sieving properties of Con A–'Sepharose'

Because Con A is bound to a matrix which by itself acts as a molecular sieve, some separations can be made on the basis of size. Solling and Wang (17) reported that glycogen synthetase could be eluted by α-MM before other leukocyte glycoproteins due to a gel filtration effect.

2.3 Preparation of Adsorbent for Use

2.3.1 Removal of Unbound Con A

It is necessary to wash Con A–'Sepharose' extensively before use in order to remove unbound Con A. This can be accomplished most easily by washing the adsorbent packed in a column with the starting buffer and then with each successive eluting solution until no 280 nm absorbing material (Con A) is found in the effluent. A final wash with the starting buffer should then be carried out before the adsorbent is mixed with the experimental material. In our experience this procedure removed the bulk of the unbound Con A. However, some Con A was always found in every eluted fraction of experimental material (10).

2.3.2 Column Choice and Packing

The dimensions of the column are not critical for selective elution of substances with α-MM and NaCl. However, if molecular sieving is desired, the same considerations that apply to methods involving gel filtration are valid for Con A–'Sepharose'. In purifying HSV antigens the same elution pattern (Figure 1) was observed regardless of column dimensions.

For packing the column, we suspended the gel in approximately twice the volume of buffer and applied the suspension in a single pouring to the column (as per instructions for packing 'Sepharose' 4B).

TABLE I. Isolation of viral components by adsorption to Con A–'Sepharose'

Virus	Component isolated	Con A preparation	Method of elution	Starting buffer	Eluant	Reference
Herpes simplex	(1) Glycoprotein (2) Mucopolysaccharide	Miles ('Glycosylex')	Affinity	0.01 M PB, pH 7.2	0.2 M α-MM in H_2O 2 M NaCl	10 and this paper
Friend leukaemia	Glycoprotein	Pharmacia	Affinity	PBS	0.1 M α-MM in PBS	19
Hepatitis B	Hepatitis B antigen	Pharmacia	Affinity	0.01 M Tris pH 7.2 + 0.001 M $CaCl_2$ + 0.001 M $MnCl_2$	5% α-MM in 0.02 M tris-maleate, pH 6.0	39
Marek's disease	Glycoprotein	Activated 'Sepharose' 4B + Con A	Batch	0.1 M Acetate buffer, pH 6.0 containing 2.5% 'Nonidet'	α-MG + α-MM in 0.1 M acetate, pH 6.0	21
Herpes virus of turkeys	Glycoprotein	Activated 'Sepharose' 4B + Con A	Batch	0.1 M Acetate buffer, pH 6.0 containing 2.5% 'Nonidet'	α-MG + α-MM in 0.1 M acetate, pH 6.0	21

359

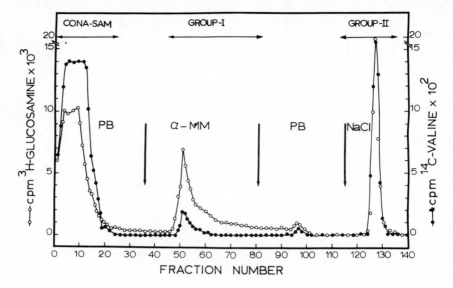

Figure 1. Affinity chromatography of HSV–SAM on Con A–'Sepharose'. The Con A column (50 ml) was prepared and developed as described in section 3.3. Fractions of 5 ml were collected (○), ³H-glucosamine label; (●), ¹⁴C-valine label. The arrows indicate the change of elution solution. Abbreviations: HSV-SAM, the soluble antigen mixture from infected cells; PB, 0.01 M phosphate buffer, pH 7.2; α-MM, 0.2 M α-methyl-D-mannoside; and NaCl, 2 M NaCl. [Reprinted with permission of ASM Publications from Ponce de Leon, M., H. Hessle and G. H. Cohen, *J. Virol.*, **12**, 766–74, (1973).]

2.3.3 Capacity of column

To purify HSV antigens, we generally employed a column with a bed volume in the ratio of 1 ml packed 'Glycosylex'-A (Miles laboratories) to 2 mg of protein [determined by the method of Lowry (18)] .

2.3.4 Flow Rates and Temperature

For HSV-infected cell extracts, we mixed the sample with 'Glycosylex'-A by gentle swirling for 2 hours at 36°C. The slurry was then repacked and elution was begun immediately after the bed had settled.

The flow rate can be speeded up considerably once elution is begun. We routinely employed a flow rate of 60 ml per hour.

The column can be packed and eluted at room temperature if the substance to be eluted is not temperature sensitive. Otherwise, the entire procedure including the adsorption step can be carried out at 4°C. It is essential in this case to increase the adsorption period from 2 hours to overnight since the rate of adsorption is temperature dependent (20).

2.3.5 Re-use and Storage of Con A—'Sepharose'

Con A—'Sepharose' can be re-used several times with reproducible results. After each use the adsorbent should be washed extensively with the starting buffer and stored as described in section 2.2. Con A—'Sepharose' should not be frozen, as the adsorbent will be irreversibly destroyed under these conditions. Because Con A slowly dissociates from the matrix during storage even at 4°C, the adsorbent should be thoroughly washed immediately before the next use to remove unbound Con A.

2.4 Separation of Virus-induced Antigens by Con A—'Sepharose' Chromatography

A list of viral components purified utilizing Con A—'Sepharose' is shown in Table I. Although the list is small, we are confident that glycoproteins from many viral sources can be effectively separated by this technique. Most of the methods listed involve column chromatography but at least one batch method has been reported (21).

3. USE OF CON A—'SEPHAROSE' TO PURIFY HERPES SIMPLEX VIRUS ANTIGENS

3.1 Choice and Preparation of Biological Material

In the studies to be described, baby hamster kidney cells (BHK) infected with herpes simplex virus strain HF (oral, type 1) were employed. Propagation of the cells, titrations and preparation of virus stock were performed as previously described (22).

To obtain the virus-specific soluble antigen preparations, monolayer cultures of BHK cells were infected at an input multiplicity of 5—10 plaque-forming units/cell. The cells were harvested after 18 hours, washed 3 times with phosphate-buffered saline (PBS) pH 7.2 and concentrated by centrifugation (10 min at 800g). The cell pellet was suspended in PBS to a final concentration of 2×10^7 cells/ml and disrupted by sonication. The preparation was centrifuged for 2 hours at 100,000g after which the supernatant fluid was collected, dialysed exhaustively against PB and concentrated in a Diaflo ultrafiltration chamber with an XM-50 membrane (Amicon Corp.). The soluble antigen mixture (SAM) from infected cells will be referred to as HSV-SAM.

Isotopically labelled SAM was obtained from infected or uninfected BHK cells by incubating the culture for 15 hours in medium containing ^{3}H-glucosamine (1.0 μCi/ml; sp. act. 1—2 Ci/mmole), ^{14}C-valine (0.5 μCi/ml; sp. act. 35 mCi/mmole) or ^{35}S-sulphate (1.0 μCi/ml; sp. act. 1000 mCi/mmole). The radioactive label was added 3 hours after infection. All radioisotopes were purchased from New England Nuclear Corp.

3.2 Adsorption of Sample to Con A—'Sepharose'

A 50 ml (settled volume) column of 'Glycosylex'-A was washed successively with PB, 0.2 M α-MM, PB, 2 M NaCl and PB until no further 280 nm absorbing material was eluted. The washed adsorbent was mixed for 2 hours at $36°C$ with HSV-SAM (1.5—3.0 mg protein/ml) and the column was repacked.

3.3 Elution of HSV Antigens from Con A—'Sepharose'

Figure 1 illustrates a typical elution profile of HSV-SAM glycoprotein removed from the column with different developing solutions. In this experiment the elution of protein label (^{14}C-valine) was closely correlated with elution of carbohydrate (^{3}H-glucosamine). The first peak fraction designated Con A-SAM corresponds to the unbound protein washed from the column with PB. Analysis of this material by SDS polyacrylamide gel electrophoresis (SDS-PAGE) (23) indicated that it contained a complex mixture of protein-staining bands (Figure 2a) and closely resembled the pattern of HSV-SAM (not-shown). The second fraction, designated Group I contained the peak of radioactivity eluted from the column with α-MM. SDS-PAGE analysis of this fraction (Figure 2b) indicated the presence of at least 2 major protein-staining bands and a number of minor bands as well. A considerable amount of Con A was present in this fraction (compare Figure 2a with 2d).

The column was washed free of α-MM with PB and the third fraction, designated Group II was eluted with 2 M NaCl. This fraction appears to contain a number of protein-staining bands (Figure 2c). Most of these bands do not correspond in position to those found in the Group I fraction (Figure 2b). These results indicate that Group I and Group II contain different components. It is also important to note the reduction in the number of bands seen by SDS-PAGE in the Group I and Group II fraction compared with Con A-SAM or HSV-SAM (Figures 2a, b and c). This indicates the Con A—'Sepharose' chromatography is useful in the purification of both Group I and Group II antigens. On the basis of specific activity (c.p.m. ^{3}H-glucosamine per mg protein) Con A—'Sepharose' chromatography purified the Group I and Group II fractions tenfold compared to HSV-SAM.

3.3.1 Recovery of Radioactive Counts from Con A—'Sepharose'

More than 90% of the input radioactive counts were routinely recovered in these experiments. No further radioactivity was removed by treatment of the column with α-MG nor with 0.1 M HCl (which destroys the binding capacity of Con A.) This indicates that all of the radioactive glycoproteins were removed by α-MM and NaCl treatment.

3.3.2 Presence of Con A in Eluants

Even though the column was exhaustively prewashed, small amounts of Con A were found in each of the eluted fractions. Con A was detected by gel diffusion employing anti-Con A serum (Miles). The radial gel diffusion method of Mancini (24) was performed to quantitate the concentration of Con A which was eluted. The amount of Con A eluted in each fraction varied from 1—10% of the total protein.

Con A was also seen as a heavy band at the bottom of SDS-polyacrylamide gels of the eluted fractions (Figure 2b).

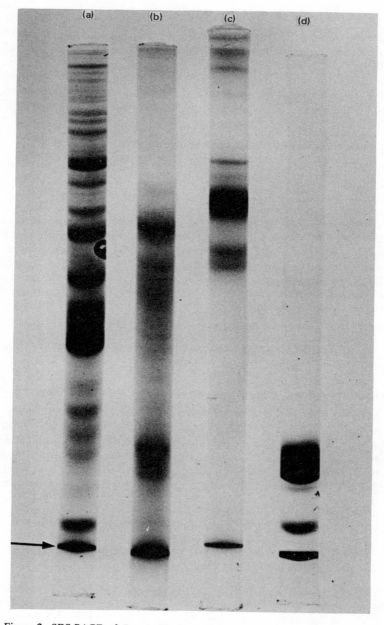

Figure 2. SDS-PAGE of Con A—'Sepharose' fractions. Samples were treated with SDS, subjected to electrophoresis in 10% gels according to the technique of Laemlii (23) and stained for protein with Coomassie brilliant blue. (a) Con A-SAM; (b) Group I; (c) Group II; (d) concanavalin A (Miles). Arrow indicates position of marker dye (brom phenol blue).

3.3.3 Efficiency of Adsorption

When the Con A-SAM fraction was collected, concentrated and recycled through the insolubilized Con A column, approximately 20% of the counts in this fraction were bound to the column. When the remaining 80% of the counts were recycled again, about 20% of these counts were bound. We conclude that the column did not have the capacity to adsorb all of the ^{3}H-glucosamine-labelled glycoprotein from SAM in a single run.

3.4 Use of Con A in the Presence of Detergents

We tested the effect of the non-ionic detergent NP-40 on the antigenic activity of the Group I and Group II fractions. Gel diffusion analysis (25) showed that the antigenic activity of these fractions was not destroyed by NP-40 treatment.

When ^{3}H-glucosamine-labelled HSV-SAM and ^{14}C-valine-labelled HSV-SAM were solubilized with 0.5% NP-40 (37°C for 1 hour) and co-chromatographed on Con A–'Sepharose' in the presence of NP-40 (0.1%) the radioactive elution profile was virtually identical to that observed in Figure 1 (10). In this experiment all of the developing solutions were prepared in 0.1% NP-40 to prevent aggregation of proteins or reassociation of lipid and protein. Thus the adsorption and elution steps can be carried out in the presence of NP-40 with no apparent effect on the binding properties of Con A–'Sepharose'.

4. BIOLOGICAL ANALYSIS OF HSV ANTIGENS ELUTED FROM CON A–'SEPHAROSE'

4.1 Complications caused by the Presence of Con A in Column Fractions

It has been shown that Con A forms precipitates with serum proteins (26,27). This property of Con A complicates immunodiffusion and serum blocking assays. Furthermore, HSV has receptor sites for Con A, and upon exposure to the lectin, the virus becomes non-infectious (28). This phenomenon complicates serum blocking assays. To overcome these problems, the immunodiffusion plates were washed with 0.2 M α-MM in saline after development of the precipitin pattern. Serum blocking assays were performed in the presence of 0.04 M α-MM. This concentration was sufficient to prevent attachment of Con A to serum proteins or to HSV. Higher concentrations of α-MM were found to be toxic to cells.

4.2 Biological Properties of Group I and Group II Antigens

The group I fraction, eluted by mannoside, contained two viral-induced precipitin bands. One of these corresponds to a glycoprotein antigen CP-1 which stimulates the production of virus-neutralizing antibody (22). This antigen appears to be a structural component of the virus envelope (10,22). The group II fraction, eluted with NaCl, contained three bands. Further gel diffusion analysis (10) indicated that none of the three bands in Group II showed identity with either band in Group I. This fraction contained no serum-neutralizing activity.

5. CHARACTERIZATION OF HSV ANTIGENS ELUTED FROM CON A—'SEPHAROSE'

5.1 Analysis of the Carbohydrates in the Group I and Group II fractions

The fact that the two groups of antigens are eluted from Con A—'Sepharose' by different agents is probably a reflection of differences in carbohydrate composition. Table II shows that Con A chromatography enriched dramatically for carbohydrate-containing (anthrone-positive) molecules in both groups. However, Group I contained 10-fold more carbohydrate per mg of protein than Group II. Sialic acid (29) and uronic acid (30) were not detected in either fraction. Both fractions contained significant amounts of ^{3}H-glucosamine label. However, it is possible that the label may have been metabolically converted to amino acids and then to protein during the 15 h incubation period. In order to determine whether ^{3}H-glucosamine label was a useful probe for oligosaccharides we measured the amount of label incorporated specifically into the carbohydrate moieties of Group I and Group II.

Samples were hydrolysed in 6 N HCl at 110°C *in vacuo* for 18 h. Acid was removed by evaporation and the basic amino acids, acidic amino acids and hexosamine fractions were isolated in an amino acid analyser (Joel, Inc) (10). Ninety-two percent of the label from Group I and 88 percent from Group II was found in the hexosamine fraction. These results assured us that ^{3}H-glucosamine is a useful probe of carbohydrate.

TABLE II. Quantitation of carbohydrate present in the Con A fractions

Sample	Carbohydrate[a] (µg/mg protein)	Specific activity (c.p.m. ^{3}H-glucosamine/mg carbohydrate)
HSV—SAM	10	
Group I[b]	1450	248
Group II	180	500

[a]Total carbohydrate was determined by the anthrone method (40).
[b]Sample was extensively dialysed to remove residual α-MM.

5.2 Pronase digestion of Group I and Group II components

Group I and Group II antigens were digested extensively with pronase for 4 days in 0.01 M Tris buffer, pH 7.8 containing 0.1% SDS and 0.001 M EDTA (31). We determined the extent of digestion by measuring the amount of acid-precipitable material remaining after pronase treatment. For this experiment, HSV-SAM protein was labelled with ^{14}C-amino acids and carbohydrate was labelled with ^{3}H-glucosamine Table III shows that virtually all of the amino acid label in HSV-SAM was solubilized by pronase. This indicated that conditions were adequate for complete proteolysis. Secondly, 94% of the carbohydrate label was precipitated by trichloroacetic acid (TCA) and of this amount, 17% remained acid precipitable after pronase digestion. This suggests that HSV-SAM contains a carbohydrate molecule resistant to pronase digestion but precipitable by TCA. The data in Table III also indicate

TABLE III.   Extent of pronase digestion of Con A fractions

Sample	Label employed	% of label precipitated by TCA	% of TCA-precipitable label after pronase[a]
HSV–SAM	^{14}C-Amino acids	100	0
HSV–SAM	^{3}H-Glucosamine	94	17
Group I	^{3}H-Glucosamine	60	3
Group II	^{3}H-Glucosamine	100	65

[a]Calculated as the percent of the original acid-precipitable label remaining after pronase digestion.

that this molecule is present in the Group II fraction of HSV-SAM. The acid-precipitable Group I molecules were solubilized completely by pronase treatment. In contrast, 65% of the acid-precipitable ^{3}H-glucosamine label in the Group II fraction was resistant to pronase digestion.

5.3 'Sephadex' Gel Filtration of Group I and Group II Carbohydrates

HSV-SAM, Group I and Group II were pronase digested and chromatographed on 'Sephadex' G-50 in order to characterize these fractions on the basis of molecular weight. Figure 3A shows the elution patterns of ^{3}H-glucosamine- and ^{14}C-amino acid-labelled SAM. A large fraction of the glocosamine label eluted with the void volume; a second fraction of lower molecular weight glucosamine-labelled material eluted next; and a third fraction of glucosamine label eluted just before the low molecular weight marker dye. All of the pronase treated ^{14}C-amino acid-labelled SAM eluted in this last fraction. In contrast to this pattern, only a small amount of the Group I carbohydrate (^{3}H-glucosamine label, Figure 3B) was eluted with the void volume. Most of the label was distributed in the middle region of the column. Molecular weight determinations employing insulin and fetuin as markers (Figure 3A) indicate that greater than 90% of the Group I carbohydrate molecules have a molecular weight distribution around 5000. Figure 3C shows that virtually all of the glucosamine label of Group II eluted in the void volume, indicating the presence of a carbohydrate with a molecular weight of greater than 10,000. When the ^{14}C-amino acid-labelled Group II material was pronase digested, all of the label eluted from the column in the V_t region. Thus the Group II carbohydrate is not linked to a pronase-resistant protein that might contribute to the observed high molecular weight.

These results show that the carbohydrates found in the two groups of antigens differ significantly in molecular weight. Most of the Group I carbohydrate consists of a heterogeneous collection of low molecular weight compounds. Group II carbohydrates have a higher molecular weight. These differences in size may reflect differences in the binding properties of Group I and Group II molecules.

5.4 Chemical Nature of the Carbohydrate of the Group II Fraction

Con A binds to neutral polysaccharides and these complexes are stable over a wide range of salt concentrations which negate any electrostatic interaction (9).

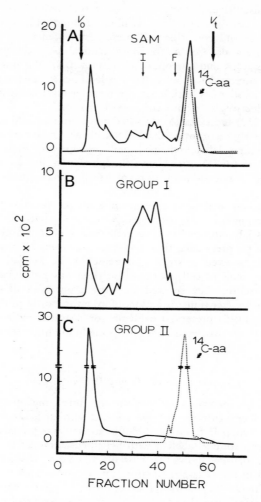

Figure 3. Gel filtration chromatography of the Group II fraction after digestion with pronase. Chromatography was on a 'Sephadex' G-50 column (1.6 x 85 cm) at 25°C. Blue dextran 2000 and phenol red were used as markers for void volume (V_0) and total volume (V_t) respectively. (1 = insulin marker, molecular weight 5000; F = fetuin marker, molecular weight 4400. Fractions of 1.4 ml were collected and radioactivity determined in Aquasol. (New England Nuclear). The dotted line is ^{14}C-amino acid label and the solid line is ^{3}H-glucosamine label. Panel (A) HSV-SAM; (B) Group I; (C) Group II

However, Con A also binds to and may precipitate polyelectrolytes and poly-saccharides which are devoid of terminal mannopyranoside (or related) residues (7—9). Doyle and coworkers (9) presented evidence for the existence of common binding sites on Con A for both neutral polysaccharides and poly-electrolytes such as RNA and mucopolysaccharides. DNA did not bind to Con A. Hydrogen bonding appeared to play a major role in neutral polysaccharide—Con A precipitate formation whereas both hydrogen bonding and electrostatic forces were implicated in polyelectrolyte—Con A complex formation.

A recent report (32) indicates that labelled glucosamine could be converted into ribose and subsequently incorporated into RNA. However, our studies show that all of the ^{3}H-glucosamine label remained acid precipitable even after extensive treatment with RNase (or DNase). Thus although the Group II fraction contained significant amounts of nucleic acid, none of this material contained label derived from glucosamine.

Mucopolysaccharides represent a reasonable choice for the Group II carbo-hydrate. These compounds have been associated closely with HSV infection. For example, numerous laboratories have reported that sulphated polyanions inhibited virus multiplication by preventing the virus from adsorbing to the cell (33). Satoh and coworkers (34) showed a striking increase in the rate of hyaluronic acid synthesis and the amount of cell-associated polymer observed after transformation by HSV-2. Furthermore, they showed that the proportion of heparin sulphate increased significantly after transformation. More importantly to the present study, Erickson and Kaplan (35) presented evidence for a virus-specific sulphated muco-polysaccharide that is excreted into the medium by pseudorabies-infected mammalian cells.

5.4.1 Incorporation of Sulphur into the Group II Fraction

Since $^{35}SO_4$ is incorporated exclusively into mucopolysaccharides by mammalian cells in cell culture (36), we asked if sulphur is incorporated into the Con A Group I or Group II fractions. Soluble antigen was prepared from infected BHK cells that were doubly labelled with ^{3}H-glucosamine and $^{35}SO_4$ or labelled with $^{35}SO_4$ alone. Figure 4 shows that virtually all of the $^{35}SO_4$ that bound to the column was eluted in the Group II fraction. These results suggest that the Group II fraction binds to Con A—'Sepharose' at least in part by a polyelectrolyte.

These results suggest also that Con A affinity chromatography not only selects carbohydrate-containing molecules out of a complex mixture, but also under the proper conditions can distinguish between neutral and charged molecules.

5.4.2 Molecular Weight of the $^{35}SO_4$-Labelled Group II Carbohydrate

The doubly labelled Group II fraction from Figure 4 was digested with pronase and chromatographed on a series of molecular weight sieves to determine whether the $^{35}SO_4$ and ^{3}H-glucosamine label were associated in the same molecule(s) and to determine the apparent molecular weights of these components. Both labels were excluded from 'Sephadex' G-50 and G-100. However, there was a hint of separation of the two labels on G-150. Figure 5 shows the radioactive elution profile of the pronase-digested doubly labelled Group II carbohydrates on 'Sephadex' G-200. The

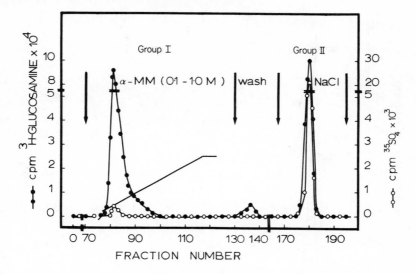

Figure 4. Affinity chromatography of ³H-glucosamine-(●) and ³⁵SO₄-(○) labelled HSV-SAM on Con A–'Sepharose'. The column was prepared and developed as in Figure 1 except that a gradient of α-MM (0.1—1.0 M) was utilized. The arrows indicate the change of elution solutions

³H-glucosamine-labelled molecules appeared to be partially excluded at this gel porosity, suggesting a molecular weight of greater than 200,000. However, the ³⁵S-containing molecules appeared to be more heterogeneous in size. Much of the label had a molecular weight of less than 200,000 and appeared to be separated from the ³H-glucosamine label. These results suggest that the Group II fraction contains a complex mixture of high molecular weight components, some of which are, in part, mucopolysaccharide.

What is the mechanism of attachment of Group II carbohydrates to Con A– 'Sepharose'? There are several reasonable alternatives: a) the antigens may be highly charged compounds (e.g. phosphorylated proteins) that bind directly to Con A– 'Sepharose'; b) the antigens may be bound non-covalently to a molecule, e.g. a mucopolysaccharide, which in turn binds to Con A–'Sepharose'; c) the antigen may be a glycoprotein which binds to Con A–'Sepharose'. Thus our studies showed that the Group II fraction is highly complex in nature, containing detectable amounts of mucopolysaccharide, nucleic acids (probably RNA) and protein.

5.5 DEAE Chromatography of Group II Components

The undigested sulphate-labelled Group II components were chromatographed on DEAE cellulose (Figure 6). No sulphur-labelled components were detected at salt concentrations lower than 0.4 M. Three peaks of ³⁵SO₄-labelled material were obtained after stepwise elution of the column with 0.5 M, 0.75 M and 1.0 M NaCl. Six fractions were pooled, dialysed and concentrated.

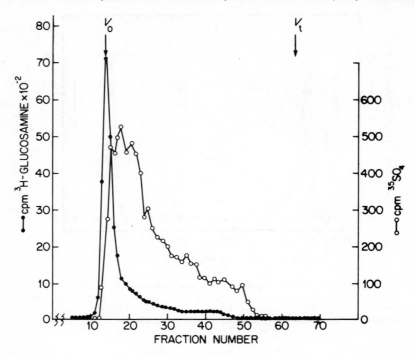

Figure 5. Gel filtration chromatography of the pronase-digested Group II fraction labelled with $^{35}SO_4$ (o) ^{3}H-glucosamine (•). Chromatography was on a 'Sephadex' G-200 column (1.6 x 85 cm) at 25°C. Blue dextran 2000 and phenol red were used as markers for void volume (V_0) and total volume (V_t) respectively. Fractions of 1.4 ml were collected and radioactivity determined in Aquasol.

5.5.1 Antigenic Properties of DEAE Fractions

The pooled fractions 1—6 were analyzed by the Ouchterlony immunodiffusion technique (25). The major Group II viral antigen was found in DEAE fraction 2 and the 2 minor Group II antigens were detected in the $^{35}SO_4$-containing peaks (#5 and #6). Since the major antigen was not covalently bound to the $^{35}SO_4$-containing molecules, it may have been bound directly to Con A–'Sepharose', or indirectly, by means of an ionic association with charged mucopolysaccharides. Further studies are needed to clarify the mechanism by which the major Group II antigen binds to Con A–'Sepharose'.

It is interesting to note that extensive pronase treatment of peaks 5 and 6 did not alter the immunoprecipitin pattern observed when these pooled fractions were reacted against specific antisera. This latter observation suggests that the antigens may be carbohydrate in nature.

5.5.2 Ultraviolet Absorption Properties of DEAE Fractions

Next we characterize the $^{35}SO_4$-containing macromolecules found in DEAE peaks 4, 5 and 6 (Figure 6). A study of the u.v. absorption properties of the 3 peaks

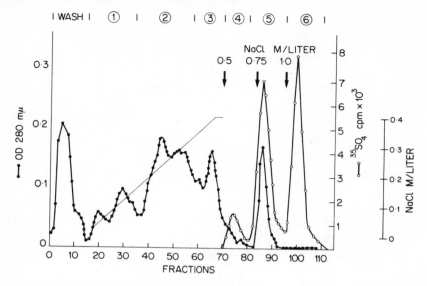

Figure 6. DEAE-cellulose (Whatman DE52) chromatography of $^{35}SO_4$-labelled Group II fraction. 5.5 mg of protein (200,000 c.p.m. of $^{35}SO_4$) were applied to a column previously equilibrated with PB until no further ultraviolet-absorbing material was eluted. The column was washed with PB and a gradient of NaCl (0.02–0.4 M) was applied (fractions 1–3). The column was then eluted stepwise with 0.5 M, 0.75 M and 1 M NaCl (fractions 4–6). The 6 pooled fractions were dialysed against PB and concentrated. Fraction volume 2.5 ml. Approximately 87% of the $^{35}SO_4$ counts and 58% of the protein were recovered from the column. Closed circles (●) O.D. 280; open circles (○) $^{35}SO_4$ label

indicates that peak 4 contained protein, while peak 5 contained nucleic acid and protein and peak 6 was almost free of ultravoilet absorbing material. Furthermore, when ^{3}H-amino acid-labelled Group II was fractionated on DEAE, insignificant amounts of label were found in DEAE fraction 6.

5.5.3 Ethanol Precipitation of Acidic Mucopolysaccharides and Sulphated Glycoproteins (37)

We found that the amount of $^{35}SO_4$-labelled material that was ethanol precipitable increased in each successive peak isolated from the DEAE column. That is, 14% of the $^{35}SO_4$ label in peak 4, 58% of the peak 5 label and 90% of the peak 6 label was ethanol precipitable. The reason for these differences is not known. However, it may be related to differences in both size and charge.

5.5.4 'Sephadex' G-200 Chromatography of $^{35}SO_4$-labelled DEAE Fractions

The G-200 elution patterns of pronase-digested DEAE fractions 5 and 6 differed significantly (Figure 7). Peak 6 constitutes the high molecular weight portion of the Group II fraction (see Figure 5) and peak 5 makes up the trailing edge of the Group II fraction.

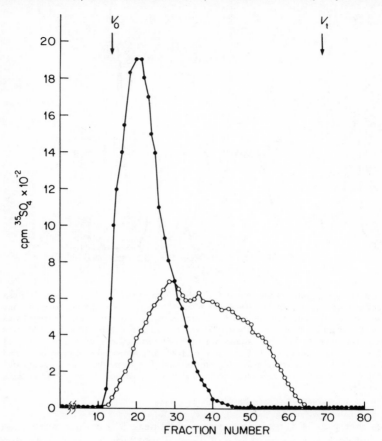

Figure 7. Gel filtration chromatography of pronase-digested DEAE peaks 5 and 6. Chromatography was on a G-200 column (1.6 × 85 cm) at 25°C. Blue dextran 2000 and phenol red were used as markers for void volume (V_0) and total volume (V_t) respectively. (●), $^{35}SO_4$-labelled peak 5; (○), $^{35}SO_4$-labelled peak 6. Fractions of 1.4 ml were collected and radio-activity determined in aquasol

5.5.5 Chondroitinase Treatment of DEAE Fractions

In an effort to characterize further the $^{35}SO_4$-containing molecules, we treated the Group II fraction as well as DEAE peaks 5 and 6 with the enzymes chondroitinase ABC (38), chondroitinase AC and hyaluronidase. All three fractions were completely resistant to hyaluronidase. After treatment of the Group II fraction with chondroitinase ABC (Table IV) 50% of the label was converted to a dialysable product, but only 18% was digested by chondroitinase AC. This suggests that the remaining 32% of the total sulphate in the Group II fraction is chondroitin

TABLE IV.   Digestion of $^{35}SO_4$-labelled Group II components by chondroitinase ABC and AC

Sample	Percent of total $^{35}SO_4$ digested by enzymes[a]		Percent of total $^{35}SO_4$ that is chondroitin B
	chondroitinase ABC	chondroitinase AC	(calculated)[b]
Group II fraction	50	18	32
DEAE peak 5	35	9	26
DEAE peak 6	63	18	45

[a]In the controls (no enzyme) fewer than 1% of the counts were dialysable. Peak 4 did not contain enough counts to examine.
[b]Calculated by subtracting percent of $^{35}SO_4$ counts digested by chondroitinase AC from the percent of $^{35}SO_4$ counts digested by chondroitinase ABC.

B. Chondroitin B also appeared to make up a significant portion of the sulphated molecules in DEAE peaks 5 and 6.

Treatment of the Group II fraction with chondroitinase ABC did not destroy antigenic activity suggesting that none of the Group II antigenic determinants contain chondroitin B.

5.5.6 Binding and Elution of DEAE Fractionated Group II Components to Con A–'Sepharose'

DEAE peaks 5 and 6 (labelled with $^{35}SO_4$) were chromatographed on Con A–'Sepharose' columns as described previously (paragraph 4.3). Figures 8A and B show the radioactive profile obtained. No radioactivity was eluted from the column by addition of α-MM or α-MG at the concentrations used. However, in both cases, greater than 90% of the $^{35}SO_4$ label was eluted with NaCl. This is direct evidence that the $^{35}SO_4$-containing molecules do in fact bind directly to Con A–'Sepharose' and are eluted by an ionic mechanism.

When DEAE peak 5 was treated with chondroitinase ABC, 35% of the $^{35}SO_4$ label was rendered dialysable. (Table IV). Of the remaining non-dialysable $^{35}SO_4$ label, 50% bound to Con A–'Sepharose' (Figure 8C) and the remaining material came through in the PB wash. Thus, 50% of the $^{35}SO_4$-containing molecules appear to have been altered by enzyme treatment in such a way that they no longer bind to Con A–'Sepharose'. Our data suggest that chondroitin contributes significantly to the binding of Group II sulphated compounds to Con A–'Sepharose'.

6. CONCLUSIONS

By utilizing Con A–'Sepharose' affinity chromatography we have been able to separate HSV antigens into two groups. Table V summarizes the known characteristics of the components of these groups. Our results suggest that Con A–'Sepharose' chromatography not only selects carbohydrate-containing molecules out of a complex mixture, but also, under the proper conditions, can distinguish between neutral and charged molecules. The Group I fraction, eluted by α-MM, contained 2 viral antigens. One of these corresponds to a glycoprotein antigen, CP-1

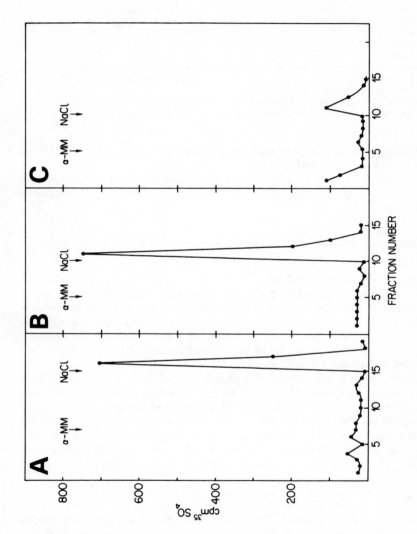

Figure 8. Affinity chromatography of $^{35}SO_4$-labelled DEAE peaks 5 and 6 on Con A–'Sepharose'. The Con A column (2 ml in a Pasteur pipette) was prepared and developed as described in paragraph 4.3, except that 1 M α-MM was utilized. Fractions of 1 ml were collected. (A) DEAE peak 5; (B) DEAE peak 6; (C) DEAE peak 5 treated with chondroitinase ABC (see Table IV)

TABLE V.   Characteristics of Group I and Group II molecules

Characteristic	Group I	Group II
Number of HSV antigens	2	3
Biological activity	Serum blocking, envelope glycoprotein	Not known
Mechanism of binding, to Con A—'Sepharose'	Specific (saccharide)	Ionic
Molecular weight of carbohydrate	Heterogeneous, approximately 5000	Homogeneous, 10,000
Presence of glucosamine	+	+
Presence of sulphated molecules		+
Presence of chondroitin		+

which stimulates the production of virus-neutralizing antibody. This antigen appears to be a structural component of the virus. The carbohydrate components of the Group I fraction comprise a heterogeneous collection of low molecular weight, glucosamine-containing molecules. The Group II fraction, eluted by NaCl contains a complex mixture of high molecular weight carbohydrate components, some of which contain sulphur and glucosamine. Our data suggest that two antigens in the Group II fraction bind directly to Con A—'Sepharose' and are closely associated with sulphated mucopolysaccharides. The third antigen appears to bind to Con A—'Sepharose' directly by an ionic mechanism or by a non-covalent mechanism. The biological function of the Group II antigens is not yet known.

7. ACKNOWLEDGEMENTS

This investigation was supported by Public Health Service grants DE-02623 from the National Institute of Dental Research (U.S.A.). G. H. Cohen was supported by A Public Health Service Career Development Award A1-23801 from the National Institute of Allergy and Infectious Diseases. R. J. Eisenberg was supported by a Public Health Service Career Development Award DE-70160 from the National Institute of Dental Research.

We acknowledge the excellent assistance of D. Weeks and M. Schieken.

8. REFERENCES

1. Goldstein, I. J., Hollerman, C. E. and Smith, E. E. (1965) 'Protein—carbohydrate interaction. II. Inhibition studies on the interaction of concanavalin-A with polysaccharides', Biochemistry, 4, 876—83.
2. Sumner, J. and Howell, B. (1936) 'The identification of the hemagglutinin of the jack bean with concanavalin A', J. Bacteriol., 32, 227—37.
3. Sharon, N. and Lis, H. (1972) 'Lectins: cell-agglutinating and sugar-specific proteins', Science, 177, 949—59.
4. Goldstein, I. J. and So, L. L. (1965) 'Protein—carbohydrate interaction. III. Agar gel-diffusion studies on the interaction of concanavalin-A, a lectin isolated from jack bean, with polysaccharides', Arch. Biochem. Biophys., 111, 407—14.
5. Portez, R. D. and Goldstein, I. J. (1970) 'An examination of the topography of the saccharide binding sites of concanavalin-A and the forces involved in complexation', Biochemistry, 9, 2890—6.

6. So, L. L. and Goldstein, I. J. (1968) 'Protein—carbohydrate interaction. IV. Application of the quantitative precipitin method to polysaccharide—concanavalin A interaction', *J. Biol. Chem.*, **242**, 1617—22.

7. Cifonelli, J. A., Montgomery, R. and Smith, F. (1956) 'The reaction of concanavalin-A with mucopolysaccharides', *J. Amer. Chem. Soc.*, **72**, 2488—9.

8. Doyle, R. J. and Tze-Jou Kan (1972) 'Interaction between concanavalin-A and heparin', *FEBS Lett.*, **20**, 22—4.

9. Doyle, R. J., Woodside, E. E. and Fishel, C. W. (1968) 'Protein—polyelectrolyte interactions: the concanavalin-A precipitin reaction with polyelectrolytes and polysaccharide derivatives', *Biochem. J.*, **106**, 35—40.

10. Ponce de Leon, M., Hessle, H. and Cohen, G. H. (1973) 'Separation of herpes simplex virus-induced antigens by concanavalin-A affinity chromatography', *J. Virol.*, **12**, 766—74.

11. Spear, P. G. and Roizman, B. (1972) 'Proteins specified by herpes simplex virus. V. Purification and structural proteins of the herpes virion', *J. Virol.*, **9**, 143—59.

12. Wilcox, W. C. and Cohen, G. H. (1969) 'The pox virus antigens', in *Current Topics in Microbiology and Immunology*, Vol. 47, (Springer-Verlag, Berlin, Heidelberg, New York), pp. 1—19.

13. Axen, R., Porath, J. and Ernbach, S. (1967) 'Chemical coupling of peptides and proteins to polysaccharides by means of cyanogen halides', *Nature*, **214**, 1302—4.

14. Lloyd, K. O. (1970) 'The preparation of two insoluble forms of the phytohemagglutinin, concanavalin A, and their interactions with polysaccharides and glycoproteins', *Arch. Biochem. Biophys.*, **137**, 460—8.

15. Agrawal, B. B. L. and Goldstein, I. J. (1968) 'Protein—carbohydrate interaction. XV. The role of bivalent cations in concanavalin A—polysaccharide interaction', *Canad. J. Biochem.*, **46**, 1147—50.

16. Sumner, J. B. and Howell, S. F. (1936) 'The role of divalent metals in the reversible inactivation of jack bean hemagglutinin', *J. Biol. Chem.*, **115**, 583—8.

17. Solling, H. and Wang, P. (1973) 'A rapid method for the purification of glycogen synthetase I and D utilizing concanavalin-A bound to agarose', *Biochem. Biophys. Res. Commun.*, **53**, 1234—9.

18. Lowry, O. H., Rosebrough, N. J., Farr, A. L. and Randall, R. J. (1951) 'Protein measurement with the Folin phenol reagent', *J. Biol. Chem.*, **193**, 265—75.

19. Moening, V., Frank, H. Hunsmann, G., Schneider, I. and Schafer, W. (1974) 'Properties of mouse leukemia viruses. VII. The major viral glycoprotein of Friend Leukemia Virus. Isolation and physicochemical properties', *Virology*, **61**, 100—11.

20. Huet, Ch., Longchampt, M., Huet, M. and Bernadac, A. (1974) 'Temperature effects on the concanavalin-A molecule and on concanavalin A binding', *Biochim. Biophys. Acta.*, **365**, 28—39.

21. Ross, L. J. N. (1974) 'Comparison of antigenic glycoproteins and glycoprotein receptors of concanavalin A isolated from duck embryo cells infected with Marek's disease virus and a herpes virus of turkeys (strain FC126)', *J. Gen. Virol.*, **24**, 549—62.

22. Cohen, G. H., Ponce de Leon, M. and Nichols, C. (1972) 'Isolation of a herpes simplex virus-specific antigenic fraction which stimulates the production of neutralizing antibody' *J. Virol.*, **10**, 1021—30.

23. Laemlii, U. K. (1970) 'Cleavage of structural proteins during the assembly of the head of bacteriophage T4', *Nature*, **227**, 680—5.

24. Mancini, G., Carbonara, A. O. and Heremans, J. F. (1965) 'Immunochemical quantitation of antigens by single radial immunodiffusion', *Immunochem.*, **2**, 235—54.

25. Crowle, A. J. (1961) *Immunodiffusion*, (Academic Press, Inc., New York).

26. Nakamura, S., Tanaka, K. and Murarawa, S. (1960) 'Specific proteins of legumes which react with animal proteins', *Nature*, **188**, 144—5.

27. Weinstein, Y., Givol, D. and Strausbach, P. H. (1972) 'The fractionation of immunoglobulins with insolubilized concanavalin-A', *J. Immunol.*, **109**, 1402—4.

28. Okada, Y. and Kim, J. (1972) 'Interaction of concanavalin-A with enveloped viruses and host cells', *Virology*, **50**, 507—15.

29. Warren, L. (1959) 'The thiobarbituric acid asssay of sialic acids', *J. Biol. Chem.*, **234**, 1971—5.

30. Galambos, J. T. (1967) 'The reaction of carbazole with carbohydrates. I. Effect of borate and sulfanate on the carbazole color of sugars', *Anal. Biochem.*, **19**, 119—32.

31. Lai, M. M. C. and Duesberg, P. H. (1972) 'Differences between the envelope glycoproteins of avian tumor viruses released from transformed and from nontransformed cells', *Virology*, **50**, 359—72.
32. Burness, A. T. H., Pardoe, I. U. and Fox, S. M. (1973) 'Evidence for the lack of glycoproteins in the encephalomyocarditis virus particle', *J. Gen. Virol.*, **18**, 33—49.
33. Roizman, B. (1969) 'The herpes viruses — a biochemical definition of the group', *Curr. Top. Microbiol. Immunol.*, **49**, 1—79.
34. Satoh, C., Duff, R., Rapp. F. and Davidson, E. A. (1973) 'Production of mucopolysaccharides by normal and transformed cells', *Proc. Nat. Acad. Sci. U.S.A.*, **70**, 54—6.
35. Erickson, J. S. and Kaplan, A. S. (1973) 'Synthesis of proteins in cells infected with herpes virus. IX. Sulfated proteins', *Virology*, **55**, 94—102.
36. Lie, S. O., McKusick, V. A. and Neufeld, E. F. (1972) 'Simulation of genetic mucopolysaccharidoses in normal human fibroblasts by alteration of pH in the medium', *Proc. Nat. Acad. Sci. U.S.A.*, **69**, 2361—3.
37. Gottschalk, A. (1972), *Glycoproteins* (Elsevier Publishing Co., Amsterdam).
38. Saito, H., Yamagata, T. and Suzuki, S. (1968) 'Enzymatic methods for the determination of small quantities of isomeric chondroitin sulphates', *J. Biol. Chem.*, **243**, 1536—42.
39. Neurath, A. R., Prince, A. M. and Lippin, A. (1973) 'Affinity chromatography of hepatitis B antigen on concanavalin-A linked to "Sepharose", *J. Gen. Virol.*, **19**, 391—5.
40. Herbert, D., Phipps, P. J. and Strange, R. E. (1971) in *Methods in Microbiology*, Vol. 2, Chapter 3, eds. Norris, J. R. and Ribbon, D. W. (Academic Press, New York).

C

Separation of Membrane Glycoproteins and Solubilization Procedures

CHAPTER 41

Solubilization of Brain Membranes for Affinity Chromatography of Glycoproteins

GIORGIO GOMBOS

Heteropolysaccharides (glycans) are the compounds most peripheral on the cell surface of most cells, there included brain cells (both neurons and glial) (1). A number of experimental data suggests that these glycans, in particular those of plasma membrane glycoproteins, might be involved in processes of cell recognition, regulation of cell growth (for review see ref. 2) and establishment of intercellular connexions (3). Specificity of cell surface is a necessary requirement for all these processes, and cell surface glycans might provide the molecular basis for this specificity (2) in view of their potential, albeit hypothetical, coding capacity.

The results obtained with cells other than brain cells have encouraged speculations that glycans might be responsible for brain cell surface specificity (not only of that specificity which differentiates a glial cell from a neuron but also of a finer specificity such as that which differentiates one type of neuron from another). In this way neuronal surface glycans may play an active role in determining the specificity of interneuronal connexions (4, 5), and in the adult brain, they may contribute to maintenance of specific synapses.

These stimulating hypotheses on the roles of cell surface glycans in the nervous system, however, do not yet have a solid experimental basis, since our knowledge on neural cell plasma membrane glycoproteins is very limited (for review see ref. 6).

We have thus started a systematic analytical study on glycoproteins from well-defined brain membrane fractions, particularly from synaptosomal plasma membrane (SPM).

In fact at the present time, the only plasma membrane of acceptable purity (for review on nerve cell plasma membrane preparations see ref. 7) which can be isolated from the central nervous system is that derived from synaptosomes, i.e. from pinched-off nerve endings (for review see refs. 8 and 9).

However all our techniques were developed by using an easy to prepare, albeit heterogeneous membrane fraction: the brain microsomal fraction* which contains most of neuronal perikarya, axons and glial cell plasma membranes, as indicated by the presence of plasma membrane markers such as gangliosides and Na^+-K^+-activated ouabain-inhibited ATPase.

In addition we have applied the techniques to synaptic vesicles (SV) prepared by the technique of Morgan and coworkers (10) and very recently to myelin prepared by two different methods (11, 12).

Glycoproteins from these membranes were separated by affinity chromatography on four lectins† (13—16). The techniques used with immobilized Con A are described elsewhere in this volume (Zanetta and Gombos, Chapter 42, and Reeber and Gombos, Chapter 43).

Here we describe our solubilization procedures which are preliminary to separation of membrane glycoproteins by affinity chromatography on different lectins.

1. SOLUBILIZATION OF MEMBRANES FOR AFFINITY CHROMATOGRAPHY OF GLYCOPROTEINS

Analytical studies of membrane proteins have been limited for many years by the absence of a method of 'solubilization' of these compounds. With the exception of a few proteins‡, the bulk of membrane proteins, imbedded in the lipid matrix, are not soluble in aqueous solvents (unless detergents are also present) not only because of the presence of lipids but also because of the particular nature of membrane proteins.

Extraction of membrane lipids by organic solvents, is generally followed by irreversible precipitation of denatured proteins [myelin proteolipid-apoprotein (17) and basic proteins (18, 19), however, are the exception to this rule].

Solubilization of membrane proteins has been obtained by treatment of membranes with phenol and formic or acetic acid in the presence of urea (20—23). The solubilized proteins were electrophoresed and, the first electrophoresis of brain

---

*Fraction which sediments in 0.32 M sucrose between 11,500g x 25 min (which sediments mitochondria, synaptosomes, lysosomes and myelin) and 100,000 g for 60 min.

†Concanavalin A, *Ulex europeus* lectin-specific for L-fucose, wheat germ agglutinin and Ricinus lectin.

‡In most cases they appear to be 'extrinsic' proteins or, exceptionally, proteins that become soluble in aqueous media after destruction of membrane structure.

membrane (myelin) proteins was obtained in this way (24—26). At the present time, this method is of limited interest particularly for solubilizing membrane glycoproteins since they can be solubilized by milder treatments which will not destroy part of their sugar moiety (glycans) as is probably the case for the acid extraction technique cited above.

Lithium diiodosalicylate has also been used to solubilize glycoproteins from red blood cell ghosts (27). Contrary to the results of some authors, claiming solubilization of macromolecules from whole brain pellet (28) and SPM (29), we have never been successful in solubilizing proteins from SPM or brain microsomal fractions with this compound.

1.1 Solubilization by Detergents

In the recent years, detergents have been widely used for membrane solubilization. Those most commonly used for nervous tissue membranes have been 'Triton' X-100 and other neutral detergents*, deoxycholate (DOC) and a strongly anionic detergent: sodium dodecylsulfate (SDS).

Two uses of detergents can be considered. One is the isolation of junctional areas of plasma membranes from different tissues. These specialized plasma membrane zones under precisely defined conditions remain insoluble in some detergents, while the rest of the plasma membrane is solubilized. 'Triton' X-100, or sodium lauroyl-N-sarcosinate or DOC do not solubilize desmosomes from a variety of tissues (30—32) or synaptic junction or postsynaptic webs in the nervous tissues (30—35). The various structures obtained are then purified by gradient centrifugation. It should be pointed out, however, that although these structures are morphologically intact, selective extraction by the detergent of certain components must occur, even if these structures remain morphologically intact. For example acetylcholinesterase, which seems to be preferentially associated with postsynaptic material in synaptosomes (36) is 'Triton' X-100 soluble.

The other use of detergents for membrane solubilization is for analytical studies on purified macromolecules. An extensive review on membrane solubilization by detergents has been published recently (37). Here we will only discuss a few points relevant to our own experience with solubilization of brain membranes.

1.2 Solubilization of Brain Membranes by 'Triton' X-100 and by SDS

We have used only two detergents, 'Triton' X-100 and SDS. Each detergent presents some advantages over the other. We will briefly summarize the positive and negative aspects of the use of 'Triton' X-100 and of SDS relative to three parameters: nature of the solubilization process, biological activity of the solubilized molecules and possibility of purification of individual solubilized molecules.

1.3 Nature of the Solubilization Process

Practically all proteins from brain membranes (SPM, SV, microsomal membranes and myelin) are solubilized by SDS. Almost complete solubilization (95% of

*'Lubrol', 'Cemulsol', etc.

proteins) of brain microsomal fraction by 'Triton' X-100 can be obtained only at very low protein concentrations and high detergent/protein ratio (Landi, F. and coworkers, in preparation). In addition only SDS is effective in solubilizing lipid-free membrane protein residues after extraction of lipids by organic solvents.

'Solubilization' is an operational term and it corresponds simply to the presence of proteins in the supernatant (rather than in the pellet) after centrifugation at 100,000 g for 120 min and not necessarily to the presence, in the solution, of separated protein molecules. It appears that the bulk of proteins solubilized in 'Triton' X-100 are in 'solution' as molecular aggregates (probably micellar aggregates of lipid, detergent and proteins, formed after destruction of membrane structure).

In fact, polyacrylamide gel electrophoresis in the presence of 'Triton' X-100 of proteins derived from membranes directly solubilized in the same detergent shows protein bands of extremely high molecular weight (38). It is possible that the elimination of lipids by ion exchange chromatography or gradient centrifugation (39) might allow a separation of individual proteins.

Electrophoresis in the presence of SDS of proteins derived from brain membranes (SPM, SV, myelin and microsomal membranes) directly solubilized in SDS, and those from lipid-free protein residues derived from the same membranes dissolved in SDS (see Figure 1) show, for the same membrane, the same complex pattern of proteins with a wide range of molecular weights.

SDS, in spite of the contrary claim of Katzman (40) appeared to monomerize proteins when lipids were extracted and sulphydryl groups blocked by alkylation*. We base this affirmation on the following evidence. We have fractionated brain microsomal membranes by preparative polyacrylamide gel electrophoresis in the presence of SDS. Each of the fractions, electrophoresed under the same conditions on analytical polyacrylamide gels, showed one or few tightly packed bands with a migration distance in the gels which corresponded to that of the fraction in the first electrophoresis. We never observed additional electrophoretic bands of molecular weight larger or smaller than that expected from the elution from the preparative gel which could have suggested subsequent aggregation or disaggregation of the proteins (Marchesini, S. and coworkers, in preparation).

In conclusion, at present, only SDS appears to monomerize membrane proteins and the term 'solubilization' does not have the same meaning when used for membranes directly solubilized in 'Triton' X-100 or for membranes directly solubilized in SDS; or for lipid-free membrane protein residues, solubilized in SDS.

1.4 Biological Activity of Solubilized Molecules

Since 'Triton' X-100 is not a denaturing agent, many enzymatic activities are intact and even activated in its presence (for review see ref. 37). However, this fact can be a disadvantage because glycosidases and proteases remain active, which can result in degradation of glycoproteins during the preparation. These enzymes should become inactive when SDS is used for solubilization since SDS is a denaturing agent. Some authors, however, have shown that proteases can remain

---

*Aggregations due to formation of disulphydryl bridges were thus avoided.

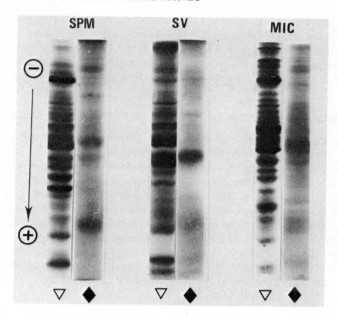

Figure 1. Polyacrylamide (12%) gel electrophoresis, in the presence of SDS (0.1%), of adult rat brain synaptosomal plasma membranes (SPM), synaptic vesicles (SV) and microsomal fraction membranes (MIC) solubilized in SDS. ▽, Protein stain, Coomassie brilliant blue; ♦, carbohydrate stain, PAS

active in the presence of SDS (for references see ref. 37). In contrast to these data and probably because the treatment with organic solvent, followed by high SDS concentration (4%) (see paragraph 3.2), has destroyed all enzymatic activity, we have never observed proteolysis in our proteins dissolved in SDS with the exception of a few cases when microorganisms (evidently resistant to the detergent) developed in the samples in the presence of low (0.07—0.08%) SDS concentration (see Zanetta and Gombos, Chapter 42).

1.5 Possibilities of Fractionation of Solubilized Membrane Proteins

Classical methods of protein separation, such as ion exchange chromatography, can be used in the presence of 'Triton' X-100. However the presence of protein micellar aggregates remains a problem also for separation of proteins solubilized in 'Triton'.

The major limitation in the separation of membrane proteins solubilized in SDS is the strong negative charge of the protein—SDS complexes, which does not allow separation of proteins by ion exchange chromatography.

At present only analytical and preparative polyacrylamide gel electrophoresis (41) and, in the case of glycoproteins, affinity chromatography on lectins, have separated brain membrane proteins solubilized in SDS (13—16).

In spite of all its limitations, we have chosen as a first approach, solubilization of brain membrane in SDS because of (a) monomerization of proteins, (b) absence of lipids (particularly glycolipids which might interfere with separation by affinity chromatography on lectins), (c) total solubilization of glycoproteins of brain membranes (see § 3.2), (d) inactivation of lytic enzymes, in the presence of SDS.

2. AFFINITY CHROMATOGRAPHY IN THE PRESENCE OF DETERGENTS

Affinity chromatography, in presence of 'Triton' X-100 or deoxycholate, of membrane proteins solubilized in these detergents has been used for purification of several membrane receptors. Affinity chromatography on Con A or other lectin in the presence of deoxycholate (DOC) has been used to separate lectin-binding glycoproteins derived from different membranes (42—44) [there included brain membranes (45)] solubilized in DOC. Akedo and coworkers (46) have precipitated, with Con A, glycoproteins from erythrocytes stroma solubilized in SDS.

Preliminary experiments had shown that affinity chromatography on Con A and other lectins (see Zanetta and Gombos, Chapter 42) in the presence of SDS was possible if the SDS concentration was kept below 0.1%. Complete solubilization of brain membrane proteins at a concentration of SDS lower than 0.1% was obtained with the technique described in § 3.2.

3. METHOD FOR MEMBRANE SOLUBILIZATION

3.1 Lipid Extraction

Lipids were extracted from the membrane fraction by the following method: the membrane pellet was homogenized in methanol (1 ml for 8—10 mg proteins in the pellet), stirred (with a magnetic stirrer) for ten minutes at room temperature, then an equal volume of chloroform was added, followed, ten minutes later, by another volume of chloroform. The mixture was stirred for 20 minutes and then centrifuged (30,000 g for 30 min). The pellet was resuspended in methanol—chloroform v/v 2:1 (same total volume as in the first extraction), stirred for 20 minutes and centrifuged as before. The lipid extraction was repeated once more.

The lipid-free pellet was then washed with methanol, to eliminate the residual chloroform, and finally with water. No proteins were released in the water from the lipid-free residue of SPM or SV while about 1% of protein was released from the lipid-free residue of microsomal fractions (14).

The washed pellet was either digested with pronase (see Reeber and Gombos, Chapter 43) or solubilized in SDS as described in the following paragraph.

3.2 Solubilization in SDS

The lipid-free pellet was left overnight in saturated EDTA to eliminate the maximum of divalent cations and to release proteins from aggregates possibly formed by interactions between cations and proteins. The EDTA solution contained also 2-mercaptoethanol (10%) to break disulphide bonds. After centrifugation the pellet (100,000 g for 120 min) derived from SPM or SV contained all the proteins of the original fraction, while 1.3% of proteins were released from the lipid-free microsomal fraction (14).

The pellet was dissolved in the minimum possible volume of 4% SDS containing 1% 2-mercaptoethanol. The high SDS concentration was used with the aim of obtaining solubilization in a small volume. At the same time, SDS bound to proteins represented a negligible percentage of the total SDS added and thus the concentration of free SDS was not significantly lowered.

The protein solution was brought to pH 8 with NaOH, and a 10-fold excess of iodacetamide was added. Alkylation took place, in the dark, for 12 hours, under nitrogen, with occasional shaking of the samples (15).

The following step, consisted of lowering the SDS concentration to 0.08–0.07%, and the elimination of excess iodoacetamide. This was achieved by a 48 h dialysis against 49 volumes of water, followed by overnight dialysis against 0.07–0.08% SDS. The sample was then centrifuged (100,000 g for 120 min). The pellet contained no glycoproteins and always less than 5% of proteins. The electrophoretic profiles of polyacrylamide gels in the presence of SDS of various brain membrane fractions are shown in Figure 1.

The alkylated lipid-free proteins were separated in the presence of SDS, by affinity chromatography on Con A (Zanetta and Gombos, Chapter 42) (15,16) or on other lectins (13, 14).

4. REFERENCES

1. Rambourg, A. and Leblond, C. P. (1967) 'Electron microscope observations on the carbohydrate-rich cell coat present at the surface of cells in the rat', *J. Cell Biol.*, **32**, 27–53.
2. Cook, G. M. W. and Stoddardt, R. W. (1973) 'Functional importance of surface heterosaccharides in cellular behavior', in *Surface Carbohydrates of the Eukaryotic Cell* (Academic Press), pp. 257–270.
3. Roseman, S. (1970) 'The synthesis of complex carbohydrates by multiglycosyl transferase systems and their potential function in intercellular adhesion', *Chem. Phys. Lipids*, **5**, 270–97.
4. Brunngraber, E. G. (1969) 'Possible role of glycoproteins in neural fraction', *Perspect. Biol. Med.*, **12**, 467–70.
5. Barondes, S. H. (1970) 'Brain glycomacromolecules and interneuronal recognition', in *The Neurosciences: Second Study Program*, ed. Schmitt, F. O. (Rockefeller University Press, New York), pp. 747–60.
6. Morgan, I. G., Gombos, G. and Tettamanti, G. 'Glycoproteins and glycolipids of the nervous system', in *Mammalian Glycoproteins and Glycolipids*, eds. Pigman, W. and Horowitz, M. I. (Academic Press, New York), in press.
7. Morgan, I. G. and Gombos, G. 'Biochemical studies of synaptic macromolecules: are there specific synaptic components?' in *Macromolecules and Neuronal Recognition*, ed. Barondes, S. (publisher), in press.
8. Whittaker, V. P. (1969) 'The synaptosome' in *Handbook of Neurochemistry*, Vol. II, ed. Lajtha, A. (Plenum Press, New York), pp. 327–64.
9. De Robertis, E. and Rodriguez De Lores Arnaiz, G. (1969) 'Structural components of the synaptic region', in *Handbook of Neurochemistry*, Vol. II, ed. Lajtha, A. (Plenum Press, New York), pp. 365–92.
10. Morgan, I. G., Vincendon, G. and Gombos, G. (1973) 'Adult rat brain synaptic vesicles. I. Isolation and characterization', *Biochim. Biophys. Acta*, **320**, 671–80.
11. Norton, W. T. and Poduslo, S. E. (1973) 'Myelination in rat brain: method of myelin isolation', *J. Neurochem.*, **21**, 749–57.
12. Waehneldt, T. V. and Mandel, P. (1972) 'Isolation of rat brain myelin, monitored by

polyacrylamide gel electrophoresis of dodecyl sulfate-extracted proteins', *Brain Res.*, **40**, 419—36.

13. Gombos, G., Reeber, A., Zanetta, J-P. and Vincendon, G. (1974) 'Fractionation of nervous tissue membranes glycoproteins', in *Méthodologie de la Structure et du Métabolisme des Glycoconjugués*, Colloque International n° 221 du Centre National de la Recherche Scientifique, Villeneuve d'Ascq, 20—27 juin 1973, (Editions du CNRS, Paris), Tome II, pp. 829—44.

14. Gombos, G., Zanetta, J. P., Reeber, A., Morgan, I. G. and Vincendon, G. (1974) 'Affinity chromatography of brain membrane glycoproteins', *Biochem. Soc. Trans.*, **2**, 627—30.

15. Zanetta, J-P., Morgan, I. G. and Gombos, G. (1975) 'Synaptosomal plasma membrane glycoproteins; fractionation by affinity chromatography on concanavalin A', *Brain Res.*, **83**, 337—48.

16. Zanetta, J-P. and Gombos, G. (1974) 'Affinity chromatography on Con A—"Sepharose" of synaptic vesicle membrane glycoproteins', *FEBS Lett.*, **47**, 276—8.

17. Tenenbaum, D. and Folch-pi, J. (1966) 'The preparation and characterization of water soluble proteolipid protein from bovine brain white matter', *Biochim. Biophys. Acta*, **115**, 141—7.

18. Eylar, E. H., Salk, J., Beveridge, G. and Brown, L. (1969) 'Experimental allergic encephalomyelitis. An encephalitogenic basic protein from bovine myelin', *Arch. Biochem. Biophys.*, **132**, 34—48.

19. Oshiro, Y. and Eylar, E. H. (1970) 'Allergic encephalomyelitis: preparation of the encephalitogenic basic protein from bovine brain', *Arch. Biochem. Biophys.*, **138**, 392—6.

20. Bagdasarian, M., Matheson, N. A., Synge, R. L. M. and Youngson, M. A. (1964) 'New procedures for isolating polypeptides and proteins from tissues. Metabolic incorporation of L-(^{14} C) valine into fractions of intermediate molecular weight in broadbean (*Vicia faba L.*) leaves', *Biochem. J.*, **91**, 91—105.

21. Work, T. S. (1964) 'Electrophoretic separation of the proteins of ribosome subunits and of encephalomiocarditis virus', *J. Mol. Biol.*, **10**, 544—5.

22. Takayama, K., McLennan, D. H., Tzagoloff, A. and Stoner, C. D. (1966) 'LXVII. Polyacrylamide gel electrophoresis of the mitochondrial electron transfer complexes', *Arch. Biochem. Biophys.*, **114**, 223—30.

23. Thorun, W. and Mehl, E. (1968) 'Determination of molecular weights of microgram quantities of protein components from biological membranes and other complex mixtures: gel electrophoresis across linear gradients of acrylamide', *Biochim. Biophys. Acta*, **160**, 132—4.

24. Mehl, E. and Wolfgram, F. (1969) 'Myelin types with different protein components in the same species', *J. Neurochem.*, **16**, 1091—7.

25. Mehl, E. and Halaris, A. (1970) 'Stoichiometric relation of protein components in cerebral myelin from different species', *J. Neurochem.*, **17**, 659—68.

26. Roboz-Einstein, E., Dalal, K. B. and Csej Tey, J. (1970) 'Biochemical maturation of the central nervous system. II. Protein and proteolytic enzyme changes', *Brain Res.*, **18**, 35—49.

27. Marchesi, V. T. and Andrews, E. P. (1971) 'Glycoproteins: isolation from cell membranes with lithium diiodosalicylate', *Science*, **174**, 1247—8.

28. Margolis, R. K. and Margolis, R. U. (1973) 'Extractability of glycoproteins and mucopolysaccharides of brain', *J. Neurochem.*, **20**, 1285—8.

29. Javaid, J. I., Herts, C. R. and Brunngraber, E. G. (1974) 'Proteins of synaptosomal membranes from whole rat brain', 5th Meeting American Soc. Neurochem. New-Orleans, Abstracts p. 110.

30. Emmelot, P., Feltkamp, C. A. and Vaz Diaz, H. (1970) 'Fractionation of the ATPase of deoxycholate solubilized rat liver and hepatoma plasma membranes and the morphological appearance of the preparations', *Biochim. Biophys. Acta*, **211**, 45—55.

31. Evans, W. H. and Gurd, J. W. (1972) 'Preparation and properties of nexuses and lipid enriched vesicles from mouse liver plasma membranes', *Biochem. J.*, **128**, 691—700.

32. Cotman, C. W. and Taylor, D. (1972) 'Isolation and structural studies on synaptic complexes from rat brain', *J. Cell Biol.*, **55**, 696—711.

33. Davis, G. A. and Bloom, F. E. (1973) 'Isolation of junctional complexes from rat brain', *Brain Res.*, **62**, 135—53.

34. Cotman, C. W., Levy, W., Banker, G. and Taylor, D. (1971) 'An ultrastructural and chemical analysis of the effects of "Triton" X-100 on synaptic plasma membranes', *Biochim. Biophys. Acta*, 249, 406—18.
35. Cotman, C. W., Banker, G., Churchill, L. and Taylor, D. (1974) 'Isolation of postsynaptic densities from rat brain', *J. Cell Biol.*, 63, 441—55.
36. McBride, W. J. and Cohen, H. (1972) 'Cytochemical localization of acetylcholinesterase on isolated synaptosomes', *Brain Res.*, 41, 489—93.
37. Helenius, A. and Simons, K. (1975) 'Solubilization of membranes by detergents', *Biochim. Biophys. Acta*, 415, 29—79.
38. Grossfeld, R. M. and Shooter, E. M. (1971) 'A study of the changes in protein composition of mouse brain during ontogenic development', *J. Neurochem.*, 18, 2265—77.
39. Simons, K., Helenius, A. and Garoff, H. (1973) 'Solubilization of the membrane proteins from Semliki forest virus with "Triton" X-100', *J. Mol. Biol.*, 80, 119—33.
40. Katzman, R. L (1972) 'The inadequacy of sodium dodecyl sulphate as dissociative agent for brain proteins and glycoproteins', *Biochim. Biophys. Acta*, 266, 3259—68.
41. Waehneldt, T. V. (1971) 'Preparative isolation of membrane proteins by polyacrylamide gel electrophoresis in the presence of ionic detergent (SDS): proteins of rat brain myelin', *Anal. Biochem.*, 43, 306—12.
42. Allan, D., Auger, J. and Crumpton, M. J. (1972) 'Glycoprotein receptors for concanavalin A isolated from pig lymphocyte plasma membrane by affinity chromatography in sodium deoxycholate', *Nature (New Biol.)*, 236, 23—5.
43. Hayman, M. J. and Crumpton, M. J. (1972) 'Isolation of glycoproteins from pig lymphocyte plasma membrane using *Lens culinaris* phytohemagglutinin', *Biochem. Biophys. Res. Commun.*, 47, 923—30.
44. Hayman, M. J., Skehel, J. J. and Crumpton, M. J. (1973) 'Purification of virus glycoproteins by affinity chromatography using *Lens culinaris* phytohemagglutinin', *FEBS Lett.*, 29, 185—8.
45. Susz, J. P., Hof, H. I. and Brunngraber, E. G. (1973) 'Isolation of concanavalin A-binding glycoproteins from rat brain', *FEBS Lett.*, 32, 289—92.
46. Akedo, H., Mori, Y., Tanigaki, Y., Shinkai, K. and Morita, K. (1972) 'Isolation of concanavalin A binding protein(s) from rat erythrocyte stroma', *Biochim. Biophys. Acta*, 271, 378—87.

CHAPTER 42

Affinity Chromatography of Brain Membrane Glycoproteins on Concanavalin A–'Sepharose' in the Presence of SDS

JEAN-PIERRE ZANETTA
GIORGIO GOMBOS

Synaptosomal plasma membranes (SPM) are rich in glycoproteins. Concanavalin A (Con A) binding material (1—3) is present on the surface of these membranes, particularly in the synaptic junction. Con A binding material appears also to be present in myelin (1, 4) which possibly contains only one glycoprotein (5, 6). Synaptic vesicles (SV) contain Con A binding glycoproteins (7) which appear not to be accessible since they do not bind ferritin—Con A (1, 2).

We have used affinity chromatography on Con A and other lectins, in the presence of SDS (sodium dodecylsulphate) for the separation of glycoproteins of synaptosomal plasma membranes (8), synaptic vesicle (7) and, more recently, of myelin (Zanetta and coworkers, in preparation). Brain microsomal fraction, a heterogeneous membrane fraction containing a large amount of nerve cell body and glial cells plasma membrane were used to set up this technique. We have described (Gombos, Chapter 41) our technique for solubilizing brain membrane proteins and keeping them soluble at low concentrations of detergent.

1. PRINCIPLE

Here we summarize the principles of the procedure: after lipid extraction, membrane proteins were solubilized in SDS, proteins were present as individual polypeptide chains. Aggregations appeared to be absent if precautions were taken to avoid the formation of disulphide bridges between different polypeptides: all sulphydryl groups were blocked by alkylation with iodoacetamide.

Another advantage of solubilization in SDS over other types of solubilization is that SDS inactivates proteases and glycosidases, presumably present in the membranes and sometimes activated by other detergents.

2. AFFINITY CHROMATOGRAPHY IN THE PRESENCE OF SDS

2.1 Utilization of Columns of Con A Bound to 'Sepharose' 4B in the Presence of SDS

The first step in setting up this method, was to ascertain that, in the presence of SDS, immobilized Con A kept, at least partially, its binding capacity for glucose- and mannose-containing polysaccharides.

Ovalbumin and Dextran blue which, in the absence of SDS, bind to Con A, were used as test material. Both compounds, dissolved in 0.1% SDS were adsorbed in the presence of SDS on the immobilized lectin column. The binding efficiency in the presence of SDS, however, was 25% of that of an identical column run in identical conditions of flow rate, amount and concentration of the test material. Thus affinity chromatography on Con A in the presence of SDS appeared to be possible in spite of the reduced binding capacity of Con A, since we had succeeded in keeping cerebral membrane glycoproteins soluble in 0.08% SDS (see Gombos, Chapter 41).

2.2 Factors Affecting Adsorption

Factors which influenced chromatography can be schematically divided into three groups. In the first group are conditions which affect the activity or the integrity of the column, in the second group are factors which affect the sample, finally in the third group are those which depend on the operating conditions of chromatography.

2.3 Factors Related to the Column

2.3.1 SDS Concentrations

The binding capacity of the Con A columns greatly decreased at SDS concentration higher than 0.1%. But the inactivation of immobilized Con A, even at high SDS concentrations was at least partially reversible. For example Con A—'Sepharose' stored 6 months in 4% SDS at 4°C, was reactivated when excess SDS was eliminated and the whole procedure of regeneration (see §3.2) was carried out. Insolubilized lectins (not only Con A but also wheat germ agglutinin, *Ricinus communis* agglutinin and *Ulex europeus* lectin specific for L-fucose), unlike the same lectins in solution, are remarkably resistant to denaturing agents (9, 10).

2.3.2 Leakage of Con A from the Column

Commercial Con A—'Sepharose' (Pharmacia), contains large amounts of non-covalently bound Con A, which must be eliminated before chromatography (whether the chromatography was carried out in presence or in absence of detergent). This was achieved by prolonged and repeated washes at alkaline and acid pH in the absence and in the presence of a haptenic inhibitor (for economical reasons we used sucrose). It was necessary to replace the ions necessary for Con A activity which might have been lost during the repeated washes with acid buffer (11, 12). The whole procedure is described in paragraphs 3.1 and 3.2.

Besides unbound Con A present in the purchased Con A—'Sepharose' a certain amount of degradation products is formed during storage of the gel as with all carbohydrate polymer supports and these must be eliminated before each chromatography.

Evidently this degradation occurs also during chromatography (usually lasting for 4 hours) but it can be very much reduced by using the minimum quantity of gel necessary for binding all specific glycoproteins present in each sample. If an excess of gel was used, traces of material, corresponding electrophoretically to Con A and not staining with PAS and which were not present in the original sample, were detected in the fraction adsorbed on the lectin and eluted by the haptenic inhibitor. The degradation material represented however a very small part of the Con A—'Sepharose' since we have gels still active after 4 years of use.

2.4 Factors Affecting the Protein Sample

2.4.1 Divalent Cations

Divalent cations are necessary to stabilize the active form of concanavalin A (11, 12). These cations can be eliminated by acid pH(11) or by EDTA. The addition of divalent cations to Con A solutions reactivates Con A(11). Divalent cations, however, precipitated SDS thus chromatography in the presence of SDS had to be carried out in the absence of divalent cations, and the activation of Con A had to be carried out by adding the cations in the absence of SDS. The detailed technique is described in paragraph 3.2.

2.4.2 Sodium and Potassium Chloride

Sodium and potassium chloride are usually used in the buffer and in the samples chromatographed to avoid non-specific adsorption. However, these salts could not be used since at high concentrations they precipitated SDS and at low concentration they precipitated most of brain membrane proteins solubilized in SDS. This precipitation did not occur when salts (0.1 M final concentration) were added to solutions in SDS of bovine serum albumin or ovalbumin.

2.4.3 Non-specific Adsorptions

Since all proteins adsorbed on Con A appeared to be glycoproteins particularly rich in mannose (8), we assumed that in the presence of SDS non-specific adsorptions are negligible and thus chromatography was carried out in the absence of sodium or potassium chloride. Since diluted protein solutions gave unsatisfactory recovery, we tried to use highly concentrated protein solutions (see §2.9). However at high protein concentration long tailing of the peak of non-adsorbed material were observed and large volumes of solution were needed to wash out this material. This could be avoided by using more diluted solutions. The highest concentration to be used without encountering this problem depends on the nature of the sample and it must be determined in each individual case. The optimal concentration was of 5 mg protein/ml for SPM or microsomal proteins, of 2 mg/ml for synaptic vesicles and of 1 mg/ml for myelin.

2.5 Operating Conditions of Adsorption

2.5.1 Temperature

We have routinely operated at room temperature since membrane proteins at concentration higher than 2 mg protein/ml precipitated at +4°C. However, proteins redissolved immediately when the sample was warmed at room temperature. There is no risk of sample degradation since protease and glycosidase activities are absent in buffered SDS solutions at the concentrations used. However it should be pointed out that samples in SDS should not be kept at room temperature for a prolonged period of time (several days or weeks) since microorganisms occasionally grow even in the presence of the detergent.

2.5.2 pH

At a pH of around 7, membrane proteins are soluble in SDS at higher protein concentration than in unbuffered SDS solutions (pH around 5). Since the optimum pH for Con A activity is reported to be between pH 6.5 and pH 8 (13), it should have been expected that chromatography at pH 5 was less efficient than at pH 7. On the contrary, the same amount of glycoproteins (from SPM or from brain microsomal fractions) with the same electrophoretic profile were adsorbed on Con A when chromatography was carried out in the presence of unbuffered SDS, as when chromatography was carried out at pH 7. However, differences between the two chromatographic conditions were observed in the fractions not adsorbed or weakly adsorbed on Con A. We will deal with these fractions in paragraph 2.7.

2.5.3 Flow Rate

We have always regulated the flow in such a way that the sample remained in contact with the gel for at least 1 hour. Longer contact times did not increase the amount of adsorbed glycoproteins. Rechromatography of the non-adsorbed fractions demonstrated that the adsorption was complete.

2.5.4 Size and Capacity of Columns

Maximal adsorption is obtained when Con A binding sites are in excess of Con A binding material. Evidently, the optimal conditions change with the type of sample (that is with the nature and the concentrations of the Con A-binding molecules) and they should be empirically found (also by rechromatography experiments) for each type of sample*. The excess of Con A bound to 'Sepharose' should not be too great if leakage of large amounts of Con A is to be avoided. In the case of the membranes that we have analysed we found that 30 ml of gel packed in a column of 1 cm diameter, adsorbed 30 mg of glycoproteins. The amount of gel and the height of the column were also important. Columns of 3 cm diameter were less efficient than those of 1 cm diameter. However when small samples (1 mg protein) were chromatographed on an equivalent amount of gel (1 ml packed volume), even if the sample contained only Con A-binding glycoproteins, the adsorption was less than quantitative. Evidently in these conditions, the probability of contact between Con A and glycoproteins was too low to ensure complete adsorption. An excess of gel (10 ml of packed gel/mg protein) was necessary for complete adsorption. In this case however leaking Con A became a serious contaminant. This is why we are pessimistic about the possibility of miniaturizing this method.

The conditions routinely used for affinity chromatography of central nervous system membrane fractions in the presence of SDS are summarized in paragraph 3.3.

2.6 Elution of Adsorbed Material

For desorption of Con A-bound material we have routinely used the specific haptenic inhibitors of Con A. α-Methyl glucoside (α-MG) was preferred to α-methyl mannoside (α-MM) to avoid erroneous estimate (due to possible contamination by α-methyl mannoside) of mannose levels of Con A-binding glycoproteins. A concentration of 0.25 M of α-methyl glucoside was sufficient to elute all the adsorbed

---

*In the case of the central nervous system membranes that we have investigated the concentrations used are given in §2.4.

material from the column. Slow flow rates were also necessary to avoid tailing of the peaks during elution.

Con A-bound material could also be desorbed by chelators and basic pH. Acid pH could not be used since it precipitated proteins from SDS solutions.

2.7 Specificity of Binding

Brain membrane glycoproteins were separated on Con A into two fractions called C0 (not absorbed) and C1 (adsorbed and eluted with α-MG) (7, 8). Mannose and glucose were present in both fractions although both sugars were more abundant in C1 than in C0. We assumed that mannose was terminal in the glycan chains of C1 while it was internal in the glycan chain in C0 protein. In fact, Con A-binding glycoproteins from microsomal and SPM fractions (fraction C1) carry on most of the polypeptides at least two types of glycans, one of which, of small molecular weight, contains practically only N-acetylglucosamine and mannose (Reeber and Gombos, Chapter 43). Since mannose in these glycans is present in larger amount than N-acetylglucosamine, it is probably a terminal sugar and it determines the affinity for Con A.

Weaker interaction between the lectins and other glycoproteins was also detected. When appropriate volumes of samples (50 ml for a column 50 x 1 cm) were chromatographed at pH 7, two peaks of proteins eluted by the buffer were present (7,8) (Figure 1). The first called C0 was eluted in the void volume of the column. The second peak, called CR (retarded) followed and partially overlapped C0. This peak contained glycoproteins weakly stained with PAS, it was rich in N-acetylglucosamine, and it disappeared (its material was eluted with C0) when chromatography was carried out with unbuffered SDS or with buffer containing α-MG. We interpreted all these results as due to a specific weak interaction of the N-acetylglucosamine of these glycoproteins with Con A.

Even glycoproteins in fractions C0 weakly interact with Con A. In fact these glycoproteins are not uniformly eluted in this fraction but they are concentrated in the last half of the peak and the subfraction within C0 had different electrophoretic profiles (Figure 1).

2.8 Precipitation and Resolubilization of the Samples after Chromatography

The samples after chromatography were concentrated by precipitation with methanol. α-MG was eliminated at the same time. The procedure (described in §3.4) was as follows: the proteins were precipitated by adding methanol after which the pH of the solution was lowered to 5. The sample was left overnight at $-20°C$. Because of the methanol density at this temperature it was necessary to bring the sample temperature at least to $0°C$ before centrifugation. Previous to the adoption of this method we had attempted to concentrate the sample by ultrafiltration. These attempts were not successful since SDS gradually destroyed the ultrafiltration membranes (Diaflo).

Acetone (3–4 volumes) immediately precipitated the protein out of the samples. However salts from the buffer were also precipitated and these salts interfered with the resolubilization of the sample.

The precipitated samples were dissolved in concentrated SDS and brought by

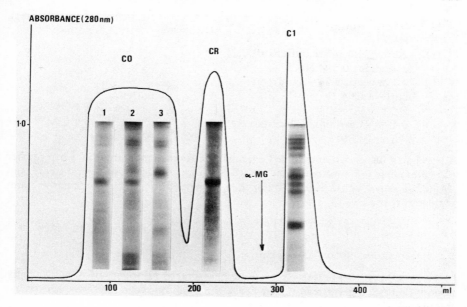

Figure 1. Proteins from adult rat brain synaptosomal plasma membranes solubilized in 0.08% SDS. Affinity chromatography on concanavalin A bound to 'Sepharose' 4B in the presence of 0.08% SDS—20 mM Tris-HCl (pH 6.7). Glycoproteins adsorbed on Con A were desorbed with 0.25 M α-methyl glucoside in the same buffer. Column: 60 cm × 1 cm; sample: 360 mg protein in 90 ml; flow rate: 0.5 ml/min. Twelve percent polyacrylamide gel electrophoregrams (8), stained for carbohydrates (PAS) of fractions CR and C1, and subfractions 1, 2 and 3 of the CO peak are also shown

dialysis, to 0.07% SDS. Some SDS in fact probably coprecipitated with the proteins and electrophoresis of precipitated and redissolved samples were blurred when the theoretical concentration of SDS was 0.08%. Sharp electrophoresis bands were obtained by slightly lowering the concentration of the SDS solution to 0.07%.

2.9 Recovery

We have obtained recovery around 100% with concentrated protein samples. Diluted protein solution gave lower recoveries. We have not further investigated this point.

3. METHOD

3.1 Elimination of Non-covalently Bound Con A from Con A—'Sepharose'

50 ml of Con A—'Sepharose' (Pharmacia) (swollen gel) were filtered in a sintered-glass filter and sequentially washed (by resuspending each time in 100 ml of the washing solution) with the following solutions.

Total volume (litres)

	H$_2$O	2 to 3
(a)	20 mM phosphate buffer (pH 7.2) containing 0.5 M NaCl	2
(b)	0.1 M acetic acid — Na acetate buffer (pH 4.5)	2
	H$_2$O	2
(c)	0.1 M Na carbonate—bicarbonate buffer (pH 9)	2

The gel was suspended and stirred overnight at room temperature in 1 litre buffer (a) containing 0.3 M sucrose and 0.02% sodium azide. The gel was filtered and washed as above with 2 litres of water then suspended and stirred overnight at room temperature in:

(d) 20 mM Tris-HCl buffer (pH 6.7) containing 0.08g% of SDS.

3.2 Regeneration of Con A—'Sepharose'

Fresh Con A—'Sepharose' gels cleaned, as described above, of not covalently bound Con A or gel already used for affinity chromatography were treated as follows.

Gels in suspension in (d) buffer were decanted and finer particles removed. The gels were filtered and SDS was washed out with:

(e) 20 mM Tris-HCl buffer (pH 7.2) total volume: about 4 litres. The gel was then suspended, and gently stirred for 2 h at room temperature in: 500 ml of buffer (e) containing CaCl$_2$, MgCl$_2$ and MnCl$_2$ each 10 mM. The gel was filtered, washed with water, resuspended and stored at +4°C in buffer (e).

3.3 Operating Conditions for Affinity Chromatography in the Presence of SDS

Immediately before use Con A—'Sepharose' (treated as described in §3.1 and 3.2) was filtered on glass filter, washed with water, resuspended in 20 mM Tris-HCl buffer (pH 6.7) containing 0.08% SDS and packed in a column of 1 cm diameter. The packing was carried out by letting the buffer percolate through the column at slow flow rate. When the column was packed, buffer was passed through the column until the base line of the u.v. recorder (280 nm) was stabilized.

The sample of defatted, alkylated brain membrane protein solubilized in 0.08% SDS (see Gombos, Chapter 41) was equilibrated at room temperature. 1/100 volume of 2 M Tris-HCl buffer pH 7.2 was added and the sample was applied to the column. Protein concentration in the applied samples was between 1 and 5 mg/ml depending on the nature of the sample.

The flow rate was regulated in such a way that the sample was in contact with the gel for at least 1 h and it was kept constant for the whole chromatography.

The elution of the non-adsorbed material was continued by washing the column with the buffer until the base line reached the initial value.

0.25 M α-methyl glucoside in the same buffer was then added and elution of adsorbed material was continued, until the base line remained constant.

3.4 Precipitation and Resolubilization of Samples

Pooled fractions were brought to pH 5.0 with glacial acetic acid. One to 2 volumes of analytical grade methanol were added, and, after shaking, the mixture was let stand overnight at $-20°C$. The suspension was allowed to stand at room temperature for 1/2 h and then centrifugated for 30 min at 30,000 g in metal centrifuge tubes. The pellet was washed twice with methanol. Traces of methanol were eliminated under a light stream of nitrogen.

Four percent SDS, enough to rapidly dissolve the sample at room temperature was added (1 ml for 1 to 10 mg protein depending on the sample). The samples were dialysed for 48 h at room temperature against 49 volumes of H_2O then against a large volume of 0.07% SDS.

3.5 Preparation of Samples for Analysis

Samples were electrophoresed on 12% polyacrylamide gels in the presence of SDS as described (8). No special preparation of the sample was necessary (see §2.8 and 3.4).

Sugars were determined by gas—liquid chromatography of the trifluoroacetate derivatives of the O-methyl glycosides obtained by methanolysis of the glycoproteins (14,15). Amino acids were determined on the same samples, after acid hydrolysis of the insoluble pellet after methanolysis and gas—liquid chromatography of the N(O)-heptafluorobutyrates of the isoamyl esters of the amino acids (16).

The samples for these analyses were prepared as follows: the glycoprotein fraction (100 to 1000 μg of proteins) was pipetted into the conic reaction vessel and then precipitated as described (§3.4) with 1 ml methanol. The pellet after centrifugation was washed with 2 ml chloroform: methanol 2:1 v/v and 2 ml 1:2v/v. The procedure was repeated and then followed by two washes with double-distilled methanol. All SDS strongly adsorbed to the proteins was removed by this technique. The pellet was dried under a light stream of nitrogen.

N-Terminal amino acids were determined according to the method of Zanetta and coworkers (17) in some protein fractions, separated by preparative polyacrylamide gel electrophoresis of microsomal C1 fractions. The samples for this analysis were simply precipitated twice with methanol, dried and dissolved in dimethyl formamide.

4. REFERENCES

1, Matus, A., De Petris, S. and Raff, M. C. (1973) 'Mobility of Concanavalin A receptors in myelin and synaptic membranes', *Nature (New Biol.)*, 244, 278—9.
2. Bittiger, H. and Schnebli, H. P. (1974) 'Binding of concanavalin A and ricin to synaptic junctions of rat brain', *Nature*, 249, 370—1.
3. Cotman, C. W. and Taylor, D. (1974) 'Localization and characterization of concanavalin A receptors in the synaptic cleft', *J. Cell Biol.*, 62, 236—42.
4. Wood, J. G. and McLaughlin, B. J. (1975) 'The visualization of concanavalin-A binding sites in the interperiod line of rat sciatic nerve myelin', *J. Neurochem.*, 24, 233—5.
5. Quarles, R. H., Everly, J. L. and Brady, R. O. (1972) 'Demonstration of a glycoprotein which is associated with a purified myelin fraction from rat brain', *Biochem. Biophys. Res. Commun.*, 47, 491—7.

6. Quarles, R. H., Everly, J. L. and Brady, R. O. (1973) 'Evidence for the close association of a glycoprotein with myelin in rat brain', *J. Neurochem.*, **21**, 1177—91.
7. Zanetta, J-P. and Gombos, G. (1974) 'Affinity chromatography on Con A—'Sepharose' of synaptic vesicle membrane glycoproteins', *FEBS Lett.*, **47**, 276—8.
8. Zanetta, J-P., Morgan, I. G. and Gombos, G. (1975) 'Synaptosomal plasma membrane glycoproteins: fractionation by affinity chromatography on concanavalin A', *Brain Res.*, **83**, 337—48.
9. Gombos, G., Reeber, A., Zanetta, J-P. and Vincendon, G. (1974) 'Fractionation of nervous tissue membranes glycoproteins', in *Méthodologie de la Structure et du Métabolisme des Glycoconjugués*, Colloque International n° 221 du Centre National de la Recherche Scientifique, Villeneuve d'Ascq, 20—27 juin 1973, (Editions du CNRS, Paris), Tome II, pp. 829—44.
10. Gombos, G., Zanetta, J-P., Reeber, A., Morgan, I. G. and Vincendon, G. (1974) 'Affinity chromatography of brain membrane glycoproteins', 547th Meeting of the Biochemical Society, London, *Biochem. Trans.*, **2**, 627—30.
11. Yariv, J., Kalb, A. J. and Levitzki, A. (1968) 'The interaction of concanavalin A with methyl α-D-glucopyranoside', *Biochim. Biophys. Acta*, **165**, 303—5.
12. Kalb, A. J. and Levitzki, A. (1968) 'Metal-binding sites of concanavalin A and their role in the binding of α-methyl-D-glucopyranoside', *Biochem. J.*, **109**, 669—72.
13. Agrawal, B. B. L. and Goldstein, I. J. (1967) 'Physical and chemical characterization of concanavalin A, the hemagglutinin from jack bean *(Canavalia ensiformis)*', *Biochim. Biophys. Acta*, **133**, 376—9.
14. Zanetta, J-P., Breckenridge, W. C. and Vincendon, G. (1972) 'Analysis of monosaccharides by gas—liquid chromatography of the O-methyl-glycosides as trifluoroacetate derivatives. Application to glycoproteins and glycolipids', *J. Chromatogr.*, **69**, 291—304.
15. Zanetta, J-P. and Vincendon, G. (1974) 'Determination of the carbohydrate composition of glycolipids and glycoproteins by gas chromatography of O-methyl glycosides as trifluoroacetate derivatives', in *Méthodologie de la Structure et du Métabolisme des Glycoconjugués*, Colloque International n° 221 du Centre National de la Recherche Scientifique, Villeneuve d'Ascq, 20—27 juin 1973, (Editions de CNRS, Paris), Tome I, pp. 47—61.
16. Zanetta, J-P. and Vincendon, G. (1973) 'Gas—liquid chromatography of the N(O)-heptafluorobutyrates of the isoamyl esters of aminoacids. I. Separation and quantitative determination of the constituent amino acids of proteins', *J. Chromatogr.*, **76**, 91—9.
17. Zanetta, J-P., Vincendon, G., Mandel, P. and Gombos, G. (1970) 'The utilization of 1-dimethylaminonaphthalene-5-sulfonyl chloride for quantitative determination of free amino acids and partial analysis of primary structure of proteins', *J. Chromatogr.*, **51**, 441—58.

CHAPTER 43

Affinity Chromatography of Glycopeptides Derived from Brain Membrane Proteins on Immobilized Concanavalin A

ANDRÉ REEBER
GIORGIO GOMBOS

Studies on structure of the sugar moieties (glycans) of glycoproteins make large use of enzymatic proteolysis of glycoproteins, for releasing the glycans bound only to a small peptide fragment (glycopeptides).

Glycopeptides from membrane glycoproteins can also be released by enzymatic proteolysis of lipid-free membrane residues. This technique was largely used for studies on membrane glycoproteins [for review of brain membrane glycopeptides see Brunngraber (1), Margolis (2)] before the development of methods of solubilization of membrane proteins.

Glycopeptides can be separated, on the basis of their sugar composition, by affinity chromatography on lectins (3,4). This technique can also by useful for the detection of glycan microheterogeneity (by using lectins of different specificity) and for isolating, from cell plasma membranes, the 'receptors' for some lectins of biological interest (5,6).

We have separated glycopeptides from brain microsomal fractions* (7) by adsorption on and desorption from a gel of concanavalin A obtained by polymerization with glutaraldehyde. The gel was prepared as described by Avrameas and coworkers (8).

More recently, in a study complementary to that of glycoproteins of brain membranes (9,10) we have used affinity chromatography on Con A–'Sepharose' to separate glycopeptides derived from these membranes (11).

1. MATERIALS AND PREPARATIONS

1.1 Preparation of Glycopeptides

The membranes used were those of the synaptosomal plasma membrane (SPM) or of brain microsomal fractions (see Zanetta and Gombos, Chapter 42; Gombos, Chapter 41). In some experiments we have labelled brain glycans by injecting $U-^{14}C$-glucosamine† intracerebrally (10 μl containing 12 μCi, in each hemisphere) according to the method of Dutton and Barondes (12).

The preparation of glycopeptides, already described (11), was as follows.

Lipids were extracted from the membranes as described (see Gombos, Chapter 41).

The lipid-free pellet was suspended in 50 mM Tris-HCl buffer (pH 7.8) containing $CaCl_2$ 1.5×10^{-3}M. The final concentration was about 15 mg protein per ml. Pronase (B Grade, 45,000 PUK units/g protein, Calbiochem.) was added (450 PUK units/g membrane protein). A few drops of toluene were added to prevent bacterial growth. Complete proteolysis was obtained after 5 days of incubation in a stoppered vessel at $37°C$ with constant agitation. Additional pronase (0.5%), was added at 12, 36 and 84 hours. At the same time, the pH was adjusted if necessary, to pH 7.8 with Tris. After incubation the suspension was centrifuged (3000g for 20 min) and the small residual pellet (which contained less than 5% of the hexoses of the membrane fraction) was discarded. The OD value at 260 nm (optical path = 1 cm) of the soluble proteolysate was determined.

Mucopolysaccharides and nucleic acids were eliminated by precipitation with N-cetylpyridinium chloride (CPC) according to the following procedure (13,14): 0.01 M sodium sulphate was slowly added, with stirring, to an equal volume of the proteolysate, followed by the same volume of CPC solution the concentration of which was calculated (in mg/ml) by multiplying the optical density (at 260 nm) of the soluble proteolysate by a factor of 0.34. The mixture was incubated under continuous agitation at $30°C$ for 30 min and centrifuged (3000g/20 min). The

---

*This extremely heterogeneous fraction is very rich in neuronal cell plasma membranes; at least 30% of its proteins are plasma membrane proteins as indicated by the comparison of its content in gangliosides and Na^+-, K^+-dependent ATPase with that of synaptosomal plasma membranes, a pure neuronal plasma membrane fraction.

†We have used D-glucosamine HCl $[^{14}C-(U)]$ specific activity 220 mCi/mM, purchased from New England Nuclear.

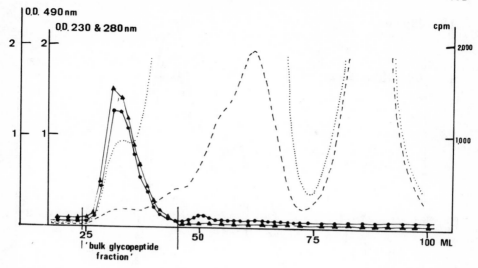

Figure 1. Adult rat brain microsomal glycopeptides after CPC treatment (see S1.1). Gel filtration on a column (30 x 2 cm) of 'Sephadex' G-25 (fine grade). Absorbance at 230 nm ($\cdots$); 280 nm ($----$). Absorbance at 490 nm ($\bullet$——$\bullet$) following phenol—sulphuric acid reaction for hexoses. Radioactivity: ▲———▲. Sample dissolved in and column equilibrated and eluted with 0.02% NaN$_3$. Sample applied in 2 ml. Flow rate: 20 ml/h

pellet was discarded. CPC was extracted from the supernatant by partitioning twice with isoamyl alcohol (0.2 ml/mg CPC).

The resulting aqueous phase was lyophilized, dissolved in water containing 0.02% sodium azide (to prevent bacterial growth) and chromatographed on a column of 'Sephadex' G-25* (fine grade) equilibrated and eluted with the same solution (Figure 1). More than 90% of the initial hexoses were in the void volume of the column.

This fraction: 'bulk glycopeptide fraction' was lyophilized and dissolved in a small volume of 20 mM ammonium formate buffer (pH 7.0) containing 0.02% sodium azide and chromatographed on a column of 'Bio-Gel' P-30 (25 x 2 cm) equilibrated and eluted with the same buffer. The elution profile (given in Figure 2) shows the presence of three hexose-containing peaks. The only sugar present in the void volume peak was glucose (probably glycogen in the microsomal fraction and, in the case of SPM, contaminating ficoll, used for the preparation) and was discarded. The other two, partially overlapping, peaks contained all the sugars present in brain glycoproteins. The material from these two fractions was pooled and this was our 'partially purified glycopeptide fraction'.

The similarity of elution profiles of hexoses and of the U-^{14}C label, indicates

---

*Depending on the volume of sample (which in turns depends on the amount of material and the viscosity of the solution), we have used columns of 50 x 5 cm, 30 x 2 cm and 20 x 1 cm.

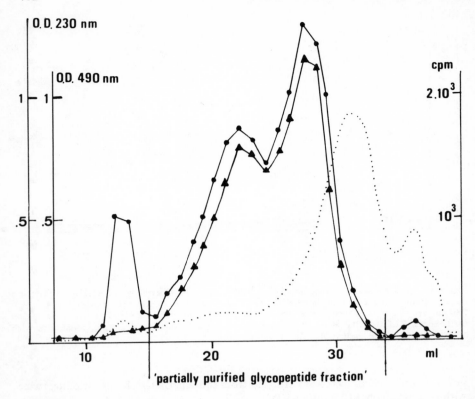

Figure 2. 'Bulk glycopeptide fraction' from adult rat brain microsomal fraction (see S 1.1). Gel filtration on a column (25 x 2 cm) of Bio-Gel P-30 (100–200 mesh). Symbols as in Figure 1. Sample dissolved in and column equilibrated and eluted with 50 mM ammonium formate buffer (pH 7) containing 0.02% NaN$_3$. Sample applied in 2 ml. Flow rate 5 ml/h

that labelling with U-^{14}C-glucosamine is a good generalized marker of these glycans, and thus it can be utilized to follow glycopeptide fractionation.

1.2 Analysis of Glycopeptide Fractions

Column eluates were monitored by u.v. adsorption (230 nm), by the ninhydrin reaction, by the phenol–sulphuric acid reaction (15) or by radioactive counting. Portions of each fraction were lyophilized for each of the following analyses. Carbohydrates were analysed by gas-liquid chromatography of the trifluorocetate derivatives of the O-methyl glycosides obtained after methanolysis of the glycopeptides (16). Amino acids were determined after acid hydrolysis of the glycopeptides by gas–liquid chromatography of the N(O)-heptafluorobutyrates of the isoamyl esters of the amino acids (17). N-terminal amino acids were analysed by dansylation (18).

2. AFFINITY CHROMATOGRAPHY OF GLYCOPEPTIDES ON CON A–'SEPHAROSE'

2.1 Pretreatment of Con A–'Sepharose'

Commercial Con A–'Sepharose' (Pharmacia) was treated as described (see Zanetta and Gombos, Chapter 42). In addition the following procedure was repeated before each chromatography, to eliminate degradation products formed during storage. Con A already utilized was also processed in the same way.

For each wash the gels were suspended in about 300 ml of buffer per 100 ml of gel (packed volume) and filtered on sintered-glass filters.

The first two steps consisted of washing the gel with:

(a) 0.1 M Tris-HCl–Acetic acid buffer (pH 5.0) containing 1 M NaCl.

followed by

(b) 0.1 M Tris-HCl buffer (pH 9.0) containing 1 M NaCl.

Steps (a) and (b) were repeated three times; then the gel was washed three times with

(c) 0.1 M Tris-HCl buffer (pH 7.0) containing 0.1 M α-methyl glucoside (α-MG).*

The washes with buffer (a) and (b) were repeated twice more. Followed by one wash with:

(d) 0.1 M Tris-HCl buffer (pH 7.0)

followed by

(e) 0.01 M Tris-HCl (pH 7.2) containing $CaCl_2$, $MnCl_2$, $MgCl_2$, each 1 mM and NaN_3 0.02%.

The gel was stored, in this buffer at +4°C.

Columns were packed in buffer (e) immediately before use. Sharper peaks however were obtained when NaCl was also present in the buffer (see §2.3).

2.2 Chromatographic Conditions

All operations were carried out at room temperature. NaN_3 was always present to prevent bacterial growth. Flow rate was kept constant at 20 ml/h for a column 30 cm x 1 cm by using a peristaltic pump.

2.3 Adsorption and Elution of Non-adsorbed or Weakly Adsorbed Material

Sample volume had not great importance for column efficiency. Generally, samples were in a volume less than or equal to half the volume of the packed Con A–'Sepharose' gel. The size of the column was chosen as described in §2.5.

---

*α-MG was used rather than α-methyl mannoside to avoid possible erroneous results in the mannose determination of glycopeptides.

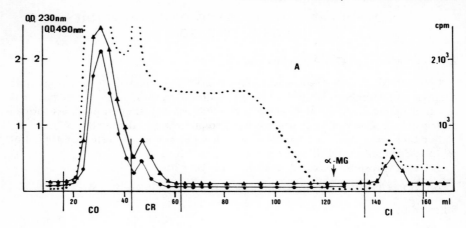

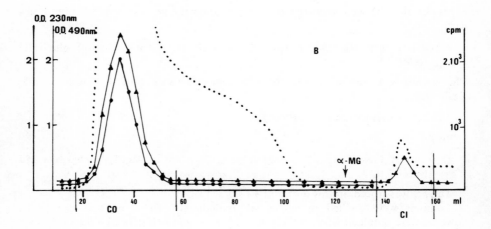

Volumes larger than these mask the peak of weakly adsorbed material (retarded fraction or fraction CR) (Figure 3A).

Buffer (e) was used for dissolving the sample and for equilibrating and eluting the column.

The presence or the absence of divalent cations in the chromatographic buffer did not affect chromatography. However the presence of sodium chloride modified the chromatographic profile.

As shown in Figure 3B the addition of NaCl (0.1 M, 0.5 M or 1 M) to buffer (e) reduced the tailing of the non-adsorbed fraction (CO fraction). At the same time, peak CR disappeared since its material was eluted with peak C0. This result indicates that interaction between Con A and CR glycopeptide is very weak since it depends on the ionic strength of the medium. It must however, be specific since CR glycopeptides did not interact with Con A in the presence of 0.1 M α-MG.

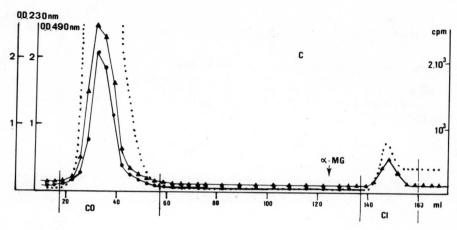

Figure 3. 'Partially purified glycopeptide fraction' from adult rat brain microsomal fraction (see S 1.1). Affinity chromatography on Con A—'Sepharose' column. Symbols as in Figure 1.

Figure	Volume of applied sample	Sample dissolved in and column equilibrated and eluted
(A)	8 ml	With buffer (e) (see text § 2.1)
(B)	2 ml	With buffer (e) (see text § 2.1)
(C)	8 ml	With buffer (e) + 0.5 M NaCl

Adsorbed glycopeptides were desorbed by 0.25 M α-MG in 0.01 M Tris-HCl buffer (pH 7.2) containing 0.02% NaN_3. Flow rate 20 ml/h

2.4 Desorption of Specifically Adsorbed Material

The material adsorbed on Con A (fraction C1) was eluted by 0.01 M Tris-HCl (pH 7.2) containing 0.25 M α-MG and 0.02% NaN_3 (Figure 3).

2.5 Column Capacity

The capacity of the column was established by rechromatography of the material not adsorbed by the first chromatography. We had determined that 1 ml of packed gel was sufficient to adsorb all the material interacting with Con A, present in a SPM or microsomal sample containing 0.3 mg of hexoses.

2.6 Recovery

Recovery from Con A—'Sepharose' could accurately be determined only with glycopeptides labelled with U-^{14}C-glucosamine (see § 1.1), in view of the several steps necessary to separate glycopeptides from α-MG (see § 1.4) with the possibility of losses of material at each step.

In the presence of 0.5 M NaCl (Figure 3C), recovery was 92% ± 8%, 59% ± 5% being in CO glycopeptides, and 33% ± 3% in C1 glycopeptides.

2.7 Purification of Glycopeptides Desorbed from Con A

Glycopeptides adsorbed on Con A (fraction C1) were eluted with 0.25 M α-MG. Glycopeptides and α-methyl glucoside were separated by repeated gel filtration on a 'Sephadex' G-25 column.

TABLE I.   Affinity chromatography on Con A-sepharose of glycopeptides derived from adult rat brain microsomal fraction. Sugar composition of Con A binding glycopeptides (C1) and glycopeptides not adsorbed on Con A (CO)

| | Molar ratio relative to N-Acetylglucosamine | | Sugar distribution expressed as % of | | | |
| | | | Total Sugars | | Each Sugar | |
	CO	C1	CO	C1	CO	C1
Fucose	0.357	0.187	7.96	0.73	92	8
Galactose	0.940	0.237	20.95	0.93	96	4
Mannose	0.702	3.439	15.65	13.42	54	46
N-Acetylglucosamine	1.000	1.000	22.29	3.90	85	15
N-Acetylgalactosamine	0.122	0.186	2.72	0.73	79	21
N-Acetylneuraminic acid	0.468	0.072	10.43	0.28	97	3
SUM			80	20		

3. SPECIFICITY OF BINDING TO CON A

An amount of 20.4% of glycopeptide sugars adsorbed on Con A is N-acetyl-glucosamine and 72.4% is mannose (see Table I). Glucose varied between experiments, and it is probably present at least in part in material other than glycoproteins. The excess of mannose indicates that some mannose is terminal.

When a sample was sequentially passed on several Con A columns, (each with an adsorption capacity smaller than that necessary to adsorb all the Con A binding material in the sample) until no material was adsorbed on Con A (see §2.5), the material adsorbed on the first column had a mannose to N-acetylglucosamine ratio higher than that of the material adsorbed on the second column indicating that the affinity for Con A of C1 glycopeptides was proportional to their mannose content.

Mannose and glucose were also present on glycopeptides not adsorbed on Con A; the mannose of these glycopeptides did not react with Con A either because it was not accessible or because its linkages with other sugars in the polysaccharide chain did not allow mannose (or glucose) to interact with Con A (see §2.7 in Zanetta and Gombos, Chapter 42). The affinity for Con A of material weakly adsorbed on Con A (CR) could be due to the same reasons as for those glycoproteins weakly bound to Con A (see §2.7 in Zanetta and Gombos, Chapter 42).

4. REFERENCES

1. Brunngraber, E. G. (1972) 'Chemistry and metabolism of glycopeptides derived from brain glycoproteins', in *Glycolipids, Glycoproteins, and Mucopolysaccharides of the Nervous*

Tissue, Advances in Experimental Medicine and Biology, Vol. 25, eds. Zambotti, V., Tettamanti, G. and Arrigoni, M. (Plenum Press, New York), pp. 17–49.

2. Margolis, R. U. and Margolis, R. K. (1972) 'Mucopolysaccharides and glycoproteins', in *Research Methods in Neurochemistry*, Vol. 1, eds. Marks, N. and Rodnight, R. (Plenum Press, New York), pp. 249–84.

3. Kornfeld, S. and Kornfeld, R. (1969) 'Solubilization and partial characterization of a phytohemagglutinin receptor site from human erythrocytes', *Proc. Nat. Acad. Sci. U.S.A.*, 63, 1439–46.

4. Smith, D. F., Neri, G. and Walborg, E. F., Jr. (1973) 'Isolation and partial characterization of cell-surface glycopeptides from AS-30D rat hepatoma which possess binding sites for wheat germ agglutinin and concanavalin A', *Biochemistry*, 12, 2111–8.

5. Burger, M. M. (1968) 'Isolation of a receptor complex for a tumor specific agglutinin from the neoplastic cell surface', *Nature*, 219, 499–500.

6. Allan, D., Auger, J. and Crumpton, M. J. (1972) 'Glycoprotein receptors from concanavalin A isolated from pig lymphocyte plasma membrane by affinity chromatography in sodium deoxycholate', *Nature (New Biol.)*, 236, 23–5.

7. Gombos, G., Hermetet, J. C., Reeber, A., Zanetta, J-P. and Treska-Ciesielski, J. (1972) 'The composition of glycopeptides, derived from neural membranes, which affect neurite growth *in vitro*', *FEBS Lett.*, 24, 247–50.

8. Avrameas, S. and Guilbert, B. (1971) 'Biologically active water-insoluble protein polymers. Their use for the isolation of specifically interacting proteins', *Biochimie*, 53, 603–14.

9. Gombos, G., Reeber, A., Zanetta, J-P. and Vincendon, G. (1973) 'Fractionation of nervous tissue membranes glycoproteins', in *Méthodologie de la Structure et du Métabolisme des Glycoconjugués*, Colloque International n° 221 du Centre National de la Recherche Scientifique, Villeneuve d'Ascq, 20–27 juin 1973, (Editions du CNRS, Paris), Tome II, pp. 829–44.

10. Zanetta, J-P., Morgan, I. G. and Gombos, G. (1975) 'Synaptosomal plasma membrane glycoproteins: Fractionation by affinity chromatography on concanavalin A', *Brain Res.*, 83, 337–48.

11. Reeber, A., Zanetta, J-P., Morgan, I. G. and Gombos, G. (1973) 'Purification and analysis of glycopeptides derived from nervous tissue membranes', in *Méthodologie de la Structure et du Métabolisme des Glycoconjugués*, Colloque International n° 221 du Centre National de la Recherche Scientifique, Villeneuve d'Ascq, 20–27 juin 1973, (Editions du CNRS, Paris), Tome II, pp. 815–28.

12. Dutton, G. R., Haywood, P. and Barondes, S. H. (1973) '[^{14}C-] Glucosamine incorporation into specific products in the nerve ending fraction *in vivo* and *in vitro*', *Brain Res.*, 57, 397–408.

13. Scott, J. E. (1960), in *Methods of Biochemical Analysis*, Vol. VIII, ed. Glick, D. (Interscience, New York), p. 145.

14. Brunngraber, E. G. and Brown, B. D. (1964) 'Fractionation of brain macromolecules. II. Isolation of protein-linked sialomucopolysaccharides from subcellular particulate, fractions from rat brain', *J. Neurochem.*, 11, 449–59.

15. Dubois, M., Gilles, K. A., Hamilton, J. K., Rebers, P. A. and Smith, F. (1956) 'Colorimetric method for determination of sugars and related substances', *Anal. Chem.*, 28, 350–6.

16. Zanetta, J-P., Breckenridge, W. C. and Vincendon, G. (1972) 'Analysis of monosaccharides by gas-liquid chromatography of the O-methyl glycosides as trifluoroacetate derivatives. Application to glycoproteins and glycolipids', *J. Chromatogr.*, 69, 291–304.

17. Zanetta, J-P. and Vincendon, G. (1973) 'Gas–liquid chromatography of the N(O)-heptafluorobutyrates of the isoamyl esters of amino acids. I. Separation and quantitative determination of the constituent amino acids of proteins', *J. Chromatogr.*, 76, 91–9.

18. Zanetta, J-P., Vincendon, G., Mandel, P. and Gombos, G. (1970) 'The utilization of 1-dimethylaminonaphthalène-5-sulphonyl chloride for quantitative determination of free aminoacids and partial analysis of primary structure of proteins', *J. Chromatogr.*, 51, 441–58.

CHAPTER 44

Application of Concanavalin A-'Sepharose' to the Purification of Glycoproteins from the Lymphocyte Plasma Membrane

DAVID ALLAN
MICHAEL J. CRUMPTON

Concanavalin A (Con A) bound to an insoluble support has been used fairly extensively to purify water-soluble carbohydrate-containing macromolecules (1). It is, however, only recently that this and related matrices have been employed for the affinity chromatography of membrane glycoproteins (2—10). The major reason for this slow rate of progress has been the lack of solvents which combine efficient solubilization and dissociation of cell membranes with the preservation of specific interactions of the antigen—antibody type. This difficulty has now been overcome with the aid of non-ionic and weakly anionic detergents, especially 'Nonidet' P-40 and sodium deoxycholate. This article describes the use of one of these detergents, sodium deoxycholate (DOC), in the affinity separation of lymphocyte plasma membrane glycoproteins on Con A–'Sepharose'.

The glycoproteins of the lymphocyte surface membrane are of particular interest since some have been implicated as mediators of well-defined physiological responses. Thus there is evidence that histocompatibility antigens (5), the receptor for antigen on B lymphocytes (11,12) and the enzymes 5'-nucleotidase and nucleotide pyrophosphatase (13) are glycoproteins. However, the best evidence for the direct involvement of membrane glycoproteins in lymphocyte behaviour comes from the stimulation by plant lectins, such as Con A and *Phaseolus vulgaris* phytohaemagglutinin, of lymphocyte transformation into blast cells that subsequently undergo mitosis. Lectins usually have an antibody-like specificity for certain carbohydrate groupings, which in the case of Con A appears to be for α-D-glycopyranosyl, α-D-mannopyranosyl, α-D-glucosaminyl and sterically related sugar residues (14).

Our original intention was to use Con A–'Sepharose' to isolate the membrane glycoprotein which mediates Con A-induced lymphocyte transformation. We initially assumed that it was a single molecular species but it rapidly became clear that if transformation is mediated by a unique glycoprotein then in the above circumstances it copurified with a considerable number of other glycoproteins, which account for the majority of the glycoproteins present in the original lymphocyte plasma membrane (2). This situation probably arises because most of the glycoproteins contain at least one carbohydrate grouping which is bound by Con A; mannosyl and glucosaminyl groups in particular are common to many glycoproteins (15). The limited data available from other workers (6,8,12) tends to confirm the idea that Con A–'Sepharose' is capable of binding many membrane glycoproteins.

1. PRINCIPLE

Theoretically, the selective solubilization of the intact cell surface membrane by using detergents, especially 'Nonidet' P-40 and 'Triton' X-100, represents a very convenient source of membrane glycoprotein. This source has, however, proven practically not to be ideal. In particular, these agents induce cell lysis and consequent contamination of surface membrane glycoproteins with intracellular carbohydrates and glycoproteins. Concomitant release of lysosomal enzymes may additionally lead to degradation of membrane proteins. A more successful approach has been to isolate initially a purified preparation of the plasma membrane. The latter approach was used in the present work.

2. SOURCE OF MEMBRANE GLYCOPROTEIN

Lymphocyte plasma membrane was isolated from pig mesenteric lymph node as previously described (16). The preparation consisted predominantly of a collection of small, smooth membrane vesicles which according to various morphological and biochemical criteria were not grossly contaminated with other subcellular elements. The yield of plasma membrane was 15% of theoretical and the degree of purification was 12- to 28-fold as assessed using various markers for plasma membrane. The method used for the preparation of pig lymphocyte plasma membrane is applicable to other soft tissues such as thymus, spleen and placenta (ref. no. 17 and D. Snary and M. J. Crumpton, unpublished observations). Alternative methods are available for the preparation of plasma membrane from single-cell suspensions (18,19).

3. SOLUBILIZATION OF MEMBRANE

3.1 Criteria for Solubilization

In most conventional methods for separation of biological macromolecules it is necessary to achieve proper solubilization (dispersion) of the components to be purified. The definition of what is truly soluble is to some extent arbitrary, especially with regard to biological membranes, and some confusion exists in the literature due to the use of various assessments of what constitutes solubilization. We have chosen to define a soluble membrane preparation as one which is not sedimented by centrifuging at $100,000\,g_{av}$ for 1 h in a solution of unit density and viscosity, and/or which contains no components excluded from 'Sepharose' 6B (exclusion limit corresponds to about 2×10^6 molecular weight) (20).

3.2 Choice of Solubilizing Agent

The choice of solubilizing agent is primarily based on its capacity to dissociate macromolecular complexes especially lipid-protein interactions with minimum alteration of conformation and destruction of biological activity. In addition, when it is proposed to exploit affinity chromatography as the basis of separation it is essential that the solvent does not interfere with specific molecular interactions such as lectin—carbohydrate, antigen—antibody and enzyme—substrate. These requirements have in the past proven difficult to satisfy. Thus, the detergent sodium dodecylsulphate while being a very efficient solubilizing agent for cell membranes, also promotes conformational alterations, loss of biological function and dissociation of specific interactions. The best compromise between the above requirements has generally been obtained by using non-ionic and weakly anionic detergents such as 'Triton' X-100, 'Lubrol' PX, 'Nonidet' P-40, sodium cholate and sodium deoxycholate (21).

3.3 Advantages and Disadvantages of Sodium Sodium Deoxycholate

We chose to use sodium deoxycholate (DOC) since we had previously found this to be a very effective solubilizing and dissociating agent for lymphocyte plasma membrane (20). Thus 1% DOC at room temperature dissolved over 85% of the membrane protein in 30 min with preservation of various biological activities such as $5'$-nucleotidase, histocompatibility antigens (5) and the capacity to interact with *Phaseolus vulgaris* phytohaemagglutinin (22). DOC also possesses the advantages that purified preparations at a concentration of 1% have a negligible adsorption at wavelengths greater than 220 nm so that protein can be estimated spectrophotometrically, and that, as it forms fairly small micelles at pH 8 and at low ionic strength about 4000 molecular weight (23)], it can readily be removed by dialysis or by elution from 'Sephadex' G-50. Alternative methods of removal include elution from a column on 'Zerolit' FFIP, SRA 62 (The Permutit Co. Ltd., U.K.) (24) or treatment with 'Amberlite' XAD-2 (Rohm & Haas Co., U.S.A.). A disadvantage of this detergent is its tendency to gel at ph <7.5 and at raised ionic strength, but this problem can be avoided by including 10 mM Tris-HCl buffer, pH 8.2. Another drawback is its affinity for low concentrations of divalent metal ions (e.g. Ca^{2+} and Mn^{2+}) which are often essential for lectin (e.g. Con A) and certain enzyme activities.

4. CON A LINKED TO 'SEPHAROSE' 4B

4.1 Preparation

Con A covalently attached to 'Sepharose' 4B is commercially available from Pharmacia Fine Chemicals AB, Sweden, but can be readily prepared in the laboratory. The latter source has the advantages that the degree of cross-linkage of the Con A to 'Sepharose' and the amount of Con A attached per unit of volume can be readily altered. It is convenient to be able to vary these parameters since they influence the affinity of the bound Con A for glycoprotein and the ease of dissociation of the adsorbed glycoprotein.

Our procedure (2) was a modification of that described by Lloyd (25). 'Sepharose' 4B (10 ml settled volume) was activated by addition of 200 mg of cyanogen bromide in 10 ml of distilled water at $2°C$; the pH was maintained at 10.5—11.0 by addition of 4 M sodium hydroxide dropwise with stirring. After 10 min the 'Sepharose' was washed at the water pump several times with water and then 0.1 M sodium bicarbonate pH 8.5, at $2°C$. The activated 'Sepharose' was resuspended in a solution of Con A [100 mg of 3x crystallized (Miles-Yeda Ltd., Israel) in 10 ml of 0.1 M sodium bicarbonate] and the slurry was gently agitated overnight at $2°C$. Ethanolamine hydrochloride (0.1 ml of a 1 M solution pH 8) was added to block any remaining active groups and agitation was continued for 30 min. The product was packed in a small column and washed sequentially by elution with 0.1 M sodium bicarbonate, 1 M sodium chloride, distilled water, 1% DOC, 5% (w/v) methyl α-D-glucopyranoside or methyl α-D-mannopyranoside in 1% DOC and 1% DOC alone until no further 280 nm adsorbing material was eluted. About 90% of the added Con A was bound to the column as judged from the optical density of the combined washings.

4.2 Stability

The Con A—'Sepharose' should be washed with 2% (w/v) methyl α-D-mannopyranoside in 1% DOC and then 1% DOC immediately prior to use for the isolation of membrane glycoproteins in order to remove soluble Con A which would otherwise contaminate the glycoprotein fraction. The origin of this soluble material is not known but it may depend either on the intrinsic proteolytic activity of many Con A preparations or on the dissociation of the non-covalently bonded subunits (26). The biological activity of Con A and the stability of subunit interaction are dependent on the presence of bound Ca^{2+} and Mn^{2+} (26). These cations are bound very strongly at pH 7 and above, but are lost more readily at pH's below 3. Some commercial manufacturers recommend that Ca^{2+} and Mn^{2+} are included in the chromatographic solvents (27), but we have found no advantage in adding these metal ions even in the presence of 1% DOC which would be expected to complex divalent cations. Some batches of Con A—'Sepharose' have been stored in distilled water containing 1 mM Ca^{2+} and Mn^{2+} at 2°C for up to 2 years without any noticeable decrease in biological activity. Similarly, Con A— 'Sepharose' suspended in 1% DOC, pH 8.2, was stable for at least 2 months at 2°C and the same sample has been re-used through six cycles of binding and elution without any apparent decrease in specificity and glycoprotein-binding capacity.

5. SEPARATION OF MEMBRANE GLYCOPROTEIN

5.1 Affinity Chromatography

Figure 1 illustrates the application of Con A—'Sepharose' to the separation of lymphocyte membrane glycoproteins. In Figure 1a pig lymphocyte plasma membrane (20 mg of protein in 2.5 ml of 10 mM Tris-HCl buffer, pH 7.5) was solubilized by incubating at 20°C for 30 min with 2.5 ml of 2% DOC in 10 mM Tris-HCl buffer, pH 8.2 and centrifuging at 100,000 g for 1 h. The supernatant, which contained 85% of the protein, was eluted from a column (13 cm x 1 cm diam.) of Con—A'Sepharose' at 2°C until no further 280 nm-absorbing material was detected. The unretarded fraction accounted for about 75% of the added protein. The column was then washed with 2% methyl α-D-glucopyranoside in 1% DOC, pH 8.2, when a small peak of 280 nm-absorbing material (equivalent to about 5% of the added protein) was eluted. Although the poor overall recovery of protein (80—85%) suggested that some material had been irreversibly adsorbed under the above conditions, no additional protein was released by elution with 50% (w/v) methyl α-D-glucopyranoside. A similar pattern (Figure 1b) was obtained by using the plasma membrane of the mouse lymphoblastoid cell line WEH1 22 that had been radioactively labelled by lactoperoxidase-catalysed iodination prior to cell breakage (M. J. Crumpton and J. J. Marchalonis, unpublished observations). Various results suggest (11) that the latter technique preferentially labels the surface membrane glycoprotein. In this case, 62% of the added radioactivity was not adsorbed by Con A—'Sepharose' and 23% was eluted with sugar. The remainder (15%) was bound by the Con A—'Sepharose' and could not be displaced by increasing the sugar concentration and/or raising the temperature to 20°C. An identical distribution of radioactivity was obtained by using 1% 'Nonidet' P-40 in place of the DOC to solubilize the WEH1 22 plasma membrane and as the solvent for affinity chromatography.

5.2 Yield and Irreversible Adsorption

It is apparent from the examples shown in Figure 1 that not all of the material added to Con A—'Sepharose' was recovered and that some material was bound irreversibly under the above conditions. Furthermore, electrophoretic analysis of the unretarded and eluted fractions from pig lymphocyte plasma membrane (see below) provided evidence that one major polypeptide chain (indicated by * in

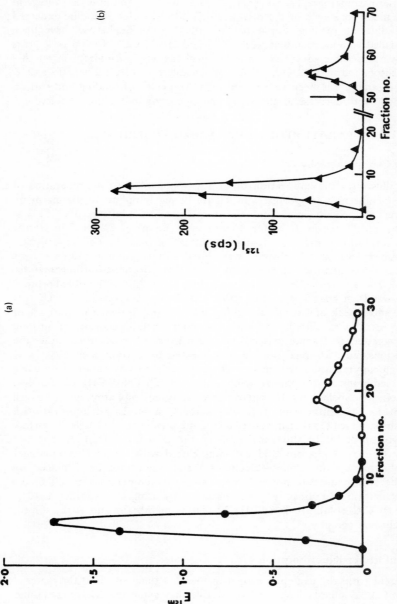

Figure 1. Fractionation of lymphocyte plasma membrane using Con A–'Sepharose'; (a) pig lymphocyte plasma membrane (20 mg of protein) and (b) plasma membrane of mouse thymoma cell line WEH1 22 that had been labelled with ^{125}I by lactoperoxidase-catalysed iodination (2 mg of protein). Membrane solubilized in 1% DOC (5 ml and 0.5 ml respectively) was eluted from a column (13 cm × 1 cm diam. and 15 cm × 0.5 cm diam. respectively) of Con A–'Sepharose' with 1% DOC until the absorbance at 280 nm (●) or the radioactivity (▲) of the eluate returned to the base line. The column was then washed (arrow) with 2% (w/v) methyl α-D-glucopyranoside in 1% DOC. ○, indicates absorption at 230 nm

Figure 2B) was irretrievably lost during the procedure. The most logical explanation for the low recoveries of material is that some of the adsorbed glycoproteins were not displaced by washing with sugar due to their high affinity for Con A—'Sepharose'. Direct evidence in support of the view that the irreversibly adsorbed material represents glycoprotein was obtained in two ways. First, when *Lens culinaris* lectin—'Sepharose' was used instead of Con A—'Sepharose' to separate the glycoproteins of pig lymphocyte plasma membrane, the sugar-eluted fraction accounted for 10—15% of the added protein compared with 5—8% and the overall recovery was increased from 80—85% to 90—95% (3). Second, using the ^{125}I-labelled WEHl 22 plasma membrane, the proportion of radioactivity irreversibly adsorbed by Con A—'Sepharose' was decreased from 15 to 3% by prior treatment of the solubilized membrane with *Lens culinaris* lectin—'Sepharose'. At the present time it is not known whether the glycoproteins which remained attached to Con A—'Sepharose' were representative of the total glycoprotein population or corresponded to a unique fraction. Also, the possibility has not been explored that different membrane glycoproteins differ in their affinities for Con A—'Sepharose' and may be eluted sequentially by using a gradient of increasing sugar concentration (see however, ref. no. 8).

5.3 Recovery of Glycoprotein

Removal of sugar and DOC from the eluted fraction and the concentration of the glycoprotein was achieved in three ways. First, by pressure dialysis either against 10 mM Tris-HCl buffer, pH 8.2 or against 0.1% DOC in 10 mM Tris-HCl buffer and subsequent removal of the remaining DOC by using an ion exchange resin such as 'Zerolit' FFIP, SRA 62 or a resin with an affinity for water-soluble organic substances, such as 'Amberlite' XAD-2 (see above). Second, by coprecipitating the DOC and glycoprotein with 0.1 vol. of 2% (v/v) acetic acid followed by removal of the DOC and any remaining sugar by extracting the pellet three times with 10 ml of 95% ethanol. Third, precipitation of the glycoprotein by adding 2 vol. of 95% ethanol at − 20°C and standing overnight at − 20°C. The residues obtained in the last two cases were dispersed in distilled water by mild sonication. The two latter procedures have the advantages that they were more rapid than dialysis and generally gave higher recoveries especially with small amounts (< 1 mg) of glycoprotein. The former technique is however, less liable to induce protein denaturation and loss of biological activity.

5.4 Composition of Eluted Fraction

The composition of the eluted fraction was assessed by assaying for protein (26), lipid phosphorus (29), neutral sugar (30) and sialic acid (31), and by polyacrylamide gel electrophoresis in the presence of 0.1% sodium dodecylsulphate (32). Chemical analyses indicated that the fraction contained primarily protein and carbohydrate, but failed to reveal any lipid phosphorus. The absence of phosphorus suggests that the membrane phospholipids were either not bound by Con A—'Sepharose' or were lost during the recovery of the eluted macromolecular species. A comparison of the carbohydrate contents of the eluted fraction and the lipid-free plasma membrane (Table I) shows that the eluted fraction was enriched two-fold in neutral sugar.

TABLE I.   Comparison of the carbohydrate contents of lipid-depleted pig lymphocyte plasma membrane and the fraction eluted from Con A–'Sepharose' with methyl -α-D-glucopyranoside. The lipid-depleted membrane was prepared by exhaustive extraction of whole plasma membrane (12 mg of protein) with aqueous n-butanol (three times with 10 ml) at $0°$C. This treatment removed 98% of the phosphorus that was extracted by chloroform–methanol (2:1, v/v). [Reproduced by permission of the copyright holder from reference 2.]

Carbohydrate	Lipid-free membrane	Eluted fraction
Neutral sugar[a] (μg of glucose/mg of protein)	23	68
Sialic acid[b] (μg of N-acetylneuraminic acid/mg of protein)	10	19

[a]Determined using the anthrone method (30).
[b]Determined by the method of Warren (31).

These results suggest that the membrane glycoprotein had been preferentially adsorbed and eluted from Con A–'Sepharose'. This interpretation is endorsed by the results of polyacrylamide gel electrophoresis (see below). On the other hand, as the unretarded fraction contained some carbohydrate, it seems likely that not all of the membrane glycoprotein was bound by Con A–'Sepharose'.

Polyacrylamide gel electrophoresis of the eluted fraction (Figure 2) showed about ten components. As expected, the pattern of the carbohydrate-staining bands (Figure 2F) resembled that of the protein-staining bands (Figure 2D), indicating that the components were glycoproteins. The protein pattern of the eluted material (Figure 2D) differed markedly from that of the whole membrane (Figure 2B) whereas the carbohydrate patterns (Figure 2E and 2F) showed marked similarities. In contrast, the unretarded fraction showed a few faint carbohydrate-staining bands only (not shown), whereas its protein pattern (Figure 2C) resembled closely that of whole membrane (Figure 2B). One prominent protein band of the whole membrane (indicated by * in Figure 2B) was not detected in either the unretarded or eluted fractions. This component did not stain for carbohydrate. The reason for its apparent loss is not known but it seems likely it was due to non-specific adsorption. The eluted fraction gave a prominent protein band in an identical position with that given by Con A (Figure 2A; mobility corresponding to a molecular weight of 27,000). This band does not represent Con A entirely since Con A does not stain for carbohydrate. However, the band stained more weakly for carbohydrate, relative to protein, than the other glycoprotein bands and, as a result, it is possible that it partly represents Con A. Contamination with Con A is not, however, a problem since it can be readily removed by treatment of the eluted fraction with a little 'Sephadex' G-25 prior to the recovery of the glycoprotein.

The above results indicate that the eluted fraction was composed predominantly of glycoprotein which accounted for the majority of the glycosylated polypeptide chains of the lymphocyte plasma membrane. Although we have not examined the eluted fraction for the presence of particular membrane glycoproteins, it appears likely from the results of other workers and by analogy with the results obtained using Lens culinaris lectin–'Sepharose' that it contained 5'-nucleotidase (3), histocompatibility antigens (5) and immunoglobulin M (12).

Figure 2. Densitometer tracings of polyacrylamide-gel electrophoresis patterns. (A) Con A (10 μg); (B) and (F) pig
lymphocyte plasma membrane (200 and 400 μg of protein
respectively; (C) unretarded fraction from Con A—'Sepharose'
(200 μg of protein); (D) and (E) eluted fraction from Con A—
'Sepharose' (50 and 100 μg of protein respectively). Samples
were dissolved by heating (100°C for 5 min) in 5% sodium
dodecylsulphate and were electrophoresed on 7.5% polyacrylamide gels in 0.1 M Na phosphate buffer (pH 7.2) containing
0.1% sodium dodecylsulphate for 17 h at 4.5 mA/tube (32);
electrophoresis was from right to left. Gels (A)—(D) were stained
for protein with 'Naphthalene Black 10B' in methanol—water—
acetic acid (3:6:1, by vol.), whereas gels (E) and (F) were stained
for carbohydrate with periodate—Schiff reagents (33). Patterns
were scanned using a Joyce-Loebl double-beam recording
microdensitometer. The large absorption at the left of the
tracings of gels (B) and (F) is due to lipid. The arrows indicate
the positions of pig immunoglobulin G (IgG), bovine serum
albumin (BSA) and ovalbumin (Ova) [Reproduced by permission
of the copyright holder from reference 2.]

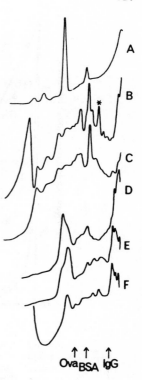

5.5 Biological Activity

Any glycoprotein that is bound by Con A would be expected to inhibit
Con A—stimulated lymphocyte transformation and thus each of the glycoproteins
in the eluted fraction may contribute towards its inhibitory capacity. Since the
inhibitory capacity of the eluted fraction (5—8% of the total protein) was increased
18-fold relative to the plasma membrane (Figure 3) it appears that the eluted
fraction accounted for most of the total inhibitory activity of the membrane. At
present we are unable to say which of the glycoprotein components of the
membrane are likely to be significant in mediating the mitogenic response of the
cell to Con A under normal culture conditions.

Antisera raised in rabbits by injecting the eluted fraction in complete Freund's
adjuvant agglutinated pig lymphocytes and erythrocytes, and induced pig lymphocytes to transform into blast cells (five- to tenfold increase in DNA synthesis). It
seems likely that these antisera would also prolong the survival of skin grafts
between non-identical pigs, by analogy with the behaviour of similar antisera raised
against the eluted fraction of mouse thymocyte plasma membrane obtained using
Lens culinaris lectin-'Sepharose' (A. J. Edwards, Clinical Research Centre,
Middlesex HA1 3UJ, U.K., personal communication).

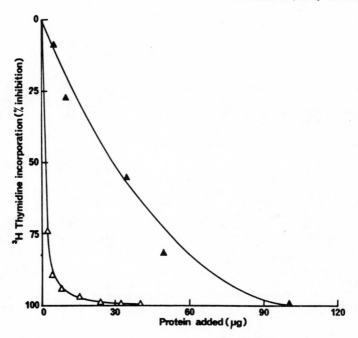

Figure 3. Inhibition of Con A-induced lymphocyte transformation by pig lymphocyte plasma membrane (▲) and the fraction eluted from Con A–'Sepharose' with methyl α-D-glucopyranoside (△). Increasing amounts of membrane or eluted fraction were incubated at 25°C for 30 min with 1 μg of Con A in 0.1 ml of saline prior to the addition of 10^6 pig lymphocytes in 1 ml of culture medium. The degree of lymphocyte transformation was assessed as previously described (22) in terms of incorporation of ^{3}H-thymidine into DNA. Amounts of ^{3}H-thymidine incorporated in the presence of inhibitor were expressed as a percentage of that incorporated in the absence of inhibitor. [Reproduced by permission of the copyright holder from reference 2.]

6. CONCLUSIONS

Con A–'Sepharose' bound the majority of the glycoproteins of lymphocyte plasma membrane which had been solubilized in either 1% DOC or 1% 'Nonidet' P-40. Approximately half of the bound glycoprotein was displaced by elution with 2% methyl α-D-glucopyranoside or 2% methyl α-D-mannopyranoside dissolved in the detergent used for solubilization. The remaining adsorbed glycoprotein was not eluted by increasing the concentration of sugar and/or by raising the temperature. Poor recoveries of membrane glycoprotein could be circumvented by using *Lens culinaris* lectin in place of Con A. The eluted fraction comprised about ten glycosylated polypeptide chains of different molecular weight. It retained essentially all of the capacity of the original plasma membrane to interact with Con A, as judged by the inhibition of Con A-stimulated lymphocyte transformation. Also, according to various criteria, antisera against the eluted fraction interacted with the surface of whole lymphocytes.

7. REFERENCES

1. Lis, H. and Sharon, N. (1973) 'The chemistry of plant lectins', *Ann. Rev. Biochem.*, **42**, 541—74.
2. Allan, D., Auger, J. and Crumpton, M. J. (1972) 'Glycoprotein receptors for concanavalin A isolated from pig lymphocyte plasma membrane by affinity chromatography in sodium deoxycholate', *Nature (New Biol.)*, **236**, 23—5.
3. Hayman, M. J. and Crumpton, M. J. (1972) 'Isolation of glycoproteins from pig lymphocyte plasma membrane using *Lens culinaris* phytohaemagglutinin', *Biochem. Biophys. Res. Commun.*, **47**, 923—30.
4. Hayman, M. J., Skehel, J. J. and Crumpton, M. J. (1973) 'Purification of virus glycoproteins by affinity chromatography using *Lens culinaris* phytohaemagglutinin', *FEBS Lett.*, **29**, 185—8.
5. Snary, D., Goodfellow, P., Hayman, M. J., Bodmer, W. F. and Crumpton, M. J. (1974) 'Subcellular separation and molecular nature of human histocompatibility antigens (HL-A)', *Nature*, **247**, 457—61.
6. Winqvist, L., Eriksson, L. C. and Dallner, G. (1974) 'Binding of concanavalin A—"Sepharose" to glycoproteins of liver microsomal membranes', *FEBS Lett.*, **42**, 27—31.
7. Gurd, J. W. and Mahler, H. R. (1974) 'Fractionation of synaptic plasma membrane glycoproteins by lectin affinity chromatography', *Biochemistry*, **13**, 5193—8.
8. Schmidt-Ullrich, R., Wallach, D. F. H. and Hendricks, J. (1975) 'Concanavalin A-reactive protein of rabbit thymocyte plasma membranes: analysis by crossed immune electrophoresis and sodium dodecylsulfate/polyacrylamide gel electrophoresis', *Biochim. Biophys. Acta*, **382**, 295—310.
9. Zanetta, J. P. and Gombos, G. (1974) 'Affinity chromotography on Con A—"Sepharose" of synaptic vesicle membrane glucoproteins', *FEBS Lett.* **47**, 276.
10. Zanetta, J. P, Morgan, I. G. and Gombos, G. (1975) 'Synaptosomal plasma membrane glycoproteins: fractionation by affinity chromatography on Con A', *Brain Res.*, **83**, 337—48.
11. Vitteta, E. S. and Uhr, J. W. (1974) 'Cell surface immunoglobulin. IX. A new method for the study of synthesis, intracellular transport and exteriorization in murine splenocytes', *J. Exp. Med.*, **139**, 1599—620.
12. Hunt, S. M. and Marchalonis, J. J. (1974) 'Radioiodinated lymphocyte surface glycoproteins: Concanavalin A binding proteins include surface immunoglobulin', *Biochem. Biophys. Res. Commun.*, **61**, 1227—33.
13. Evans, W. H. (1974) 'Nucleotide pyrophosphatase, a sialoglycoprotein located on the hepatocyte surface', *Nature*, **250**, 391—4.
14. So, L. L. and Goldstein, I. J. (1967) 'Protein—carbohydrate interaction. IX. Application of the quantitative hapten inhibition technique to polysaccharide—concanavalin A interaction. Some comments on the forces involved in concanavalin A—polysaccharide interaction', *J. Immunol.*, **99**, 158—63.
15. Marshall, R. D. and Neuberger, A. (1970) 'Aspects of the structure and metabolism of glycoproteins', *Advan. Carbohydr. Chem. Biochem.*, **25**, 407—78.
16. Allan, D. and Crumpton, M. J. (1970) 'Isolation and characterization of pig lymphocyte plasma membrane', *Biochem. J.*, **120**, 133—43.
17. Allan, D. and Crumpton, M. J. (1972) 'Isolation and composition of human thymocyte plasma membrane', *Biochim. Biophys. Acta*, **274**, 22—7.
18. Wallach, D. F. H. and Lin, P. S. (1973) 'A critical evaluation of plasma membrane fractionation', *Biochim. Biophys. Acta*, **300**, 211—54.
19. Crumpton, M. J. and Snary, D. (1974) 'Preparation and properties of lymphocyte plasma membrane', in *Contemporary Topics in Molecular Immunology*, Vol. 3, ed. Ada, G. L. (Plenum Press, New York/London), pp. 27—56.
20. Allan, D. and Crumpton, M. J. (1971) 'Solubilisation of pig lymphocyte plasma membrane and fractionation of some of the components', *Biochem. J.*, **123**, 967—75.

21. Crumpton, M. J. and Parkhouse, R. M. E. 'Comparison of the effects of various detergents on antigen-antibody interaction', *FEBS Lett.*, **22**, 210—2.
22. Allan, D., Auger, J. and Crumpton, M. J. (1971) 'Interaction of phytohaemagglutinin with plasma membranes of pig lymphocytes and thymus cells', *Exp. Cell Res.*, **66**, 362—8.
23. Small, D. M. (1971) 'The physical chemistry of cholanic acids', in *The Bile Acids: Chemistry, Physiology and Metabolism*, Vol. 1, eds. Nair, P. P. and Kritchevsky, D. (Plenum Press, New York/London), p. 318.
24. Hardwicke, P. M. D. and Green, N. M. (1974) 'The effect of delipidation on the adenosine triphosphatase of sarcoplasmic reticulum: electron microscopy and physical properties', *Eur. J. Biochem.*, **42**, 183—93.
25. Lloyd, K. O. (1970) 'The preparation of two insoluble forms of the phytohaemagglutinin, concanavalin A, and their interactions with polysaccharides and glycoproteins', *Arch. Biochem. Biophys.*, **137**, 460—8.
26. Cunningham, B. A., Wang, J. L., Pflumm, M. N. and Edelman, G. M. (1972) 'Isolation and proteolytic cleavage of the intact subunit of concanavalin A', *Biochemistry*, **11**, 3233—9.
27. *Affinity Chromatography: Principle and Methods* (1974)' Pharmacia Fine Chemicals AB, Sweden, pp. 41—2.
28. Lowry, O. H., Rosebrough, N. H., Farr, A. L and Randall, R. J. (1951) 'Protein measurement with the Folin phenol reagent', *J. Biol. Chem.*, **193**, 265—75.
29. Bartlett, G. R. (1959) 'Phosphorus assay in column chromatography', *J. Biol. Chem.*, **234**, 466—68.
30. Scott, T. A. and Melvin, E. H. (1953) 'Determination of dextran with anthrone', *Anal. Chem.*, **25**, 1656—61.
31. Warren, L. (1963) 'Thiobarbituric acid assay of sialic acids', in *Methods in Enzymology* Vol. 6, eds. Colowick, S. P. and Kaplan, N. O. (Academic Press, London/New York), pp. 463—5.
32. Maizel, J. V. (1969) 'Acrylamide gel electrophoresis of proteins and nucleic acids', in *Fundamental Techniques in Virology* eds. Habel, K. and Salzman, N. P. (Academic Press, London/New York), pp. 334—62.
33. Zacharius, R. M., Zell, T. E., Morrison, J. H. and Woodlock, J. J. (1969) 'Glycoprotein staining following electrophoresis on acrylamide gels', *Anal. Biochem.*, **30**, 148—52.

CHAPTER 45

Isolation of Insulin Receptors

PEDRO CUATRECASAS
KWEN-JEN CHANG

As described elsewhere (1) in this volume, the plant lectins, concanavalin A (2—5) and wheat germ agglutinin (2,3), simulate the action of insulin in isolated fat cells.

There is considerable evidence that these plant lectins can interact directly with insulin receptors in intact cells, isolated plasma membrane preparations and isolated receptor macromolecules (2,3).

The studies described above indicate that the insulin receptor (or complex) is a protein of complex carbohydrate composition which has several chemically distinct sites capable of binding plant lectins in a manner that perturbs the insulin-receptor interaction. The fact that the detergent-solubilized, isolated receptor can still be perturbed by these plant lectins with respect to insulin binding has served as the rational basis for the use of these lectins for purification of the receptor structures (2). Although it is possible that the insulin-like activity of wheat germ agglutinin may be related to direct perturbation of insulin receptor macromolecules, this explanation is unlikely for concanavalin A since tryptic digestion of cells destroys the insulin receptor (and response) without affecting the biological activity of concanavalin A (3).

Wheat germ agglutinin enhances the specific binding of insulin to isolated fat cells and to liver cell membranes at a concentration of about 1 μg/ml (Figure 1). Wheat germ agglutinin increases insulin binding by increasing the rate of insulin-receptor complex formation; the protein does not alter the rate of dissociation of the insulin-membrane complex or the total number of binding sites for insulin. Higher concentrations of wheat germ agglutinin, as well as concanavalin A at various concentrations (Figure 2), block the binding of insulin to fat cells when the plant lectins are added to the cells before insulin. If the insulin-receptor complex is

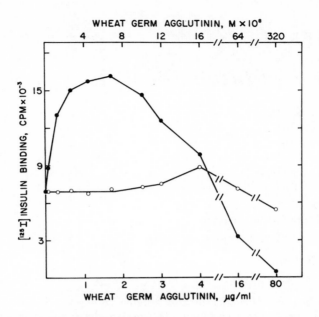

Figure 1. Effect of wheat germ agglutinin on the specific binding of ^{125}I-insulin to isolated fat cells. The adipocytes from the epididymal fat pads of 10 rats were suspended in 23 ml of Krebs–Ringer bicarbonate buffer containing 0.1% (w/v) albumin. Cell suspensions (0.2 ml containing about 3×10^5 cells) were incubated for 60 min at 24° with various concentrations of wheat germ agglutinin (●); in some cases (○) the wheat germ agglutinin had been preincubated (20 min, 24°) with 0.1 M N-acetylglucosamine before addition to the cells. The final concentration of the sugar in the cell incubation media was 10 mM. After incubation with the plant lectin, ^{125}I-insulin $(2.2 \times 10^5$ c.p.m.; 8×10^{-10} M) (7) and 2 μg of native insulin (for control tubes only) were added. The specific binding of insulin was determined by 'Millipore' membrane filtration (8) after incubating for another 50 min at 24°. N-Acetylglucosamine at a concentration of 50 mM does not by itself alter the binding of ^{125}I-insulin to fat cells. This concentration of sugar completely abolishes the effects of wheat germ agglutinin on insulin binding. It has been established independently (9) that wheat germ agglutinin binding to fat cells or membranes reaches equilibrium in 30 to 40 min with all concentrations of the plant lectin used in these experiments. Wheat germ agglutinin does not modify the non-specific binding of ^{125}I-insulin. Data from ref. 3

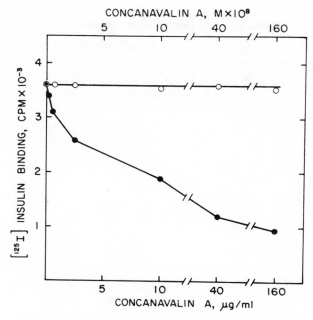

Figure 2. Effect of concanavalin A on the specific binding of ^{125}I-insulin to isolated fat cells in the absence (●) and presence (○) of α-methyl-D-mannopyranoside. The experiments were performed essentially as described in Figure 1. The fat cells from five rats were suspended in 11 ml of Krebs—Ringer bicarbonate buffer containing 0.1% (w/v) albumin. After incubating the cells (0.2 ml containing about 3×10^5 cells) at 24° for 60 min with the indicated concentration of concanavalin A ^{125}I-insulin (1.8×10^5 c.p.m.; 4×10^{-10} M) was added and the incubation was continued for another 50 min (24°). In some cases concanavalin A was preincubated with 0.5 M α-methyl-D-mannopyranoside (○). In these cases the final sugar concentration in the cell suspension was 50 mM; concentrations of the sugar as high as 0.1 M do not alter insulin binding. Under the conditions described here the binding of concanavalin A to fat cells approaches equilibrium (9). Data from ref. 3

formed before addition of the plant lectin, only an increase in insulin binding with wheat germ agglutinin and virtually no effect of concanavalin A are demonstrable. Studies using combinations of both plant lectins suggest that these proteins are binding to different regions of the insulin receptor and that some lectin molecules can bind to the cell in a way which is without effect on insulin binding unless the other lectin is also present.

Wheat germ agglutinin and concanavalin A modify insulin binding in membrane preparations (Table I) in a manner similar to that described for intact fat cells. Similar alterations of insulin binding occur with soluble preparations of the insulin

TABLE I.   Effect of wheat germ agglutinin and concanavalin A
on the specific binding of insulin to liver and fat cell membranes.
Fat and liver cell membranes were incubated for 50 minutes at
24° in 0.2 ml of Krebs—Ringer bicarbonate buffer containing
0.1% (w/v) albumin and the plant lectin at the indicated
concentration. ^{125}I-Insulin (1.1 x 10^5 c.p.m.) was added, and
specific binding was determined after incubating for 50 minutes
at 24°C. The liver and fat cell membrane concentrations were
680 and 41 μg of protein per ml, respectively. Data from
reference 3

Addition	Specific ^{125}I-insulin bound[a]	
	fat membranes	liver membranes
None	830 ± 60	18,200 ± 810
Wheat germ agglutinin (μg/ml)		
0.3	1500 ± 110	18,300 ± 220
1.2	2010 ± 160	21,900 ± 180
5	820 ± 40	32,800 ± 290
20	590 ± 30	29,500 ± 510
80	510 ± 40	16,000 ± 320
300	400 ± 20	10,100 ± 600
Concanavalin A (μg/ml)		
2.5	880 ± 40	16,600 ± 310
40	760 ± 50	15,300 ± 440
160		14,300 ± 390
600	440 ± 20	4900 ± 90

[a]c.p.m.; average ± standard error of the mean of three replications.

receptor of liver and fat cell membranes (Table II). These results indicate that the
changes in the binding of insulin occur by direct binding of the plant lectins to the
insulin-receptor macromolecule. All of the effects of the plant lectins are reversed
rapidly by adding simple sugars having selective specificity for the proteins,
N-acetyl-D-glucosamine   (for   wheat   germ   agglutinin)   or α-methyl-
D-mannopyranoside (for concanavalin A).

1. MATERIALS

1.1 Preparation of Plant Lectin Affinity Adsorbents

Agarose derivatives of the lectins are prepared by the CNBr procedure or by
reaction with activated N-hydroxysuccinimide esters of diaminodipropylaminosuccinyl-agarose (2,12). Forty millilitres of 'Sepharose' 4B are activated with
6 g of CNBr and allowed to react with 60 ml of ice-cold 0.1 M sodium phosphate at
pH 7.4 containing 500 mg of concanavalin A and 0.1 M α-methyl-D-mannopyrano-
side. After 16 hours at 4°, 2 g of glycine are added, and the incubation is continued
for 8 hours at room temperature. This adsorbent contains 5.5 mg of protein per
millilitre of gel. Wheat germ agglutinin (1.4 mg/ml) is similarly coupled to activated
agarose in the presence of 0.1 M N-acetyl-D-glucosamine; 1.1 mg of protein is
coupled per millilitre of gel. ^{125}I-labelled plant lectins may be used during the

TABLE II.   Effect of wheat germ agglutinin and concanavalin A on the binding of insulin to the solubilized receptor of liver and fat cell membranes. Liver and fat cell membranes were extracted (10, 11) with 1% (v/v) 'Triton' X-100 and centrifuged for 45 minutes at 300,000 g. The liver (210 μg of protein per ml) and fat (26 μg of protein per ml) supernatants were incubated for 50 minutes at 24° in 0.2 ml of Krebs–Ringer bicarbonate buffer containing 0.1% (w/v) albumin, 0.4% (v/v) 'Triton' X-100, and the indicated lectin. ^{125}I-Insulin (1.4 x 10^5 c.p.m.) was then added, and the samples were incubated for 50 minutes at 24°. Specific insulin binding was determined by the polyethylene glycol assay (10, 11). Data from reference 3

Addition	Specific ^{125}I-insulin bound (c.p.m.)	
	fat	liver
None	1030	12,660
Wheat germ agglutinin (μg/ml)		
0.3	2800	16,600
1.2	4800	25,400
5	5250	31,100
20	3200	30,600
80	880	13,710
300	250	6700
Concanavalin A (μg/ml)		
100	810	11,400
500	620	7400

coupling procedures. By phase-contrast microscopy, the lectin-coupled beads become heavily coated with erythrocytes when these are mixed at 24° for 1–2 hours. The cells can be rapidly desorbed with 50 mM N-acetyl-D-glucosamine.

2. METHOD

2.1 Affinity Chromatography (Figure 3)

Columns (1 ml, Pasteur pipettes) containing wheat germ agglutinin–agarose and concanavalin A–agarose are equilibrated at room temperature for 3 hours with Krebs–Ringer bicarbonate buffer containing 0.1% (v/v) 'Triton' X-100. Two millilitres of liver membrane extract (14 mg of protein per millilitre) are applied to each column, and the columns are washed with 10 ml of the buffer. The adsorbed proteins are eluted with 0.1 M sodium bicarbonate at pH 8.4 containing 0.1% 'Triton' X-100, 0.1% (w/v) albumin, and either 0.3 M α-methyl-D-mannopyranoside (for concanavalin A column) or 0.3 M N-acetyl-D-glucosamine (for wheat germ agglutinin column); the column flow is stopped for 3 hours after addition of the eluting buffer. About 60–80% of the insulin-binding activity of the sample is extracted by the columns. Wheat germ agglutinin appears to be more effective than concanavalin A. If the specific sugar is added to the crude sample before chromatography, the receptor does not adsorb to the column. The protein content

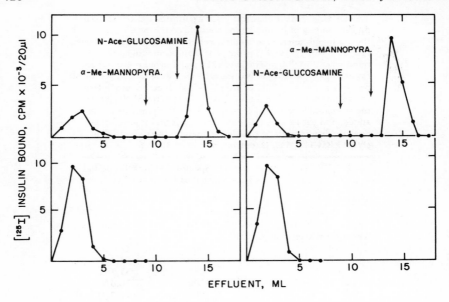

Figure 3. Affinity chromatography of detergent-solubilized (10, 11) insulin receptors of liver cell membranes on columns containing wheat germ agglutinin—agarose (left) and concanavalin A—agarose (right). The columns (1 ml, Pasteur pipettes) were equilibrated at 24° for 3 h with Krebs—Ringer bicarbonate buffer containing 0.1% (v/v) 'Triton' X-100, 2 ml of membrane extract (14 mg of protein per ml) was applied (at 24°) to each column, which was eluted with 0.1 M sodium bicarbonate buffer (pH 8.4) containing 0.1% (v/v) 'Triton' X-100, 0.1% (w/v) albumin, and either 0.3 M α-methyl-D-mannopyranoside (M) or 0.3 M N-acetyl-D-glucosamine (G); the column flow was stopped for 3 h after addition of the eluting buffer. In control columns (bottom), the specific sugar was added to the crude sample before chromatography. The protein content of the pooled material of the void volume was the same as that of the control column, indicating that only a very small fraction of the total protein had actually adsorbed. The binding protein was purified about 3000-fold, and the recovery of binding activity was greater than 90%. Data from ref. 2

of the pooled material of the void volume is the same as that of the control column, indicating that only a very small fraction of the total protein is actually adsorbed. The binding protein is purified about 3000-fold and the recovery of binding activity has been greater than 90% (2).

Concanavalin A and wheat germ agglutinin bind quite well to the insulin-binding protein in the presence of 0.2% 'Triton' X-100 which is used in the chromatography buffers. Fortunately, most of the lectin-binding glycoproteins of the membrane bind very weakly in the presence of this detergent. Thus very small quantities of protein adsorb to these affinity columns, and it is possible to achieve substantial purification (3000-fold) of the insulin-binding protein by these procedures. Since the receptor appears to bind to both types of lectin columns, greater and more selective purification should be achieved if both columns are used in a sequential manner.

3. COMPARISON WITH OTHER METHODS

The use of lectin-containing affinity adsorbents for purification has special advantages, compared to insulin-containing adsorbents (6), for purification of insulin receptors. For example, with lectin adsorbents elution of the receptor can be affected easily with simple sugars and without denaturation while elution from insulin adsorbents requires conditions (e.g. use of the denaturant, urea) which can adversely affect the protein. Furthermore, the capacity of lectin-containing adsorbents is much greater because the amount of insulin used in affinity gels must be kept to a minimum to reduce the quantity of insulin 'leakage' to the point that it does not interfere with the chromatographic adsorption of the receptor. With plant lectin-containing adsorbents, there is no possibility of contamination with insulin. On the other hand, the specificity of the lectin columns is far inferior to that of insulin columns since many other glycoproteins will be purified as well. This will, of course, also tend to decrease the potential capacity of the lectin columns for insulin receptors. As expected, the purification (3000-fold) achieved with the lectin columns is very much less than that which can be achieved (250,000-fold) with insulin columns. However, because the yields from lectin columns are so much better, these can be most useful as assessory steps in the overall purification, and they can in addition be convenient tools for concentration of dilute samples of the receptor.

4. REFERENCES

1. Chang, K.-J. and Cuatrecasas, P. (1975) 'Insulin-like effect of concanavalin A', this volume, Chapter 58, p. 599—604.
2. Cuatrecasas, P. and Tell, G. P. E. (1973) 'Insulin-like activity of concanavalin A and wheat germ agglutinin — direct interactions with insulin receptors', *Proc. Nat. Acad. Sci. U.S.A.*, **70**, 485—9.
3. Cuatrecasas, P. (1973) 'Interaction of concanavalin A and wheat germ agglutinin with the insulin receptor of fat cells and liver', *J. Biol. Chem.*, **248**, 3528—34.
4. Czech, M. P. and Lynn, W. S. (1973) 'Stimulation of glucose metabolism by lectins in isolated white fat cells', *Biochim. Biophys. Acta*, **297**, 368—77.
5. Czech, M. P., Lawrence, J. C. and Lynn, W. S. (1974) 'Activation of hexose transport by concanavalin A in isolated brown fat cells', *J. Biol. Chem.*, **249**, 7499—505.
6. Cuatrecasas, P. (1972) 'Affinity chromatography and purification of the insulin receptor of liver cell membranes', *Proc. Nat. Acad. Sci. U.S.A.*, **69**, 1277—81.
7. Cuatrecasas, P. (1971) 'Insulin-receptor interactions in adipose tissue cells: Direct measurement and properties', *Proc. Nat. Acad. Sci. U.S.A.*, **68**, 1264—8.
8. Cuatrecasas, P. (1971) 'Perturbation of the insulin receptor of isolated fat cells with proteolytic enzymes', *J. Biol. Chem.*, **246**, 6522—31.
9. Cuatrecasas, P. (1973) 'Interaction of wheat germ agglutinin and concanavalin A with isolated fat cells', *Biochemistry*, **12**, 1312—23.
10. Cuatrecasas, P. (1972) 'Properties of the insulin receptor isolated from liver and fat cell membrane', *J. Biol. Chem.*, **247**, 1980—91.
11. Cuatrecasas, P. (1972) 'Isolation of the insulin receptor of liver and fat-cell membranes', *Proc. Nat. Acad. Sci. U.S.A.*, **69**, 318—22.
12. Cuatrecasas, P. and Parikh, I. (1972) 'Adsorbents for affinity chromatography. Use of *n*-hydroxysuccinimide esters of agarose', Biochemistry, **11**, 2291—9.

CHAPTER 46

Concanavalin A–Agarose in the Study of Rhodopsin and its Derivatives

ROBERT RENTHAL
JORDAN S. POBER
ADRIAN STEINEMANN
LUBERT STRYER

Rhodopsin contains a covalently attached carbohydrate moiety consisting of three *N*-acetylglucosamine and three mannose residues (1). We have shown that Con A binds to disc membranes with a dissociation constant of 2×10^{-7} M (2). At saturation, one Con A monomer is bound per rhodopsin. The binding of Con A to disc membranes is specific since it is inhibited by α-methyl-D-mannoside and by D-glucose. We have used this specific interaction as the basis for the affinity chromatography of rhodopsin and some of its derivatives on Con A—agarose. Three applications are discussed here:

1. Rhodopsin solubilized by a variety of detergents can be separated from some other constituents of the rod outer segment (2).
2. Periodate-oxidized rhodopsin can be separated from unmodified rhodopsin (3).
3. Rhodopsin can be cleaved by a variety of proteolytic enzymes into two interacting fragments, F1 and F2, which can be purified as a complex by Con A—agarose chromatography. These fragments can be separated by photoelution from a Con A—agarose column (4).

1. CHROMATOGRAPHY OF RHODOPSIN ON CON A—AGAROSE

1.1 Preparation of Retinal Disc Membranes

Rod outer segments isolated from bovine retinas were suspended in deionized water, vortexed for about 15 sec, and centrifuged at 27,000g for 40 min. This procedure was repeated three times, yielding disc membranes in the form of closed vesicles having an average diameter of 0.4 μ. These washed disc membranes were then suspended in the standard buffer (10^{-3} M $CaCl_2$ and 10^{-3} M $MnCl_2$ in 0.05 M sodium acetate buffer, pH 5.0). All chromatographic procedures involving rhodopsin were carried out at $23°$ under dim red light unless otherwise noted.

1.2 Chromatography

Disc membranes containing 70 nmol of rhodopsin were solubilized by the addition of 2 ml of detergent buffer (1.4% cetyltrimethylammonium bromide in standard buffer). This solution was loaded on a 1 x 9 cm column of Con A—agarose equilibrated with detergent buffer. The elution profile is shown in Figure 1. Almost no rhodopsin emerged when the column was washed with detergent buffer. These fractions contained about 14% of the amino acid residues and nearly all of the phosphorus that was loaded on the column. The amino acid composition of these

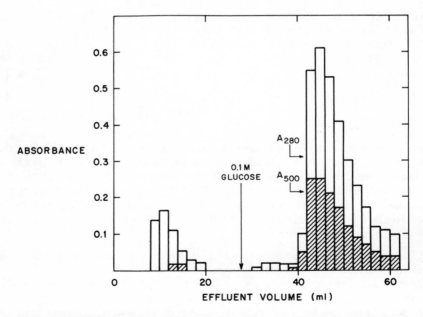

Figure 1. Affinity chromatography of rhodopsin on a Con A—agarose column. The rhodopsin bound to the column was eluted by 0.1 M D-glucose in detergent buffer (effluent from 38 to 62 ml). The absorbance of the effluent was measured at 280 nm (top of unfilled bars) and at 500 nm (top of crossed bars)

fractions was distinctly different from that of rhodopsin. Nearly all of the rhodopsin that was loaded on the column remained bound until it was eluted by the addition of 0.1 M D-glucose (or 0.1 M α-methyl-D-mannoside) in detergent buffer. These sugars displaced rhodopsin from the saccharide binding site of Con A. The recovery of rhodopsin from the column, calculated from the integrated 500 nm absorbance of the second peak, was 90%.

1.3 Choice of Detergents

A variety of detergents can be used instead of cetyltrimethylammonium bromide. We have obtained similar results with 3% dodecyltrimethylammonium bromide at 3°, 1% 'Triton' X-100 at 23° and 1% lauryl dimethylamine oxide ('Ammonyx' LO) at 23°. An advantage of using 'Triton' X-100 is that it does not interfere with subsequent SDS—acrylamide gel electrophoresis. However, 'Triton' X-100 is unsuitable for some spectroscopic studies because it absorbs and fluoresces strongly when excited at 280 nm.

2. CHROMATOGRAPHY OF PERIODATE-OXIDIZED RHODOPSIN

The presence of a carbohydrate unit in rhodopsin raises interesting questions concerning its function. Is an intact carbohydrate unit required for the 500 nm absorption band of rhodopsin and for the regeneration of this band after bleaching?

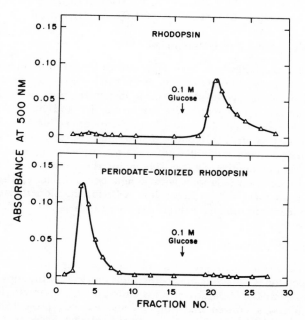

Figure 2. Comparison of the elution profiles of rhodopsin (top) and periodate-oxidized rhodopsin (bottom) from Con A—agarose columns

The carbohydrate moiety was modified by oxidation with sodium periodate to answer this question. Also, the resulting aldehydes were reacted with fluorescent hydrazides for spectroscopic studies. We monitored the extent of oxidation by chromatographing the periodate-treated rhodopsin on Con A—agarose.

Disc membrane vesicles containing 60 nmol of rhodopsin were suspended in 2 ml of 0.01 M $NaIO_4$ buffered to pH 5.0 with 0.1 M acetate. At the end of the reaction period, this suspension was centrifuged at 27,000g. The pellet was washed three times with 10 ml distilled water, and then dissolved in 2 ml of detergent buffer (1% cetyltrimethylammonium bromide in standard buffer). After centrifugation, 1 ml of the supernatant was applied to a 1 x 9 cm column of Con A—agarose equilibrated with the detergent buffer. Two ml fractions were collected, and their absorbance at 500 nm was measured. The column was washed with detergent buffer for about 15 fractions, followed by detergent buffer containing 0.1 M D-glucose (or 0.1 M α-methyl-D-mannoside). The elution profiles of rhodopsin solubilized from untreated disc membranes and from membranes exposed to sodium periodate for 4.5 h are compared in Figure 2. It is evident that the carbohydrate unit of nearly all rhodopsin molecules was modified after 4.5 h of oxidation by periodate. About half of the rhodopsin was completely oxidized after 1 h of periodate treatment. We found that periodate-oxidized rhodopsin retains the 500 nm absorption band and is regenerable after bleaching, provided that the sulphydryl groups are protected from oxidation (3).

3. SEPARATION OF PROTEOLYTIC FRAGMENTS OF RHODOPSIN BY PHOTOELUTION FROM CON A—AGAROSE

3.1 Proteolysis of Retinal Disc Membranes

Retinal disc membranes have been subjected to limited proteolysis by a variety of enzymes to ascertain the accessibility of the rhodopsin molecule (5—7). We have found that thermolysin cleaves rhodopsin in disc membranes or in 'Triton' X-100 detergent solution into two different fragments, F1 and F2 (4). The apparent molecular weights of opsin, F1, and F2 on SDS—acrylamide gels depend on the particular gel system used: 38,000, 30,000, and 18,000, respectively, in one system (8), and 37,000, 26,000 and 13,000, respectively in another (9). F2 contains the retinyl binding site, as shown by the formation of a fluorescent N-retinyl derivative upon reduction with sodium borohydride. F2 also contains the sulphydryl group that is alkylated by N-(iodoacetamidoethyl)-1-aminonaphthalene-5-sulphonate to produce a fluorescent derivative.

3.2 Separation of Fragments

The location of the carbohydrate group was determined by carrying out affinity chromatography on Con A—agarose. Thermolysin-cleaved rhodopsin, like native rhodopsin, binds to Con A—agarose and is eluted by 0.1 M α-methyl-D-mannoside. SDS—acrylamide gels showed that both F1 and F2 were present in the fractions eluted by α-methyl-D-mannoside. In contrast, only F1 binds to Con A—agarose when thermolysin-cleaved rhodopsin is bleached before it is applied to the column. If thermolysin-cleaved rhodopsin was loaded on Con A—agarose in the dark and then bleached while it was bound to the column (Figure 3a), light eluted 25% of

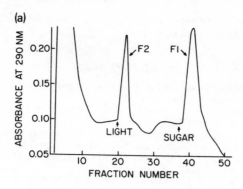

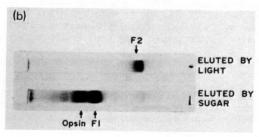

Figure 3. (a) Elution profile of thermolysin-cleaved rhodopsin from a Con A—agarose column. The absorbance of the effluent was measured at 290 nm to maximize the absorbance of tryptophan relative to that of 'Triton' X-100. The presence of 2 mM dithiothreitol does not change this elution pattern. (b) SDS—acrylamide gels stained with Coomassie blue show that the two peaks in the elution profile are F2 (top) and F1 (botton). A small amount of undigested opsin is evident in the bottom gel

the bound amino acid residues. The remainder could then be eluted by α-methyl-D-mannoside. SDS—acrylamide gel patterns (Figure 3b) showed that light had released F2, whereas F1 remained bound to the column until it was eluted by α-methyl-D-mannoside. Thus, only F1 contains the carbohydrate binding site for Con A. It seems likely that F2 adheres to the column prior to illumination because it is non-covalently associated with F1, and that the effect of light is to dissociate this complex.

The availability of large amounts of F1 and F2, which can readily be prepared by the above procedure, may aid the elucidation of the amino acid sequence of rhodopsin. Spectroscopic and other physical—chemical studies of these interacting fragments may provide insight into the conformational basis of visual excitation.

4. ACKNOWLEDGEMENTS

This work was supported by grants from the National Eye Institute (EY-01070) and the National Institute of General Medical Sciences (GM-16708). R. R. was an

NIH Postdoctoral Fellow (EY-54064). J. S. P. is a predoctoral fellow of the Medical Scientist Training Program (GM-02044).

5. REFERENCES

1. Heller, J. and Lawrence, M. A. (1970) 'Structure of the glycopeptide from bovine visual pigment 500', *Biochemistry*, **9**, 864–8.
2. Steinemann, A. and Stryer, L. (1973) 'Accessibility of the carbohydrate moiety of rhodopsin', *Biochemistry*, **12**, 1499–502.
3. Renthal, R., Steinemann, A. and Stryer, L. (1973) 'The carbohydrate moiety of rhodopsin: lectin-binding, chemical modification, and fluorescence studies', *Exp. Eye. Res.*, **17**, 511–5.
4. Pober, J. S. and Stryer, L. (1975) 'Light dissociates enzymatically-cleaved rhodopsin into two different fragments', *J. Mol. Biol.*, **95**, 477–80.
5. Bonting, S. L., DeGrip, W. J., Rotmans, J. P. and Daemen, F. J. M. (1974) 'Use of photoreceptor membrane suspension for the study of rhodopsin and associated enzyme activities', *Exp. Eye Res.*, **18**, 77–88.
6. Trayhurn, P., Mandel, P. and Virmaux, N. (1974) 'Removal of a large fragment of rhodopsin without changes in its spectral properties by proteolysis of retinal rod outer segments', *FEBS Lett.*, **38**, 351–3.
7. Saari, J. C. (1974) 'The accessibility of bovine rhodopsin in photoreceptor membranes', *J. Cell Biol.*, **63**, 480–91.
8. Weber, K. and Osborn, M. (1969) 'The reliability of molecular weight determinations by dodecyl sulfate – polyacrylamide gel electrophoresis', *J. Biol. Chem.*, **244**, 4406–12.
9. Fairbanks G., Steck, T. L. and Wallach, D. F. H. (1971) 'Electrophoretic analysis of the major polypeptides of the human erythrocyte membrane', *Biochemistry*, **10**, 2606–17.

D

Separation of Subcellular Fractions

CHAPTER 47

Separation of Subcellular Fractions from Rat Brain

HELMUT BITTIGER

Established methods for separation of subcellular components, e.g. sedimentation and density-gradient centrifugation, take advantage of differences in physical properties, such as size, form, mass and density of the moieties to be separated (1). More specific methods based on receptor–ligand interactions, which are successfully applied in the isolation of proteins and other macromolecules (2), have very rarely been used to separate subcellular fractions (3,4).

1. PRINCIPLE

The separation of subcellular components by affinity binding on Con A— 'Sepharose' introduces several new aspects not encountered in the isolation of macromolecules.

Since only macromolecules with molecular weights below 20×10^6 daltons can penetrate 'Sepharose' 4B gel beads (5), subcellular membranous particles are expected to bind exclusively to the surface of the gel particles.

In densely packed, sedimented gels (diameter of 'Sepharose' 4B beads $40-190 \mu m$) subcellular particles of sizes ranging from several hundred Ångstroms to some microns will be sterically hindered in movement, flow, binding and release. Batch procedures under conditions in which the gel particles are freely floating and separated from each other are therefore preferable to column methods.

Subcellular membranous particles of unfractionated homogenates are generally very heterogeneous in size, form and flexibility and in the number, distribution, density and accessibility of their receptors (Figure 1). The degree of binding on the gel surface will largely depend on the number of actually interacting ligand— receptor pairs, which is obviously not identical with the total number of receptors. Bound particles can be released by the application of shearing forces. The ease of detachment depends mainly on the size, form and mobility of the particles, and on the number of actual attachment sites. The susceptibility of particles to shearing forces obviously increases with their size. The likelihood of a particle being released in this way is tentatively rated, in diminishing order, by the numbers 1 to 7 in Figure 1. Small particles with a high density of surface receptors (No. 7) are presumably the most difficult to detach.

In an attempt to solve these foregoing problems, the following technique was developed.

The material to be separated is incubated in a 10—20% (v/v) Con A—'Sepharose' suspension and simultaneously rotated slowly (1—2 r.p.m.) end-over-end, to minimize shearing forces. Under these conditions, the gel particles are freely floating and their surface is completely accessible.

The suspensions are then filtered and washed on nylon gauze (35 μm mesh) and placed into a large-diameter Buchner funnel to obtain a thin layer of sedimented

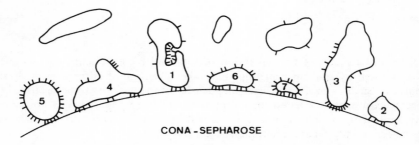

Figure 1. Scheme for the binding of subcellular particles to Con A—'Sepharose'. Probability of release of particles by shearing forces is rated from 1 to 7 (7 = release most difficult)

gel. This minimizes unspecific trapping of particles in the gel during filtration and washing. The gel is washed and shearing forces are applied in a very simple manner by squirting buffer from a polyethylene rinsing bottle onto the filter. Depending on the number and strength of washings fractions of different compositions can be obtained.

The release of strongly bound particles by addition of the competitive inhibitor α-methyl mannoside is aided by *vigorous* (about 60 r.p.m.) end-over-end rotation of the gel suspensions.

2. SEPARATION OF POSTSYNAPTIC DENSITIES* FROM RAT BRAIN

As has been shown in Chapter 10, the binding sites for Con A are unequally distributed among the components of a crude mitochondrial fraction from rat brain. PSD's* revealed by far the highest density of Con A receptors, whereas they are almost completely lacking on the mitochondria (6—8). A dispersed staining pattern for Con A receptors is found on synaptosomes, axons and myelin.

PSD's can thus be regarded as small particles carrying a high density of Con A receptors (Figure 1, No. 7). A necessary condition for the binding of PSD's to Con A–'Sepharose' is the opening of the synaptic cleft. This is achieved to a sufficient extent by osmotic-shock treatment of a crude mitochondrial fraction, removal of Ca^{2+} ions from the medium and incubation in salt solutions of relatively high ionic strength (9,10). PSD's are purified by incubating a crude mitochondrial fraction treated as described in a suspension of Con A–'Sepharose'. The portion still adhering after exhaustive washings is finally eluted with α-methyl mannoside under vigorous end-over-end rotation of the suspension. The released material is called the PSD fraction.

2.1 Preparation of a Crude Mitochondrial Fraction and Opening of the Synaptic Cleft

For the separation of the PSD's a crude mitochondrial fraction from the cerebral cortices of 20 adult male Sprague Dawley rats (average weight 350 g) is prepared according to standard procedures (11). This fraction is exposed to hypotonic conditions (100 ml 1 mM Tris buffer, pH 7, containing 2 mM EDTA) for 1 hour at $4°C$; the suspension is subsequently centrifuged at 100,000g (Beckman rotor SW-27 at 27,000 r.p.m.) for 40 minutes. The resultant pellet is suspended in 60 ml 100 mM acetate buffer, pH 6.4, containing 1 M NaCl, 1 mM $MgCl_2$, and kept at $20°C$ for 1 hour.

2.2 Affinity Binding

To this suspension is added 40 ml of sedimented Con A–'Sepharose' gel (Pharmacia, Uppsala, Sweden) and the volume is made up to 200 ml with 100 mM acetate buffer, pH 6.4, containing 1 M NaCl, 1 mM $CaCl_2$, 1 mM $MnCl_2$ and 1 mM

---

*Postsynaptic densities are defined in this paper as the structures comprising the subsynaptic web and the adhering postsynaptic membrane. Abbreviation: PSD.

$MgCl_2$. The suspension is gently agitated by slow end-over-end rotation (1—2 r.p.m.) for 2—3 hours at $20°C$, filtered through nylon gauze (35 μm mesh, Schweizerische Seidengazefabrik, Thal, SG, Switzerland) which is placed on a broad Buchner funnel (about 10 cm diameter) and washed with the same buffer solution (2—3 litre) until the optical density of the eluates at 280 nm is less than 0.005.

During the washing procedure, fractionation of the crude mitochondrial fraction is achieved. In the 'void volume', i.e. the first filtrate (about 200 ml), predominantly mitochondrial structures are observed and the pelleted material is of a brownish colour typical of mitochondria. The amount of protein in that fraction is usually 15—20% of the total. The following fractions obtained by further washing of the gel are virtually devoid of mitochondria and are similar to synaptic membrane fractions slightly contaminated with myelin. Complete removal of bound material requires large volumes of washing solutions (at least 3 litre), i.e. 15 times the volume of the incubation medium. More details of this type of fractionation are given in §3. To release the strongly bound material, the gel is suspended in 150 ml 100 mM acetate buffer, pH 6.4, containing 0.3 M NaCl, 1 mM $CaCl_2$, 1 mM $MgCl_2$ and 0.2 M α-methyl mannoside (Fluka, Buchs, Switzerland), and shaken by vigorous end-over-end rotation (60 r.p.m.) for several hours at $4°C$. The suspension is again filtered through nylon gauze, and the gel washed with 50—100 ml of the α-methyl mannoside solution. The eluate is centrifuged at 75,000g (Beckman SW-27 at 24,000 r.p.m.) for 1 hour. The yield of PSD fractions from 20 adult rat cerebral cortices ranges from 1.5 to 3 mg protein.

2.3 Characterization of the Separated Material

2.3.1 Electron Microscopy

Electron microscopical examination is an important part of the characterization of subcellular fractions. PSD's are relatively difficult to identify in conventional osmium/uranyl acetate/lead citrate-stained preparations. This difficulty can be overcome to a certain extent by using Con A—ferritin conjugates (see Chapter 8) or ethanolic phosphotungstic acid (EPTA) (12) as specific stain for PSD's.

It is recommended, however, that the contaminating membranous material should be stained lightly by adding a few more drops of water than prescribed (12) to the ethanolic phosphotungstic acid solution.

In Figures 2 and 3 electron micrographs of the PSD fractions purified according to the above affinity procedure are shown. Many of the EPTA-stained structures in Figure 2 can be identified as typical bar-shaped PSD's (arrows), as they appear if cut perpendicularly to the main plane of the discoid PSD's. Sections cut at other angles reveal only part of the geometrical characteristics of PSD's (Figure 2, arrowheads). The remaining EPTA-positive material is ill defined, and generally appears in the form of thick membranes, which are vesiculated, coiled or aggregated to polymorphous structures. In preparations doubly stained with uranyl acetate and lead citrate (Figure 3) the PSD's are more difficult to identify than in EPTA preparations, but membrane-like structures are more easily recognized. If exposed to Con A—ferritin conjugates many of the PSD's are labelled with ferritin particles (see insert in Figure 3) and in addition it can be seen that some of the PSD's appear

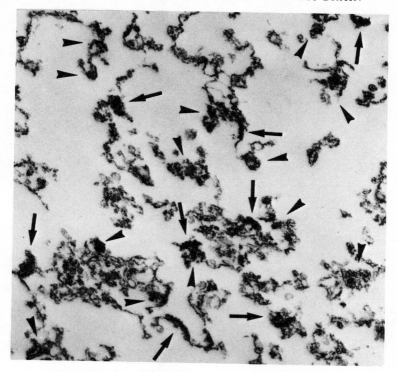

Figure 2. Postsynaptic density (PSD) fraction stained with ethanolic phospho-tungstic acid. Identified PSD's are marked by arrows and suspected PSD's by arrowheads (26,500 ×)

in the form of closed 'microsacs'. Virtually all the PDS's are completely separated from the presynaptic membranes. Remnants of lysed synaptosomes are not detectable and structures of possible mitochondrial origin are rarely found. Besides identifiable PSD's most of the isolated material appears to consist of membranous structures covered by a 'fuzzy coat'. It is difficult, if not impossible, to arrive at a representative estimate of the purity of the PSD fraction by electron microscopy, as PSD's having thin subsynaptic webs — especially those originating from Gray type II synapses (13) — can assume any structure owing to the flexibility gained through separation from their presynaptic counterparts: their only characteristic feature is the thin 'fuzzy coat' on the membranes.

2.3.2 *High-affinity Binding of Neurotransmitter-receptor Agonists or Antagonists*

Since electron microscopy has drawbacks as far as recognition of PSD's in subcellular fractions is concerned complementary biochemical methods have to be used. At present, however, there is no specific enzymatic assay for PDS's available. Specific high-affinity binding of neurotransmitter-receptor agonists or antagonists

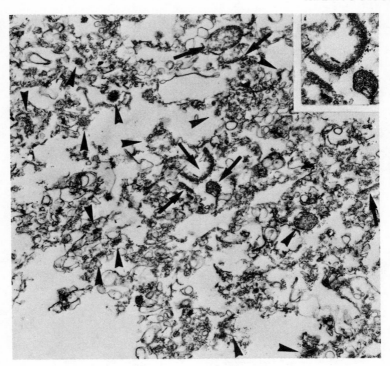

Figure 3. Postsynaptic   density   fraction (PSD)   reacted   with   concanavalin A—ferritin conjugates and stained with uranyl acetate and lead citrate
(26,500 x). Insert (45,000 x) shows concanavalin   A—ferritin conjugates
bound to PSD's

appears to be the most promising approach. The enrichment of PSD's achieved by
the method outlined above has been estimated by means of high affinity binding
(K_d = 8.1 x 10^{-9} M) (14) of the cholinergic muscarinic agonist pilocarpine and of
GABA.

For this purpose membrane fractions (300—900 μg of protein (15)) are
incubated in 2 ml 50 mM Na,K-phosphate buffer, pH 7.5, containing 9 nM
^{3}H-labelled pilocarpine (6.95 Ci/mmol, New England Nuclear, Boston) for 1 hour
at 20°C. Unspecific binding is determined in the presence of 70 μM cold pilocarpine
or 200 μM carbachol (Merck, Darmstadt, Germany). Incubation is terminated by
centrifugation at 45,000g for 15 min at 4°C. The supernatant is carefully removed
from the pellet and the walls of the tube. The pellets are dissolved in suitable
solvents (e.g. 'Soluene'-100, Packard) and the radioactivity determined in a liquid
scintillation counter.

High-affinity binding is assayed at about half-maximum saturation (9 nM) of the
pilocarpine binding sites.

Specific binding (5 separate preparations of PSD's) as expressed in counts per
minute per mg protein was found to be 1860 ± 320 for the PSD fraction and
480 ± 42 for the lysed, salt-treated crude mitochondrial fraction, which was reacted

with Con A—'Sepharose'. This difference suggests that the enrichment of PSD's, in comparison with the crude mitochondrial fraction, was about 4-fold. The specific binding of pilocarpine to the fractions representing synaptosomal membranes was somewhat less than in the crude mitochondrial fraction. Binding studies of GABA-receptors (16) resulted in a 7-fold enrichment of PSD's.

2.4 Problems in the Separation of PSD's

2.4.1 Purity of Fractions

As affinity binding to Con A—'Sepharose' separates particles according to *relative* differences in their Con A-receptor density and size, pure fractions of PSD's cannot be expected. Nevertheless this method yields fractions highly enriched in PSD's, i.e. in postsynaptic parts of synaptic junctional complexes devoid of presynaptic membranes. Further purification by density-gradient centrifugation or differential centrifugation was not possible, as PSD's banded at different densities between 1.0 M and 1.5 M sucrose.

Starting from synaptosomal membranes instead of a crude mitochondrial fraction has not been found to be advantageous with regard to the separation of PSD's, as in the above procedure mitochondrial membranes in any case separated during the washing procedure and myelin was removed by shearing forces. Since PSD's exhibit a wide range of densities, a significant loss of PSD's would be inevitable during centrifugation procedures to obtain synaptosomal membranes. As mitochondria account for about 20% (at the most 30%) of the protein content of the crude mitochondrial fraction, the effect of such a prepurification procedure would not significantly influence the enrichment.

2.4.2 Effect of Temperature

In investigating special properties of PSD's, the exposure of the material to $20°$C for several hours may be a drawback. Experiments performed at $10°$C and $4°$C have shown that the present purification procedure is in principle feasible, although the purification of PSD's is less good than at $20°$C.

2.5 Comparison with Other Methods

Purification of intact PSD's cannot be achieved by conventional centrifugation methods, since there are no exploitable differences in density between PSD's and other components of crude mitochondrial or synaptosomal fractions. Only by solubilizing synaptosomal membranes with detergents and subsequently separating the synaptic junctional complexes (SJC) — which are more resistant to that treatment — by density-gradient centrifugation, can SJC's and PSD's be isolated (17—21). Exposure to detergents solubilizes 50—97% of the membrane-bound proteins. At comparable detergent concentrations, the acetylcholine receptors of electric fish are already completely solubilized (22,23). Therefore, the structural and functional integrity of PSD's obtained by detergent procedures is likely to be severely impaired. By contrast PSD's purified by affinity binding on Con A—'Sepharose' in the absence of detergents appear to be largely intact, because they display a high neurotransmitter-receptor binding capacity for pilocarpine and GABA; in addition most of the protein (95%) and virtually all phospholipids remained particle bound during the isolation procedure. As neurotransmitter receptors can be expected to be fully accessible in PSD fractions due to the complete removal of the presynaptic membranes, further purification based on interactions of neurotransmitter receptors and agonists or antagonists should be possible.

3. SEPARATION OF MITOCHONDRIA FROM CRUDE MITOCHONDRIAL FRACTIONS

Extrasynaptosomal mitochondria do not bind any Con A on their outer membranes (6) (see Chapter 10). They are, therefore, not attached to Con A—'Sepharose' and can be isolated from a crude mitochondrial fraction in the 'void volume' of the gel suspension by filtration. The material still adhering to Con A—'Sepharose' thus represents a 'synaptosomal' fraction devoid of mitochondria and can be obtained by release with α-methyl mannoside. Alternatively, for further fractionation of the material according to the size of the particles and the density of Con A receptors, the gel is washed with acetate buffer.

3.1 Preparation of a Crude Mitochondrial Fraction

A crude mitochondrial fraction is obtained according to the standard procedure (11) from the cerebral cortices of 5 adult rats.

3.2 Separation of Fractions

A pellet obtained from the crude mitochondrial fraction is suspended in 70 ml 50 mM acetate buffer, pH 6.4, containing 0.15 M NaCl, 1 mM $CaCl_2$, 1 mM $MnCl_2$, 1 mM $MgCl_2$ and 10—15 ml sedimented Con A—'Sepharose'. The suspension is gently agitated by slow end-over-end rotation for 2—3 hours at 20°C and filtered through nylon gauze (35 μm mesh size, Schweizerische Seidengazefabrik, Thal, SG, Switzerland) which is placed on a broad (7 cm) Buchner funnel. The extrasynaptosomal mitochondria are recovered in the first filtrate corresponding to the 'void volume'. The 'synaptosomal' fraction can be obtained by incubating the sedimented gel in acetate buffer, pH 6.4, containing 0.1—0.2 M α-methyl mannoside under vigorous end-over-end rotation. The released material is again recovered by filtering the suspension through nylon gauze.

Alternatively, the gel is washed with acetate buffer as above.

3.3 Characterization of Fractions

The material is prepared for electron microscopy by conventional staining and thin-sectioning procedures. To check the purification of the mitochondria, the inner mitochondrial membrane marker cytochrome c oxidase (cytochrome c: O_2 oxidoreductase, E.C. 1.9.3.1.) can be used (24). As judged by electron microscopy, the fraction corresponding to the 'void volume' is highly enriched in mitochondria

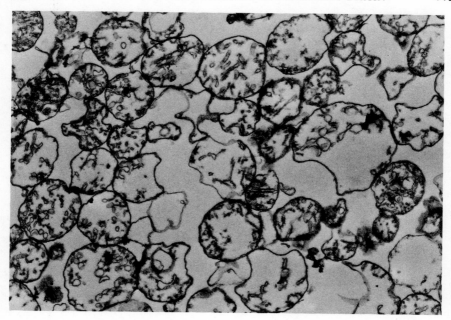

Figure 4. Fraction corresponding to the 'void volume' of a crude mitochondrial fraction which was incubated with concanavalin   A—'Sepharose' (17,250 ×)

(Figure 4). In the 'retained' fraction, synaptosomes, axons, myelin, but rarely extrasynaptosomal mitochondria are detected (Figure 5). This composition does not change significantly in the fractions obtained by washing the gel with additional buffer solution, but a general tendency towards smaller particles with a higher Con A-receptor density (detected by Con A-ferritin conjugates) is observed, increasing with the number of washings. In Table I the specific activity of

TABLE I.   Specific activity of cytochrome c oxidase in different fractions (50 ml) obtained by washing bound components of a crude mitochondrial fraction from Con A—'Sepharose'

Fraction	Protein (mg)	Specific activity of cytochrome c oxidase (μmoles substrate consumed per mg protein per hour)
1 ('void volume')	6.19	23.40
2	7.75	12.50
3	7.63	6.08
4	6.45	6.07
5	4.13	4.20
6	2.15	5.50
7	1.56	5.40

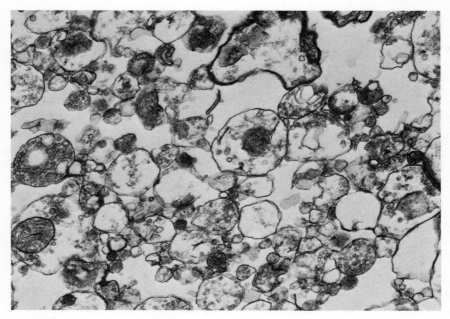

Figure 5. Fraction No. 3 (Table I) obtained from a crude mitochondrial fraction which was incubated with concanavalin A—'Sepharose' (24,375 x)

cytochrome c oxidase and the amount of protein in the individual fractions is indicated.

The fraction corresponding to the 'void volume', containing about 15% of the total protein of the crude mitochondrial fraction, has a high specific activity. From the third fraction on, the specific activity of cytochrome c oxidase remained constant, probably representing the activity exhibited by the enzymes of intra-synaptosomal mitochondria.

3.4 Comparison with Other Methods

In our experience the extent of separation of mitochondria from crude mitochondrial fractions by the affinity method is comparable to that achieved by centrifugation procedures. However, this method allows the separation of mito-chondria under strictly isotonic conditions.

4. GENERAL REMARKS

Affinity binding of subcellular components permits the separation of structures not exhibiting the differences in such physical properties as size, mass and density which are prerequisites for separation by centrifugation methods. A further advantage of the method is the possibility of separating fractions under isotonic conditions. So far, the method has only been applied to fractionate rat brain. In principle, components of any cell type differing in Con A—receptor density and distribution should be separable on the basis of affinity binding. The first step in such cases would be the study of Con A binding sites with the aid of Con A-ferritin or any other marker.

5. REFERENCES

1. 'Biomembranes part A' (1974), in *Methods in Enzymology*, Vol. 31, eds. Fleischer, S. and Packer, L. (Academic Press, New York, San Francisco, London).
2. 'Affinity techniques, enzyme purification, part B' (1974) in *Methods in Enzymology*, Vol. 34, eds. Jakoby, W. B. and Wilcheck, M. (Academic Press, New York, San Francisco, London).
3. Zachowski, A. and Paraf, A. (1974) 'Use of concanavalin A polymer to isolate right side-out vesicles of purified plasma membranes from eukariotic cells', *Biochim. Biophys. Acta.*, **57**, 787—92.
4. Flanegan, S. D., Taylor, P. and Barondes, S.H. (1975) 'Affinity partitioning of acetyl-choline receptor enriched membranes and their purification', *Nature*, **254**, 441—3.
5. 'Beaded "Sepharose" 2B-4B-6B', Pharmacia Fine Chemicals AB, Uppsala, Sweden, p. 3.
6. Matus, A., De Petris, S. and Raff, M. C. (1973) 'Mobility of concanavalin A receptors in myelin and synaptic membranes', *Nature (New Biol.)*, **244**, 278—80.
7. Bittiger, H. and Schnebli, H. P. (1974) 'Binding of concanavalin A and ricin to synaptic junctions of rat brain', *Nature*, **249**, No. 5455, 370—1.
8. Cotman, C. W. and Taylor, D. (1974) 'Localization and characterization of concanavalin A receptors in the synaptic cleft', *J. Cell Biol.*, **62**, 236—42.
9. Pfenninger, K. H. (1973) 'Synaptic morphology and cytochemistry', in *Progress in Histochemistry and Cytochemistry*, Vol. 5, (Gustav Fischer Verlag, Stuttgart, Portland), pp. 1—86.
10. Pfenninger, K. H. (1971) 'The cytochemistry of synaptic densities. II. Proteinaceous components and mechanism of synaptic connectivity', *J. Ultrastruct. Res.*, **35**, 451—75.
11. Whittaker, V. P. (1969) 'The synaptosome' in *Handbook of Neurochemistry*, Vol. 2, ed. Lajtha, A. (Plenum Press, New York, London), pp. 327—64.
12. Bloom, F. E. and Aghajanian, G. K. (1968) 'Fine structural and cytochemical analysis of the staining of synaptic junctions with phosphotungstic acid', *J. Ultrastruct. Res.*, **22**, 361—75.
13. Gray, E. G. (1959) 'Axosomatic and axo-dentritic synapses in the cerebral cortex. An electron microscope study', *J. Anat.*, **93**, 420—33.
14. Schleifer, L. S. and Eldefrawi, M. E. (1974) 'Identification of the nicotinic and muscarinic receptors in subcellular fractions of mouse brain', *Neuropharmacol.*, **13**, 53—63.
15. Lowry, O. H., Rosebrough, N. J., Farr, A. L. and Randall, R. J. (1951) 'Protein measurement with the Folin phenol reagent', *J. Biol. Chem.*, **193**, 265—75.
16. Zukin, S. R., Young, A. B. and Snyder, S. H. (1974) 'Gamma-aminobutyric acid binding to receptor sites in the rat central nervous system', *Proc. Nat. Acad. Sci. U.S.A.*, **71**, 4802—7.
17. De Robertis, E., Azcrurra, J. M. and Fiszer, S. (1974) 'Ultrastructure and cholinergic binding capacity of junctional complexes isolated from rat brain', *Brain Res.*, **5**, 45—56.
18. Cotman, C. W., Levy, W., Banker, G. and Taylor, D. (1971) 'An ultrastructural and chemical analysis of the effect of "Triton" X-100 on synaptic plasma membranes', *Biochim. Biophys. Acta.*, **249**, 406—18.
19. Cotman, C. W. and Taylor, D. (1972) 'Isolation and structural studies on synaptic complexes from rat brain', *J. Cell Biol.*, **55**, 696—701.
20. Davis, G. A. and Bloom F. E. (1973) 'Isolation of synaptic junctional complexes from rat brain', *Brain Res.*, **62**, 135—53.
21. Cotman, C. W., Banker, G., Churchill, L. and Taylor, D. (1974) 'Isolation of postsynaptic densities from rat brain', *J. Cell Biol.*, **63**, 441—55.
22. Raftery, M. A. (1974) 'Isolation of acetylcholine receptor — α-bungarotoxin complexes from *Torpedo californica* electroplax', *Arch. Biochem. Biophys.*, **154**, 270—6.
23. Chang, H. W. (1974) 'Purification and characterization of acetylcholine receptor I from *Electrophorus electricus*', *Proc. Nat. Acad. Sci. U.S.A.*, **71**, 2113—7.
24. Cooperstein, S. J. and Lazarow, A. (1951) 'A microspectrophotometric method for the determination of cytochrome oxydase', *J. Biol. Chem.*, **189**, 665—70.

E

Separation of Cells

CHAPTER 48

Fractionation and Manipulation of Cells with Concanavalin A-coated Fibres

URS RUTISHAUSER
GERALD M. EDELMAN

447

1. AFFINITY FRACTIONATION OF CELLS

There is an obvious need in cell biology for methods that utilize chemical differences at the cell surface as a basis for either cell fractionation or cell manipulation. The main requirements of such methods are specificity, wide applicability, high yield and maintenance of cell viability. In addition, it is necessary to define some criteria for purity. Because of their complexity, cell populations cannot be analysed by many of the procedures routinely used for molecules. It is therefore necessary to define purity in terms of a single feature that can be assayed by specific methods. Of course, cells homogeneous with respect to one property will usually be heterogeneous for many others, and repeated selection by many different criteria would be necessary to obtain cells that are closely similar. Usually, the fragility of cells makes the realization of such extensive procedures unlikely. Nevertheless, the isolation of cells having even one property in common would greatly facilitate the study of that property and its relationship to other aspects of the cell phenotype. Although the fragility of cells as compared with molecules limits the conditions and extent of fractionation procedures, the size and metabolic capabilities of a cell can be a distinct advantage.

The chemical separation of cells according to their affinity for a biospecific reagent on a solid support has been achieved in a variety of systems. Usually beads of plastic (1), glass (2), polyacrylamide (3) or agarose (4) have been used by adsorbing or covalently attaching substances such as antigens, antibodies, lectins or hormones to provide the requisite specificity. In most instances, this matrix is packed into a cylinder through which the cells are passed. The properties of the retained and effluent cells are then compared to those of the unfractionated cells. Unfortunately, many cells are retained non-specifically by bead columns, and because specific elution has not in general been achieved, it is not possible to study the properties of the specifically bound cells or to isolate enriched populations of cells with specific receptors.

2. PRINCIPLE OF FIBRE FRACTIONATION (5,6)

The method to be described here separates cells according to their ability to bind specifically to strung nylon fibres coated with molecules such as antigens, antibodies or lectins. In this book the primary focus will be on the use of fibres coated with the lectin concanavalin A (Con A). The particular geometry and mixing conditions were designed to avoid many of the difficulties encountered in the use of columns, and at the same time to keep the procedure simple and inexpensive. Fractionation can be achieved according to the presence of a unique cell surface component, or by differences in the properties of the same component on different cells.

The basic principle of fibre fractionation is illustrated in Figure 1. The molecule or macromolecule to be used for fractionation is attached to smooth nylon fibres that are held under tension in a frame without crossing each other, thus avoiding non-specific trapping of cells. Cells are shaken in suspension with the fibre, after which unbound cells are washed away. The bound cells may be transported to other media for further characterization or released by plucking the taut fibre with a needle. This serves to shear a cell from its point of attachment to the fibre.

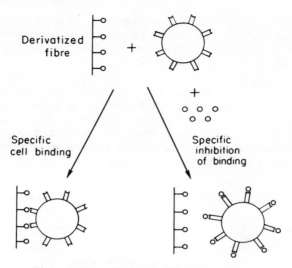

Figure 1. General scheme of fibre fractionation

Specificity in the system can be tested by inhibition of cell-fibre binding with a competitive inhibitor for either the ligand on the fibre or on the cell.

3. ADVANTAGES AND DISADVANTAGES OF THE FIBRE FRACTIONATION TECHNIQUE AND COMPARISON WITH OTHER METHODS

The fibre technique has certain advantages over previous affinity fractionation procedures. The use of a simple spatial arrangement for the fibres permits direct observation and quantitation of the cells. The fractionated cells can also be manipulated on the fibres under a variety of conditions, and individual cells can be identified throughout an experiment.

Most cells have a natural tendency to adhere to surfaces, and this poses a serious difficulty in many affinity fractionation methods. With columns of beads, physical trapping also occurs, and 5—50% of the cells loaded are non-specifically retained. In the isolation of cells that occurs at low frequencies, this amount of non-specific binding is unsatisfactory. With fibre fractionation, the simple configuration of the solid support minimizes non-specific binding of cells. In the presence of a competitive inhibitor of binding, less than 0.01% of the cells in the incubation mixture are bound to the fibres or the accompanying support. Adequate specificity can therefore be achieved in the isolation of even a minor subpopulation of cells.

Although the initial binding of cells to a surface may be specific, elution of bound cells by a competitive inhibitor can rarely be achieved. Studies with concanavalin A-derivatized beads (5) suggest that after being specifically bound, the cell membrane interacts with the surface to form secondary adhesions that can be broken only by physical shearing or when the cell is distorted by hypotonic shock. Unfortunately, osmotic elution can be achieved with erythrocytes, but not with nucleated cells. As with bead column methods, cells bound to nylon fibres cannot in general be removed by addition of a competitive inhibitor. They can, however, be rapidly and quantitatively released by plucking the taut fibre with a needle. This mechanical removal of cells has the additional advantage that it is not limited to cases in which a competitive inhibitor of binding is available.

A disadvantage of all solid-phase methods, including fibre fractionation, is the possible perturbing effect of the ligand or surface on cell metabolism or function. This stricture also extends to the method of removing the cells from the surface. Our experience with the method indicates that a number of variables are critical in obtaining good results, and these may vary considerably in different situations. A careful study of each system in terms of these parameters is therefore essential for success in a particular application.

4. PREPARATION OF CON A-COATED FIBRES

4.1 Selection of the Fibre

The major criteria used in selecting a fibre are its chemical reactivity, appearance, strength and its tendency to bind cells non-specifically. Nylon is particularly suitable in that it has free amino and carboxyl groups, is available as a smooth transparent monofilament in a range of sizes, has exceptional strength and does not bind cells non-specifically after derivatization.

Monofilament is generally used in two size ranges, 125–250 μm diameter for routine fractionation and manipulation and 20–50 μm diameter for certain morphological and histological observations. The larger fibre is available commercially as size 50 transparent sewing nylon (Dyno Merchandise Corp., Elmhurst, New York, or at notions stores). The smaller fibres can be obtained directly from the manufacturer (Dupont, Wilmington, Delaware).

4.2 Stringing of the Fibre

In order to derivatize the nylon, and to bind, observe and manipulate the cells, it is useful to immobilize the fibre by stringing it under tension in a supporting frame. Although a variety of designs can be used, an inexpensive and reusable frame (Figure 2) can be cut from from plastic stoppers (size 5 or 6 polyethylene or polypropylene stopper, Mallinckrodt, New York, New York). These collars fit snugly into conical 35 x 10 mm petri dishes (NUNC, Vanguard International, Red Bank, New Jersey). For routine assays, a single parallel row of 10–15 fibres is sufficient; for preparative fractionation, up to two rows of 24 fibres each can be strung in a single frame. Fresh fibres are usually used for each experiment.

4.3 Coating the Fibre with Con A

Prior to coating of the fibres, it is necessary to remove surface contaminants by 10-minute extractions first with petroleum ether and then with carbon tetrachloride. If desired, the number of free amino and carboxyl groups may then be increased by partial hydrolysis of the nylon with 3 N HCl for 30 minutes at room temperature. The fibres must be rinsed with water for at least 30 minutes after

Figure 2. Petri dish (35 mm x 10 mm) containing polyethylene frame strung with nylon monofilament

hydrolysis. The hydrolysis procedure is useful in the covalent coupling of ligands to the fibres, but often can be eliminated in coating them with proteins.

4.3.1 Adsorption of Con A to Fibres

Proteins are adsorbed onto the surface of many plastics, including nylon fibres. During a 30-minute incubation at room temperature, a nylon fibre will adsorb enough Con A molecules for most applications. This non-covalent binding is stable in physiological media. Because the unbound protein can be recovered intact, this procedure is particularly useful in cases where only a small amount of the protein is available.

4.3.2 Cross-linking of Protein to the Fibre by Carbodiimide

A water-soluble carbodiimide, 1-cyclohexyl-3-(2-morpholinoethyl) carbodiimide metho-p-toluenesulphonate (Aldrich), can be used to couple Con A (at 0.25—2.5 mg/ml) covalently to the nylon. Four millilitres of these reagents in 0.15 M NaCl, pH 6—7, at a carbodiimide:protein ratio of 5:1 (w/w) is added to petri dishes containing the fibres, and the reaction mixture incubated at room temperature for 30 minutes. The derivatized fibres are washed in 0.15 M NaCl and used the same day.

4.3.3 Activation of Carboxyl Groups on Fibres

For activation of fibre carboxyl groups, thionyl chloride or phosphorus pentachloride in benzene or dioxan can be used to convert the nylon carboxyl groups to acid chlorides. The fibres can then be reacted with free amino groups on proteins or other molecules. Mild conditions (1:500 dilution of $SOCl_2$ in dioxan, 10 minutes at 25°) must be used in order to avoid extensive damage to the fibre during the activation.

TABLE I.   Coating of fibres with concanavalin A. Formation of non-covalent and covalent bonds

Protein (mg/ml)	WSC[a] (mg/ml)	Wash[b]	C.p.m. of washed fibre[c]
[125]I-Con A[d]			
1.25	0	PBS	18,250
1.25	0	6 M Guanidine	7140
1.25	6.25	PBS	22,724
1.25	6.25	6 M Guanidine	18,715

[a]WSC, water-soluble carbodiimide; PBS, phosphate-buffered saline; Con A, concanavalin A.
[b]Incubation at 25° for four hours.
[c]C.p.m. for 25 cm fibre.
[d]2.4×10^7 c.p.m./mg.

In order to determine the stability of the protein coat, the fibre to which the ^{125}I-labelled protein (7) has been attached is incubated with phosphate-buffered saline (PBS) or 6 M guanidine hydrochloride. The amount of radioactivity bound before and after treatment with these solutions is then determined (Table I). Using Con A without the addition of carbodiimide, over 50% of the bound protein may be removed by guanidine. This non-covalent binding is, however, stable in physiological media. With the addition of carbodiimide, less than 20% of the bound protein is extracted by the guanidine wash. The use of carbodiimide is also useful in achieving a high degree of derivatization (Figure 3).

4.4 Controlling the Density of Con A Molecules on the Fibre

Two methods are available for preparing fibres coated with a different average density (molecules per linear cm of fibre) of Con A on their surface. One involves changes in the concentration of Con A during derivatization with carbodiimide, whereas the other involves adsorption of protein from solutions containing different ratios of Con A and some 'spacer' protein such as bovine serum albumin.

The derivatization with protein and carbodiimide proceeds smoothly over a wide

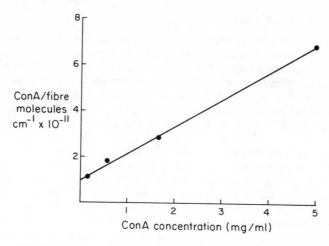

Figure 3. Coupling of Con A to nylon fibre (0.125 mm diameter) using a water-soluble carbodiimide at a 5:1 (w/w) ratio of carbodiimide to protein. The vertical axis indicates the number of Con A molecules per linear cm of fibre, and the horizontal axis indicates the concentration of Con A in the reaction mixture

concentration range (Figure 3). In order to determine the extent of coupling at a particular protein concentration, protein labelled with [125]I (7) is used for derivatization.

To prepare fibres with different densities of adsorbed Con A, a series of protein solutions containing the lectin and bovine serum albumin are used in which the concentration of lectin is decreased by factors of two from 500 μg/ml to 4 μg/ml and enough BSA is added to obtain a total protein concentration of 500 μg/ml (8). The density of lectin and/or BSA on the fibres is determined by using ^{125}I-labelled proteins. In this way the average density of Con A molecules can be controlled by interspacing them with albumin molecules. Given the number of protein molecules of known size adsorbed to a fibre with a given surface area, the average distance between adjacent molecules can be estimated. Using a cylindrical fibre with a diameter of 0.125 mm, one cm of fibre has enough surface area to be coated by approximately 17 ng of BSA or 20 ng of Con A. The amount of Con A and/or BSA actually bound per cm of fibre is near these values, so it appears that essentially the entire surface of the fibre is covered by Con A and/or BSA by this procedure.

5. BINDING OF CELLS TO FIBRES

With the fibre method certain variables must be adjusted to each application. The important variables include the method of cell preparation, the temperature and mixing conditions during incubation, the composition of the separation medium, the choice of ligand, the degree of derivatization and the number of cells added.

5.1 Preparation of Cell Suspension

In order to obtain good results, it is necessary to use cell suspensions that are free of aggregated material, and conditions under which the cells will not aggregate or release large amounts of DNA. Otherwise, clumps of cells will adhere firmly to the fibres and seriously reduce the quality of the fractionation. Cell populations from various sources will require different methods of preparation. In cases where enzymes or chemicals are used for cell dissociation, it is important to consider the possible effect of these reagents on the receptors by which the cell will bind to the fibre.

Lymphoid cell suspensions, for example, can be prepared by teasing a minced spleen, lymph node or thymus through a wire mesh into cold phosphate-buffered saline (PBS), Hank's balanced Salt solution (HBSS, Grand Island Biological Co., Grand Island, New York) or minimal essential medium (MEM, Microbiological Associates, Bethesda, Maryland) containing 20 μg/ml DNase. The presence of DNase prevents aggregation and non-specific binding caused by the release of DNA from damaged cells. Low-speed centrifugation is used to remove aggregates and the cells are washed three times in the medium. After washing, the cell suspension is passed through a fine-gauge nylon mesh to remove residual aggregates. The viability of the cells treated in this manner exceeds 90% for thymocytes and lymph node cells, and 70% for spleen cells.

5.2 Mixing Conditions, Temperature and Medium

The specific binding of cells to the fibres involves a balance between the rate and angle of collision of the fibre and cell, the dwell time of the cell at the fibre surface, and the settling of the cell due to gravity. If the cells are allowed to settle by gravity onto the fibres, non-specific binding may result. On the other hand, rapid shaking results in no binding at all. The best results are obtained by incubating the petri dish containing strung fibres for 15—60 minutes with 4 ml of a cell suspension on a reciprocal shaker having a 3—4 cm horizontal stroke at 70—90 oscillations per minute. The fibres must be aligned perpendicular to the direction of shaking, and it is essential that the medium covers the fibres at all times during shaking.

Many cell types that bind well to Con A-coated fibres at 25° (Figure 4) bind poorly at 4° (8). Incubation at 37°, however, may produce non-specific binding, and in some applications it may be desirable to keep the cells at 4° to minimize cell damage and possible effects of the fibre on the membranes and metabolism of the bound cells. In some cases, metabolic energy may be required for binding with Con A-coated fibres inasmuch as the presence of 10 mM NaN_3 or 1 mM dinitrophenol can decrease the number of cells bound (8).

Except for the addition of DNase or a specific inhibitor, the composition of the medium during the incubation does not in general affect the amount of binding. Any physiological medium can be used provided that the cells do not aggregate. Specific inhibitors are added to the cell suspension to test the specificity of the binding or to fractionate cells with respect to their affinity for the soluble inhibitor. With Con A fibres, 10 mM α-methyl mannoside is highly effective as an inhibitor of binding. The amount of glucose present in HBSS or MEM will decrease but not prevent binding of most cell types to Con A-coated fibres.

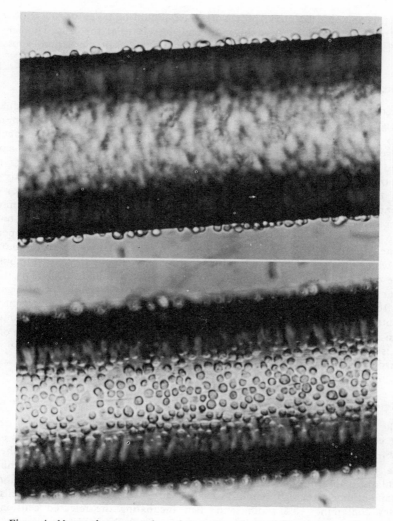

Figure 4. Mouse thymocytes bound to a Con A-coated fibre. Top: the field was focused on the edge of the fibre. Bottom: the field was focused on the face of the fibre. The fibre was derivatized with 5 mg/ml Con A and 25 mg/ml carbodiimide (see Figure 3) and incubated with 2×10^7 thymocytes in 4 ml phosphate-buffered saline

5.3 Removal of Unbound Cells

In many experiments it is necessary to remove all unbound cells from the fractionation apparatus. Unbound cells are removed by complete immersion of the dish in a series of larger vessels containing medium. It is also necessary to replace the petri dish in order to remove additional cells, which adhere to the dish bottom.

During this and all subsequent procedures care should be taken not to remove the fibres from the liquid because this will result in release of the cells. The cells are otherwise firmly bound, and extensive washing, mixing or manipulation is possible.

5.4 Quantitation of Bound Cells

The number of cells attached to the fibres can usually be determined by directly counting those bound to a segment of the fibre at 100—200 × magnification, most suitably with an inverted microscope. This procedure is simplified by the fact that the edge of the fibre can be sharply focused and cells in this focal plane may be counted over a reasonable length of fibre. Using ^{51}Cr-labelled cells, it has been found that the total number of bound cells is linearly proportional to this edge count. The factor to convert edge to total counts should be determined for each different experimental context. For analytical experiments, it is useful to limit the number of cells along one fibre edge to 400—800 cells per cm (depending on cell size) so that the cells do not overlap each other. At 800 cells/cm the cells appear to lie quite near to each other, but are actually spread out over an area about five times wider than a single cell. Therefore, only about 10% of the available binding surface on the fibre is occupied, and the number of cells bound is still proportional to their frequency in the cell suspension (8). With the fibres shown in Figure 4, about 10^4 thymocytes are bound per cm of fibre. Generally several different fibre segments of known length are counted per dish, and the mean of these cell counts is used in calculating the total number of bound cells. The standard deviation of this mean is 5—10% and the number of cells bound to five identical dishes also has a standard deviation of the mean of 5—10%.

5.5 Removal of Cells from the Fibre

Cells can be released from the fibres by plucking the taut fibres with a needle. Direct visual inspection of this process shows that transverse components of the fibres' mechanical vibration project the bound cells into the medium. This shearing of the cell—fibre bond may produce a lesion in the cell surface membrane, particularly if the fibre has a high surface density of Con A. However, fractionated cell populations with high viabilities can be obtained by removing the cells into medium containing 10% serum and then incubating the released cells in this medium for 30 minutes at 37°. The ability of cells so obtained to survive has been established in both *in vivo* transfer experiments with lymphocytes (9) and in culture experiments with tumour cell lines (10). In the absence of serum, however, as few as 5% of the cells may survive rupture of the cell—fibre bond.

With the apparatus described above, cells obtained by fibre fractionation are removed into a minimum volume of about 2.5 ml per dish. This results in relatively dilute cell suspensions ($10^4 - 10^6$ cells/ml). Counting the cells by haemacytometer is laborious (several chambers may have to be scanned) but satisfactory. Alternatively, the cell suspension may be counted directly with a Coulter counter (Coulter Electronics) or Cytograph (Bio/Physics Systems). In both cases, care must be taken to be sure that all unbound cells have been washed away before removing the cells from the fibres.

Concentration of the cells by conventional centrifugation and resuspension can result in substantial losses. Although the addition of carrier particles or cells will eliminate this difficulty, it may defeat the purpose of the fractionation. In our experience, good results have been obtained by lifting out the strung fibres after plucking, placing the petri dish into the top of a 50 ml clinical centrifuge cup, and centrifuging at 1500 r.p.m. for 10 minutes. This concentrates the cells into a monolayer, where they can be used *in situ* for cytological studies or resuspended into a small volume.

6. EXTENSION OF THE METHOD TO OTHER SOLID MATRICES

In addition to fixation on fibres, cells may be attached to surfaces of different geometries, including meshes and flat surfaces such as plastic petri dishes. Nylon meshes can increase the capacity of the fractionation considerably, although some non-specific binding can occur and certain features of the fibre system, including the removal of cells by plucking and easy visualization of the bound cells, are hindered by the interwoven geometry.

Plastic petri dishes may be coated directly by incubation with solutions of Con A in PBS. The protein adheres very tenaciously to the plastic and does not wash off even after several days of incubation at 37°. Flat surfaces may be useful for immobilization of cells, but are often unsatisfactory for fractionation because of non-specific adherence of cells. In addition, derivatized flat sufaces are not very useful for recovery of specifically bound cells, for it is difficult to remove the cells quantitatively and without damage. There are, however, advantages to using petri dishes rather than nylon fibres when the object of the experiment is to observe the effects of various surface-bound ligands on cell migration, growth or cell functions during long-term culturing. One example of the use of derivatized petri dishes comes from the work of Andersson and coworkers (11). They showed that when Con A is bound to a plastic petri dish, it stimulates mitosis of bone marrow-derived lymphocytes but not thymus-derived lymphocytes, whereas the reverse is true for Con A in solution.

7. APPLICATIONS

The fibre method has been used in three major experimental contexts: 1) the isolation and characterization of cell subpopulations; 2) the immobilization and subsequent manipulation of cells for an analysis of cell—cell interactions; and 3) studies on the effects of fibre-bound reagents on the morphology, motility and surface behaviour of cells attached to these fibres.

7.1 Differential Binding of cells to Con A-coated Fibres (5,8)

Fibre fractionation can be used to isolate cells according to differences in the amount of a particular component on the cell surface, and/or in the properties (distribution, mobility, etc.) of the same component on different cells.

Mouse thymocytes and erythrocytes have been used as a model system to demonstrate fractionation by differential binding using fibres derivatized with different densities of Con A by the carbodiimide method (6) (Figure 5). The

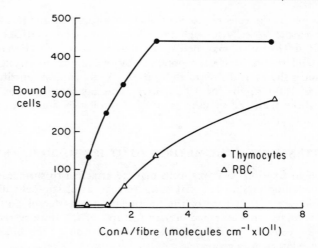

Figure 5. Binding of mouse erythrocytes and thymocytes to different fibres as a function of the number of Con A molecules per cm of nylon fibre (see Figure 3). The vertical axis represents the number of cells bound to one edge of a 1 cm fibre segment when 10^6 cells in 4 ml phosphate-buffered saline are incubated with the fibres for 30 minutes at $25°$

binding of these cells to Con A fibres is inhibited by over 95% by the presence of 10 mM α-methyl-mannoside, but not by galactose. Both thymocytes and erythrocytes have carbohydrate receptors for Con A, and will bind to fibres with a high density of Con A molecules. Erythrocytes, however, do not bind to fibres with a low Con A density. Since the thymocytes will bind to these fibres, they can be preferentially isolated. This means of fractionation has been used to characterize villus and crypt epithelial cells from intestine (12).

Similarly, the binding of various normal and transformed cells to fibres coated with Con A—serum albumin mixtures has been investigated under a variety of conditions (8). These cell types often differ in their ability to bind to the Con A fibres (Table II), and the differences become more evident as the ratio of Con A to albumin on the fibres is decreased. It has been shown that cells fixed with glutaraldehyde do not bind to the fibres, suggesting that mobility of cell surface receptors for the lectin is required for cell—fibre binding. Because a requirement for receptor mobility should increase as the density of Con A on the fibre decreases, it has been suggested that a titration of cell—fibre binding versus Con A density can be used to measure the relative mobility of receptors for the lectin on different cell types. In addition to Con A, a number of other lectins including soy bean agglutinin, wheat germ agglutinin and wax bean agglutinin have been used in these experiments.

7.2 Characterization of Isolated Cells

The ability to visualize, manipulate and remove the cells bound to a fibre facilitates their subsequent characterization. This includes the identification of cell

CON A COATED FIBRES

459

TABLE II.   Binding of mouse cells to fibres coated with concanavalin A[a]

Cell type	Fibre-bound cells/cm[b]
Thymocytes	712
Erythrocytes	15
Lymph node lymphocytes	603
Lymphomas:	
YAC ascites tumour	693
L 1210 culture line	618
Myeloid leukaemia cells:	
clone 21 D(D$^+$IR$^+$)	502
clone 1 M(D$^-$IR$^-$)	263
Normal fibroblasts	127
SV40 transformed fibroblasts	502

Data from references 5 and 8.
[a] Fibres coated with Con A by incubation with a 0.5 mg/ml solution of the lectin. Cells (5 x 10^5/ml in PBS) were incubated with fibres for 30 min at 25°.
[b] Number of cells bound along one edge of a 1 cm fibre segment.

surface receptors by cytotoxicity, immunofluorescence and rosette assays, the quantitation and localizaton of receptors with radiolabelled or fluorescent reagents and by rosette formation, and the study of cell function by *in vivo* transfer or cell culture techniques.

a. A number of antisera can be used to identify cells with a particular cell surface antigen by killing them in the presence of complement. These cytotoxicity assays can be carried out directly on the fibre-bound cells (13). The viability of cells bound to the fibres exceeds 95%, even though the unfractionated cell population may include up to 35% dead cells. Although the reason for this selection is not known, it greatly facilitates subsequent cytotoxicity assays by eliminating the usual background of non-viable cells. In the fibre cytotoxicity assay, fibre-bound cells are incubated with the appropriate antiserum at 37° for 45 minutes. The serum is then washed away, and the cells are incubated with a 1:10 dilution of guinea pig complement for 30 minutes at 37° with gentle shaking. When strong antiserum and complement are used, cells bearing the target antigen are severely damaged and no longer remain bound to the fibres; with less potent reagents, some dead cells may not be released, but can be detected by a trypan blue viability test. The number of viable bound cells is counted before and after treatment with complement.

b. Antibodies to a cell surface component may also be used to estimate the amount of this component on fibre-bound cells (14). This is carried out by incubating fibre-bound cells with increasing amounts of the appropriate ^{125}I-labelled antibody (7) for 30 minutes at room temperature, until saturation is achieved. After washing away the unbound antibody, the cells are removed from the fibres by plucking, and 10^7 carrier cells are added to facilitate further washing by centrifugation. The radioactivity of the cells is then determined using a gamma spectrometer.

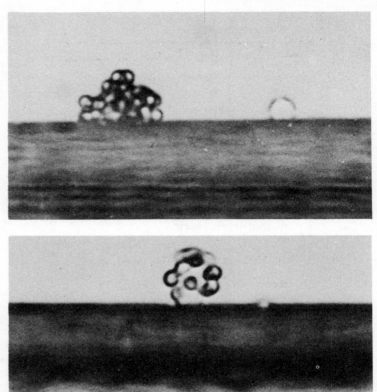

Figure 6. Top: two lymphocytes bound to a section of fibre, one of which forms an erythrocyte–antibody–complement (EAC) rosette and one which does not. Bottom: fibre-bound lymphocyte which forms a rosette with antigen-coated erythrocytes

    c. Rosette assays may be carried out directly on the fibres to identify a variety of cell surface receptors (13,14). For example, B lymphocytes have receptors for antigen–antibody-complement complexes. When sheep erythrocytes are used as the antigen the B cells form visible erythrocyte–antibody-complement (EAC) rosettes (Figure 6). T lymphocytes do not have these receptors, and B cells on a fibre can be distinguished from T cells after incubating the fibre-bound cells with 4% EAC in MEM with gentle shaking for 30 minutes at $37°$. Similar methods have been used to detect specific antigen-binding receptors on lymphocytes (Figure 6).

    d. Fluorescence microscopy can be employed to permit the detection of receptors on fibre-bound cells (15,16). In this case, care must be taken that the nylon fibres have not been treated with fluorescent additives. It is also useful to use monofilament with a small diameter (less than 50 μm) in order to minimize fluorescence by the fibre and to facilitate observation of the cells at high magnification.

e. Although functional studies have been carried out using lymphoid cells fractionated with antigen-coated fibres (9), experience with the function of cells fractionated on Con A-coated fibres is more limited. It has been observed, however, that lymphoma tumour cells removed from Con A-coated fibres can be repeatedly passaged in ascites form, and that myeloid leukaemia cell lines can survive in culture after being bound to and removed from these fibres (10).

7.3 Studies of Cell—Cell Interactions

In addition to being a useful reagent to select cell populations, Con A-coated fibres can also serve as an anchor to immobilize cells in a variety of experimental contexts. One of these is the study of cell—cell binding (6,16). Binding among cells induced by Con A has been widely used as a probe to study changes in the cell surface membrane. This agglutination process is generally studied by adding the lectin to a cell suspension, shaking the suspension and recording the degree of cell aggregation. Although such an assay is rapid and convenient, it simultaneously measures the binding of lectin to the cells and of several cells to one another. In contrast, use of the fibre method to analyse cell—cell binding induced by Con A (16,17) allows the quantitative study of binding between single cells, in which the binding of Con A to a cell and the binding of that cell to another cell can be carried out as successive steps.

7.3.1 Cell—Cell Binding Assay

Nylon fibres with a high density of Con A molecules are prepared as described above. The assay (Figure 7) involves three steps: (1) binding of cells to Con A-coated (2.5×10^6 cells/ml in PBS) for 30 minutes at $25°$ and removal of unbound cells by washing; (2) binding of Con A ($10–200$ μg/ml) for 30 minutes at $25°$ to the fibre-bound (FB) cells, followed by three washes with PBS; and (3) incubation of FB cells with a cell suspension (10^6 cells/ml in PBS) at $25°$ for 30 minutes with gentle shaking, as in the binding of cells to fibres. In each step, a new petri dish is used to prevent carry-over of cells or Con A. After unbound cells are washed away, the number of cell-bound (CB) cells are counted along one edge of a 1 cm segment of fibre.

7.3.2 Studies of Cell—Cell Binding Induced by Con A among Normal and Transformed Cells (16, 17)

The fibre system has been used to analyse the binding of normal or tumour cells to themselves or to each other, and to examine the effects of fixation, enzymatic treatments and metabolic inhibitors on this process. The ability to quantitate single binding events between two different cell types, or cells treated with different reagents has permitted a number of new observations to be made. For example, although no binding is obtained when both cells are fixed with glutaraldehyde, fixation of a cell before coating it with Con A greatly enhances its ability to bind an unfixed, uncoated cell. This indicates that mobility of receptors for Con A is

Figure 7. Assay for cell–cell binding induced by Con A (16). Top fibre: lymphoma tumour cells bound to a Con A–coated fibre. These immobilized cells are then coated with Con A and incubated with cells in suspension. This results in cell–cell binding as shown on the bottom fibre. The binding is quantitated by counting the number of cell-bound cells along one edge of a 1 cm fibre segment

required in only one of the two cells. Fixation of the uncoated cell, however, decreases its ability to bind to an unfixed, Con A-coated cell. Together these results indicate that clustering of receptors for Con A is not required for, and may actually hinder the cell–cell binding induced by this lectin. In particular, the inability of normal lymphocytes and fibroblasts to agglutinate with Con A appears to result largely from cross-linking of receptors on the same cell by the lectin. Another significant result obtained by the fibre method was that no binding occurs between two cells that have been coated with Con A. The fact that very high concentrations of Con A do not inhibit agglutination has sometimes been interpreted as evidence that cell–cell binding induced by Con A does not involve the direct formation of cell–lectin–cell bridges. However, the requirement for unoccupied receptors in the fibre assay, in which saturation of receptors can occur prior to cell contact, suggests that lectin bridging is the initial step in cell–cell binding.

7.4 Studies on the Morphology, Motility and Surface Behaviour of Cells Bound to Fibres (15)

The relationship of cell movement to cell surface events is poorly understood. In studying this problem, a number of factors can be distinguished: local movement of cell surface receptors (diffusion and patch formation), global movements of these receptors (capping), local morphological changes (microvillus formation, blebbing and ruffling), global morphological changes altering cell shape and translocation of the whole cell.

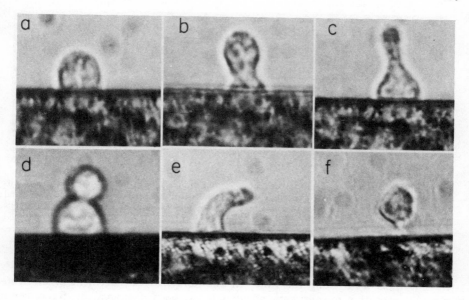

Figure 8. Various shapes of lymphocytes bound to antigen-coated fibres. Cells bound to Con A—coated fibres appear as in (a)

It is experimentally difficult to analyse these events at both the microscopic and molecular level in a single cell type. To reduce this difficulty, it is useful to make the cells sessile by attaching them to nylon fibres via their cell surface receptors. The bound lymphocytes do not translocate on the fibre; for this reason, they may be identified and readily observed *in situ*. The method, may, therefore, be used to study the movement and surface behaviour of single cells as well as populations. It is particularly useful for exploration of the polar effects of various ligands and for molecular analysis of cell surface events.

The morphological changes undergone by fibre-bound lymphocytes, the identification of cellular elements affecting their movement and shape, and the connexion between these factors and those controlling the local movement of cell surface receptors have been studied using antigen-coated and Con A-coated fibres (15). In these studies it was observed that mouse B lymphocytes bound through their immunoglobulin receptors to antigen-coated fibres exhibit continuous morphological changes (Figure 8). Fibre-bound cells are incubated in MEM at 21° or 37° for one hour and the changes in cell shapes are scored immediately after incubation. A water-immersion lens is used to examine the cells at 400x magnification, and time-lapse cinematomicrography can be used for continuous monitoring of cell shapes. For example, an Arriflex motion picture camera has been used in our studies with a Universal Zeiss microscope and a 40x water-immersion lens. A time lapse of two seconds, between two-second exposures with Plus-X film gave a final speed-up of 96 times at 24 frames/second. Cinematography shows that these shape changes are associated with local and global movements, although the attached cells

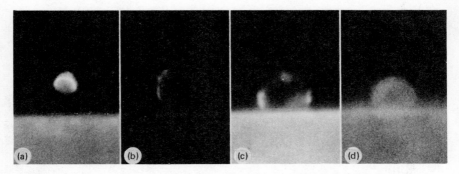

Figure 9. Redistribution of surface immunoglobulin induced by anti-immunoglobulin on cells bound to fibres. (a) Capped distribution on cell bound to antigen fibre. Note the distal orientation of the cap relative to the fibre. (b) Cap with random orientation found on cells incubated with colchicine after binding to antigen- or Con A—coated fibres. (c) Patched distribution on cell incubated with sodium azide after binding to antigen-coated fibre. (d) Diffuse distribtion on cell bound to Con A—coated fibre

do not translocate along the fibre. In addition, the rearrangement of immuno-globulin receptors by anti-immunoglobulin has been examined by fluorescence microscopy (Figure 9). Anti-immunoglobulin induces cap formation (18) on B cells bound to antigen-coated fibres and the cap is always located opposite to the point of attachment between the cell and the fibre.

In contrast, B lymphocytes bound to Con A-coated fibres remain more or less spherical in shape, and cannot be induced to form caps with anti-immunogobulin. The inhibition of cap formation by Con A-coated fibres can be reversed by the presence of colchicine, although the position of these caps is random with respect to the fibre. Colchicine also inhibits the movement of cells bound to antigen-coated fibres. From these and other studies, it has been suggested that colchicine-sensitive structures within a cell regulate the mobility of various receptors (19) and that these structures are altered by binding to Con A-coated fibres (15). The fibre experiments also suggest the important conclusion that interactions of *local* areas of the lymphocyte surface with certain ligands and substrates can strongly effect the movement and morphology of the entire cell.

8. ACKNOWLEDGEMENTS

This work was supported by USPHS grants from the National Institutes of Health and by grants from the National Science Foundation.

9. REFERENCES

1. Wigzell, H. and Mäkelä, O. (1970) 'Separation of normal and immune lymphoid cells by antigen-coated columns', *J. Exp. Med.*, **132**, 110—26.
2. Abdou, N. I. and Richter, M. (1969) 'Transfer of antibody-forming capacity to irradiated rabbits by antigen-reactive cells isolated from normal allogeneic rabbit bone marrow after passage through antigen-sensitized glass bead columns', *J. Exp. Med.*, **130**, 141—64.
3. Wofsy, L., Kimura, J.and Truffa-Bachi, P. (1971) 'Cell separation on affinity columns: The

preparation of pure populations of anti-hapten specific lymphocytes', *J. Immunol.*, **107**, 725—9.

4. Davie, J. M. and Paul, W. E. (1970) 'Receptors on immunocompetent cells. I. Receptor specificity of cells participating in a cellular immune response', *Cell. Immunol.*, **1**, 404—18.

5. Edelman, G. M., Rutishauser, U. and Millette, C. F. (1971) 'Cell fractionation and arrangement on fibers, beads and surfaces', *Proc. Nat. Acad. Sci. U.S.A.*, **68**, 2153—7.

6. Edelman, G. M. and Rutishauser, U. (1974) 'Specific fractionation and manipulation of cells with chemically derivatized fibers and surfaces', *Meth. Enzymol.*, **34**, 195—225.

7. Williams, C. A. and Chase, M. W. (eds.) (1967) *Methods in Immunology and Immunochemistry*, Vol. 1, (Academic Press, New York).

8. Rutishauser, U. and Sachs, L. (1975) 'Receptor mobility and the binding of cells to lectin-coated fibers', *J. Cell Biol.*, in press.

9. Rutishauser, U., D'Eustachio, P. D. and Edelman, G. M. (1974) 'Immunological function of lymphocytes fractionated with antigen-derivatized fibers', *Proc. Nat. Acad. Sci. U.S.A.*, **70**, 3894—8.

10. Rutishauser, U. and Sachs, L., unpublished observations.

11. Andersson, J., Edelman, G. M., Möller, G. and Sjöberg, O. 'Activation of B lymphocytes by locally concentrated concanavalin A', *Eur. J. Immunol.*, **2**, 233—5.

12. Podolsky, O. K. and Weiser, M. M. (1973) 'Specific selection of mitotically active intestinal cells by concanavalin A-derivatized fibers', *J. Cell Biol.*, **58**, 497—500.

13. Rutishauser, U. and Edelman, G. M. (1972) 'Binding of thymus- and bone marrow-derived lymphoid cells to antigen-derivatized fibers', *Proc. Nat. Acad. Sci. U.S.A.*, **69**, 3774—8.

14. Rutishauser, U. and Edelman, G. M. (1972) 'Specific fractionation of immune cell populations', *Proc. Nat. Acad. Sci. U.S.A.*, **69**, 1596—600.

15. Rutishauser, U., Yahara, I. and Edelman, G. M. (1974) 'Morphology, motility and surface behaviour of lymphocytes bound to nylon fibers', *Proc. Nat. Acad. Sci. U.S.A.*, **70**, 1149—53.

16. Rutishauser, U. and Sachs, L. (1974) 'Receptor mobility and the mechanism of cell—cell binding induced by concanavalin A', *Proc. Nat. Acad. Sci. U.S.A.*, **71**, 2456—60.

17. Rutishauser, U. and Sachs, L. (1975) 'The mechanism of cell—cell binding induced by different lectins', *J. Cell Biol.*, **65**, 247—57.

18. Taylor, R. B., Duffus, W. P. H., Raff, M. C. and de Petris, S. (1971) 'Redistribution and pinocytosis of lymphocyte surface immunoglobulin molecules induced by anti-immunoglobulin antibody', *Nature (New Biol.)*, **233**, 225—9.

19. Edelman, G. M., Yahara, I. and Wang, J. L. (1973) 'Receptor mobility and receptor—cytoplasmic interactions in lymphocytes', *Proc. Nat. Acad. Sci. U.S.A.*, **70**, 1442—6.

CHAPTER 49

Binding and Release of Tissue Culture Cells on Lens culinaris Lectin Immobilized on 'Sepharose' Beads

VOLKER KINZEL
DIETER KÜBLER
JAMES RICHARDS
MICHAEL STÖHR

1. PRINCIPLE

Techniques taking advantage of different cell surface properties for the separation of cells starting with the work of Wigzell and Andersson (1) have been proved to be of great value.

In another approach concanavalin A (Con A) has been covalently linked to different solid supports. Thus cells could be bound to solid matrices (2), but the specific release of cells from immobilized Con A upon addition of the specific

467

sugars is, without mechanical aid, difficult if not impossible (2). Similar difficulties which might be due to the same property of Con A have been reported on the fractionation of cell surface glycoproteins (3). The yield of specifically eluted material from immobilized Con A was found to be rather low.

In general, separation of cells by binding on Con A beads in column procedures is not satisfactory, as steric hindrance and unspecific binding occurs in a system of densely packed sedimented gel beads. These problems were overcome by the use of lentil agglutinin instead of Con A, selecting large 'Sepharose' beads with diameters over 200 μ (diameter of commercially available 'Sepharose' 2B is 60—250 μ) and by employing a batch technique.

The lectin isolated from *Lens culinaris* binds the same sugar residues as Con A. Its affinity, however, is 50 times lower (4). The *Lens culinaris* lectin (LCL) also resembles the Con A in its lymphocyte-stimulating activity (5). 'Sepharose' 2B beads ranging in size from 200 to 250 microns were separated from beads of smaller size by using a gauze having a pore size of 200 microns. The advantages of using large 'Sepharose' beads in the batch procedure described are: (i) a rather uniform sedimentation behaviour, (ii) a large size difference of bound and free cells in suspension, (iii) enough space between the settled beads to allow the free flow of cells (which seems to be also important for application in a column procedure). For the separation of free cells (unbound or released) and of bound cells a batch procedure has been devised using a small apparatus which allows an easy and fast handling of the samples. Suspended cells and uncovered or covered beads are simply separated by means of a gauze of defined pore size (100 microns).

2. MATERIALS AND EQUIPMENT

2.1 Isolation of the *Lens culinaris* Lectin

LCL was isolated from lentils mainly following Hayman and Crumpton (3) as modified by Rohrschneider (L. R. Rohrschneider, personal communication) and coupled to selected 'Sepharose' 2B beads which had been activated according to March and coworkers (6).

Unless otherwise stated all steps were carried out at $4°C$. For buffers and solutions see Table I.

Common lentil seeds (500 g) are soaked overnight in 1 litre buffer no. 1.

swollen seeds are homogenized in several batches (2 min each) in a Waring blender at highest speed with the addition of buffer no. 1 to a final volume of about 1.5 litre, homogenate is centrifuged at about 12,000 g for 15 min;

the supernatant fluid is slowly mixed with 170 g $(NH_4)_2SO_4$ per litre under constant stirring for 30 min, the solution is centrifuged at about 24,000 g for 30 min;

to the supernatant fluid 350 g $(NH_4)_2SO_4$ per litre are added and centrifuged as before;

precipitate is dissolved in 200 ml buffer no. 1 by rapidly pipetting with a large pipette, the solution is dialysed against running tap water for 30 min followed by an overnight dialysis against buffer no. 1 (4—6 litre) with 3 changes, precipitates are removed by centrifugation;

TABLE I.   Buffers and solutions

No.	Strength	Substance	pH
1.	75 mM	Sodium phosphate	7.4
	75 mM	NaCl	
2.	No. 1 together with		
	10^{-4} M	$MnCl_2$	
	10^{-4} M	$CaCl_2$	
3.	No. 2 together with		
	0.1 M	MαGP	
4.	0.1 M	$NaHCO_3$	7.4
	0.15 M	NaCl	
	0.1 M	MαGP	
	10^{-4} M	$MnCl_2$	
	10^{-4} M	$CaCl_2$	
5.	0.1 M	$NaHCO_3$	8.2
	15 mM	NaCl	
	10^{-4} M	$MnCl_2$	
	10^{-4} M	$CaCl_2$	
6.	No. 5 together with		
	5 mM	Ethanolamine HCl	8.2
7.	330 mOsM	PBS	7.2
	0.1%	BSA	
	2 mM	EDTA	
8.	330 mOsM	PBS	7.2
	0.1%	BSA	
	10^{-4} M	$MnCl_2$	
	10^{-3} M	$CaCl_2$	
	5×10^{-4} M	$MgCl_2$	
9.	0.3 M	Sugar	7.2
	0.1%	BSA	
	10^{-4} M	$MnCl_2$	
	10^{-3} M	$CaCl_2$	
	5×10^{-4} M	$MgCl_2$	

clear supernatant is mixed with 20 g 'Sephadex' G-75 preswollen and washed in buffer no. 2;

after 30 min the sediment is washed 3 times with buffer no. 2 on a filter and drained by gentle suction;

the 'Sephadex' cake is resuspended in buffer no. 2, filled into a column and washed with buffer no. 2;

the lectin is eluted with buffer no. 3 (yield: about 200 mg LCL).

The lectin activity is correlated with the peak recorded at 280 nm as determined by coprecipitation with human serum in an Ouchterlony test (7). The material separated on dodecylsulphate gels gave two major and one minor band.

The peak fractions are pooled and dialysed against buffer no. 4 which is also used for coupling to the activated beads.

2.2 Isolation and activation of Large 'Sepharose' Beads

'Sepharose' 2B beads (60–250 μ in diameter) were used for the isolation of the large fraction (> 200 to 250 microns) by the use of a gauze of defined pore size (200 microns: Swiss silk bolting Cloth Mfg. Co. Ltd., Zürich/Switzerland,: 'Stabiltex' polyester multifilament Cat. number 7-200). With this gauze a bag is formed, filled with 50 ml batches of 'Sepharose' 2B, closed at the top and dipped several times into a large beaker with distilled water, until the smaller-sized beads have left the bag as checked microscopically.

Cyanogen bromide activation was according to March and coworkers (6).

To one volume of settled beads suspended in an equal amount of distilled water one volume of 2 M Na_2CO_3 is added with stirring;

0.05 volume of a solution containing 2 g cyanogen bromide in 1 ml acetonitrile is added all at once while vigorously stirring the reaction mixture for 2 min;

beads are washed on a glass filter funnel with 10 volumes each of 0.1 M $NaHCO_3$, pH 9.5, distilled water and coupling buffer no. 4. To prevent beads from aggregating only the last washing step is carried out with suction until the beads are drained to a cake.

Temperature of the reaction was not allowed to exceed room temperature. The activitation of the 'Sepharose' beads was performed immediately before use.

2.3 Coupling of *Lens culinaris* Lectin to 'Sepharose'

Usually about 100 mg of LCL was reacted with about 30 ml of settled beads.

The drained beads are transferred to a glass-stoppered vial, suspended in the LCL solution and rotated slowly end-over-end on a Bellco rocking platform for about 20 h;

beads are drained (volume of the first drainage is measured and protein is determined) and washed with about 100 ml of buffer no. 5. Coupling efficiency varied from 50 to 85%;

beads are drained and suspended in buffer no. 6 and rotated on the rocking platform for 90 min at room temperature;

beads are transferred to a column and washed free of ethanolamine with buffer no. 5, distilled water and finally with PBS, pH 7.2 (and 0.1% azide for longer storage);

beads are stored at $4°C$.

2.4 Apparatus

The apparatus for the batch procedure to separate suspended cells and covered or uncovered beads is shown in Figure 1. Erlenmeyer flask (A) with a possibility for equilibration of pressure (D) and a top part (B) which facilitates addition and removal of solutions. Part A is connected to part B by the double screw fitting (C) but separated by a gauze (pore size 100 microns) to keep beads in part A. A gasket

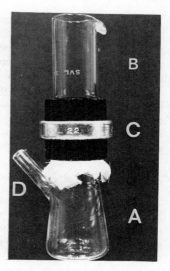

Figure 1. Binding and separation apparatus (height 11.5 cm). Erlenmeyer flask (A) connected to the top part (B) by a double screw fitting (C). A gasket tightens part B against the gauze and part A. Opening D is for equalizing pressure

between part B and the gauze tightens the system. If it is intended to handle larger amounts of cells and beads an accordingly larger sized part A might be used.

The apparatus is made of commercially available parts by a glassblower:

(Sovirel/France

glass tube with winding	cat. no. 703—02
double screw fitting	cat. no. 701—12
gasket	cat. no. 701—02

Gauze ('Stabiltex' polyester multifilament) is supplied by: Swiss silk bolting Cloth Mfg. Co. Ltd. Zürich/Switzerland, cat. no. 16-100.)

2.5 Recycling of Used LCL 'Sepharose'

LCL beads can be reused several times. For recycling, beads are suspended in about 20 times their settled volume of distilled water containing 0.5 M MαMP and agitated on a rocking platform for about 1 hour at room temperature. Most conveniently the beads are then filled into a small column and washed at 4°C with large amounts of cold distilled water.

Storage should be in PBS at 4°C. Storage for a longer period of time should be in the presence of antibacterial substances.

Before reuse the beads should be washed 3 to 4 times in solution no. 8.

3. METHODS

3.1 Preparation of Cell Suspensions

The following cells were investigated: 1 to 3 days old subconfluent monolayers of either a cloned HeLa cell line grown in MEM with 10% calf serum or of a SV-3T3 line grown in BME with 10% foetal calf serum.

Single-cell suspensions were obtained as follows: Cells are washed with 2 to 4 changes of prewarmed (37°C) buffer no. 7 (1 min each). If available, a warming plate is useful to have the dishes permanently warmed. When the cells have nearly detached from the dish surface as checked microscopically they are harvested as single cells by a gentle stream of buffer no. 7. To prevent cells from initial aggregation a small amount of DNAase was added routinely during the preparation procedure. The cell suspensions now handled at room temperature are transferred to a plastic centrifuge tube. The cells are sedimented (200 g, 5 min) and washed free of EDTA with buffer no. 8 in two cycles. The cell pellet is then resuspended in a small volume of buffer no. 8.

3.2 Binding of Cells

For the binding usually about 0.5 ml settled LCL beads are washed twice in buffer no. 8 and mixed with about 4 to 6 x 10^6 cells suspended also in buffer no. 8 to about 5 ml final volume in a 5 cm petri dish (or directly in the lower part of the apparatus shown in Figure 1). The vessel is placed on a slowly rocking platform (15 r.p.m.) to bring the cells into contact with the beads. This is usually done at room temperature to decrease the metabolism of the cells. For the addition of sugars, aliquots from 0.3 M sugar solution no. 9 are simply mixed with buffer no. 8 in different ratios to obtain the concentrations desired. The binding is readily appreciated microscopically.

LCL 'Sepharose' beads bind tissue culture cells as illustrated in Figures 2 and 3. In contrast, untreated 'Sepharose' 2B and CNBr-activated 'Sepharose' without LCL failed to bind any cells. About 0.5 ml of settled LCL beads bind 6 to 8 x 10^6 cells. Up to 70% of both HeLa and SV-3T3 cells are bound to LCL beads within 20 min at room temperature as well as at 37°C. Decrease of the temperature to 4°C reduces the rate of binding. HeLa cells were never observed to form aggregates during the binding procedure. SV-3T3 cells, however, tend to aggregate either with or without LCL beads present upon being brought into motion on the rocker platform. This phenomenon could not be prevented by omitting or increasing the BSA amount or by the addition of serum or of higher amounts of DNAase. The binding of SV-3T3 cells to beads, however, was so fast that the aggregation phenomenon did not cause any major interference (Table II, control panel).

3.3 Release of Cells

After the binding of cells to the beads in part A of the separation apparatus unbound cells remaining in suspension are simply decanted into a centrifuge vial in such a way that no fluid enters opening D. The beads now packed underneath the gauze are rinsed off by a slow stream of about 5 ml solution passing through part B. Depending on the amount of beads and cells employed usually 2 to 3 washings are enough to remove most of the free cells. In the same way sugar solutions are added and decanted.

For the specific release of cells from LCL beads the preceding binding time should be about 45 to 60 min in order to increase the mechanical stability of the cell — bead unit. Free cells are removed as described above. Most conveniently the release is done in the lower part (A) of the separation apparatus at room

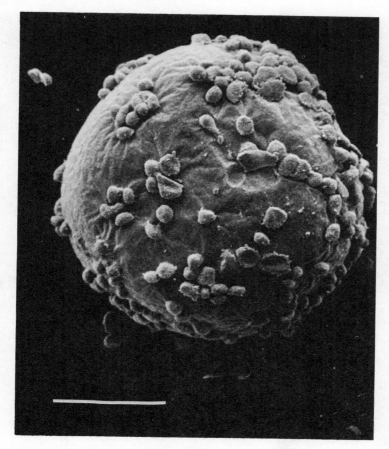

Figure 2. Scanning electronmicrograph of HeLa cells bound to a LCL 'Sepharose' bead (which has been recycled 3 times already). Measure indicates 100 microns

temperature. The concentrations of the specific sugar necessary and the time schedule have to be evaluated for each particular cell line. The release is improved by a rocking speed of about 30—40 r.p.m. to facilitate the chemical release by mild mechanical shearing.

The fractionated release of cells from LCL beads may be approached either by applying the same concentration of a specific sugar for different periods of time or by different sugar concentrations (or sugars).

4. EXAMPLE OF APPLICATION

The incubation of cells with LCL beads together with sugars known to inhibit LCL such as MαMP and MαGP prevented the binding of the cells to different degrees as shown in Figures 3 and 4 and Table II in which typical experiments are

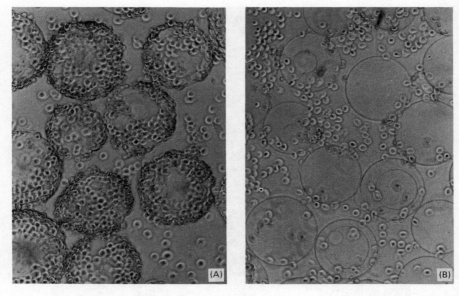

Figure 3. Binding of HeLa cells to immobilized LCL in the presence of 0.1 M D-galactose (A) and of 0.1 M MαMP (B) during 20 min at room temperature

TABLE II. Binding of SV-3T3 cells to LCL beads in the presence of different sugars. Three portions of about 4×10^6 SV-3T3 cells were incubated each in 7 ml total volume with 0.5 ml settled LCL beads in the presence of 0.1 M concentrations of different sugars[a] at room temperature. At the times indicated the vessels were scored microscopically

Time (min)	0.1 M D-Galactose		0.1 M MαMP		Control (without beads) aggregation
	binding	aggregation	binding	aggregation	
0	−	−	−	−	−
5	+	− +	−	+	+ +
10	+ +	− +	−	+ + +	+ + +
20	+ + + +	− +	−	+ + + +	+ + + +

[a]MαGP interferes also with the binding of SV-3T3 cells but not as well as MαMP.
Explanation:

Binding		*Aggregation*
No binding	−	No aggregation
Few cells bound per bead	− +	Some 2 to 3 cell aggregates
More than 8 cells per bead	+	About 50% small aggregates, rest single cells
Up to about half of cells bound	+ +	Many small aggregates
More than half cells bound	+ + +	Many small and few large aggregates
Nearly all cells bound	+ + + +	Large aggregates

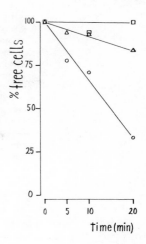

Figure 4. Binding of HeLa cells to LCL beads in the presence of different sugars. Three portions of about 7×10^6 HeLa cells were incubated each with 0.5 ml LCL beads in the presence of 0.1 M concentrations of different sugars ($\circ$ = D-galactose; $\triangle$ = MαGP; $\square$ = MαMP) at room temperature in 3.5 ml total volume. At the times indicated 50 μl aliquots were withdrawn and counted. Each point represents the mean of two determinations

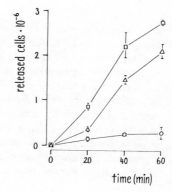

Figure 5. Release of HeLa cells from LCL beads by different sugars. Three portions of about 4×10^6 HeLa cells were allowed to bind each to 0.5 ml settled LCL beads in the presence of 0.2 M D-galactose for 60 min at room temperature in 5 ml total volumes. From each supernatant 3 ml was removed and immediately replaced by 4 ml of the particular sugar solutions (no. 9) to give the final concentration 0.2 M (in the case of the D-galactose group 4 ml of 0.2 M D-galactose was added. $\circ$ = D-galactose; $\triangle$ = MαGP; $\square$ = MαMP). The rocking speed was increased and at the times indicated 100 μl aliquots were withdrawn and counted. Each point represents the mean of two determinations

depicted. D-Galactose, however, which is not bound by LCL does not prevent the cells from being bound to LCL beads (Figures 3 and 4 and Table II). D-Glucose interfered also to some degree with the binding (data not shown).

A typical release experiment with HeLa cells is depicted in Figure 5. It is clearly shown that MαMP is more powerful in displacing cells than MαGP at the same concentration, a result which agrees with data from binding studies such as those demonstrated in Figure 4 where it was found that MαGP is less capable of preventing the cells from being bound. It may be mentioned that these data are in accord with the lower binding constant of LCL for MαGP. From Figure 5 it is also noticeable that a small number of cells are removed unspecifically. HeLa cells and SV-3T3 cells (Table III) are both specifically released as single cells. Unlike HeLa cells, however, SV-3T3 cells after removal exhibit the same aggregation phenomenon as described above. Therefore cell counts were not obtainable. The MαGP (data not shown) seems to be also less efficient at releasing SV40-3T3 cells. The sugar concentrations necessary for a substantial release of cells during 1 h

TABLE III.   Release of SV-3T3 cells from LCL
beads in the presence of different sugars. Two portions
of about 4 to 5 x 10⁶ SV-3T3 cells were incubated
each in 7 ml total volume with 0.5 ml settled LCL
beads in the presence of 0.2 M D-galactose for 40 min
at room temperature. From each the supernatant was
removed and immediately replaced by 5 ml of the
particular sugar solution. The rocking speed was
increased and at the times indicated the vessels were
scored microscopically

Time (min)	0.2 M D-Galactose binding	0.2 M MαMP binding
0	+ + + +	+ + + +
5	+ + + +	+ + + +
20	+ + +	+ + +
30	+ + +	+ +
60	+ + +	+

For explanation of binding see Table II.

appear to be higher than the concentration which is able to prevent the cells from
being bound to LCL beads.

The integrity of released cells was studied using different approaches. The less
sensitive but nevertheless rather valuable method utilizes fixed cells which are
prepared according to Trujillo and Van Dilla (8). Using a flow microfluorometer
(Cytofluorograph from Biophysics) frequency distributions of cellular DNA content as measured by emission intensity of DNA-binding ethidium bromide were
plotted versus the light-scattering equivalent to the cell size on the monitor of a
storage oscillograph. A pattern obtained from a normal culture is shown in
Figure 6A. A comparison with the picture received from a partly impaired culture
(Figure 6B) indicates that a significant number of cells have a reduced cellular

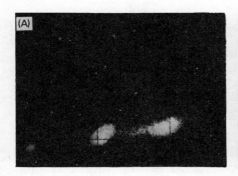
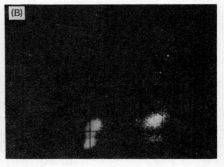

Figure 6. Volume/DNA plot as obtained by a storage oscillograph of a microfluorometer.
Comparison of non-impaired (A) and impaired (B) HeLa cells showing the lower 'cloud'
phenomenon. For further explanations see text. Ordinate: light-scattering equivalent to cell
volume. Abscissa: fluorescence intensity equivalent to cellular DNA amount

volume thus generating the lower 'clouds'. By adding BSA to the solutions for instance the lower 'cloud' phenomenon could be eliminated. The more sensitive method using trypan blue showed a slight increase in positively stained cells from 6.7 to 7% in control cells kept for 2 h in binding solution no. 8 (with 0.2 M D-galactose) versus 11.7% in cells after binding (1 h) and release (1 h) which is still in the range reported for suspended monolayer cells studied with fluorescein diacetate (9).

For reasons of sterility only short-term experiments on viability were performed. We determined the ability of released cells to attach to culture dishes. It could be shown that about 75% of the released cells attach within 5 h showing only a few mitotic figures whereas the control cells kept for 2 h in binding solution no. 8 (with 0.2 M D-galactose) showed many mitotic figures; the recovery was almost 100%.

The increase of trypan blue-positive cells and the lower attachment rate in the release group as compared with the control could be either due to the mechanical action of the beads rolling with the cells on the surface of the dish during the binding and release procedure or to the chemical influence of the immobilized LCL acting on the cell surface.

5. SUGGESTIONS AND PROBLEMS

Several factors must be considered when employing an immobilized lectin system as opposed to a system using soluble lectins: the amount of lectin coupled to a given batch of beads, steric hindrance and batchwise differing lectin activities at the bead surface.

The results demonstrate that LCL induces binding of the cells to derivatized 'Sepharose' beads, and that this binding can be prevented to different degrees by specific hapten sugars. Once a cell is bound the binding seems to be strong enough to prevent a major mechanical removal of the cells during the washings in the batch procedure described. As reported for Con A (2) the amount of LCL coupled to the beads strongly influences the binding of the cells to the beads. About 2.5 mg LCL per ml settled beads (the concentration mostly used in this study) induces a good binding which occurs so fast that the release procedure can be started probably before secondary interactions between cell and bead take place as suggested by Edelman and coworkers (2). With high, but with regard to osmolarity still physiological concentrations of the specific hapten sugars, the removal occurs sufficiently fast for the integrity of the cells which seems to be mainly unimpaired as determined by the methods described.

The differences in the results obtained with the two cell lines used indicate that the use of other cell lines would require particular adaptations in methods of preparing and handling cell suspensions, and reevaluation of the binding procedure and the sugar concentrations for the removal of bound cells. We have not tested how much LCL can be maximally coupled to a certain amount of 'Sepharose' 2B. It might be the case, however, that particular cell lines may only bind to a minor degree to the amounts of immobilized LCL used in this study. For this reason it is probably preferable to couple as much LCL as possible. A differential binding may then be titratable by specific sugar.

6. ABBREVIATIONS

BSA	bovine serum albumin
BME	basal medium Eagle's
Con A	concanavalin A
LCL	*Lens culinaris* lectin
MαGP	methyl α-D-glucopyranoside
MαMP	methyl α-D-mannopyranoside
MEM	minimum essential medium (Eagle)
PBS	phosphate-buffered saline

7. ACKNOWLEDGEMENT

We thank Dr. L. R. Rohrschneider for communicating the isolation procedure of LCL, Dr. N. Paweletz for the scanning electronmicrograph, Dr. H. P. Schnebli for helpful suggestions in preparing cell suspensions, and Mr. E. Fey for technical assistance.

This work was supported by the Deutsche Forschungsgemeinschaft and the Verein zur Förderung der Krebsforschung in Deutschland.

8. REFERENCES

1. Wigzell, H. and Andersson, B. (1969) 'Cell separation on antigen-coated columns', *J. Exp. Med.*, **129**, 23–36.
2. Edelman, G. M., Rutishauser, U. and Millette, C. F. (1971) 'Cell fractionation and arrangement on fibers, beads and surfaces' *Proc. Nat. Acad. Sci. U.S.A.*, **68**, 2153–7.
3. Hayman, M. J. and Crumpton, M. J. (1972) 'Isolation of glycoproteins from pig lymphocyte plasma membrane using *Lens culinaris* phytohemagglutinin' *Biochem. Biophys. Res. Commun.*, **47**, 923–30.
4. Stein, M. D., Howard, I. K. and Sage, H. J. (1971) 'Studies on a phytohemagglutinin from the lentil. IV. Direct binding studies of *Lens culinaris* hemagglutinin with simple saccharides' *Arch. Biochem. Biophys.*, **146**, 353–5.
5. Young, N. M., Leon, M. A., Takahashi, T., Howard, I. K. and Sage, H. J. (1971) 'Studies on a phytohemagglutinin from the lentil. III. Reaction of *Lens culinaris* hemagglutinin with polysaccharides, glycoproteins, and lymphocytes', *J. Biol. Chem.*, **246**, 1596–601.
6. March, S. C., Parikh, I. and Cuatrecasas, P. (1974) 'A simplified method for cyanogen bromide activation of agarose for affinity chromatography', *Anal. Biochem.*, **60**, 149–52.
7. Howard, I. K., Sage, H. J., Stein, M. D., Young, N. M., Leon, M. A. and Dyckes, D. F. (1971) 'Studies on a phytohemagglutinin from the lentil. II. Multiple forms of *Lens culinaris* hemagglutinin', *J. Biol. Chem.*, **246**, 1590–5.
8. Trujillo, T. T. and Van Dilla, M. A. (1972) 'Adaptation of the fluorescent Feulgen reaction to cells in suspension for flow microfluorometry', *Acta Cytol.*, **16**, 26–30.
9. Rotman, B. and Papermaster, B. W. (1966) 'Membrane properties of living mammalian cells as studied by enzymatic hydrolysis of fluorogenic esters', *Proc. Nat. Acad. Sci. U.S.A.*, **55**, 134–41.

F

Purification by Agglutination

CHAPTER 50

Purification of Oncornaviruses by Concanavalin A

MARGARET L. STEWART

Generally, the viruses studied in most detail are those that are most easily purified. Part of the difficulty in studying oncornaviruses is that high concentrations of particles are difficult to obtain in purified form.

Murine leukaemia viruses are enveloped, spherical particles 100–150 nm in diameter (sedimentation coefficient approximately 600S). The viral RNA, approximately 1–2% of the particle weight, is a 60–70S molecule composed of two or three subunits 3–4 x 10^6 daltons molecular weight, each in turn having a sedimentation coefficient of about 36S (1). Physically, oncornaviruses are representative of most enveloped RNA viruses. They are related to progressive pneumonia-visna, and foamy viruses of the genus 'leukovirus'. They are also similar to mammary tumour viruses, paramyxoviruses, some arboviruses, and orthomyxoviruses (2). Representative of the last three groups are measles virus and the more extensively studied Vesicular Stomatitis virus (VSV) and Influenzae viruses (2). These viruses have a common mode of maturation in that they 'bud' from the surfaces of infected cells, and in this process, they take with them a large proportion of cellular material. Most of this is incorporated into the viral envelope and it is largely representative of the membranes of the cells in which the viruses are grown. The similarity between virus envelopes and cell membranes probably presents the main difficulty in definitive purification of these viruses.

Often much is known about virus-specified events in the cell because like poliovirus (a small, 25 nm diameter RNA-containing virus) and VSV, many animal viruses are very efficient at shutting off host macromolecular synthesis. In these instances most of the material being synthesized in the cell is virus specific late in the infectious cycle. Of course, infection is destructive to the cells. Leukaemia viruses are able to infect and replicate in appropriate cells without causing any readily observable effects on the cells. Morphological evidence for productive infection is by observation of C-type particles in thin sections of cells examined in the electron microscope. It has been estimated that mammalian oncornavirus-specific macromolecular synthesis in such cells amounts to one percent of ongoing host cell synthesis (3).

Many animal viruses replicate very efficiently and are produced in large amounts in culture. This, of course, facilitates their isolation and characterization. At maximal rates of virus formation in cultures where every cell can be infected (2), 20,000 polio- and VSV particles are produced per cell per hour. Similar estimates for ortho- and paramyxoviruses are 670 particles, and for Rous-associated virus (an avian oncornavirus), 30 particles per cell per hour.

Much is known about the structural organization of many animal viruses because they may be relatively easy to concentrate and purify. Isolates of poliovirus can contain 10^{14} particles per ml of a caesium chloride gradient fraction, and one in 200–300 of these particles is infectious (4). Isolates of VSV routinely contain 10^{11} PFU and 10^{13} particles/ml of a fraction from a sucrose velocity gradient that is used in the final step of purification (D. F. Summers, personal communication).

Friend leukaemia virus (Fr-MuLV) is produced at very low levels by continuous cell lines of mouse 3T3 fibroblasts. By electron microscopy, about one in ten cells appears to be productively infected at any one time. By assaying infectivity in the medium, we find that 0.4 PFU are released per cell in seven hours and from gel electrophoretic data, we estimate that an average of 3 particles of virus are produced per cell per hour. This is approximately one-tenth the average rate for Rous-associated virus (2). Both Friend and Rauscher strains of murine leukaemia viruses have proven to be extremely sensitive to methods of purification that are more successfully applied to other enveloped RNA viruses such as VSV (5) and Influenzae (6). While this is due, no doubt, to the low concentration of virions released by the cells, it is also probably complicated by the known susceptibility of the particles to degradation by certain factors in the medium (7). Moreover, whether related to this or not, the lifetime of an infectious virion in culture medium may be of the order of a few hours (8). Taken together, these considerations prompted us to search for a simple, specific and gentle method which would provide us with pure, infectious leukaemia virus within a day, at most, after harvest of medium from the cells.

1. PRINCIPLE

Concanavalin A binds in high concentrations to the surfaces of C-type particles newly formed by the cell (Figure 1). It also rather quickly cross-links virions that are free in the medium much like antigen—antibody complexes, and after collection of insoluble complexes by centrifugation at low speed, virions can be released by solubilization with α-methyl mannose, a competitive inhibitor of Con A. Furthermore, freed from Con A and other non-viral material by a single cycle of equilibrium centrifugation, infectious isolates of pure virions can be analysed 10—12 hours after harvest of medium if necessary (9—11).

This method provides significant purification of viral, from non-viral material. Relative to serum albumin present in parental medium, purification of the major polypeptide in the virus is at least 50,000-fold (11). The PFU/particle ratio of the purified virus is one/800—1000, but the maximum concentration of particles we can presently obtain is approximately 10^{10}/ml. This is sufficient for morphological studies and analysis of structural components of the virion, but we are not yet able to study events in a single, synchronous infectious cycle with high multiplicities of infectious particles as can be done with poliovirus, VSV and other animal viruses.

With a simple assay for virus-specific 60—70S RNA in Con A complexes, we can detect, and monitor virus production by the cells (12). For this analysis, at least 10^5 cells are required. For morphological studies of purified virions and analysis of major structural proteins by staining in SDS-acrylamide gels, at least 10^8 cells are required. Of course, many more radiolabelled structural proteins can be detected in gels by radioautography. Con A is expensive, and this, at present, will probably economically limit the maximum amount of virus that can be obtained by the method described here.

2. REAGENTS AND MATERIALS

Concanavalin A, lyophilized (3X crystallized, Miles-Yeda, Israel) (alternatively, in saturated NaCl at 40 mg/ml)
α-Methyl-D-mannopyranoside (Sigma Chemicals) 2.0 M in TSE buffer
TSE buffer: 0.01 M Tris-HCl, pH 7.4; 0.1 M NaCl, 0.002 M EDTA
Radioactive precursors:
^{3}H-uridine, sp. act. 28.5 Ci/mmol;
^{35}S-methionine, sp. act. greater than 100 Ci/mmol;
^{14}C-glucosamine, sp. act. 164 mCi/mmol.
(From New England Nuclear, Boston, Massachusetts)

2.1 Cells and Virus

Normal mouse 3T3 fibroblasts are exogenously infected with virus in spleen extracts taken from mice with Friend disease. Passed in the mouse, this virus

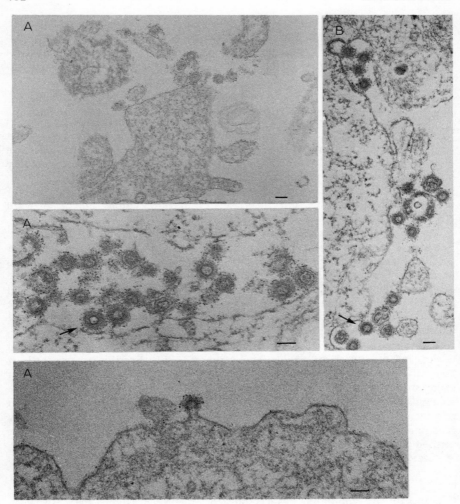

Figure 1. Ferritin-conjugated concanavalin A bound to surfaces of extracellular oncornavirions in thin sections of monolayers of infected 3T3 cells. Panels A, Friend-infected 3T3 mouse fibroblasts; B, 3T3 mouse cells producing endogenous viruses (10). The bar indicates 100 nm. Arrows mark completed particles resembling the budding particle, lower panel

complex is associated with an 'erythroleukaemia' (for review, see ref. 13). In most cases, only the leukaemia component Fr-MuLV, successfully replicates in established mouse cell lines. Culture medium incubated with infected cells is the source of virus used for study (for infection and growth conditions, see ref. 9 and 10).

Leukaemia virus infectivity is assayed by plaque formation in a monolayer of MSV-transformed, mouse (S+L−) cells (14).

2.2 Radiolabelling of Viral Components

Viral RNA is labelled by addition of ^{3}H-uridine to the cells in fresh medium at a concentration of $40-50$ μC/ml medium. When required, viral proteins can be labelled with ^{35}S-methionine, 70 μC/ml medium and in this case medium contains 1/10 the normal concentration of methionine (11). (This radioactive precursor is a satisfactory one since it is available in high specific activity. Presently, it appears there are no methionine-deficient polypeptides present in Fr-MuLV.)

Viral glycoproteins are labelled with ^{14}C-glucosamine, 6.25 μCi/ml medium. (It is possible that some viral glycoproteins do not contain glucosamine, and other carbohydrates should be used in addition for further definition of glycopolypeptides.)

For harvest of virus, and labelling of viral components produced by cells growing in roller bottles (670 cm^2), the medium volume is decreased from 100 ml used for growth, to 15 ml. For harvest of virus from cells growing in T$_{75}$ flasks (75 cm^2), the medium volume is decreased from 20 ml used from growth, to 6.0 ml. Medium always contains the normal amount of foetal calf serum. Incubation times may vary from 7.5 to 16 h and cells used for harvest and labelling of virus are discarded.

2.3 Electron Microscopic Analysis of Virus Morphology

For preparation of thin sections and analysis of uranyl acetate-stained particles adsorbed to carbon-coated grids, see ref. 16.

3. METHOD: PROCEDURE FOR ISOLATION OF LEUKAEMIA VIRUS FROM CULTURED MEDIUM INCUBATED WITH INFECTED CELLS

The optimum time to harvest virus from 3T3 cells is when they are in mid-log phase of growth. The yield of virus per cell is low within the first 12 h after passage, and cultures at high density tend to shed more cellular debris into the culture medium. This, of course should be avoided because this increases the likelihood of contamination of the virus preparation with cellular material. Cell membranes and virus have the same buoyant density (1.16 g/cm^3) in sucrose and if they are not effectively removed from the medium (see pretreatment of medium, below), they will co-band with the virus in sucrose density gradients. The decreased volume of medium used for harvest if virus increases the concentration of virus in the medium. Less Con A is required for agglutination and amounts of radioactive precursors are also reduced.

3.1 Pretreatment of the Medium: Removal of cells and subcellular debris

Conditions described in Figure 2 have been optimized to quantitatively provide infectious Friend leukaemia virus particles and viral 60—70S RNA in the shortest possible time. These conditions will undoubtedly vary with different viruses and they should be determined separately for each system. Analyses used to determine these conditions follow in subsequent paragraphs.

After incubation with cells, the medium is removed from the monolayer, divided into glass centrifuge tubes (e.g. 30 ml Corex tubes, Corning Glass, New York), and

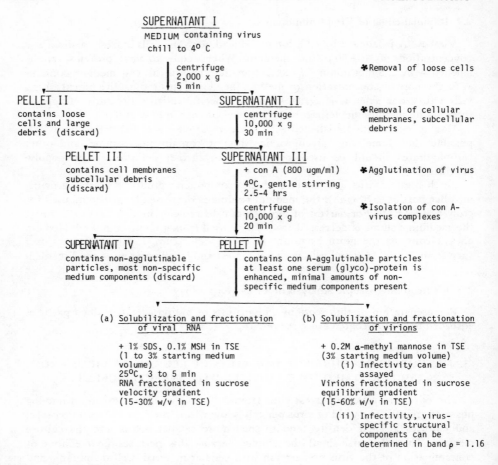

Figure 2. Procedure for isolation of leukaemia virions from culture medium incubated with infected cells. For analysis of viral RNA in Con A—virus complexes (pellet IVa), 12 ml medium containing 50 μCi/ml ^{3}H-uridine is incubated with 4×10^6 cells growing in two, T_{75} flasks. Labelling is for 16 h. For analysis of viral RNA, protein, glycoprotein and infectivity in purified virions of density 1.16 solubilized from Con A complexes with α-methyl mannoside (pellet IVb), 60 ml medium containing appropriate radioactive precursors is divided between four roller bottles containing a total of 10^8 cells. Labelling may be from 7.5 to 16 hours.

chilled to 4° C. All subsequent steps are carried out at this temperature unless otherwise specified. Loose cells are first removed by centrifugation of the medium at low speed, 2000g. Subcellular material is then removed by centrifugation at 10,000g. (Cells could also be removed by centrifugation in this step, but they could break in these conditions, and this would contaminate the medium with soluble cellular material.) The clear supernatant III is transferred to a 100 ml Erlenmeyer flask containing a small stirring bar.

3.2 Agglutination of Virus

Lyophilized Con A is added to a final concentration of 800 μg/ml medium (supernatant III). Alternatively, and for small volumes of medium, Con A in saturated NaCl can be used. We have found by SDS-acrylamide gel electrophoresis that Con A in solution can be degraded on storage at 4°C giving rise to proteolytic fragments of 13,000, 15,000 and 18,000 molecular weight (15). Though there is no noticeable effect upon agglutination of the virus, bands of this size in acrylamide gels may confuse patterns of stained, (but of course, not radiolabelled) polypeptides of similar size. For these reasons, lyophilized Con A is recommended, and solutions made from it should be discarded after use.

After gentle stirring in an ice bath for 2.5—4 h, Con A-agglutinated complexes are isolated from the cloudy medium by centrifugation at 10,000g for 20—30 minutes in 30 ml glass centrifuge tubes in the Sorvall centrifuge. The resulting supernatant, IV, is discarded. Inside surfaces of the tubes are rinsed with ice-cold TSE buffer and this too, is discarded. The pellet IV can be gently washed by resuspending it in 2—3 ml cold TSE buffer and recentrifugation without loss of viral components. This washing step, though not always necessary, reduces amounts of soluble medium components entrapped by occasionally heavy precipitates.

3.3 Assay of Viral RNA in Con A—Virus Complexes (Pellet IV)

Viral 60—70S RNA can be solubilized directly from this pellet in TSE buffer containing 1% SDS and 0.1% MSH. The pellet is gently resuspended in a volume equal to 1—3% of the starting culture medium, supernatant I. After 5 min at 25°C, the clear solution is layered onto a 14 ml, 15—30% linear sucrose gradient in TSE prepared in a Spinco SW40 centrifuge tube, or a 12 ml gradient prepared in a Spinco SW41 tube. In each case, centrifugation is for 120 min at 40,000 r.p.m. at 4°C. The radiolabelled 70S RNA is found in the middle of the gradient. (For further details, see ref. 12. Examples of this analysis appears in Figures 6 and 7 below.)

3.4 Separation of Virions from Con A Complexes

Virions are released by solubilization of Con A complexes with α-methyl mannoside (α-MM). The pellet IV is gently resuspended with a Pasteur pipette in TSE buffer (3% of volume of supernatant I) containing 0.2 M α-methyl mannoside and, within 5 min, with gentle mixing at 4°C, it will dissolve to an opalescent, slightly yellowish solution. Infectivity of the solubilized virions can be assayed at this point after passing the solution through a sterile membrane filter (0.45 μ pore size, Swinney filter, Millipore). Free virus particles can be observed in the electron microscope after a tenfold dilution of the α-MM-solubilized pellet in TSE buffer. The major components visible, however, are small spheres 12 nm in diameter that probably contain a serum (glyco)polypeptide, m.wt. 180,000 (11) that reacts with Con A.

Virus is purified from this, and other soluble material by fractionation in a sucrose (15—60% w/v) equilibrium gradient prepared in TSE. Centrifugation is in the SW40 or SW41 rotor for a minimum of 4 h, and a maximum of 16 h at 16,000 r.p.m. in the Spinco ultracentrifuge. Depending on the amount of material isolated in pellet IV, fractionation in the SW27 rotor may provide a cleaner virus preparation.

The position of virus in the gradient can often be determined by location of a single light-scattering band in the centre of the gradient. (For details on fractionation of the gradients, and assay of TCA-precipitable radioactivity, see ref. 10.) Density of isolated gradient fractions is determined by refractive index. For an example of density gradient centrifugation of Con A-agglutinated virus see Figure 4 below.

4. APPEARANCE AND INFECTIVITY OF LEUKAEMIA VIRUS ISOLATED FROM MEDIUM BY AGGLUTINATION WITH CON A

Many viruses are routinely collected by centrifugation to concentrate them in various purification procedures. We have found that doing this affects the physical structure, and greatly decreases infectivity of murine leukaemia viruses. The morphological integrity of Fr-MuLV is preserved under the gentler conditions of purification with Con A, and infectivity seems to be similarly unaffected. The appearance and infectivity of Con A-isolated virus is shown in Figure 3 and this is compared to virus isolated directly from medium under centrifugation conditions necessary to pellet free virions in a reasonable time, 80,000g for 90 min.

Extracellular virions in thin sections of infected cell monolayers are shown in the upper panel of Figure 3. The particles are spherical with a diameter of approximately 100 nm. Individuals in these populations are often quite characteristic of newly formed particles (see also, Figure 1), in that, like particles observed budding from the cell, they possess electron-lucent centres. With age, the centres appear to condense to uniformly electron-dense cores in the middle of the virus particles (18).

In pellet III which is usually discarded, membranous material, vesicles and subcellular organelles are observed. Cell membranes (1000S) are generally larger than virions and they should pellet under these conditions used to isolate pellet III in pretreatment of the medium. Occasionally an electron-dense particle resembling a virus particle is also observed in this fraction and from the size classes of uridine-labelled RNA (4S, 18S and 28S, ref. 12), it contains ribosomes, and sometimes as much as 20% (see also Figure 7 below) of the viral 60–70S RNA eventually isolated in pellet IV. This is not too surprising given the affinity of this virus for cell membranes. (We find that extracellular virus is very difficult to wash free from cell surfaces.) Also, some virus in the medium will probably become associated with the membranes in centrifugation.

For practical reasons, infectivity is not usually assayed in this fraction.

Virions in thin sections of Con A-complexed aggregates, pellet IV, appear identical to cell-associated virus in the starting medium above. They have, however, acquired an outer coating of Con A that resembles antibody bound to virus surfaces. The inset, at higher magnification, shows an example of a 'young' particle in pellet IV that appears with a frequency of about 1 in 50 particles. It has a rather beautiful, and organized internal structure similar to particles in Figure 1.

Compared to particles in the starting material, those in a thin section of pellet IVa collected in the absence of Con A, show some distortion of the outer viral membrane giving some particles a 'tailed' appearance. The central cores of these particles are also more uniformly electron dense. After resuspension in buffer of the particles in pellet IVa and examination by negative staining, 'tailed' particles are visible amid other amorphous material. Distortion of these viruses is known to occur through preparation on the grid (21) but it is usually minimal with uranyl acetate stain as used here and it could not have occurred in the fixed pellet prepared for thin sectioning. It thus appears that the particles *are* physically affected by centrifugation out of the medium. Particles collected onto a 60% sucrose cushion appear very similar to these.

Infectivity of α-methyl mannose-solubilized virions in pellet IV amounts to 63% of that in the starting medium. Some infectivity remains in supernatant IV and we have shown that this is associated with certain, presently uncharacterized particles of density 1.18 that are non-agglutinable by Con A. These particles are present only in medium incubated with newly infected cells (a method for determining this is illustrated in Figure 4, below).

Infectivity of particles in pellet IVa is reduced to 1/10,000th of that in the starting material. Part of this loss in infectivity may only be apparent since these particles tend to aggregate easily. In contrast, particles in pellet IV are more homogeneously resuspendable. Though at this point, infectivity is not severely affected by purification with Con A, subsequent centrifugation of the virus to equilibrium in sucrose causes a 100-fold reduction in infectivity. This is due simply to centrifugation in sucrose and is not a secondary affect of agglutination with Con A (see Figure 3). Con A-purified particles appear to be physically intact in the electron microscope. They are not permeable to stain as are broken particles and they retain their spherical shape. Almost all of the particles are at least partially covered with small knobs 8 nm in diameter. By negative staining, they are slightly larger (150 nm) than they appear in thin sections. This may be due to some collapsing of the envelope in preparation on the grid.

Virions in gradient fractions of density 1.16 should be the only material visible in the electron microscope. Evidence of membrane-like, or other matter means that the preparation may be contaminated with cellular components, or that damage has occurred to the virus during purification. Occasionally, we observe stranded material that we presume to be ribonucleoprotein released from a few particles in preparation of the grid.

Within 3–4 days, at 4°C, Con A-purified virus acquires the appearance of particles isolated in pellet IVa and at the same time infectivity decreases to similar levels. Keeping the virus at room temperature hastens these effects. Dialysis, or recentrifugation similarly affects the virus and further manipulations while attempting to preserve its morphological and biological integrity are not presently very successful. Therefore, infectivity, morphological studies and analyses of macromolecular components are performed immediately without storage of the virus.

4.1 Selective Agglutination of Virions

The selective isolation of infectious virions from among any other particles and soluble components in culture medium can be demonstrated by comparing the profiles of labelled particles in whole medium with those isolated in pellet IV and

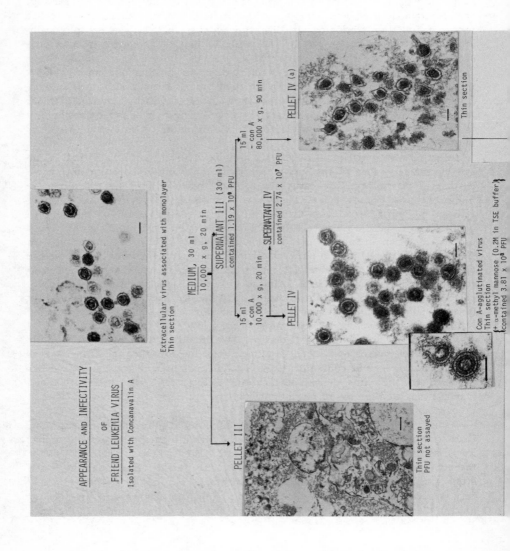

APPEARANCE AND INFECTIVITY

OF

FRIEND LEUKEMIA VIRUS

Isolated with Concanavalin A

Extracellular virus associated with monolayer
Thin section

MEDIUM, 30 ml
10,000 × g, 20 min

SUPERNATANT III (30 ml)
contained 1.19 × 10⁹ PFU

15 ml
+ con A
10,000 × g, 20 min

PELLET IV

15 ml
– con A
80,000 × g, 90 min

SUPERNATANT IV
contained 2.74 × 10⁷ PFU

PELLET IV (a)

Thin section

Con A-agglutinated virus
Thin section
(+ α-methyl mannose (0.2M in TSE buffer)
contained 3.81 × 10⁸ PFU

PELLET III

Thin section
PFU not assayed

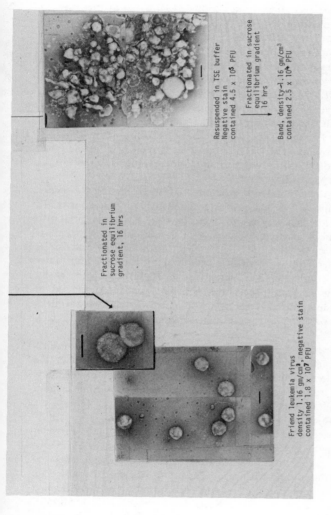

Friend leukaemia virus
density 1.16 gm/cm³ negative stain
contained 1.8 x 10⁷ PFU

Fractionated in
sucrose equilibrium
gradient, 16 hrs

Resuspended in TSE buffer
Negative stain
contained 4.5 x 10⁵ PFU

Fractionated in sucrose
equilibrium gradient
16 hrs

Band, density ~1.16 gm/cm³
contained 2.5 x 10⁶ PFU

Figure 3. Appearance and infectivity of leukaemia virus isolated by (a) agglutination with con-canavalin A, and high-speed centrifugation of free virions in medium in the absence of Con A. The surface of a monolayer of cells passaged five times after infection was rinsed with 10.0 ml TSE buffer, and the buffer was discarded. The cells were fixed with glutaraldehyde and prepared for thin sectioning (upper photograph). Medium (30 ml) incubated with 2×10^8 cells from the same culture but growing in two roller bottles was 'pretreated' as described in Figure 2. Con A was added to one half (15 ml) of supernatant III (+ Con A), and the other half, (− Con A), received nothing. Conditions for collection of pellet IVa are the minimum required to isolate free virions (600S) within a reasonable time. Small duplicate aliquots amounted to 1 percent of the volume of virus-containing fractions taken for assay of infectivity. Initial dilutions were sterilized by passing them through a membrane filter, 0.45 μ pore size, (Swinney filters, Millipore). Infectivity in virions, density 1.16, represents the total recovered in sucrose gradient fractions of density between 1.14 and 1.17 (for an example, see Figure 4, below)

solubilized with α-methyl mannose (Figure 4). This assay, together with others described below can be used to optimize conditions described in Figure 2.

[3]H-Uridine is used as the most specific radiolabelled marker for virus. It is available in high specific activity, and of TCA-precipitable radioactivity in supernatant III, as much as 10 percent is present in virions of density 1.16 g/cm³. Of TCA-precipitable ^{35}S-methionine and ^{14}C-glucosamine present in supernatant III, 3–5%, and 1–2%, respectively is present in virions (11). Possibly because of instability of the virus, in medium at 37°C (7), these amounts decrease with increased incubation times with the cells. For these experiments, centrifugation is done in the SW27 rotor (36 ml tubes) to provide enough capacity to handle 15 ml medium on top of an 18 ml gradient. Solubilized pellet IV is fractionated in a 36 ml gradient, and supernatant III, in an 18 ml gradient (both 15–60% sucrose in TSE). Sixty percent sucrose is added to supernatant III prior to fractionation to a final concentration of 15% to prevent dilution of the upper part of the sucrose gradient. After centrifugation at 16,000 r.p.m., thirty, 1.2 ml fractions were collected from the gradient containing pellet IV, and thirty, 0.6 ml fractions from the lower half of the gradient containing supernatant III. The upper 18 ml is collected in one pool. Refractive indexes of selected fraction are determined, and radioactivity in (50 μl) aliquots from each fraction is assayed.

In addition to virions of density 1.16, newly infected 3T3 cells release several kinds of particles into the medium that are differentiated by buoyant density. In early passage medium, virus is selectively agglutinated by Con A, and the presently uncharacterized, non-agglutinable particles of density 1.18, 1.20 and 1.24 disappear from the medium with passage of the cells. After the fourth passage and for as many as fifteen passages afterward, only virus is produced by the cells and under optimal conditions, their consistent and quantitative agglutination is easily demonstrated by this analysis. As a check that the particles are produced in response to infection, a comparable analysis of medium from uninfected cells is done and no radiolabelled particles are present in such medium (note FLV(−), lower left panel).

For morphological studies, the gradient fractions containing maximal amounts of Con A-agglutinable radioactivity are analysed in the electron microscope (see Figure 2). It is not possible to similarly analyse particles in gradient fractions containing whole medium because of the excessive amounts of serum protein present in all of the gradient fractions.

The specific infectivity (proportion of infectivity/radiolabel, ref. 9) in Con A-agglutinated particles of density 1.16 is equivalent to that in the same particles fractionated directly from medium. By this analysis, Con A affects neither infectivity nor buoyant density of the virus. As Figure 2 describes, a single cycle of centrifugation drastically reduces infectivity recovered in both cases, for it amounts to only one percent of that in the starting medium. Fractionation of the virus in a medium such as 'Renografin', (E. R. Squibb & Sons, Inc., Princeton, N.J. 08540), or a non-ionic solute such as 'Metrizamide' (Nyegaard and Co., A/S, Oslo, Norway), may improve recovery of biological activity.

It is possible that some, or all of the particles produced by infected cells are infectious, but *not* agglutinable by Con A. This can be demonstrated by assaying infectivity across the gradient containing whole medium and comparing profiles with infectivity in Con A-isolated particles (see fifth passage). This panel shows that the infectivity which remains in supernatant IV in fifth passage medium

(Figure 3), is associated with certain non-agglutinable particles of density slightly greater than virus (0.02 g/cm^3). Though they were just as infectious as virus of density 1.16 (9) on a label/infectivity basis, they must differ from virus by the absence, or inaccessibility of appropriate surface carbohydrate.

The selective agglutination of virus by Con A can also be demonstrated by first fractionating supernatant III in a gradient, and adding Con A to each of the isolated fractions (see inset, Figure 4). Radioactivity recovered in pellets IV from these fractions should coincide with that in virus. Radioactivity in supernatant fractions is due to 'trapping' of soluble radiolabel. If desired, this can be reduced by washing the pellets on TSE as described above. By adding appropriate competitive inhibitors of Con A to the fractions, a measure of the affinity of Con A for viral surface carbohydrate could also be obtained. This technique can also be used to collect virions for analysis of viral RNA (see below).

5. CHARACTERIZATION OF THE METHOD

5.1 Optimizing This Method for Your System

The minimum amount of Con A that is required to quantitatively agglutinate virus should be determined since we have found that excess Con A may contaminate virus-containing fractions in equilibrium gradients. This could promote non-specific contamination of viral proteins with glycoproteins adventitiously bound to the outside of virions (see ref. 12) and may also promote agglutination of purified virions causing an apparent drop in infectivity of the virus.

As a preliminary step, TCA-precipitable ^{3}H-uridine content of complexes formed with varying concentrations of Con A added to the medium can be determined (Figure 5). In this experiment, the minimum amount of Con A required to provide maximal agglutination in 2.5 h at $35°$C was $250-500$ μg/ml medium. At $4°$C, a higher concentration, 800 μg/ml is required (data not shown). This number can be checked by measuring radioactivity in particles of density 1.16 in α-methyl mannose-solubilized pellets IV as described in Figure 4 above, or as radioactivity in viral 60—70S RNA of pellets IV (Figure 7, below). In Figure 5, pellet III was not removed from the medium before addition of Con A, and the amount of radioactivity collected in the *absence* of Con A (1.5×10^4 c.p.m.) amounts to 43% of the maximum radioactivity isolated in this case. It is this baseline which increases with longer incubation (labelling) times with the cells. It also increases as the cells reach confluency, and it may be promoted by extended incubation of the cells in suboptimal volumes of medium.

The ^{3}H-uridine-labelled 60—70S RNA content of pellet IV is a relatively quick and more specific way to not only detect virus in culture medium (10), but also to obtain viral RNA in a relatively undegraded form for other biochemical and morphological analyses. We have used this assay (illustrated in Figure 6) as a separate measurement of the quantitative agglutination of virus. It shows that 60—70S RNA is quantitatively isolated from culture medium by collection of large Con A—virus complexes under gentle centrifugation conditions. Probably because the particles are intact (Figure 3, pellet IV), the yield of RNA obtained in the presence of Con A (Figure 6, A and B) is four and often as much as tenfold more than that collected in particles isolated under more physically damaging conditions (Figure 3, pellet IVa, and Figure 6, C). By analysis of the proteins in the pelleted

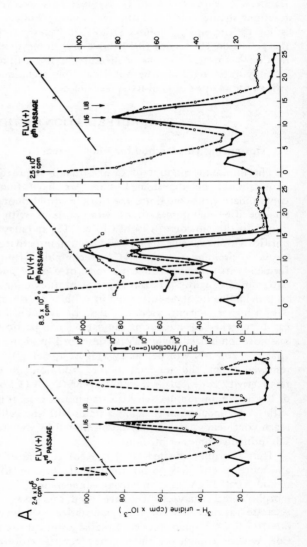

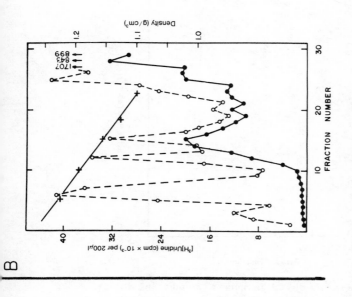

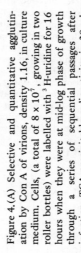

Figure 4. (A) Selective and quantitative agglutination by Con A of virions, density 1.16, in culture medium. Cells, (a total of 8×10^7, growing in two roller bottles) were labelled with ^{3}H-uridine for 16 hours when they were at mid-log phase of growth through a series of sequential passages after infection. TCA-precipitable radioactivity in 30 ml supernatant III was routinely between 4 and 6×10^6 c.p.m. Con A was added to one half and the other, received nothing. TCA-precipitable radioactivity in (●——————●), α-methyl mannose-solubilized pellet IV, and (○ – – – ○), supernatant III. Infectivity in (■——————■), pellet IV; (□ – – – □), supernatant III. (X——————X), density, g/cm^3, determined by refractive index. FLV(−), lower left panel, radioactivity in comparable fractions isolated from (1.2×10^8) uninfected cells. TCA-precipitable radioactivity in supernatant III, 8×10^6 c.p.m. (B) Selective agglutination of virions in supernatant III prefractionated in a sucrose equilibrium gradient: inset lower right. Medium with 10 μCi/ml ^{3}H-uridine was incubated for 16 hrs with 4×10^7 cells, third passage. Supernatant III from this medium, containing 2×10^5 c.p.m. TCA-precipitable ^{3}H-uridine, was fractionated in an 18 ml gradient as in (A) above. Centrifugation was from right to left. Fractions (0.5 ml) were collected and chilled immediately. TCA-precipitable radioactivity in 50 μl aliquots was determined and radioactivity is corrected to represent 200 μl from each fraction. Con A (1.5 mg/ml) was added to 200 μl from each fraction and radioactivity in SDS-solubilized pellets IV was counted directly. (○ – – – ○), Radioactivity in supernatant III; (●——————●), radioactivity in pellet IV. (X——————X), density, g/cm^3, determined by refractive index

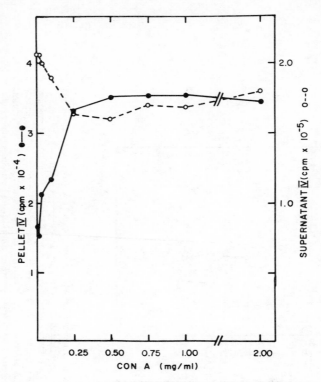

Figure 5. Recovery of TCA-precipitable ^{3}H-uridine in pellet IV with different concentrations of Con A. Infected cultures (third passage) were labelled with 50 μCi/ml of ^{3}H-uridine for 16 h. Medium contained 1.12×10^5 TCA-precipitable c.p.m./ml. In this experiment, pellet II was not removed prior to addition of Con A. Con A was added to 2.0 ml aliquots of culture medium at the indicated final concentrations. After isolation of pellets IV, supernatants were saved and acid-insoluble counts remaining in 50 μl aliquots were determined. Pellets were washed in 400 μl of TSE and resuspended in 200 μl of TSE containing 0.1 M α-methyl mannoside. After dissociation, TCA-precipitable radioactivity in 25 μl aliquots was determined. Counts are corrected to represent the total fractions assayed. ($\bullet$———$\bullet$), pellet IV; ($\circ$ — — — — $\circ$), supernatant IV

free virus (C), we find there is almost no purification of viral proteins from medium components. RNA in broken, and SDS-solubilized virions would be susceptible to degradation by any nucleases still present in this pellet, and this may explain the relatively low recovery and heterogeneous sedimentation in velocity gradients.

Often as much as 60% of the ^{3}H-uridine in pellet IV is in viral 60—70S RNA (Figure 7). If pellet III is not removed from medium before addition of Con A, radioactivity in pellet IV is divided between 4S, 18S, 28S and 60—70S components.

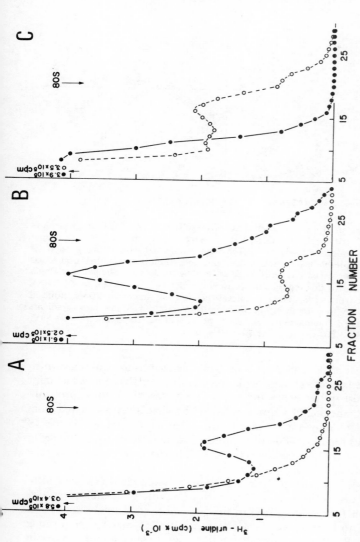

Figure 6. Isolation of 60–70S RNA by centrifugation of Con A–virus complexes. Supernatant III (20 ml) incubated for 16 hours with 2×10^7 cells growing in one roller bottle contained a total of 2.66×10^6 c.p.m. TCA-precipitable uridine. Similarly to the outline in Figure 3, the medium was divided into two 10 ml aliquots and Con A was added to one of them (supernatant III). Pellets IV (+ Con A) and IVa (– Con A) were collected from both aliquots after centrifugation first, at 5000 g for 15 min (A), then 16,000 g for 20 min (B) and finally, 80,000 g for 30 min (C). ³H-Uridine-labelled RNA and SDS-solubilized pellets were analysed in sucrose velocity gradients. Samples were collected directly into scintillation vials and counted in 10 ml aquasol (New England Nuclear) containing 10% water by volume. Direction of sedimentation is from left to right. Position of rabbit reticulocyte (80S), ribosomes (arrows) fractionated in parallel gradients containing 0.01 M Tris-HCl, pH 7.4; 0.05 M KCl and 0.015 M MgCl₂ was located by absorbance at 260 nm. RNA solubilized from pellet IVa directly from supernatant IIIa (omitting steps A and B above) sediments exactly as in C. (●————●), ³H-Uridine in pellets IV; (○— — —○), ³H-Uridine in pellets IVa

495

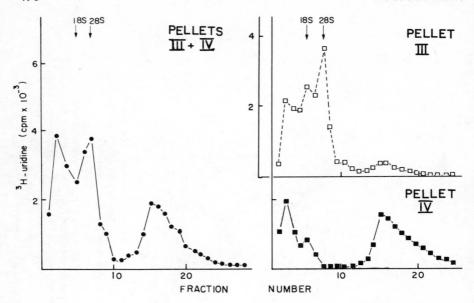

Figure 7. The distribution of ³H-uridine-labelled RNA in pellets III and IV. Cells growing in two T$_{75}$ plastic flasks (2 × 10⁶ cells per flask, 14th passage), were labelled with ³H-uridine (40 μCi/ml, 6.0 ml medium per flask for 16 h). Supernatant III containing a total of 4.6 × 10⁵ c.p.m. TCA-precipitable uridine was divided into two aliquots and chilled. Con A was added to one aliquot and the other was centrifuged at 16,000g for 15 min. The pellet (III) was saved and Con A was added to this supernatant to isolate pellet IV. After 2.5 h at 4°, Con A aggregates were isolated from each. TCA-precipitable uridine in the three SDS-solubilized pellets was determined. TCA-precipitable radioactivity in RNA fractionated in sucrose gradients as shown. TCA-precipitable radioactivity in each pellet before fractionation: III + IV, 3.54 × 10⁴ c.p.m.; III, 1.43 × 10⁴ c.p.m.; IV, 2.70 × 10⁴ c.p.m.

Most of the 18S and all of the 28S RNA is probably derived from ribosomes which can be easily removed in pretreatment of the medium (pellet III). A small amount of viral RNA is also lost in this pellet and as previously described this is probably associated with membrane-bound virions. Most of the viral RNA is isolated from medium in pellet IV in response to added Con A (pellet IV). Much of the 4S material present in this fraction is due to trapping of soluble radiolabel in the pellet and this can be reduced by washing the pellet in TSE buffer before solubilizing the virions in SDS as described in section 3.2.

Oncornaviral RNA is sensitive to heat and dimethylsulphoxide (1). The SDS-solubilized pellet IV can be heated at 95°C for one minute (or dissolved in 95% dimethylsulphoxide) before layering it onto the gradient. Radioactivity in 60–70S material is converted to homogeneously sedimenting 30–35S material confirming the viral origin of the RNA. Under these conditions, some concomitant increase in 4S material is also observed.

In newly formed Rous sarcoma virions, viral RNA is apparently found in the 35S form, and it subsequently 'matures' to 60–70S aggregates in the particle (16,17). Thus, harvests of a few minutes contain 35S RNA, and medium incubated with the

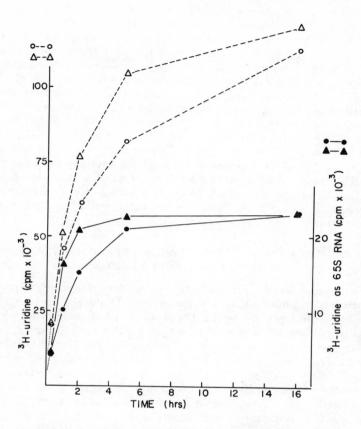

Figure 8. The effect of temperature on the rate of agglutination of 65S RNA in virions. Medium (20 ml containing 1 mCi ^{3}H-uridine) was incubated with 2×10^7 cells (14th passage) in one roller bottle. After labelling, supernatant III containing 3.46×10^6 c.p.m. TCA-precipitable uridine was divided into two 10 ml aliquots. One was rapidly chilled to $4°$, the other was kept at $25°$. Con A was added to each, and pellets were isolated immediately for the earliest time point (15 min). Centrifugation was at the temperature of incubation for each aliquot. To minimize possible degradation of RNA at $25°$ by medium components, two sequential 1 h harvest (0—1, 1—2 h) and then one 3 h harvest (2—5 h) were taken from each of the two aliquots. The last pellet isolated was a 9 h harvest (5—16 h). RNA in pellets: 15 min, 1 h, 2 h, 5 h, was assayed after the 5 h time point; sixteen h points were assayed at 17 hours. The amount of 65S RNA present in each pellet is shown as a sum with that from previous time points. (○————○, ●————●) $4°$C; (△————△, ▲————▲) $25°$C

cells for longer times contains greater proportions of 70S RNA, Though newly formed virions are agglutinable by Con A (9, Figure 3), under conditions described here, very little labelled RNA of subunit size is detectable. This may be related to our observation that these virions are very difficult to remove from cell surfaces and some time may elapse before newly made particles are released into the medium. In any case, the incubation time used here (16 h) would permit 'maturation' of the majority of radiolabelled RNA molecules present in any 'young' particles to occur.

The 70S RNA content of pellet IV can be used to determine the rate of agglutination of virions at various temperatures. By this assay, the rate of agglutination of virus is slower at $4°C$ than it is at $25°C$ (Figure 7). Maximal amounts of 60—70S RNA are agglutinated at 1—3/4 h at $25°C$, and at 3 h at $4°C$. At these early times in these experiments, approximately 25% of the radioactivity is present as 60—70S RNA. Continued incubation promotes accumulation of TCA-precipitable radiolabel at both temperatures. That this additional label is not associated with virions of density 1.16 can be determined in equilibrium gradients as described in Figure 2. By velocity gradient centrifugation the size of this additional material is approximately 4S.

6. USE OF THIS METHOD IN DETERMINING STRUCTURAL COMPONENTS OF CON A-ISOLATED VIRIONS

We have found that the pattern in SDS-acrylamide gradient gels of polypeptides associated with Con A-purified oncornaviruses (11) differs from those of conventionally purified particles in that (1), there are many more bands present than are usually assigned to these viruses, and (2) most of the polypeptides are larger, instead of smaller than the major polypeptide called P30 (20). While this may be due to differences in the cells in which the viruses are propagated, we have with some confidence, assigned most of the polypeptides to a structural role in the virions by the following criteria: (1) Purification of the Fr—MuLV P30 relative to serum albumin, the major non-viral protein present in medium, is at least 50,000-fold (11). (Antagonistic components in the medium [7] could be similarly excluded from viral isolates.) (2) Most proteins in patterns in SDS-acrylamide gels are covariant with recognizable, intact virions (Figure 2), infectivity (Figure 4) and viral 60—70S RNA (data not shown) in Con A-purified particles of density 1.16 in equilibrium gradients. A different pattern of polypeptides appears in similarly analysed, spontaneously arising endogenous particles of density 1.18 produced by BALB/c mouse cells (10,11).

Certain reservations may apply to patterns of proteins from Fr-MuLV, and BALB/c endogenous viruses and they may apply to other viruses purifiable with Con A. Some radiolabelled bands of apparent molecular weight greater than 170,000 (four out of 24, and the same four out of 28 bands, for Fr-MuLV and endogenous virus respectively) are found continuously from supernatant to virus-containing fractions of equilibrium gradients as though they are loosely bound components of the virions which are washed from the particles during centrifugation. This is quite possible, since we, and others (21) have found that the viral surface knobs, at least, appear to be quite loosely associated with the viral envelope. However, by labelling with ^{14}C-glucosamine, we find that two of these bands are glycopolypeptides. By staining in gels, virus fractions (11) contain residual amounts of Con A that in this case, are very difficult to remove from the virus (of the order of 1 μg/10^{10} particles). While this does not interfere with infectivity of the virus, it may promote weakly adventitious association of non-viral glycoproteins with the outer surfaces of the virus, and caution should thus be observed in assignment of polypeptides with similar behaviour in equilibrium gradients containing virions.

7. APPLICATION OF THIS METHOD TO OTHER VIRUSES

In addition to Friend Leukaemia virus and BALB/c endogenous C-type particles produced by cells in culture, we have purified Rauscher leukaemia virus, and Friend virus produced by leukaemic cell lines transformed in the mouse (unpublished data). It is also possible to isolate Avian Myeloblastosis virus (an avian oncorna-virus) directly from whole serum. By analysing the viral 60—70S viral RNA content of Con A complexes we have monitored newly infected cells for new virus production and we have been able to pinpoint the time of production of spontaneous endogenous viruses by mouse cells during serial passage in culture (10). Recently, others have used this method to detect production of oncornavirus-like particles by cultures of marrow cells taken from leukaemic patients (22).

Every enveloped virus that contains appropriate Con A receptors is potentially purifiable by this method. Except for Picornaviruses and Reoviruses this may include all of the RNA viruses and one important group of sometimes oncogenic DNA viruses, the Herpes viruses. Probably the best indication that native virions can be purified by this method is by visualization of ferritin-, or otherwise conjugated Con A bound to extracellular virus in infected cultures as is shown in Figure 1. It follows that Influenza virus (23) should be suitably purified by this method. Some enveloped viruses may not be agglutinable with Con A if Con A receptors are sequestered as was originally proposed for cells (24). Agglutination of such viruses may occur after some degradation of the viral envelope. Purified viruses that are agglutinated by Con A are Fowl Plague and Influenza viruses, New Castle Disease virus and SV5 virus, Sindbis and Semliki Forest virus, apparently VSV (25—27) and Rauscher leukaemia and mammary tumour viruses (28).

Infectivity may be variably recoverable after isolation of virus with Con A. This may depend on effective removal of Con A by whichever competitive inhibitor is used for solubilization of Con A—virus complexes. Added directly to some virus preparations, Con A appears to reduce infectivity (26). In the case of Japanese Encephalitis virus (JEV) (29) this effect is ascribed to aggregation of virions rather than to virus neutralization. Infectivity is recoverable in JEV, Semliki Forest virus (25), Herpes virus and Sendai virus (30) and, of course, Friend leukaemia virus (Figure 3, pellet IV) on dissociation of Con A—virus complexes with α-methyl mannose. The 1 μg of Con A which remains associated with 10^{10} purified Fr-MuLV particles (11) apparently does not alter infectivity of the virus (Figure 4). Nevertheless, minimal amounts of Con A necessary to quantitatively agglutinate virus should be used. Recently, Con A has been used to purify infectious measles and Subacute Panencephalitis virus (SSPE) from culture medium (31). Fifty micrograms of Con A per millilitre of medium will quantita-tively agglutinate measles virus, but greater amounts of Con A apparently reduce infectivity in purified isolates (B.N. Fields, personal communication).

It cannot be assumed that all infectious particles produced by one culture are agglutinable by Con A. We have found that medium incubated with early passage Friend virus-infected 3T3 mouse cells contains infectious leukaemia virus that is *not* agglutinable by Con A (Figure 4). Thus Con A revealed a previously unsuspected heterogeneity in infectious particles with respect to inaccessible, or non-existent Con A receptors on the surfaces of some particles.

Because little is known about the molecular organization of many virions, it is probably easier to try this simple method of purification than to try to predict its suitability for your particular virus system. Whether it works or not, *concanavalin A* may *tell* you something about the molecular organization of your virus.

8. REFERENCES

1. Tooze, J. (1973) 'The RNA tumor viruses: Morphology, composition and classification', in *The Molecular Biology of Tumor Viruses*, Cold Spring Harbor Monograph Series, pp. 502–66.
2. Fenner, F., McAuslan, B. R., Mims, C. A., Sambrook, J. and White, D. O. eds. (1974) (a) 'The nature and classification of animal viruses', pp. 1–33, (b) 'The multiplication of RNA viruses', in *The Biology of Animal Viruses*. 1974 (Academic Press, New York and London), pp. 221–73.
3. Shanmugan, G., Vecchio, G., Attardi, D. and Green, M. (1972) 'Immunological studies on viral polypeptide synthesis in cell replicating murine-sarcoma-leukemia viruses', *J. Virol.*, 10, 447–55.
4. Levintow, L. and Darnell, J. E. (1960) 'A simplified procedure for purification of large amounts of polio virus: Characterization and amino acid analysis of type I poliovirus', *J. Biol. Chem.*, 235, 70–3.
5. Mudd, J. A. and Summers, D. F. (1970) 'Protein synthesis in VSV-infected Hela cells', *Virology*, 42, 328–40.
6. Compans, R. W., Klenk, H. D., Caliguiri, L. A. and Choppin, P. W. (1970) 'Influenza virus proteins. I. Analysis of polypeptides of the virion and indentification of spike glyco-proteins', *Virology*, 42, 880–9.
7. McClain, K. and Kirsten, W. H. (1972) 'Electrophoretic analysis of the RNA from a mouse leukemia virus', *Cancer Res.*, 32, 1470–5.
8. Smith, R. E. (1974) 'High specific infectivity avian RNA tumor viruses', *Virology*, 60, 543–7.
9. Stewart, M. L., Summers, D. F., Soeiro, R., Fields, B. N. and Maizel, J. V., Jr. (1970) 'Purification of oncornaviruses by agglutination with concanavalin A', *Proc. Nat. Acad. Sci. U.S.A.*, 70, 1308–12.
10. Stewart, M. L. and Maizel, J. V., Jr. (1975a) 'Friend leukemia and endogenous viruses. I. Production of Friend and endogenous C-type particles by mouse cells cultured *in vitro*', *Virology*, 65, 55–66.
11. Stewart, M. L. and Maizel, J. V., Jr. (1975b) 'Friend leukemia and endogenous viruses. II. Proteins of Friend leukemia and mouse endogenous C-type particles produced by BALB/c mouse cell cultures', *Virology*, 65, 67–76.
12. Stewart M. L. and Maizel, J. V., Jr. (1974) 'Rapid analysis of oncornaviral RNA employing agglutination of virions with concanavalin A', *Virology*, 59, 595–9.
13. Steeves, R. A. (1975) 'Spleen focus forming virus in Friend and Rauscher leukemia virus preparations', *J. Nat. Cancer Inst.*, 54, 289–97.
14. Bassin, R. H., Tuttle, N. and Fischinger, P. J. (1971) 'Rapid cell culture assay technique for murine leukemia viruses', *Nature*, 229, 564–6.
15. Edmundson, A. B., Ely, K. R., Sly, D. A., Westholm, F. A., Powers, D. A. and Liener, I. E. (1971) 'Isolation and characterization of concanavalin A polypeptide chains', *Biochemistry*, 10, 3554–9.
16. Duesberg, P. (1968) 'Physical properties of Rous sarcoma virus RNA', *Proc. Nat. Acad. Sci. U.S.A.*, 60, 1511–7.

17. Cheung, K-S., Smith, R. E., Stone, M. P. and Joklik, W. K. (1972) 'Comparison of immature (rapid harvest) and mature Rous sarcoma virus particles', *Virology*, 50, 851—64.
18. Sarkar, N. H. Nowinski, R. C. and Moore, D. H. (1971) 'Helical nucleocapsid structure of the oncogenic ribonucleic acid viruses (oncornaviruses)', *J. Virol.*, 8, 564—72.
19. Luftig, R. B. and Kilham, S. S. (1971) 'An electron microscope study of Rauscher leukemia virus', *Virology*, 46, 277—9.
20. August, J. T., Bolognesi, D. P., Fleissner, E., Gilden, R. V. and Nowinski, R. C. (1974) 'A proposed nomenclature for the virion proteins of oncogenic RNA viruses', *Virology*, 60, 595—601.
21. Nermut, M. V., Frank, H. and Schäfer, W. (1972) 'Properties of mouse leukemia viruses. III. Electron microscopic appearance as revealed after conventional preparation techniques as well as freeze drying and freeze etching', *Virology*, 49, 345—58.
22. Mak, T. W., Kurtz, S., Manaster, J. and Housman, D. (1975) 'Viral-related information on oncornavirus-like particles isolated from cultures of marrow cells from leukemia patients in relapse and remission', *Proc. Nat. Acad. Sci. U.S.A.*, 72, 623—7.
23. Klein, P. A. and Adams, W. R. (1972) 'Location of ferritin-labelled concanavalin A binding to influenza virus and tumor cell surfaces', *J. Virol.*, 10, 844—54.
24. Burger, M. M. (1969) 'A difference in the architecture of the surface membrane of normal and virally transformed cells', *Proc. Nat. Acad. Sci. U.S.A.*, 62, 944—1002.
25. Oram, J. D., Ellwood, D. C., Appleyard, G. and Stanely, J. L. (1971) 'Agglutination of an arbovirus by concanavalin A', *Nature (New Biol.)*, 233, 50—1.
26. Becht, H., Rott, R. and Klenk, H-D. (1972) 'Effect of concanavalin A on cells infected with enveloped RNA viruses', *J. Gen. Virol.*, 14, 1—8.
27. Birdwell, C. R. and Strauss, J. H. (1973) 'Agglutination of Sindbis virus by plant lectins', *J. Virol.*, 11, 502—7.
28. Calafat, J. and Hageman, P. C. (1971) 'Binding of concanavalin A to the envelope of two murine RNA tumor viruses', *J. Gen. Virol.*, 14, 103—6.
29. Yoshinaka, Y. and Shiomi, T. (1975) 'Agglutination of Japanese encephalitis virus with concanavalin A', *J. Virol.*, 15, 671—4.
30. Okada, Y. and Kim. J. (1972) 'Interaction of concanavalin A with enveloped viruses and host cells', *Virology*, 50, 507—15.
31. Miller, C. A. and Fields, B. N. (1975) 'Measles and subacute sclerosing panencephalitis (SSPE) viruses: Characterization of purified particles', *Abstr. Ann. Meeting Amer. Soc. Microbiol.*, Paper S191, p. 245.

Section VI:

BIOLOGICAL APPLICATIONS

CHAPTER 51

Lymphocyte Stimulation by Concanavalin A

JAN ANDERSSON
FRITZ MELCHERS

When the immune system encounters antigen, reactions along either of two different pathways are possible. Antigen may induce a specific immune response (immunogenic action of the antigen), or its reaction with the immune system may lead to specific unresponsiveness (tolerogenic, paralytic action of the antigen). Either reaction requires as a first step, the binding of antigen to specific receptors located on the surface membrane of lymphocytes.

Two major classes of lymphocytes, thymus-derived (T-) and bursa equivalent (B-) cells cooperate in such responses (1). Since both T- and B-cells recognize antigen specifically both must possess antigen—specific receptor structures. On B-cells, these receptors are immunoglobulin molecules, predominantly of μ-heavy-chain class (2). The nature of the receptor molecules on T-cells is not known at present.

A first encounter of T- and B-cells with antigen can lead to an increased specific responsiveness or to specific unresponsiveness for a second challenge with the same antigen (3). Increased responsiveness is most certainly a result of clonal expansion of antigen-reactive T- and B-cells which show positive cooperation for an immune response. Unresponsiveness may result from the absence or elimination of reactive cells (clonal elimination) (4). It may also emerge from the presence or clonal expansion of unreactive or suppressive cells which show specific negative cooperation (suppression) for an immune response (5).

It appears that B-cells only have the choice to be or not to be stimulated to proliferation and differentiation into immunoglobulin-secreting plasma cells making them the site of humoral antibody production (6). The magnitude of this antibody production, i.e. the number of antibody secreting plasma cells, is usually measured as the magnitude of an immune response. T-cells either amplify or suppress the antibody (B-cell) response and thus regulate the magnitude of such an immune response. It is likely that different stages of differentiation of the T-cell lineage perform these positive or negative regulatory functions on B-cells, and that B-cells have to reach a certain stage in their differentiation to become regulatable by such T-cells (7).

Mitogens stimulate lymphocytes to polyclonal proliferation and subsequent maturation (8). Soluble Con A stimulates T-cells to synthesize DNA, divide and differentiate into effector cells capable of helping or suppressing a B-cell response to antigen. As reported in this volume by Bevan (Chapter 52), Con A also induces T-cells to perform as killer cells in a cytotoxic reaction. Insolubilized Con A (c-Con A) (9,10) or insolubilized phytohaemagglutinin (11), as well as bacterial lipopolysaccharide (LPS) (12), purified protein derivative (PPD) from tuberculin bacteria (13) and components of the outer membrane of bacteria such as the lipoprotein of the outer membrane of E. coli (14) all stimulate B-cells to synthesize DNA, to proliferate and to differentiate into plasma cells, secreting large amounts of immunoglobulins.

Mitogens, therefore, mimic the action of antigens on lymphocytes. While, however, an antigenic determinant stimulates only a very small proportion of all lymphocytes, mitogens stimulate a large number of them. It is likely that different mitogens distinguish between different stages in the differentiation of T- and B-cells (15). Of a given type of lymphocyte, however, all appear to be stimulated by a given mitogen. Mitogen-stimulated lymphocytes are therefore useful for bio-chemical studies of the cellular and molecular events connected with lymphocyte stimulation since changes in a large number of cells in a population allow for the biochemical monitoring of such changes. Furthermore, they may let us distinguish between antigen-dependent (specific) and antigen-independent (unspecific) events of lymphocyte stimulation.

It is the purpose of this article to delineate the general features of T- and B-lymphocyte stimulation by Con A, and to discuss some of the cellular and molecular processes involved in proliferation and maturation of lymphocytes and thus in the immune response.

1. TECHNIQUES FOR STUDYING MITOGENIC STIMULATION OF LYMPHOCYTES

1.1 Cell Sources

Practically every lymphoid organ at almost all times of prenatal and postnatal life consists of a mixture of lymphocytes at different stages in their differentiation. Certain organs, at particular times in ontogeny, are sites where particular sub-populations may be enriched. Thus the thymus appears essentially devoid of B-cells and is therefore regarded as a source of pure T-cells. Spleen and lymph nodes from normal mice constitute a mixture of T- and B-cells and it appears that the T-cells in the spleen represent subpopulations which are different from those which are enriched in the thymus. As stated in the introduction, it appears that different functions of T-cells, such as helper-, suppressor-, killer- and delayed-type hyper-sensitivity mediator-functions, are properties of different T-cell subpopulations (7). These different subpopulations may change during life. They may be stimulated by different mitogens (7). Genetically athymic 'nude' mice lack the thymus and are devoid of the differentiated, functional forms of T-cells. In such mice spleen and lymph nodes consist of the different stages of B-cell development and possibly of precursor T-cells (16). Different stages in the B-cell development appear to be stimulated by different mitogens (15). During life the composition of the lymphoid organs, i.e. their content of cells in the various stages of B-cell development, changes. Mitogen-reactive B-cells arise at different times during ontogeny (15), are fully developed only 2 to 3 weeks after birth and decline thereafter. One year after birth very few mitogen-reactive B-cells are left in the spleen (Melchers, unpublished). We have therefore used thymus, spleen and lymph nodes of mice between 3 weeks and 3 months old, at an age when they possess a maximal number of mitogen-reactive lymphocytes. Lymphocyte heterogeneity will often make interpretations of experimental results obtained with a certain population of cells difficult, and sometimes impossible, as we will point out below.

It is obvious that at least partial purification of different lymphocyte popul-ations, such as T- and B-cells, is necessary. Mice were chosen as a species of experimental animals because murine T- and B-lymphocytes can be distinguished and can also be partially purified from each other.

1.1.1 B-Cells

Three methods are commonly used to obtain T-cell-depleted B-cell populations:

a. Spleens from genetically athymic mice lacking functional T-cells. They are either bred in the individual laboratories or can be purchased from Gr. Bomholtgård Ltd., Rye, Denmark.

b. Since T- but not B-cells display the θ-alloantigen on their surface mem-brane, T-cells can be lysed by alloantibodies directed against the θ-antigen in the presence of complement. This yields spleen and lymph node cell populations which are predominantly B-cells.

c. Adult mice are thymectomized, irradiated and 'reconstituted' with syngeneic bone marrow cells previously treated with anti-θ-serum and complement. After 6 to 8 weeks spleens from such animals contain predominantly B-lymphocytes.

1.1.2 T-Cells

Since only B-, but not T-cells contain surface-bound immunoglobulin of high density and slow turnover, T-cells can be enriched from normal spleen cells by passage of the cell suspension through columns of 'Sephadex' G-100 coated with rabbit anti-mouse immunoglobulin antibodies (17).

1.2 Preparation of Cell Suspensions

Mice are killed, dipped in 70% alcohol and the lymphoid organs are immediately taken out with sterile instruments. The organs are placed in sterile plastic petri dishes containing 10 ml balanced salt solution (BSS) (18), teased into a single cell suspension with two forceps and slowly pipetted up and down. The cell suspension is transferred to a 50 ml conical plastic tube (Falcon Plastics No. 2074) and additional 40 ml BSS are added. The cells are spun (in the cold) in a clinical table centrifuge with a swinging bucket head 1000 r.p.m. for 20 min. The supernatant is removed. The cells are carefully resuspended in BSS with a 10 ml pipette. This washing procedure is repeated once. The final cell pellet is resuspended in tissue culture medium and the number of living cells counted in a Bürker haematocyto-meter after 1/10 dilution of the cell suspension in trypan blue solution (0.01% in PBS). Washing of cells with large volumes of BSS is important since it removes suppressive components which have been found also in mouse serum and which inhibit the induction of T- and B-cells (19,20).

1.3 Lymphoid Cell Culture Techniques

1.3.1 Serum-free Cultures

Most batches of foetal calf serum (FCS) stimulate both T- and B-cells to proliferation and maturation (19–21). Therefore tissue culture systems were designed in which lymphocytes could be cultured in the absence of serum for up to 5 days (21). This reduces the influence of unknown serum factors affecting the subsequent mitogen-induced response. Most important for good mitogenic responses in serum-free cultures of mouse cells appears to be the quality of the water used in the preparation of BSS and of the tissue culture medium. In serum-containing cultures, whatever is toxic for the lymphocytes is 'buffered' away by low concentrations of FCS (0.2–0.5%). The best medium for serum-free cultures, in our hands, is fresh RPMI 1640 obtained from Microbiological Assoc-iates (Bethesda, Md., USA No. 12.126). The medium should not be older than 3 months. It is supplemented with 2 mM L-glutamine, 50 U/ml each of penicillin and streptomycin, 10 mM HEPES (N-2-hydroxyethylpiperazine-N'-2-ethanesulphonic acid) all from Microbiological Associates. Furthermore, 2-mercaptoethanol is added to a final concentration of 5×10^{-5} M.

1.3.2. Serum-containing Cultures

Cells are cultured in the medium described above (1.3.1) but supplemented with FCS. Different batches of FCS have different mitogenic effects on T- and B-lymphocytes. Consequently each new 'batch' of serum has to be tested for its intrinsic mitogenic activity with lymphocytes from different sources. In splenocytes DNA synthesis is optimally stimulated at 0.2% to 0.5% FCS, if batches of FCS from Gibco (Grand Island Biological Co.) are used. The capacity of FCS to stimulate the development into plaque-forming cells varies much more from batch to batch. Usually this 'background activity' of FCS is not desired, thus batches of FCS are selected which develop less than 25 PFC/10^6 cultured cells (for the assay see 1.6). This PFC development usually shows a dose optimum between 3% and 5% FCS. The two different dose optima for induction of DNA synthesis and for the development of PFC (22) in splenocytes are taken as an indication for different cell populations which proliferate or which mature to PFC. It is evident that the interpretation of experimental data obtained on lymphocyte stimulation by Con A will be complicated by the presence of serum in the culture medium.

1.3.3 Important Variables

'*In vitro*' mitogenic responses of lymphocytes are dependent on cell density. This is illustrated by a simple experiment, shown in Figure 1, where normal murine splenic lymphocytes are stimulated at different cell densities with a constant dose of the B-cell mitogen LPS in serum-free and in 10% FCS-containing culture medium. It is evident that different cell densities give an optimal PFC development at different times after stimulation. Thus, 5×10^6 cells/ml give an optimal response at day 3: the response in serum-free medium is one fifth of that in serum-containing cultures. On the other hand, 5×10^5 cells/ml give an optimal response at day 5. The response in serum-free medium is not significantly above background; that in FCS-containing medium is as high as the one with 10 times more cells (5×10^6 cells/ml) in the medium. The data in Figure 1 can thus be interpreted as resulting from the stimulation of two populations of B-cells in the spleen (15): one does not need serum and is frequent enough in adult spleen to be detected at

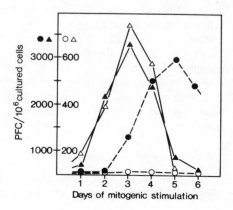

Figure 1. Dependence of the development of PFC after mitogenic stimulation on cell density. Splenocytes of C57Bl/6J × DBA/2J F$_1$ mice were cultured with 50 µg/ml LPS at 5×10^6 cells/ml (triangles) or at 5×10^5 cells/ml (circles) in serum-free (open symbols) or in serum-containing (closed symbols) medium. PFC were assayed with TNP-SRBC as described in paragraph 1.6

5×10^6 cells/ml, but not at 5×10^5 cells/ml. The other may be too frequent at 5×10^6 cells/ml and may not find enough mitogen or nutrients to be optimally stimulated. It has to be diluted to 5×10^5 total cells to be optimally stimulated.

This example is meant to illustrate how important it is to measure stimulation at different time intervals and with different cell densities. We will outline the importance of the dose of mitogen below (see paragraph 3). It could ideally be expected that a linear dependence existed between cell dose and the degree of stimulation (for quantitative measurement of the degree of stimulation, see 1.5 and 1.6). Lymphocyte heterogeneity in every lymphoid organ (outlined in 1.1) will in practice lead to a superimposition of effects on different cell populations, giving apparent non-linear dose dependencies of the degree of stimulation on cell density. Since this lymphocyte heterogeneity will be different in the various lymphoid organs at different times during life and under different physiological conditions, only qualitative interpretations of the results, but no quantitative treatment of the data will be possible.

In our studies with Con A, lymphocytes are either cultured at $5 \times 10^6 - 10^7$ cells per ml in tissue culture petri dishes (Falcon Plastics No. 3054) or at 1×10^6 cells in 0.5 ml tubes (Falcon Plastics No. 2954) in a $37°$ incubator supplied with 5% CO_2 in air.

1.4 Preparation and Standardization of Soluble and Insolublized Con A

Concanavalin A is purchased from Miles Yeda Ltd. and purified on 'Sephadex' G—100 as described by Agrawal and Goldstein (23). After extensive dialysis against distilled water the Con A solution is subjected to centrifugation at 15,000 r.p.m. for 90 min in an SS34 rotor of the Sorvall RC 2B centrifuge. The supernatant is lyophilized and the dry Con A is stored at $-20°$. Solutions of the lyophilized Con A are made fresh every day. Each new batch of Con A is tested for mitogenicity in the standard test system (see below) at twofold dilutions in the concentration range of $0.01-40$ μg/ml. For every change, such as cell source, tissue culture medium or batch of foetal calf serum introduced in the standard test system a new titration curve is made.

Insolublized Con A (c-Con A) is prepared by adding graded amounts $(0.1-10$ mg) of Con A in sterile $\overline{0.1}$ M carbonate buffer, pH 8.5 containing α-methyl-D-mannoside, 10 moles in excess over Con A to 1 gram of cyanogen bromide-activated (24) 'Sepharose' 4B or 'Sephadex' G—100 beads. The coupling is performed over night at $4°$ under very gentle agitation which leaves the beads intact. The beads are then washed ten times with a tenfold excess of sterile phosphate-buffered saline, pH 7.2, removing the wash buffer from the beads each time by centrifugation for 5 min at 3000 r.p.m.

Lymphocytes are cultured with $1/10-1/1000$ dilution of the packed c-Con A bead material to determine the amount of c-Con A beads giving optimal mitogenicity.

1.5 Determination of DNA Synthesis

Incorporation of ^{3}H-thymidine (^{3}H-TdR) into lymphocytes is performed by adding 5 μCi ^{3}H-TdR (sp. act. 5 Ci/mmol, Amersham, England) in 0.1 ml tissue culture medium supplemented with 10^{-4} M non-radioactive thymidine (Sigma Chemical Co.) to 1 ml cultures. The culture is pulsed for 1—4 hours and the cells harvested and washed with 6—10 ml cold BSS containing 1 mg/ml unlabelled thymidine. The supernatant after centrifugation (10 min at 2500 r.p.m) is decanted and the cell pellet resuspended in one drop of a 1% protein solution (gamma globulin) on a Vortex mixer. Five ml of 10% trichloroacetic acid (TCA) is added and the resulting precipitate spun at 4000 r.p.m. for 30 min. The supernatant is decanted. The pellet is dissolved in 0.5 ml 0.2 M KOH (15 min at $37°$). Five ml 10% TCA is added to the solution and the precipitate spun again (as above). The pellet is dissolved in 0.5 ml of 0.2 M KOH, the samples are transferred to scintillation vials and counted in 5 ml of a dioxan-based thixotropic gel ('Cab-O-Sil', Packard Instuments Co.) scintillator fluid.

Studies using thymidine incorporation as the only measure for mitogenic stimulation may lead to gross misinterpretations. This is not only so because of the above mentioned heterogeneity in T- and B-lymphocyte populations (see 1.1) but also because of the way in which lymphocytes use thymidine to incorporate it into DNA. First, lymphocytes are not auxotrophic for thymidine: an internal pool of TTP exists which is fed from orotic acid via UTP. This internal TTP pool is used for DNA synthesis, the precursor UTP pool for RNA synthesis. Thymidine, given in the medium, has first to enter the cell. It uses an unknown transport system and is triphosphorylated in the cell by an enyzmatic pathway different from that used for the production of endogenous TTP. All the precursor pools, transport and enzymatic activities could be different in different lymphocyte subpopulations, and the various mitogens, at different doses, may effect different changes. Furthermore this may be different in different inbred strains of mice. In view of this complexity of thymidine uptake into DNA, mitogenic activation measured by thymidine uptake may become uninterpretable, if additional information by other experimental approaches is not sought.

1.6 Determination of the Number of Plaque-forming Cells (PFC)

Sheep red blood cells (SRBC) bind SRBC-specific antibodies. The SRBC-bound antibodies (but not the free) can bind and activate complement. Activated complement, a complex mixture of lytic enzymes and cofactors (complement is usually available in high concentrations in guinea pig serum), will attack the red cell membrane and lyse the cell. An antibody-producing and -secreting cell with specificity against SRBC, immersed in a layer of SRBC fixed in agar, will lyse the SRBC layer around it, when complement is added.

Trinitrophenyl groups (TNP) can be coupled to SRBC in a chemical reaction described below. It can be coupled in different densities of TNP groups on the red cells, depending on the amount of coupling reagent (see below) added per red cell. A *densely* coupled TNP-SRBC will bind immunoglobulin molecules of *many different* specificities. This is thought to be so, since TNP groups are generally hydrophobic, smaller than the total site on the antibody molecule binding antigen and are existing in such high densities on the coupled SRBC that multivalent binding of antibodies with very low affinities can take place. This multivalent binding is particularly favoured by decavalent 19S IgM molecules forming direct haemolytic plaques. Thus, highly coupled TNP-SRBC detect a broad spectrum of antibody-secreting PFC's with *many* different specificities arising from many different precursor cells (polyclonal detection), while uncoupled SRBC detect only antibody-producing cells with high affinity for SRBC ('specific', 'mono'clonal detection).

1.6.1 Coupling of TNP groups to SRBC

In principle the original technique by Rittenberg and Pratt is followed (25). Two ml of packed SRBC in Alsevier's solution are washed 3 times with 0.28 M Na cacodylate buffer, pH 7.0 and finally suspended in 10 ml of this buffer. Under stirring any amount between 0.2 mg and 80 mg 2,4,6-trinitrobenzenesulphonic acid (TNBS) (Sigma Chemical Co.), dissolved in 2 ml cacodylate buffer are added dropwise (30 seconds) at room temperature. The cells are stirred for another 10 min, then washed six times with BSS.

Figure 2 shows the relationship between the amount of TNP coupled to SRBC

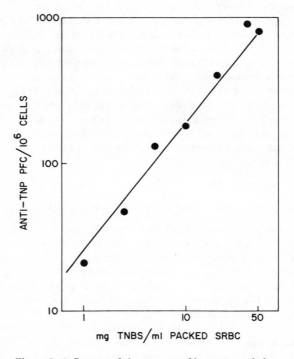

Figure 2. Influence of the amount of hapten coupled to indicator red cells for the development of plaque-forming cells in the haemolytic plaque assay. Mouse spleen cells stimulated *in vitro* for 3 days by the B-cell mitogen LPS were assayed for anti-TNP plaque-forming cells (PFC) using sheep red blood cells (SRBC) coupled with different amounts of 2, 4, 6-trinitrobenzene-sulphonic acid (TNBS)

and the number of PFC detected within one and the same population of mitogen-stimulated B-cells. It shows that higher epitope densities detect more PFC and thus illustrates that densely coupled TNP-SRBC can be used for polyclonal detection of immunoglobulin-synthesizing and -secreting cells.

Caution: Sheep cells from many donors cannot withstand the coupling of 40 mg TNBS per ml packed SRBC and tend to lyse either during the washing procedure after coupling or in the assay. Blood from different sheep, therefore, has to be screened for its resistance to the TNP-coupling reaction. It should also be emphasized that not all sheep red cells will be lysed equally well by antibodies plus complement. Selection should also be made for this.

1.6.2 Plaque Assay

This assay was developed by W. W. Bullock (19). One gram Agar (Difco) is disolved in 100 ml BSS by heating it up to boiling 3 times (take off the flame, when the foam starts coming up). Put the hot agar-BSS solution into a 43° water bath. When the solution has cooled to 43°, add under stirring or shaking 1.5 ml DEAE−dextran (Pharmacia, Sweden) 50 mg/ml in PBS, pH 7.2. The agar turns turbid. DEAE−dextran neutralizes charged groups in the agar, which otherwise would be anti-complementary. Dispense 0.6 ml amounts of the agar into 43° prewarmed tubes with a prewarmed pipette.

a) Add 50 µl 1:2−1:5 diluted packed TNP-SRBC (dilute packed TNP-SRBC in BSS). Dilution factor is determined visually so that the suspension of SRBC in agar is neither too dilute nor too concentrated to allow for easily visible plaques. This dilution factor has to be determined for each sheep blood used in the assay. Now work fast!

b) Add 200 µl lymphocytes at convenient dilution. (A culture of 5×10^6 original lymphocytes stimulated for 3 days by LPS is normally resuspended in 0.5−1.0 ml BSS and kept on ice!)

c) Add 50 µl of 1:2−1:5 diluted guinea pig complement (Difco) (diluted in BSS) (Optimal dilution is determined for each batch of complement by assaying for the number of PFC developing with a given mitogen-stimulated B- lymphocyte suspension.) Shake on Vortex; take up in 43°-prewarmed 1 ml pipette. Dispense in three 0.2 ml aliquots on a plastic petri dish (marked with the appropriate number of the lymphocyte culture. A plastic petri dish not repelling water should be used!) Streak out quickly with the pipette into 3 round areas. Let stand for *one* minute to solidify. Put the top of the petri dish on to prevent drying out. Incubate for 2 hours at 37° in a *moist* incubator. Count plaques. Dishes containing the plaques can be stored at 4° for up to 2 days without changes in the morphology or number of plaques on the plate.

The number of direct PFC in a mitogen-stimulated B-cell suspension assayed with this method correlates well with the extent of IgM secretion measured by biosynthetic labelling of such cell suspension with tritiated leucine followed by serological detection of the labelled IgM (10,15,22,26−28). The plaque assay employing densely coupled TNP-SRBC, therefore, is a fast, reliable, quantitative assay for polyclonal activation of B-cells by mitogens to differentiation into Ig-secreting cells.

2. TYPICAL RESPONSES OF LYMPHOCYTES TO CON A

2.1 Stimulation of T-Cells by Soluble Con A:

T-Cells from various sources such as thymus cells, cortisone-resistant thymus cells, normal spleen cells, normal spleen cells passed over columns of 'Sephadex' G-100 coupled with rabbit anti-mouse Ig (17), all were induced by soluble Con A to increased rates of thymidine uptake. On the other hand, purified B-cells, in the absence of T-cells, such as normal mouse spleen cells treated with anti-θ-serum and complement, spleen cells from 'nude' athymic mice or spleen cells from thymecto-mized, irradiated and bone marrow reconstituted mice did not respond to soluble Con A by increased rates of thymidine incorporation (29).

The response to Con A, measured in Figure 3 with normal spleen cells containing T- and B-cells, occurred within a narrow concentration range. In the absence of serum the maximum response was found at 0.3 μg Con A/ml (Figure 3, top). In FCS-containing medium there was comparable thymidine uptake in the same

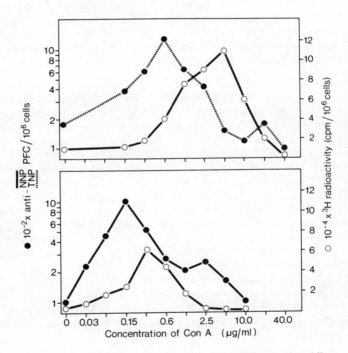

Figure 3. The influence of Con A concentration on thymidine uptake ($\circ$) and development of immunoglobulin-synthesizing and secreting cells ($\bullet$) in normal mouse spleen cells. Top: in serum-free medium. Bottom: in medium supplemented with 5% FCS. The assays were done as described in paragraphs 1.5 and 1.6, except that for the serum-free cultures PFC were determined with SRBC densely coupled with a different hapten; NNP (4-hydroxy-3,5-dinitrophenacetyl-), at 120 hours of mitogenic stimulation

concentration range, the maximum response, however, now was shifted to 5 µg Con A/ml (Figure 3, bottom). We take these results as a first indication that more cells could be activated in FCS-containing medium (at higher Con A concentrations), than were activated in serum-free medium (at the lower Con A concentrations).

Data in Figure 3 illustrate:

a) How important it is to always test a series of concentrations of Con A in order not to miss the narrow peak of maximum response.

b) The influence of foetal calf serum on the induction of thymidine uptake in lymphocytes by Con A. There may be additive effects of Con A + FCS on the same lymphocyte population. This could lead to a higher rate of thymidine uptake into DNA. The fact that it occurs at *higher* concentrations of Con A may be due to Con A binding to serum components. Alternatively the data could indicate heterogeneity in the resonding lymphocytes which either do or do not require serum to be stimulated to thymidine uptake.

2.2 Stimulation of B-Cells by Soluble Con A

When a mixture of T- and B-cells, such as normal spleen cells, were treated with Con A both types of cells were activated (21,30). This led to the development in these Con A-stimulated spleen cells of immunoglobulin-secreting PFC (Figure 3). Since B-cell populations devoid of T-cells, however, are not activated, and since T-cells devoid of B-cells are activated to thymidine uptake only (see above) we must conclude that activation of T-cells by Con A led to a subsequent activation of B-cells which is T-cell dependent.

From the data in Figure 3 it is evident that low doses of Con A activated spleen cells to PFC development. These doses were clearly lower than the doses needed for a maximal response of thymidine uptake both in serum-free and in FCS-containing media. If T-cells were activated by Con A to produce factors which subsequently, and without the action of Con A, activate B-cells, then such a T-helper-cell subpopulation could be activated maximally at these low concentrations. If, however, soluble Con A was again needed in the subsequent activation of B-cells, then this low concentration of Con A could have merely represented the *optimal* concentration at which this cooperation of T- and B-cells occurred, which led to development of PFC (B-cells).

Thymidine uptake began to increase 16–18 hours after stimulation with Con A. The B-cell response, measured by the appearance of PFC showed a lag period of around 50 to 60 hours. Thymidine uptake reached a maximum around 50 hours, while the PFC development did so much later, at 120 hours after stimulation. This was expected from the polyclonal growth of B-lymphocytes after stimulation, in which a period of proliferation precedes the differentiation of cells into PFC. Direct stimulation of B-cells by B-cell mitogens such as LPS or PPD, however, has shown an earlier PFC development with peak responses around 72 hours (12,13) (see Figure 1, cell density of 5×10^6 cells/ml). For these responses an average clone size of 4 to 8, representing 2 to 3 divisions before differentiation, has been measured (31). Responses in which the peaks of PFC development occur later, i.e. at 120 hours, should result from a longer period of proliferation. When normal spleen cells are stimulated by soluble Con A, a period of T-cell proliferation may be necessary and precede a subsequent period of B-cell proliferation before maturation of B-cells into PFC. This is, however, not the conclusion reached by Coutinho and coworkers (21) from experiments, in which pretreatment of spleen cells with mitomycin C prior to stimulation with Con A led to a 96% inhibition of thymidine uptake, but to unimpared PFC development. Their conclusion that proliferation is not preceding the differentiation of PFC may not be valid, if the low level of thymidine incorporation seen in Con A-stimulated, mitomycin C- treated spleen cells actually reflects a slowly dividing B-cell sub-population or if the stimulated B-cells represent only a minor portion of the total thymidine-incorporating lymphocytes. A more far-fetched interpretation of the data is that a subpopulation of B-cells is resistant to these concentrations of mitomycin C and does not transport thymidine into the cells even after induction.

2.3 Stimulation of Lymphocytes by Insolubilized Con A

It could be shown in two different experimental systems that insolubilized Con A probably activates B-cells in the absence of T-cells. Con A was either cross-linked to the bottom of tissue culture petri dishes (9) or covalently bound to 'Sepharose' 4B or 'Sephadex' G-100 beads (c-Con A, 10). In the first system only ^{3}H-TdR uptake was measured in cultures of spleen cells from nude mice or from thymectomized, irradiated and bone marrow-reconstituted mice. These results indicated that insolubilized Con A can act as B-cell mitogen, but were insufficient since no direct B-cell function, such as PFC development or immunoglobulin synthesis, had been determined. This system was also not suitable for an analysis which distinguishes direct B-cell activation by insolubilized Con A from indirect B-cell activation by soluble Con A, since a considerable leakage of Con A (up to 10% of the initially bound material after 72 hours in tissue culture) occurred during the culture period (32) With Con A—'Sepharose' beads (c-Con A) B-cells could be activated to increased immunoglobulin synthesis and secretion. A number of difficulties were, however, encountered:

a) Batch to batch variations in c-Con A preparations are considerable. The parameters for obtaining optimal mitogenicity have, however, never been determined.

b) It is difficult to avoid bacterial contaminations in the bead material which subsequently give rise to infections in the tissue cultures.

c) The easiest assay for induction of B-cells — the plaque assay (described in paragraph 1.6) — cannot be used with c-Con A since 'Sepharose' beads are plated with the cells in the plaque assay. The beads give the appearance of plaques. Incorporation of radioactive precursors and serological determination of immuno-globulin synthesis and secretion, therefore, must always be done as a reliable B-cell assay.

d) In the light of recent experiments with more defined B-cell mitogens it is not clear whether stimulation of B-cells by c-Con A is the result of a direct action of insolubilized lectin on B-cells. It may also be that c-Con A renders B-cells more susceptible to factors present in serum or produced by other cells, or to the agarose or dextran backbone of the bead material, all of which could act as a subsequent mitogenic signal.

Taken together these considerations makes c-Con A, at the present time, a less suitable 'B-cell' mitogen. For studies on the stimulation of B-lymphocytes mitogens more defined in their action on B-cells are recommended.

2.4 Suppression of B-Cell response by Soluble Con A

It was observed that concentrations of Con A which lead to maximal thymidine uptake in FCS-containing media (5 μg/ml; see Figure 3, bottom) often resulted in a suppression of the small PFC response induced in spleen cells by FCS. This suppressive action could more clearly be demonstrated when a known strong B-cell mitogen, PPD, was added together with various doses of soluble Con A to normal spleen cells cultured in serum-free medium (Figure 4).

A partial or complete suppression of the PPD-induced B-cell PFC response was observed at concentrations of Con A (0.3–1.2 μg/ml) which were optimal for a maximal response in thymidine uptake. When the same experiment was done with spleen cells from athymic 'nude' mice lacking functional, Con A-responsive T-cells, this suppression could not be observed (Figure 5). Thus the suppression is due to the presence of T-cells in the cultures.

In parallel cultures of normal spleen cells (Figure 4) thymidine uptake was actually increased at and above concentrations of Con A in the presence of PPD which led to suppression of B-cell PFC response. Again, the synergistic effect of Con A and PPD on thymidine uptake was due to the presence of T-cells, since spleen cells from athymic 'nude' mice did not show this effect (Figure 5).

At very low concentrations of Con A (0.0015 μg/ml) an enhancement of the PPD-induced B-cell PFC response occurred (Figure 4). This lends support to the idea that synergy can exist between B-cell mitogens and Con A-induced T-cells or T-cell products at the level of B-cell precursors capable of developing into PFC (helper effect of T-cells). Since FCS also is a B-cell mitogen (19,22), this then explains the more pronounced difference in dose optima for Con A-induced T- and B-cell responses in the presence of foetal calf serum as compared to that observed in serum-free cultures (see Figure 3, top and bottom).

The inhibition of B-cell PFC responses by Con A-activated T-cells strikingly resembles inhibition of PFC response to B-cell mitogens by anti-immunoglobulin antibodies (33). Thus when the addition of Con A was delayed for 24 hours after inactivation of the B-cell responses (induction by PPD or LPS) only very slight or no inhibition of the PFC response occurred. When Con A was added even later and at lower concentrations an enhancement of the B-cell PFC response was seen. These observations were also made in a system, where immune induction of B-cells by antigen was studied (30,34–38).

Very high concentrations of Con A (5 μg/ml and above) were inhibitory even in T-cell-depleted B-cell populations for B-cell PFC development induced by PPD (Figure 5). There was no measurable influence on thymidine uptake at this concentration in this cell system. This effect of Con A is again similar to that observed after treatment of B-cells with anti-Ig antibodies (33) and could well be

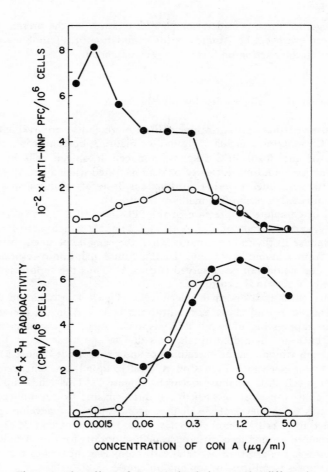

Figure 4. The effect of Con A stimulation on the differentiation of B-cells induced by a direct B-cell mitogen in a mixed
T—B-cell population. Spleen cells from normal mice were
cultured in serum-free medium with various doses of Con A
alone (○) or together with 100 μg/ml of the B-cell mitogen,
purified protein derivative from tuberculin bacteria (PPD) (●).
Proliferation was measured as thymidine uptake after 48 hours
of culture (see paragraph 1.5; lower part of figure). Development of anti-NNP PFC was also measured at 48 hours of
culture (see paragraph 1.6; upper part of figure), since the
peak B-cell response to PPD occurs at this time. At this early
time of culture a low number of anti-NNP PFC have been
induced by Con A alone

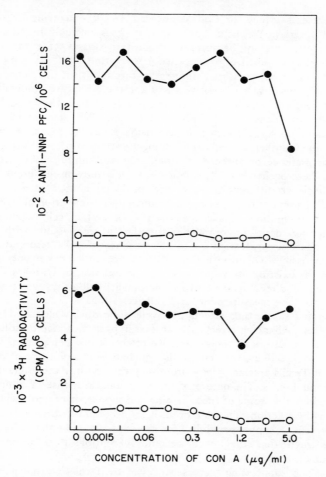

Figure 5. The effect of Con A stimulation on the differentiation of B-cells induced by a direct B-cell mitogen in a T-cell depleted B-cell population. Same experimental conditions as in Figure 4, except that spleen cells from athymic 'nude' mice were used. Note, that in the absence of Con A-reactive T-cells, only high doses of Con A are inhibitory to PPD-induced B-cell maturation

due to a direct interaction of Con A with the glycoprotein IgM (39) located in the surface membrane of mitogen-sensitive B-lymphocytes.

3. GENERAL COMMENTS

It is now well established that soluble Con A directly activates T-lymphocytes to different effector functions, such as helping and suppressing effects on B-lymphocyte stimulation, and killer and delayed-type hypersensitivity mediator functions.

Our paper presents effects of Con A-activated T-cells enhancing or suppressing B-cell PFC responses. We have outlined that it is difficult to measure T-cell activation directly, mainly because no quantitative assays for the activity of individual T-cells in positive or negative cooperation exist at present. T-Cell activation is therefore measured indirectly through their influence on B-cell performance by the only quantitative assay for the activity of individual B-cells available today, Jerne's haemolytic plaque assay (40). It is thus evident that we cannot assay how many T-cells become activated. The amount of radioactive thymidine taken up by a given cell suspension is, by no means, a quantitative measure for the number of cells responding to stimulation. Culture conditions, especially the presence or absence of stimulating and inhibiting factors in serum, cell density, heterogeneity of lymphocyte populations in different organs of different animals at different times of life, problems connected with thymidine uptake by lymphocytes and the dose dependency of stimulation all contribute to the extent of thymidine uptake measured in a certain cell population. Only qualitative measurements on mitogenic stimulation are possible with thymidine uptake; quantitative conclusions as to the number of cells responding to what extent are impossible. Quantitative conclusions are, however, erroneously drawn very often in the existing literature on Con A stimulation of lymphocytes. We are clearly in need of direct, quantitative, functional T-cell assays which will replace thymidine uptake as a measurement for T-cell activation.

Enhancing and suppressing effects of T-cells are probably mediated by different cell populations. Enhancing effects are, at least in part, mediated by soluble factors (41,42), while the nature of the suppressive effects is not yet known. In principle direct T-cell to B-cell contact could also mediate some of these effects. Con A stimulation of T-cells appears a very promising way to study the interacting factors between T- and B-cells. The nature of the polyclonal response of lymphocytes to mitogens predicts that some of these helping and suppressing factors will be antigen unspecific in their action. The role of antigen in lymphocyte stimulation remains unclear at present. A simple possibility for its action would be that it modulates the lymphocytes such that they then become susceptible to a second signal of unspecific stimulation.

Enhancing and suppressing factors conveyed by T-cells are recognized by the same precursor B-cell, the antigen- and mitogen-sensitive small, resting B-lymphocyte which is capable of proliferation and differentiation into PFC. They must therefore have a way of distinguishing between factors mediating stimulation and factors mediating inhibition for stimulation. This is reminiscent of the action of direct B-cell mitogens, such as PPD or LPS on B-cells and the interference of anti-Ig antibodies with this stimulation (33). For this situation a functional complex of structures in the surface membrane of B-cells has been postulated, which includes the antigen- and anti-Ig antibody-recognizing Ig molecules, the PPD receptors and the LPS receptors. It appears that this receptor complex can exist in two conformations: one favouring induction to stimulation, the other fixing the cell in an unstimulatable condition (suppression). Since the kinetics of induction to the suppressed state by T-cells stimulated with high doses of Con A are so similar to the direct induction of the suppressed state of B-cells by anti-Ig antibodies, it is tempting to speculate that at least suppressive factors, and possibly also enhancing factors of Con A-activated T-cells are recognized within the same functional complex of structures forming the receptor complex on B-cells which regulates their growth and differentiation.

4. ACKNOWLEDGEMENTS

J. A. is a Fogarty International Postdoctoral Fellow also receiving support from the Swedish Medical Research Council. Part of these studies were supported by two grants from N.I.H. to Dr. E. S. Lennox.

5. REFERENCES

1. Möller, G. (ed) (1969) 'Antigen sensitive cells. Their source and differentiation', *Transpl. Rev.*, 1.
2. Greaves, M. F. and Hogg, N. M. (1971) 'Immunoglobulin determinants on the surface of antigen binding T and B lymphocytes in mice', *Progr. Immunol.*, 1, 111–26.
3. Möller, G. (ed) (1971) 'Immunological tolerance – Effect of antigen on different cell populations', *Transpl. Rev.*, 8.
4. Dresser, D. W. and Mitchison, N. A. (1968) 'The mechanism of immunological paralysis', *Adv. Immunol.*, 8, 129.
5. Droege, W. and Zucker, R. (1975) 'Lymphocyte subpopulations in the thymus', *Transpl. Rev.*, 25, 3.
6. Miller, J. F. A. P. and Mitchell, G. F. (1969) 'Thymus and antigen reactive cells', *Transpl. Rev.*, 1, 3–42.
7. Möller, G. (ed) (1975) 'Separation of T and B lymphocyte subpopulations', *Transpl. Rev.*, 25.
8. Möller, G. (ed) (1972) 'Lymphocyte activation by mitogens', *Transpl. Rev.*, 11.
9. Andersson, J., Edelman, G. M., Möller, G. and Sjöberg, O. (1972) 'Activation of B-lymphocytes by locally concentrated concanavalin A', *Eur. J. Immunol.*, 2, 233–5.
10. Andersson, J. and Melchers, F. (1973) 'Induction of immunoglobulin M synthesis and secretion in bone marrow-derived lymphocytes by locally concentrated concanavalin A', *Proc. Nat. Acad. Sci. U.S.A.*, 70, 416–20.
11. Greaves, M. F. and Bauminger, S. (1972) 'Activation of T and B cells by insoluble phytomitogens', *Nature*, 235, 67–70.
12. Andersson, J., Sjöberg, O. and Möller, G. (1972) 'Induction of immunoglobulin and antibody synthesis *in vitro* by lipopolysaccharides.', *Eur. J. Immunol.*, 2, 349–53.
13. Nilsson, B. S., Sultzer, B. M. and Bullock, W. W. (1973) 'PPD tuberculin induces immunoglobulin production in normal mouse spleen cells', *J. Exp. Med.*, 137, 127–36.
14. Melchers, F., Braun, V. and Galanos, C. (1975) 'The lipoprotein of the outer membrane of *Escherichia coli*: a B-lymphocyte mitogen', *J. Exp. Med.*, in press.
15. Melchers, F., von Boehmer, H. and Phillips, R. A. (1975) 'B-Lymphocyte subpopulations in the mouse. Organ distribution and ontogeny of immunoglobulin synthesizing and of mitogen-sensitive cells', *Transpl. Rev.*, 25, 26–58.
16. Pantelouris, E. M. (1968) 'Absence of thymus in a mouse mutant', *Nature*, 217, 370–1.
17. Schlossman, C. F. and Hudson, L. (1973) 'Specific purification of lymphocyte populations on a digestable immunoadsorbent', *J. Immunol.*, 110, 313–5.
18. Mishell, R. I. and Dutton, R. W. (1967) 'Immunization of dissociated spleen cell cultures from normal mice', *J. Exp. Med.*, 126, 423–42.
19. Bullock, W. W. and Möller, E. (1972) 'Spontaneous B-cell activation due to loss of normal mouse serum suppressor', *Eur. J. Immunol.*, 2, 514–7.
20. Bullock, W. W. and Andersson, J. (1973) 'Mitogens as probes for immunocyte regulation – specific and non-specific suppression of B cell mitogenesis', in *Ciba Foundation Symposium on Immunpotentiation*, eds. Wolstenholme, G. E. W. and Knight, J. (Associated Scientific Publishers, Amsterdam), pp. 173–88.
21. Coutinho, A., Möller G., Andersson, J. and Bullock, W. W. (1973) '*In vitro* activation of mouse lymphocytes in serum-free medium. Effect of T and B cell mitogens on proliferation and immunoglobulin synthesis', *Eur. J. Immunol.*, 3, 299–306.

22. Melchers, F. and Andersson, J. (1974) 'Early changes in immunoglobulin M synthesis after mitogenic stimulation of bone marrow derived lymphocytes', *Biochemistry*, 13, 4645—53.
23. Agrawal, B. B. L. and Goldstein, I. J. (1965) 'Specific binding of concanavalin A to cross-linked dextran gels', *Biochem. J.*, 96, 23c—25c.
24. Axén R., Porath, J. and Ernback, S. (1967) 'Chemical coupling of peptides and proteins to polysaccharides by means of cyanogen halides', *Nature*, 214, 1302—4.
25. Rittenberg, M. B. and Pratt, K. (1969) 'Anti-trinitrophenyl (TNP) plaque assay. Primary response of Balb/c mice to soluble and particulate immunogen', *Proc. Soc. Exp. Biol. Med.*, 132, 575—81.
26. Melchers, F. and Andersson, J. (1973) 'Synthesis, surface deposition and secretion of immunoglobulin M in bone marrow-derived lymphocytes before and after mitogenic stimulation', *Transpl. Rev.*, 14, 76—130.
27. Andersson, J., Lafleur, L. and Melchers, F. (1974) 'Immunoglobulin M in bone marrow-derived lymphocytes. Synthesis, surface deposition, turnover and carbohydrate composition in unstimulated mouse B-cells', *Eur. J. Immunol.*, 4, 170—80.
28. Melchers, F. and Andersson, J. (1974) 'Immunoglobulin M in bone marrow-derived lymphocytes. Changes in synthesis, turnover and secretion, and in number of molecules on the surface of B-cells after mitogenic stimulation', *Eur. J. Immunol.*, 4, 181—8.
29. Andersson, J., Möller, G. and Sjöberg, O. (1972) 'Selective induction of DNA synthesis in T and B lymphocytes', *Cell Immunol.*, 4, 381—93.
30. Andersson, J., Sjöberg, O. and Möller, G. (1972) 'Mitogens as probes for immunocyte activation and cellular co-operation', *Transpl. Rev.*, 11, 131—77.
31. Andersson, J. and Melchers, F. (1974) 'Maturation of mitogen-activated bone marrow-derived lymphocytes in the absence of proliferation', *Eur. J. Immunol.*, 4, 533—9.
32. Möller, G., Andersson, J., Pohlit, H. and Sjöberg, O. (1973) 'Quantitation of the number of mitogen molecules activating DNA synthesis in T and B lymphocytes', *Clin. Exp. Immunol.*, 13, 89—99.
33. Andersson, J., Bullock, W. W. and Melchers, F. (1974) 'Inhibition of mitogenic stimulation of mouse lymphocytes by anti-mouse immunoglobulin antibodies. I. Mode of action', *Eur. J. Immunol.*, 4, 715—22.
34. Sjöberg, O., Andersson, J. and Möller, G. (1973) 'Reconstitution of immunocompetence in B cells by addition of concanavalin A or concanavalin A treated thymus cells', *Clin. Exp. Immunol.*, 13, 213—23.
35. Dutton, R. W. (1972) 'Inhibitory and stimulatory effects of concanavalin A on the response of mouse spleen suspensions to antigen. I. Characterization of the inhibitory cell activity', *J. Exp. Med.*, 136, 1445—58.
36. Dutton, R. W. (1973) 'Inhibitory and stimulatory effects of concanavalin A on the response of mouse spleen cell suspensions to antigen. II. Evidence for separate stimulatory and inhibitory cells', *J. Exp. Med.*, 138, 1496—505.
37. Rich, R. R. and Pierce, C. W. (1973) 'Biological expressions of lymphocyte activation. II. Generation of a population of thymus-derived suppressor lymphocytes', *J. Exp. Med.*, 137, 649—59.
38. Rich, R. R. and Pierce, C. W. (1974) 'Biological expression of lymphocyte activation. III. Suppression of plaque forming responses *in vitro* by supernatant fluids from concanavalin A-activated spleen cell cultures', *J. Immunol.*, 112, 1360—7.
39. de Petris, S. (1975) 'Concanavalin A receptors, immunoglobulins and θ antigen of the lymphocyte surface. Interactions with concanavalin A and with cytoplasmic structures', *J. Cell Biol.*, 65, 123—46.
40. Jerne, N. K., Nordin, A. H. and Henry, C. (1963) 'The agar plaque technique for recognizing antibody-producing cells', in *Cell-bound Antibodies*, eds. Amos, B. and Koprowski, H. (Wistar Institute Press, Philadelphia, Pa., USA), pp. 109—25.
41. Andersson, J., Möller, G. and Sjöberg, O. (1972) 'B lymphocytes can be stimulated by concanavalin A in the presence of humoral factors released by T cells', *Eur. J. Immunol.*, 2, 99—101.
42. Dutton, R. W., Falkoff, R., Hirst, J. A., Hoffmann, M., Kappler, J. W., Kettman, J. R., Lesley, J. F. and Vann, D. (1971) 'Is there evidence for a non-antigen specific diffusable chemical mediator from the thymus-derived cell in the initiation of the immune response?', *Progr. Immunol.*, 1, 355—68.

CHAPTER 52

Induction of Cytotoxic Effector T-cells by Concanavalin A

MICHAEL J. BEVAN

Lymphocytes of mammals and birds and probably other vertebrates can be broadly divided into two classes: those which pass through the thymus during their differentiation, called thymus dependent, or T-cells, and those which differentiate independently of the thymus, called B-cells. The only well-established function of B-cells after stimulation by antigen is to differentiate into cells which secrete large amounts of antibody. T-Cells on the other hand, upon induction by antigen, appear to perform many diverse functions. Under some circumstances antigen-induced T-cells help the response of B-cells to antigen, while under other circumstances they can suppress the B-cell response. T-Cells also play a role in delayed skin reactions, probably by secreting soluble mediators which recruit non-antigen-specific effector cells. Cytotoxic T-cells are those which can specifically lyse a target cell bearing a surface antigen against which the T-cell is directed. This class of effector T-cell is postulated to play a major role in the rejection of grafts, tumours and infected cells.

Soluble Con A is a T-cell mitogen, that is, it induces small, resting cells to undergo blast transformation and division (see Andersson, Chapter 51). Some of the functions attributed to antigen-induced T-cells are also induced by Con A. Among these is the ability of a subfraction, or all, of Con A-induced cells to kill various target cells. That the effector cell in Con A-induced cytotoxicity is a T-cell and not a macrophage or other non-lymphoid cell (1) has been well established. It has been shown for example that populations rich in T-cells, but not populations depleted of T-cells, become cytotoxic upon Con A treatment (2—6). Also, the effector cell in Con A-induced cytotoxicity bears Thy-1 (θ), a T-cell surface antigen in the mouse (7). *Bona fide* cytotoxic T-cells, i.e. those induced by the major histocompatibility antigens of the species, can also be assayed in an identical way to Con A-induced cytotoxic cells, suggesting that the two are probably the same, only the inductive stimulus is different (7).

I will describe first the technique used to induce mouse T-cells to become cytotoxic. The following major part of this article will deal with the assay of Con A-induced cytotoxic cells. Finally, I will return to the induction step and describe some of the culture conditions which affect the efficiency of induction.

1. INDUCTION OF MOUSE SPLEEN CELLS WITH CON A

Mice are killed by cervical dislocation and the spleens removed asceptically. Usually 5—10 week old mice are used, though spleens of mice as young as 4—7 days contain precursors of cytotoxic T-cells which can be induced to become effector cells with Con A. Adult levels of precursors seem to be reached around 3 weeks of age (Bevan and Byrd, unpublished). After removal, the spleen is rinsed through four sterile Petri dishes containing 5 ml Hank's balanced salt solution and then teased into Hank's balanced salt solution with forceps. The suspension of cells is transferred to a tube and clumps allowed to settle under gravity on ice for 5 min. The supernatant is transferred and centrifuged at 1300 r.p.m. in a model CL International centrifuge for 5 min, the pellet resuspended in culture medium and viable cell counts made in PBS — 0.05% trypan blue. The cell suspension is kept on ice.

The medium used is RPMI 1640 (no. 12-702, Microbiological Associates, Bethesda, Maryland) supplemented with penicillin, streptomycin, 5% foetal calf serum and 50 μM 2-mercaptoethanol. Cells are usually cultured in this medium at a concentration of 2—4 x 10^6 viable cells/ml. The optimum concentration of Con A both for mitogenesis and induction of cytotoxicity is 2 μg/ml (Con A from Calbiochem, San Diego, California, no. 234567). The mitogen is added at time zero to the cells. One ml of cell suspension in 35 mm tissue culture dishes, or 10—15 ml in 100 mm tissue culture dishes (Falcon Plastics, Div. of Bioquest, Oxnard, California) are cultured in an atmosphere of 83% N_2, 10% CO_2, 7% O_2 (8). Cytotoxicity is present at day 2 of culture and reaches a maximum at day 3. Background cytotoxicity in cultures of cells without Con A does develop in this medium containing FCS and 2-mercaptoethanol (9). It develops slowly and on day 3 might reach 1—5% of Con A-induced cytotoxicity. At the end of culture the cells are harvested from the dish by pipetting (scraping is not necessary), centrifuged once or twice, resuspended in assay medium and kept on ice until use.

2. ASSAY FOR CON A-INDUCED CYTOTOXIC CELLS

Much of the early work on mitogen-induced cytotoxicity was done without the realization that the mitogen was performing two roles: one, that of inducing cytotoxic effects and two, providing close contact between killer and target cell in order for lysis to occur. Thus, lymphocytes, targets, and mitogen were mixed together at time zero and the extent of lysis of the target measured at intervals thereafter (2). The favoured targets were erythrocytes and fibroblasts. The targets in common use now are non-adherent tumour cells, grown either *in vitro* or as ascites. The targets used in this laboratory are P815, a DBA/2 mastocytoma, and EL4, a C57BL/6 lymphoma. These tumours grow *in vitro*, label very well with ^{51}Cr-sodium chromate, are easier to count than fibroblast monolayers and are very sensitive to T-cell lysis. Stejskal and coworkers have shown that mitogen-induced lysis is greater on an allogeneic or syngeneic target than on a xenogeneic target, so that one should choose a target of the same species as the induced lymphocyte population (10).

2.1 ^{51}Cr-Labelling the Target

The medium used for assaying cytotoxicity is Dulbecco-modified Eagle's medium supplemented with 5% foetal calf serum (2-mercaptoethanol has no effect on the assay). P815 and EL4 are maintained in culture in this laboratory. They grow in suspension in Petri dishes (Falcon Plastics, Div. of Bioquest, Oxnard, Calif.) in Dulbecco-modified Eagle's medium containing 10% foetal calf serum in an atmosphere of 10% CO_2-air. Logarithmically growing cells are used for the assay. Cells are spun down from medium and $2-5 \times 10^6$ in 0.2 ml assay medium are labelled with $100-200 \mu Ci$ ^{51}Cr-sodium chromate (Amersham/Searle Corp., Arlington Heights, Ill.). Labelling is carried out in a tube (no. 2017, Falcon Plastics) at $37°$ in an atmosphere of 10% CO_2-air for $45-60$ min. The labelled cells are spun down from 10 ml medium two or three times before use in the assay.

With P815 and EL4 this labelling technique gives 10,000-60,000 c.p.m. (counted at about 50% efficiency) per 5×10^4 cells, i.e. about 1 cpm/cell. ^{51}Cr-Sodium chromate labels cell proteins, and the label is not reutilized by other cells after release (11). Other labels can be used, for example rubidium. which has a high rate of spontaneous release (12), or ^{125}I-iododeoxyuridine (13) or labelled amino acids (14) which have very low rates of spontaneous release.

2.2 Assay

The mouse lymphoid population which has been cultured with 2 μg/ml Con A for 2 or 3 days is washed before the assay. Serial twofold or threefold dilutions are made and 0.5 ml aliquots used. The assay can be carried out in tubes (no. 2052, Falcon Plastics) in which case they are kept upright and stationary, or in 35 mm Petri dishes (no. 1008, Falcon Plastics) when they are rocked at $5-7$ cycles/min (Bellco Glass Inc., Vineland, New Jersey). The ^{51}Cr-labelled target is usually adjusted to 1×10^5 cells/ml and just before addition to the lymphocytes, PHA-P is added to a final concentration of 20 μg/ml (PHA-P, Difco Labs, Detroit, Michigan). At the start of the assay 0.5 ml of the target plus PHA mixture is added to 0.5 ml lymphocytes. The assay usually is carried out for 4 h at 37° in 10% CO_2—air. The mixture is centrifuged at the end of the assay and aliquots of the supernatant removed for counting directly in a gamma counter. The results are plotted as percent ^{51}Cr released (linear scale) versus lymphocyte:target ratio (log scale).

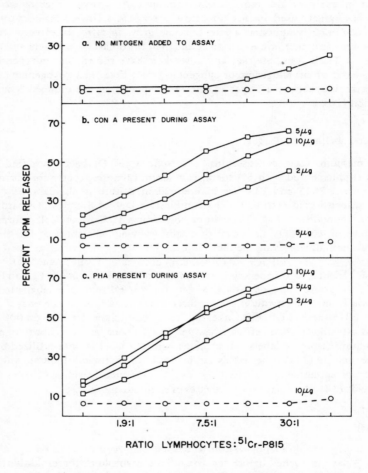

Figure 1. Assay of the lysis of ^{51}Cr-P815 by Con A-induced mouse spleen cells in the presence or absence of added agglutinant. C57BL/6 spleen cells were cultured with 2 μg/ml Con A for 48 h (□———□) or with no mitogen (○ — — — ○). Cells were washed once and serial 2-fold dilutions assayed for lysis of 5×10^4 ^{51}Cr-P815 for 4 h in tubes. (a) No mitogen added to the assay; (b) 2, 5 or 10 μg/ml Con A added; (c) 2, 5 or 10 μg/ml PHA-P added to the assay. [Reproduced by permission of the copyright holder from reference 7.]

2.3 Requirement for Agglutinants in the Assay

Figure 1 shows the amount of label released from ^{51}Cr-P815 during a 4 hour assay by C57BL/6 spleen cells which had been cultured with 2 μg/ml Con A for 48 hours and washed once by centrifugation before the assay. With no mitogen added to the assay (Figure 1a) then a 60:1 ratio of lymphocytes:P815 caused only 26% release of label. Spontaneous release of label by P815 cells alone or in the presence of non-induced spleen cells was 7% in this experiment. When various doses of Con A (Figure 1b) or PHA-P (Figure 1c) were added to the assay, lysis of P815 was greatly enhanced. The spontaneous release of ^{51}Cr remained at 7% in the presence of mitogens. Similar enhancement of the rate of mitogen-induced cytotoxicity by added agglutinants has been shown by others (3—5,10).

Since a cytotoxic T-cell probably has to bind to a target in order to cause lysis, one can postulate that the agglutinants work by providing this close contact. The agglutinants might also work by stimulating the cytotoxic cell to express its lytic function (see paragraph 2.5). The lysis caused by these C57BL/6 Con A-induced cells on P815 in the absence of added agglutinant (Figure 1a) has now been shown to be due partly to Con A carried over from the induction, but due mainly to the subfraction of Con A-induced cytotoxic T-cells which have specific receptors for the H-2^d antigens expressed by P815 (see paragraph 2.6).

The Con A-induced effector cytotoxic cell is sensitive to anti-Thy-1 serum plus complement, i.e. it is a T-cell (7). Alloimmunized cytotoxic T-cells can also lyse tumour cells when agglutinated with PHA (7,13). In Figure 2 C57BL/6 spleen cells (H-2^b) were immunized for 4 days in mixed lymphocyte culture (MLC) with BALB/c spleen cells (H-2^d) which had been inactivated with mitomycin-C. The reciprocal immunization, BALB/c anti-C57BL/6 was also made. In the absence of an agglutinant these alloimmune cytotoxic T-cells showed great specificity for targets which carry the immunizing H-2 antigens. Thus C57BL/6 anti-BALB/c cells lysed ^{51}Cr-P815 (H-2^d) much more efficiently than BALB/c anti-C57BL/6 cells (Figure 2a). The BALB/c anti-C57BL/6 cells however are the more active when assayed against ^{51}Cr-EL4 (H-2^b) (Figure 2b). In the presence of 10 μg/ml PHA however, both cytotoxic populations were able to lyse either target. This analogy with the assay for Con A-induced cytotoxic cells reassures one that the two types of induced effector cells are the same.

The dose of agglutinant which I regularly use in the assay is 10 μg/ml PHA-P (Difco Labs, Detroit, Michigan). This is for two reasons: a) the shape of the curve and the maximum percent release of label with this dose of PHA is most similar to that observed in antigen-specific T-cell-mediated lysis (Figures 1 and 2); b) PHA agglutinates some cells which are not agglutinated by Con A (15).

2.4 Choice of Target Cell

The mastocytoma P815 and the lymphoma EL4 are sensitive both to antigen-specific and antigen-non-specific (PHA-revealed) T-cell-mediated lysis (Figures 1 and 2). However not all targets which are susceptible to specific lysis can be lysed non-specifically. T-cell and B-cell blasts (prepared by stimulation with T- or B-cell mitogens) are susceptible to antigen-specific T-cell lysis but are not efficiently lysed by syngeneic cytotoxic T-cells when agglutinated with PHA (7,16). Figure 3a

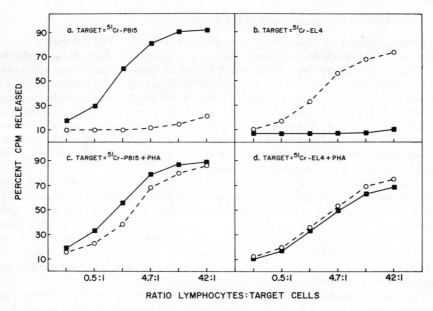

Figure 2. Assay of the cytotoxic activity of MLC-induced cytotoxic cells in the presence or absence of 10 μg/ml PHA-P. C57BL/6 spleen cells (H-2^b) were immunized with inactivated BALB/c spleen cells (H-2^d) (■———■), and BALB/c immunized with C57BL/6 (o — — — o). Cytotoxic assay was carried out in tubes for 4 h against (a) ^{51}Cr-P815 (H-2^d); (b) ^{51}Cr-EL4 (H-2^b); (c) ^{51}Cr-P815 in the presence of 10 μg/ml PHA-P; (d) ^{51}Cr-EL4 in the presence of 10 μg/ml PHA-P. [Reproduced by permission of the copyright holder from reference 7.]

shows that Con A-induced C57BL/6 killers lyse syngeneic EL4 very efficiently in the presence of PHA. ^{51}Cr-Labelled Con A blasts or LPS blasts are not readily lysed under these conditions however (Figures 3b and c). Small lymphocytes are also relatively resistant to lysis by Con A-induced cytotoxic cells (7,17). Thus small lymphocytes and their induced progeny are poor indicators for assaying non-specific T-cell lysis.

Table I shows a comparison of the rate of lysis of ^{51}Cr-P815, chicken erythrocytes, horse erythrocytes and sheep erythrocytes by Con A-induced C57BL/6 spleen cells. Because spontaneous release by erythrocyte targets varies with cell concentration, the assay was set up at a constant ratio of lymphocytes: targets and a kinetic assay was performed. As a control, non-induced C57BL/6 lymph node cells were used. At this ratio of 17:1 almost complete lysis of P815 occurred in 2 h, by which time the CRBC and HRBC showed only 2–3% release above the control. After 5–21 h the chicken and horse erythrocytes showed considerable differences in ^{51}Cr release with control and Con A-induced cells. Even after 21 h however the SRBC showed no more lysis in the presence of induced versus non-induced cells. It is important to note that all of these targets are agglutinated by PHA. The reason for the difference between sheep and horse

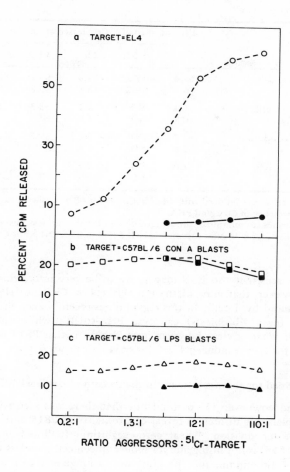

Figure 3. Lysis of various ^{51}Cr-labelled targets by Con A-induced cells. C57BL/6 spleen cells at a concentration of 2×10^6 cells/ml were induced with Con A for 3 days washed once and serial 3-fold dilutions titrated against various targets for 4 h in the presence (open symbols) or absence (closed symbols) of 10 μg/ml PHA-P. (a) 5×10^4 EL4 lymphoma cells as target (C57BL/6); (b) 5×10^4 C57BL/6 spleen cells induced with Con A for 3 days and labelled with ^{51}Cr as target; (c) 5×10^4 ^{51}Cr-labelled C57BL/6 spleen cells which had been induced with 10 μg/ml lipopolysaccharide (LPS) for 3 days as target cells

TABLE I.   Comparison of the rate of lysis of various targets by Con A-induced mouse spleen cells in the presence of PHA[a]

	Percent of total ^{51}Cr released at time[b]:					
Target	1 h	1.5 h	2 h	3 h	5 h	21 h
^{51}Cr-P815 + Con A cells	19.8	57.2	75.0	84.1	87.2	ND
^{51}Cr-P815 + NLN	6.1	9.2	13.8	15.2	21.9	ND
^{51}Cr-CRBC + Con A cells	1.4	ND	4.1	5.3	12.5	38.3
^{51}Cr-CRBC + NLN	0.7	ND	1.9	ND	2.9	12.4
^{51}Cr-HRBC + Con A cells	2.7	ND	5.6	9.5	15.6	50.3
^{51}Cr-HRBC + NLN	1.8	ND	3.4	ND	6.1	37.7
^{51}Cr-SRBC + Con A cells	2.0	ND	2.5	2.6	2.9	9.2
^{51}Cr-SRBC + NLN	2.0	ND	2.7	ND	3.9	10.3

[a]C57BL/6 spleen cells were induced with 2 μg/ml Con A for 48 h, washed and assayed for lysis of targets in the presence of 10 μg/ml PHA.
[b]Assay set up in dishes (to minimize sedimentation times for the various cells) on a rocking platform, at a ratio of lymphocytes (normal C57BL/6 lymph node (NLN) or Con A induced) to target of 17:1.

erythrocytes in sensitivity to lysis may prove to be very interesting. It should be pointed out however, that none of the lytic effects on CRBC or HRBC have been shown to be caused by T-cells. In this regard it has recently been shown that many types of cells (even fibroblasts) can lyse PHA-CRBC, while only T-cells lyse PHA-P815 (18). What is clear however from Table I is that P815 is a much more sensitive target for Con A-induced T-cell lysis than are erythrocytes.

2.5 Is there a Need for the Agglutinant in the Assay to be a T-cell Mitogen?

Asherson and coworkers (5) pointed out that there was a correlation between the ability of plant agglutinins to reveal cytotoxicity on P815 and their ability to activate T-cells for division. Wheat germ agglutinin (Calbiochem, Wheat Germ Lectin, no. 681818) is not mitogenic for mouse spleen cells, i.e. less than a fourfold stimulation of ^{3}H-thymidine incorporation over a range of 0.2–600 μg/ml. I have found that it can however serve to reveal the lytic effect of Con A-activated cells on ^{51}Cr-EL4 (Figure 4). At this optimum dose of 175 μg/ml during a 4h assay, it is not toxic to the target cells and reveals cytotoxic cells almost as efficiently as does PHA. The possibility that Con A carried over from the induction phase, might be acting synergistically with WGA in the assay (i.e. WGA providing agglutination and Con A providing a terminal stimulus for lysis) is ruled out in Figure 4, since washing out most of the Con A with α-methyl-D-mannoside before the assay had no effect on WGA- or PHA-revealed lysis.

WGA up to 300 μg/ml does not efficiently reveal C57BL/6 Con A-induced cytotoxic cells when P815 is the target, (data not shown) in aggreement with the results of Asherson and coworkers (5). This clearly speaks for some difference in the surface membranes of P815 and EL4. Another interesting feature of using WGA in the cytotoxic assay is that it works much better with 3-day Con A-induced cells than with 2-day induced cells. Figure 5 shows that C57BL/6 spleen cells cultured

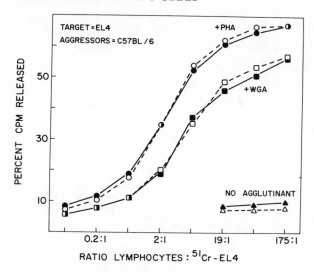

Figure 4. Lysis of ^{51}Cr-EL4 by syngeneic Con A-induced spleen cells in the presence of PHA-P (○,●), or wheat germ agglutinin (WGA) (□,■) or no added lectin (△,▲). C57BL/6 spleen cells were induced with Con A for 3 days. Half of the induced cells were incubated in 50 mM α-methyl-D-mannoside for 90 min before harvesting (open symbols) and the other half were not treated with sugar (solid symbols). The sugar-treated and untreated cells were titrated against 5×10^4 ^{51}Cr-EL4 in the presence of 10 μg/ml PHA-P (○,●), or 175 μg/ml WGA (□,■) or with no added agglutinant (△,▲)

with Con A for 65 h will lyse ^{51}Cr-EL4 in the presence of PHA or WGA. However, cells cultured only for 40 h with Con A lyse EL4 much more efficiently in the presence of PHA than in the presence of WGA. Changing the concentration of WGA in the assay does not increase lysis. This suggests that WGA binds to a structure on the cytotoxic T-cell which increases with time of Con A induction. This, in fact correlates well with recent findings on binding sites for fluorescent WGA on resting versus activated lymphocytes (19).

Figure 5 also illustrates that although PHA can reveal cytotoxic cells after 2 or 3 days induction with Con A, the lines for the plot of percent release versus log lymphocyte:target ratio for the two populations are not parallel. This makes quantitating the relative activity of the two populations difficult. It is not known whether the different slopes reflect differences in the lytic efficiency of individual cytotoxic cells or a difference in their relative frequency in the population.

2.6 Assay of H-2-specific Cytotoxic Cells Induced by Con A

Con A in the absence of added antigen, induces a large fraction of T-cells to become cytotoxic. The precursor small lymphocyte has a predetermined antigen specificity, so that if this specificity is retained after Con A induction one could

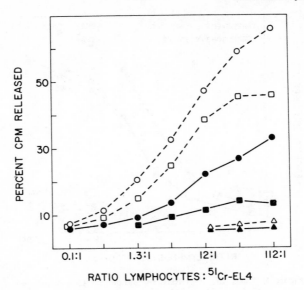

Figure 5. Lysis of ^{51}Cr-EL4 in the presence of PHA-P or WGA by spleen cells induced with Con A for 40 h (solid symbols) or 65 h (open symbols). C57BL/6 spleen cells induced with Con A were titrated versus ^{51}Cr-EL4 for 4 h in tubes. Assays carried out in the presence of 10 μg/ml PHA-P (○,●), in the presence of 175 μg/ml WGA (□,■) or no added agglutinant (△,▲).

expect to observe specific lysis (i.e. PHA independent) of an allogeneic target cell. This has recently been shown to occur (20—22). In assaying for specific lysis where the only bridge between killer and target cell should be receptor—antigen interaction, it is essential to exclude from the assay any non-specific agglutinating agent. Con A can serve to reveal cytotoxic cells (Figure 1), and Con A carried over from the induction period must be washed off the cells or neutralized by a sugar such as α-methyl-D-mannoside which binds Con A.

Figure 6 shows the result of an experiment in which C57BL/6 and DBA/2 spleen cells were induced with Con A for 3 days and assayed for lysis of ^{51}Cr-P815 (DBA/2) in the presence of 10 μg/ml PHA-P as well as in the absence of PHA-P but containing various doses of α-methyl-D-mannoside. Both induced populations lysed P815 very efficiently in the presence of PHA-P. In the absence of PHA-P the DBA/2 cells (syngeneic with P815) caused about 5% release of label above background at a ratio of 258:1. This lysis was inhibited completely by 100 mM α-methyl-D-mannoside and was probably due to residual Con A causing agglutination. Lysis by the C57BL/6 cells is much greater and is only partially inhibitable by sugar. This, and other data (20—22) demonstrate conclusively that a subfraction of specific H-2-reactive cytotoxic cells can be demonstrated in Con A-induced populations.

Usually in the assay for specific cells induced by Con A we prefer to wash out the mitogen before the assay by incubating the cells for 2—13 h in 50 mM

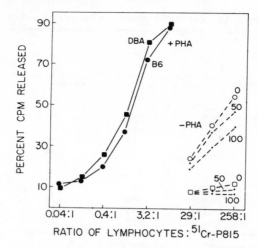

Figure 6. The effect of α-methyl-D-mannoside on the lysis of ^{51}Cr-P815 by syngeneic (DBA/2) and allogeneic (C57BL/6) Con A-induced cells. Spleen cells of DBA/2 (□,■) and C57BL/6 (○,●) mice were cultured with Con A for 65 h, washed, and assayed for lysis of P815 for 4 h in dishes. Assays carried out in the presence (closed symbols) or absence (open symbols) of 10 μg/ml PHA-P. Assays were also carried out in the absence of PHA-P with 50 mM and 100 mM α-methyl-D-mannoside present as indicated

α-methyl-D-mannoside. The only specific cytotoxic cells which have been demonstrated so far to be induced by Con A are those reactive with the species major histocompatibility antigens which may reflect the high frequency of such precursor cells (22).

3. FACTORS WHICH AFFECT THE INDUCTION

The work discussed here was done some time ago. C57BL/6 cells were used for induction with 2 μg/ml Con A, and the assays done for 4 h in tubes with PHA-P815. In paragraph 2.5 we presented data to show that cells cultured 3 days with Con A are more active than 2-day activated cells.

3.1 Cell Concentration and Cell Source

Doubling concentrations of mouse lymphoid cells were cultured with Con A for 2 days. Cells were washed and viable cell recoveries and cytotoxicity estimated (Table II). One might have expected cultures containing originally 1.0×10^6 cells to be twice as active on a per culture basis in the cytotoxic assay as cultures set up with 0.5×10^6 cells. By the same reasoning, taking cell recoveries into account, then one would have expected all the cultures to develop the same degree of

TABLE II.   The effect of initial cell concentration on the induction of cytotoxicity by Con A

Cultures[a]	TdR incorporation[b]	Cell recovery (%)	Relative cytotoxicity per culture[c]	Relative cytotoxicity per recovered cell[c]
Spleen cells				
1×10^6/ml − Con A	2039	43	0	0
0.5×10^6/ml + Con A	38,737	123	10.5	40
1×10^6/ml + Con A	55,986	116	32	64
2×10^6/ml + Con A	73,125	115	100	100
4×10^6/ml + Con A	94,038	82	480	336
Lymph node cells				
1×10^6/ml − Con A	1169	55	0	0
0.5×10^6/ml + Con A	64,204	196	6.3	14.8
1×10^6/ml + Con A	69,182	144	32	51
2×10^6/ml + Con A	85,401	157	71	52

[a]C57BL/6 spleen cells cultured with or without Con A for 48 h.
[b]A 4 h pulse with ^{3}H-thymidine, h 48−52.
[c]Assayed for 4 h against ^{51}Cr-P815 in the presence of 10 μg/ml PHA. Calculated from a comparison of the number of induced cells required to achieve the same amount of ^{51}Cr release (23).

cytotoxic activity *per recovered cell*. In fact, one reproducibly finds that doubling the cell number in the low range of initial cell concentrations the cytotoxic activity per culture more than doubles. With spleen cells the activities do not level off up to 4×10^6/ml. In two experiments with lymph node cells the relative activity per recovered cell reached a plateau at between $1−2 \times 10^6$ ml. On the other hand in terms of ^{3}H-thymidine incorporation, the increments are always less than twofold.

There are at least two possible explanations for this finding: one is that cells which are dividing quickly (i.e. seeded at low density) are relatively inefficient killer cells. The second possibility is that there is some sort of cooperative interaction among the cells in the development of cytotoxic activity.

Table II also illustrates that under these culture conditions spleen cells generate more cytotoxicity than do lymph node cells. I have compared these lymphoid organs on four other occasions at initial cell concentrations of 1×10^6/ml and spleen has given 1.4−5 times as much cytotoxic activity per culture as lymph node. This is despite the fact that lymph node is approximately twice as rich in T-cells as spleen. Cortisone-resistant thymocytes in one experiment gave about the same cytotoxic activity as did lymph node cells.

3.2 Induction of Cytotoxic Activity in the Absence of DNA Synthesis

Cytosine arabinoside (5×10^{-5} M) added to mouse spleen cultures at the same times as a mitogenic dose of Con A almost completely inhibits subsequent incorporation of labelled thymidine (Table III). The viable cell recovery in the inhibited cultures is very low, about one tenth of the recovery in cultures receiving Con A alone. But over 70% of the recovered cells in both kinds of cultures at day 2 and day 3 are blast cells. The results in Table III show that along with this

TABLE III. Con A induction of cytotoxic activity in the absence of DNA synthesis[a]

Cultures	Day 2				Day 3			
	TdR[b] incorporation	Cell[c] recovery	Relative cytotoxicity per culture[d]	Relative cytotoxicity per recovered cell[d]	TdR[b] incorporation	Cell[c] recovery	Relative cytotoxicity per culture[d]	Relative cytotoxicity per recovered cell[d]
Cells alone	2231	71	0	0	2054	70	1.1	2.3
Cells + Con A	68,350	109	100	100	85,221	151	100	100
Cells + Con A + cytosine arabinoside	465	11	17	168	280	7	31	669

[a] 2×10^6 C57BL/6 lymph node cells/ml were cultured alone or with 2 μg/ml Con A, plus or minus 5×10^{-5} M cytosine arabinoside.

[b] A 4 h pulse of ^{3}H-TdR.

[c] Viable cell recovery expressed as percentage of original.

[d] Assayed for 4 h against ^{51}Cr-P815 in the presence of 10 μg/ml PHA-P, calculated as in Table II.

535

morphological transformation of cells in the presence of Con A and cytosine arabinoside there is also a functional differentiation of some of the cells into cytotoxic effector cells. Although on a per culture basis the relative cytotoxic activity is small in the inhibited, compared to uninhibited cultures, on a recovered cell basis the cytotoxic activity is higher. By day 3 the viable cells recovered from the inhibited cultures are 6–7 times as active as the uninhibited cells. A small amount of background cytotoxicity develops in control cultures by day 3, due to stimulation of the cells by the medium.

This suggests that the first explanation offered in the preceding paragraph of the fact that dense cultures develop proportionately more cytotoxicity than sparse cultures is correct: Con A-stimulated cells which are prevented from expending energy in growth either because of limited nutrients or by an inhibitor, are more efficient in their lytic function.

4. GENERAL COMMENT

It is clear that, in the absence of added antigen, Con A can induce resting precursors of cytotoxic T-cells to differentiate into functional effector cells. Thus, among Con A-induced cells are killer cells which can specifically lyse a target which carries a foreign (non-self) major histocompatibility antigen and these can be assayed in an exactly similar way to antigen-induced cells. The mitogenic induction is polyclonal, and therefore to assay for most killers one has to artificially provide close contact of the target with an agglutinant such as PHA in order to reveal lysis. One can use Con A as an inducer to answer questions about precursor frequency, for example, especially in cases where the induction by antigen has rather stringent requirements.

Assaying populations for lytic activity against tumour targets in the presence of an agglutinant probably does not artifactually reveal T-cells as being cytotoxic effectors when under normal circumstances they might perform a quite different function. The choice of target is likely to be important here, tumour cells being preferred over erythrocytes for example since they are much more sensitive to PHA-revealed T-cell lysis and probably more resistant to lysis by non-T-cell cytotoxic effector mechanisms.

There are still many unknowns in the induction and assay of cytotoxic T-cells. For example does Con A induce all potential precursors to become cytotoxic effector cells whilst retaining their antigen-binding specificity? As mentioned above, Con A certainly induces cells with specificity for the species major histocompatibility antigens, but we do not know if it induces killers specific for other antigens, or even if it can induce primed (antigen-experienced) cytotoxic memory cells. Another unknown is, does the assay of a population of cells for lysis of a labelled tumour in the presence of PHA reveal all the cytotoxic effector cells?

5. ACKNOWLEDGEMENTS

I thank Dr. Melvin Cohn for discussion and support. Supported by National Institute of Allergy and Infectious Diseases Research Grant A105875 and National Institute of Allergy and Infectious Diseases Training Grant A100430.

6. REFERENCES

1. Cerottini, J.-C. and Brunner, K. T. (1974) 'Cell mediated cytotoxicity, allograft rejection, and tumor immunity', *Adv. Immunol.*, 18, 67—132.
2. Perlmann, P. and Holm, G. (1969) 'Cytotoxic effects of lymphoid cells *in vitro*', *Adv. Immunol.*, 11, 117—93.
3. Greaves, M. F. and Janossy, G. (1972) 'Elicitation of selective T and B lymphocyte responses by cell surface binding ligands', *Transpl. Rev.*, 11, 87—130.
4. Möller, G., Sjöberg, O. and Andersson, J. (1972) 'Mitogen-induced lymphocyte-mediated cytotoxicity *in vitro*: effect of mitogens selectively activating T or B cells', *Eur. J. Immunol.*, 2, 586—92.
5. Asherson, G. L., Ferluga, J. and Janossy, G. (1973) 'Non-specific cytotoxicity by T cells activated by plant mitogens *in vitro* and the requirement for plant agents during the killing reaction', *Clin. Exp. Immunol.*, 15, 573—89.
6. Stavy, L., Treves, A. J. and Feldman, M. (1972) 'Capacity of thymus cells to effect target cell lysis following treatment with Con A', *Cell. Immunol.*, 3, 623—8.
7. Bevan, M. J. and Cohn, M. (1975) 'Cytotoxic effects of antigen- and mitogen-induced T cells on various targets', *J. Immunol.*, 114, 559—65.
8. Mishell, R. I. and Dutton, R. W. (1967) 'Immunization of dissociated spleen cell cultures from normal mice', *J. Exp. Med.*, 126, 423—42.
9. Bevan, M. J., Epstein, R. E. and Cohn, M. (1974) 'The effects of 2-mercaptoethanol on murine mixed lymphocyte cultures', *J. Exp. Med.*, 139, 1025—30.
10. Stejskal, V., Holm, G. and Perlmann, P. (1973) 'Differential cytotoxicity of activated lymphocytes on allogeneic and xenogeneic target cells', *Cell. Immunol.*, 8, 71—81.
11. Berke, G. and Amos, D. B. (1973) 'Mechanism of lymphocyte-mediated cytolysis. The LMC cycle and its role in transplantation immunity', *Transpl. Rev.*, 17, 71—109.
12. Henney, C. S. (1973) 'Studies on the mechanism of lymphocyte-mediated cytolysis. II. The use of various target cell markers to study cytolytic events', *J. Immunol.*, 110, 73—84.
13. Forman, J. and Möller, G. (1973) 'Generation of effector lymphocytes in mixed lymphocyte reactions. I. Specificity of the effector cells', *J. Exp. Med.*, 138, 672—85.
14. Bean, M. A., Pees, H., Rosen, G. and Oettgen, H. F. (1973) 'Prelabeling target cells with ^{3}H-proline as a method for studying lymphocyte cytotoxicity', *Nat. Cancer Inst. Monogr.*, 37, 41—8.
15. Kirschner, H. and Blaese, R. M. (1973) 'Pokeweed mitogen-, Con A-, and phytohemagglutinin-induced development of cytotoxic effector lymphocytes. An evaluation of the mechanisms of T cell-mediated cytotoxicity', *J. Exp. Med.*, 138, 812—24.
16. Miggiano, V. C., Bernoco, D., Lightbody, J., Trinchieri, G. and Cepellini, R. (1972) 'Cell-mediated lympholysis in vitro with normal lymphocytes as target: specificity and cross-reactivity of the test', *Transpl. Proc.*, 4, 231—7.
17. Holm, G. (1967) 'The *in vitro* Cytotoxicity of human lymphocytes; comparison with other cells', *Exp. Cell Res.*, 48, 327—33.
18. Muchmore, A. V., Nelson, D. L. and Blaese, R. M. (1975) 'The cytotoxic effector potential of some common non-lymphoid tumors and cultured cell lines with PHA and heterologous anti-erythrocyte sera', *J. Immunol.*, 114, 1001—3.
19. Robinson, P. J. and Roitt, I. M. (1974) 'Identification of a population of mouse leukocytes using wheat germ agglutinin', *Nature*, 250, 517.
20. Clark, W. (1975) 'An antigen-specific component of lectin-induced cytotoxicity', *Cell Immunol.*, 17, 505—16.
21. Waterfield, J. D., Waterfield, E. M. and Möller, G. (1975) 'Lymphocyte-mediated cytotoxicity against tumor cells. I. Con A activated cytotoxic cells exhibit immunological specificity', *Cell Immunol.*, in press.
22. Bevan, M. J., Langman, R. E. and Cohn, M. (1976) 'H-2 antigen specific cells induced by Con A: estimation of their relative frequency', *Eur. J. Immunol.*, in press.
23. Bevan, M. J. (1975) 'Alloimmune cytotoxic T cells: evidence that they recognize serologically defined antigens and bear clonally restricted receptors', *J. Immunol.*, 114, 316—9.

CHAPTER 53

Concanavalin A as a Probe of Cell-mediated Immunity: Production of Lymphokines by Con A-Stimulated T-Cells

RONALD P. PELLEY
HOWARD J. SCHWARTZ

Much varied evidence links thymus-derived (T) lymphocytes to the induction
and expression of cellular immunity. Neonatal thymectomy inhibits graft rejection
and delayed cutaneous hypersensitivity and decreases cellular immunity (CMI) to
infections by intracellular parasites. The injection of anti-lymphocyte serum
depletes the lymphocytes in thymus-dependent areas of lymph nodes, impairs the
rejection of transplants, inhibits delayed cutaneous hypersensitivity and suppresses
immunity to infection. The cellular transfer of immunity to *Listeria* infection was
inhibited when the cells were treated with antibody specifically cytotoxic for
T-lymphocytes in the presence of complement. Finally, congenitally athymic
individuals show impaired cellular immunity to infection.

Soluble factors produced by T-lymphocytes were long suspected to be the
mediators of CMI. Dumonde (1) proposed that the term 'lymphokine' be applied to
any cell-free soluble factor generated during the interaction of immune lymphocytes with specific antigen which mediates CMI phenomena. These factors do not
possess any inherent immunological specificity of their own.

Con A stimulates normal lymphocytes from a wide variety of sources to
produce lymphokines. Some of these lymphokines activate cells: migration
inhibitory factor (MIF) and monocyte chemotactic factor (MCF) activate macrophages, ESP activates eosinophils and LM enhances lymphocyte functions. Other
lymphokines inhibit cells; lymphotoxin (LT) kills target cells, PIF inhibits target
DNA synthesis, and T-cell suppressor substance inhibits lymphocyte function (3—5,10,21,22,33,35—38, review 34).

In this chapter we will review the methods for assessing lymphokine production,
and then we will demonstrate that Con A induces T-cells, in particular, to produce
MIF. Our research has concentrated on MIF, since it was the first *in vitro* correlate
of delayed hypersensitivity discovered and it is the best characterized lymphokine (2).

1. PREPARATION OF CELL CULTURES

1.1 Rabbits and Guinea Pigs

Randomly bred New Zealand white rabbits of either sex (2—3 kg), and randomly
bred female guinea pigs (400—500 g) were used throughout these studies.

After anaesthesia animals (rabbits with 30 mg 'Diabutal' [Diamond Laboratories, Des Moines, Iowa], i.v. and guinea pigs with ether [Squibb] by inhalation) were exsanguinated by cardiac puncture. Lymph nodes were aseptically removed from the cervical, popliteal, inguinal and brachial areas (guinea pigs and rabbits) and axillary, auricular and submandibular areas (rabbit only). The nodes were trimmed of fat and the cells were released by teasing between sterile stainless-steel wire screens held by haemostats. This teasing was done in 60 x 15 mm tissue culture dishes (#3002, Falcon Plastics, Oxnard, California) in 5 ml of medium HBSS—10% serum (Hank's buffered salt solution containing 100 units/ml penicillin and 50 μg/ml streptomycin — supplemented with 10% serum). Normal rabbit serum (NRS), guinea pig serum (GPS) and foetal calf serum (FCS) were obtained from Grand Island Biological Co., Grand Island, N.Y. and were inactivated by heating at 56° for 30 minutes.

The cell suspensions were filtered through gauze and washed twice in HBSS—10% serum by alternate suspension and centrifugation of the cells for 10 minutes at 1000 r.p.m. at room temperature in an International centrifuge. The cells were resuspended to 20 ml in HBSS—10% serum and counted in a haematocytometer. The cells were again centrifuged and suspended to a concentration of either one or 2×10^7 cells per ml in tissue culture medium. Cell viability was in excess of 90%, as assessed by exclusion of trypan blue dye. Cell suspensions with lower viability at the initiation of culture did not consistently produce reasonable quantities of lymphokines and (as will be discussed later) released toxic products that interfered with the assay of certain lymphokines.

1.2 Mice

Swiss albino female CF1 mice (Carworth Farms, New City, New York) weighing 18—22 g were used. Since this is not a highly inbred strain, the lymphoid cell preparation from each mouse was cultured separately, and after culture supernates were pooled. After ether anaesthesia, mice were exsanguinated by decapitation. Their spleens were aseptically removed and disaggregated in 5 ml HBSS—10% FCS. Disaggregation was accomplished by injecting the spleens with 1 ml of HBSS—10% FCS and gently aspirating the cells in and out of 1 ml syringes with 21 gauge needles twenty times (Becton-Dickinson and Co., Rutherford, New Jersey). The cells were then gently filtered through gauze and washed in HBSS—10% FCS using the procedure described in 1.1 above and resuspended in 5 ml of medium. Cells were enumerated in a Model Z_f counter (Coulter Electronics, Inc., Hialeah, Florida), centrifuged and suspended in tissue culture medium at a cell concentration of 1×10^7/ml. By this procedure, cell viability in excess of 90% is easily obtained.

1.3 Culture Media — Rabbits

In our recent studies we have uniformly used serum-containing media. Both antigen-induced (2) and Con A-induced (10) lymphokines are generated in the absence of serum. Cell viability, however, was much better in serum-supplemented cultures. When culture media were supplemeted with serum, viability at the end of

24 hours was 70—90% for guinea pig or rabbit lymph node cell cultures and 60—70% for mouse spleen cell cultures. When studies on protein or DNA synthesis were done concomitantly with studies on lymphokine production, high viability was essential. When studies were conducted on production of skin reactive factor (SRF), toxic products were released from dying cells resulting in 'non-specific' inflammation upon injection of control cell culture supernates into the skin of a test animal (see assay section). Finally, many of our studies examined lymphokine production at time periods beyond 24 hours in culture. At these times the effect of serum supplement upon cell viability was pronounced.

For culture of rabbit lymph node cells the following medium is optimal. Eagle's minimal essential medium (MEM), in either basal or spinner modification (Grand Island Biological, or Flow Laboratories, Rockville, Maryland) was used. For experiments studying protein synthesis, MEM lacking cold leucine was utilized. We have not observed significant lot-to-lot variation in the acceptability of this medium. This medium was supplemented at the time of culture with 2 mM glutamine, 100 units/ml penicillin, 50 μg/ml streptomycin and minimal essential medium non-essential amino acids (all obtained from either GIBCO or Flow Laboratories). This medium will be referred to as Complete MEM. HEPES (10 mM) (Microbiological Associates, Bethesda, Md. or Flow Laboratories) was used as an additional buffer. When HEPES buffer was used, culture tubes could simply be tight capped and incubated in a regular 37° incubator. MEM contains 2 g/litre $NaHCO_3$, satisfying any cell requirement for bicarbonate. This procedure obviated the need for a cumbersome CO_2 incubator for control of pH.

Viability of rabbit lymph node cells appeared to be optimal when 1×10^7 cells/ml were cultured in Complete MEM supplemented with 10% NRS. Increasing the serum concentration beyond this level did not appear to improve protein synthesis or lymphokine production. Although we have reported good cell survival (80—90%) when rabbit peripheral blood lymphocytes were cultured in FCS-supplemented medium (8), lymph node cell survival was not optimal in FCS-supplemented medium. Often less than 50% viability was observed during the second 24 hours. At later times in culture this effect was even more striking. However, the study of effects of antiserum to immunoglobulin upon the immune response requires FCS in order to avoid the formation of antigen—antibody complexes (which occur when goat anti-rabbit IgG serum is added to cultures supplemented with NRS).

MIF production and DNA synthesis were studied simultaneously by a modification of the method of Fanger and coworkers (11). Four $\times 10^7$ lymph node cells were cultured for 24 hours in 2 ml of Complete MEM—10% FCS containing 3×10^{-5} M glutathione (Sigma Chemical Co., St. Louis, Missouri). The use of glutathione was essential for optimal DNA synthesis by rabbit B cells. At the end of the first 24 h of culture the supernate tissue culture fluid was removed by Pasteur pipette from the settled cells. Fresh media containing 5 μCi/ml ^{3}H-thymidine was added and the cells incubated for an additional 24 hours. At the end of culture the cells were centrifuged, the supernates removed and stored, and the cellular DNA precipitated with 10% cold TCA after it was released from freeze-thaw lysed cells. Non-radiolabelled Con A was used in tissue cultures to be assayed for DNA synthesis.

1.4 Culture Media — Guinea Pigs

Guinea pig lymphokine production has been studied only during the first 18—24 hours in culture. In general, lymphokine production was optimal when 1×10^7 lymph node cells were cultured in 1 ml of tissue culture medium (TCM-199) supplemented with 10% GPS. Raising cell numbers and increasing serum concentration did not appear to result in increased lymphokine production. When spleen cells were cultured serum supplementation was not needed. We have avoided the use of peritoneal exudate lymphocytes as a source of SRF since unstimulated cells released significant inflammatory materials into the culture. After addition of mitogen, culture tubes were tight capped and incubated in the same manner as rabbit cultures. TCM-199 appeared to contain sufficient $NaHCO_3$ and Na_2HPO_4 for the maintenance of pH during these short-term cultures.

1.5 Culture Media — Mice

Lymphokine production by murine cell suspensions during the first 24 hours in culture did not require serum or highly specialized culture conditions. However, if DNA synthesis is to be simultaneously studied, we have found RPMI-1640 culture medium and 10% FCS to be optimal. We have noted that DNA synthesis in the mouse was very dependent upon proper culture conditions and optimal mitogen dose. In order to obtain significant, reproducible results it will be well worth the time invested for each investigator to carefully determine the optimal cell and mitogen concentrations for the particular lymphoid organ of the particular stain of mouse he is dealing with. Different lots of medium and serum accounted for a tenfold variation in incorporation of ^{3}H-thymidine. This variability was not encountered in studying lymphokine production.. The culture media for DNA synthesis in the mouse (RPMI-1640) and rabbit (glutathione-supplemented Complete MEM) both contain glutathione. However the presence of this reagent was not required for Con A-induced lymphokine production in either species.

2. REAGENTS

2.1 Con A

Con A was obtained as the lyophilized powder from a commercial source (Calbiochem) and stored in this form at 4°. After reconstitution the sterile solution (10 mg/ml) can be stored at 4° for considerable periods of time (6 months) without detectable loss of biological activity. Prior to use the solution was filtered (.22 μ 'Millex', Millipore Corp., Bedford, Mass.) and dilutions made. Con A may be unstable in dilute solution. Therefore, fresh dilutions were made up for each experiment. We have not encountered difficulties with spontaneous precipitation of Con A at the stock concentration employed (10 mg/ml).

2.2 Con A Labelling

Tem mg lots of Con A were labelled with 5 mCi ^{125}I by either the chloramine T or ICl method (12). ^{125}I-Con A was desalted and purified by absorption onto

'Sephadex' G-75, followed by elution with 1 M glucose. Glucose was removed and the sample concentrated by vacuum dialysis at $4°$ for 72 h against 2 litres of buffer (0.14 M saline, 0.01 M phosphate, pH 7.0, 1 x 10^{-4} M $CaCl_2$ and 1 x 10^{-4} M $MnSO_4$). After centrifugation (30 min 20,000 r.p.m., $4°$) the protein content was determined by absorbance at 280 nm, and adjusted to 1 mg/ml with buffer. The ^{125}I-Con A was sterilized by filtration through a .22 μ 'Millex' filter and stored at $4°$. This procedure usually resulted in Con A with a specific activity of 10,000 c.p.m./μg. ^{125}I-Con A was generally used within 2—3 weeks of preparation. After longer periods of time Con A became inactive (incapable of binding to 'Sephadex' G-200) and dialysable ^{125}I-activity was observed.

2.3 Antiserum to T-Cells

Goat antiserum to rabbit thymus cells (ATG) was prepared against thymus cells of exsanguinated 6-week old animals. Cells were teased from thymuses of 6 young rabbits into HBSS containing heparin (5 units/ml), filtered through gauze, and washed three times in HBSS. The initial injection (1.8 x 10^{10} thymocytes in Complete Freund's adjuvant) was divided among six flank sites. Two weeks later a total of 0.5 x 10^{10} thymocytes in Complete Freund's adjuvant were injected into the same sites. At 4 and again at 12 months after the initial series of inoculations, booster injections of 10^{10} thymocytes were given. Two weeks after the last series of injections, the goat was exsanguinated. The goat antiserum was found to have a cytotoxic titre against thymus cells of 1:2048 to 1:4000 in the presence of complement by the method of Terasaki (13).

The complement activity in the antiserum was inactivated by addition of 5 ml 2% EDTA to 100 ml serum. The antiserum was then absorbed with packed, washed, leukocyte-poor, rabbit red cells until free of haemagglutinating activity. The globulin (ATG) was then prepared by salt fractionation with ammonium or sodium sulphate. The IgG fraction was prepared by the method of McDonough and Inman (14) by elution from DEAE-cellulose (Whatman DE-23, H. Reeve Angel Inc., Clifton, N.J.) by 0.0175 M phosphate buffer, pH 6.9. The 7S IgG fraction was obtained by gel filtration on 'Sephadex' G-200. This material is pure IgG by immunodiffusion analysis with rabbit anti-whole goat antiserum.

The evidence for the specificity of ATG can be summarized as follows:

a) Fluorescein-labelled ATG binds to over 99% of the lymphocytes in the rabbit thymus (15). Diminishing numbers of lymphocytes from other organs are stained by ATG; lymph node and peripheral blood lymphocytes show the next largest percentage of staining, and appendix and splenic lymphocytes show the least staining. Peripheral blood and lymph node lymphocytes are stained by ATG more intensely than thymus cells. Immunofluorescence studies show positive staining of subcortical areas of lymph node sections by ATG. Germinal centres are not intensely stained by ATG.

b) ATG does not inhibit putative B-cell functions. DNA synthesis induced by antiserum to rabbit immunoglobulin in cultures of peripheral blood lymphocytes (8) or lymph node cells (15) is not inhibited by ATG. ATG does not inhibit ongoing antibody synthesis and during the *in vitro* secondary response to Keyhole limpet haemocyanin (KLH), treatment with ATG at 24—48 hours in culture does not inhibit the synthesis of anti-KLH antibody seen at days 4, 5 and 6 (16). Antibody synthesis by B-cells is susceptible to inhibition by antiserum as shown by the inhibitory effect of antibody to histocompatibility antigens.

c) ATG does inhibit a number of known T-cell functions. Con A- and PHA-induced DNA synthesis in peripheral blood lymphocytes and Con A-induced DNA synthesis in lymph node cells are both inhibited by ATG (8,15). The induction of the *in vitro* antibody secondary response to the thymus-dependent antigen — KLH can be inhibited by ATG treatment during the first 24 hours of culture (16). The response to KLH in these cells can be restored by addition of lymph node cells depleted of B-cells by treatment with anti-immunoglobulin serum and complement (Cook and Stavitsky, unpublished results). Suppression of the function of T-cells is not solely due to cytotoxicity (although ATG is cytotoxic for peripheral blood lymphocytes, it does not kill lymph node cells) and may involve an effect on cyclic AMP.

d) Absorption of ATG by thymus cells, lymph node cells and peripheral blood lymphocytes results in a loss of the ability of ATG to inhibit the DNA synthesis induced in peripheral blood lymphocytes by T-cell mitogens. Bone marrow absorption is ineffective, and spleen and appendix are rather ineffective in absorbing out ATG activity (15). Brain and liver absorption are without effect on ATG activity. Thymus-absorbed ATG is incapable of suppressing the *in vitro* secondary response to KLH or of inhibiting KLH-induced MIF production (R. P. Pelley, Ph.D. Thesis, 1975, Case Western Reserve University).

e) The best characterized antiserum to T-cell antigens is anti-θ (17). This antiserum is produced by immunizing AKR mice with thymus cells from another strain of mice (usually C_3H). The differences between anti-θ and ATG can be summarized as follows: Unlike anti-θ, ATG does not require complement activity in order to effect its suppression; ATG does not kill concomitantly with suppression; ATG is not absorbed by brain cells and even without absorption contains little functional anti-histocompatibility antibody; and finally, immunofluorescent studies show more intense staining of peripheral lymphoid tissue with ATG than do thymus cells.

In summary, ATG is unlike anti-θ and appears to inhibit T-cell function with no demonstrable effect on B-cells.

3. LYMPHOKINE ASSAYS

3.1 MIF

Because several detailed descriptions of the MIF assay have been published elsewhere (18—20) our assay will be only briefly described. Peritoneal exudates were induced in normal animals by an intraperitoneal injection of sterile light mineral oil ('Klearol', Witco Chemical Co., Berwyn, Ill.). Large (4—5 kg) unanaesthetized rabbits were injected with 50 ml. We have noted radical variability in the yield of cells from rabbit to rabbit. We can only suggest the use of large animals which have been allowed to acclimatize to local facilities for several weeks before use. Guinea pigs and mice were anaesthetized with ether and then injected with 30 ml and 1.5 ml respectively. Three days after injection the animals were exsanguinated (rabbits by cardiac puncture and mice and guinea pigs by decapitation). Thorough exsanguination was essential to prevent contamination of exudates with blood. Animals were then given an i.p. injection of HBSS—10% serum (rabbits — 100 ml, guinea pigs — 30 ml and mice — 5 ml), the abdomen gently kneaded, and the exudate harvested through a peritoneal dialysis cannula using the method of David (18). The peritoneal exudate cells (PEC) were separated by centrifugation at room temperature at 1200 r.p.m., washed three times with cold HBSS—10% serum and then reconstituted to 4×10^7 cells/ml in tissue culture medium 199 (Grand Island Biological), containing all the supplements used in Complete MEM (referred to as Complete TCM-199).

Non-heparinized 75 mm capillary tubes (Dade Capiletts, American Hospital Supply) were filled with the cell suspension, sealed with 'Critoseal' (Arthur H. Thomas Co., Philadelphia, Pa.) and centrifuged at room temperature at 550 r.p.m. for 5 minutes. The tubes were cut at the cell—fluid interphase, and the portion containing the cells placed in a Mackaness-type chamber, two per chamber (Berton Plastics, Hackensack, N.J.). Half a ml of Complete TCM-199—10% serum and 0.5 ml of the sample to be assayed were added to the chamber.

Chambers were incubated for 24 h at $37°$, projected, traced and the area of migration was measured by planimetry. The average area of migration for each preparation, based on at least duplicate capillary determinations, was used to calculate the per cent inhibition as follows:

$$\frac{\%\ \text{migration}}{\text{inhibition}} = \frac{\text{Control migration} - \text{Experimental migration}}{\text{Control migration}} \times 100$$

Recently we have adapted a Millipore πMC particle measuring system interfaced with a Wang 600 programmable calculator to the measurement of cell areas. This system, although very expensive, greatly shortens the length of time required to measure migrations, calculate migration inhibition and perform necessary statistics. One hundred capillaries can be measured and the final data derived in less than one hour with this system.

3.2 Difficulties with and Modifications of the MIF Assay

Until recently, the MIF system has not been extensively applied to animals other than guinea pigs. Two major problems arise with this lymphokine assay: availability of cells and cross-species sensitivity. Macrophages are unobtainable from humans, and the lymphokine measured by an effect on the migration of human neutrophils is not MIF but the separate molecule, leukocyte inhibitory factor (LIF) (21,22). Macrophages can only sporadically be obtained in large quantities from rabbits and only small numbers are obtainable from mice. For murine MIF tests, it is advisable to use the smallest capillaries available. Disposable microsampling pipettes (20 μl, Corning catalogue #7099S) are suitable after thorough washing. The guinea pig is the only animal that consistently yields large numbers of PEC. Unfortunately, there are not a large number of defined immunological reagents developed for guinea pig systems (such as the assays for antibody synthesis, T-cells and B-cells now available for men, mice and rabbits). Guinea pig macrophages can be used to assay murine

MIF (23,24) but they are less sensitive to the lymphokine than murine macrophages. It is exceedingly difficult to measure antigen-induced rabbit MIF at all with guinea pig macrophages. However, the large amounts of MIF produced by Con A-stimulated rabbit lymphocytes does allow assay in the guinea pig cell migration system. Hopefully, further development of assays that utilize migration of cells from droplets will allow more extensive study of MIF in systems other than the guinea pig (25–27).

3.3 Lymphotoxin (LT) Assay

Complete medium for maintenance of L-cells was prepared by adding 45.6 ml tryptose phosphate broth and 114 ml foetal calf serum (heat inactivated) to each 1000 ml of MEM. Prior to use, 200 units/ml penicillin, 40 μg/ml streptomycin and 1 ml L-glutamine (200 mM) (100 x) were added to the medium. Saline A was prepared by dissolving 8 g NaCl, 0.4 g KCl, 1.0 g dextrose, 0.35 g NaHCO$_3$ in 990 ml demineralized water. Ten ml of phenol red stock solution (2 g/100 ml demineralized water), 200 units/ml penicillin, 40 μg/ml streptomycin were added. Trypsin–versene solution was prepared by mixing 4 parts 0.01% versene in phosphate-buffered saline (without calcium or magnesium) with 1 part 0.25% trypsin in saline A. The 10% foetal calf serum (in MEM), saline A and trypsin–versene were filtered through a 0.2 μ porosity 'Selas' filter after preparation as described.

Mouse L-cell fibroblast monolayers were subcultured thrice weekly for maintenance of stock lines or for assay. All medium was aspirated and the monolayer washed twice with saline A. Trypsin–versene solution was then added to the flask for ten seconds and removed by aspiration. The flasks were allowed to sit for 10–15 minutes or until the sheet of cells was freed from the flask.

For direct cytotoxicity assay 5–8 x 10^5 L-cells were added to 4 ml of medium in 30 ml plastic tissue culture flasks. One day later, concanavalin A (10 μg/ml final concentration) was added to the flasks. One x 10^7 lymphocytes were then added to each flask (control or experimental) and the cultures incubated for an additional 48 hours. The medium was aspirated off the monolayer and the cells washed twice in cord saline A and once in saline A at 37°. The remaining attached cells were removed by incubating them at 37° with 1.5 ml trypsin–versene. One ml of saline A was added to each flask and the cells suspended evenly by gentle pipetting. One ml was added to 19 ml of normal saline and counted in a model B Coulter counter, thresholds 30 to ∞, 1/amplification 1/4 and 1/aperture current 0.354. Duplicate counts were made on each culture, and the values obtained for duplicate cultures averaged.

For lymphotoxin assays L-cells were cultured as above except that instead of 4 ml of medium, 3 ml of medium and 1 ml of lymph node tissue culture supernate were added.

3.4 Skin Reactive Factor (SRF) Assay

Frozen culture supernates were thawed, clarified by centrifugation (20,000 r.p.m., 20 min in a Sorvall centrifuge) and sterilized by passage through a .22 μ

'Millipore' filter. Skin sites on the flanks of normal guinea pigs were prepared first by shaving with an Oster model A-2 animal clipper followed by depilation with 'Nair' (Carter Products, Division of Carter-Wallace, Inc., New York, N.Y.). Five minutes after application, the Nair was washed off with running tepid tap water. Samples of supernates (0.10 ml) were injected 2 hours after preparation of skin sites by the intradermal route with a 26 gauge intradermal needle (Beckton-Dickinson). Groups of 6 guinea pigs are optimal in order to assure reproducible results. Care was taken to avoid injections near areas of erythema due to depilation. Injections were evaluated for the presence of induration and erythema at 3 hour intervals. Reactions were graded according to size by a modification of the technique of Gell and Benacerraf (31) as follows: ±, 0—3 mm erythema, no induration; +, 4—7 mm erythema, no induration; 2+, 8—11 mm erythema, slight induration; and 3+, larger than 12 mm erythema with positive induration. The most frequently encountered difficulty in interpreting the results of skin reactions induced by culture supernates was non-specific inflammation in control supernates. This effect appeared to be inversely related to cell viability and was so pronounced when peritoneal exudate lymphocytes were used that studies with this population of lymphocytes have been discontinued. Finally, guinea pigs did not respond as vigorously to SRF produced in mice or rabbits as they did to SRF produced in guinea pigs. This finding was consistent with the (paragraph 3.2) cross-species sensitivity of the MIF assay.

Cell cultures of lymphoid tissues derived from rabbits and mice produced SRF. However, in these species, quantitation of SRF was not optimal. Murine SRF was quantitated by measuring (with a micrometer) the swelling induced by injection of 25 μl of tissue culture supernate into the hind footpad of an unsensitized mouse. SRF activity was expressed as the difference in thickness between the control (tissue culture medium or control culture supernate) and experimental footpads. Therefore only one variable could be examined per mouse, and a minimum of 8 mice per group were required for accurate determinations. The very large number of animals required for even simple experiments did not make this a practical assay. Furthermore, decalcification of specimens preparatory to sectioning resulted in loss of cellular detail that prevented histological analysis.

There were two technical difficulties in testing rabbit SRF. First, there were thickened, raised areas of skin on the flanks of occasional rabbits. Supernates should not be injected into these areas because of the non-specific neutrophil infiltrate that is generally present. Supernates should always be injected into skin sites of comparable thickness (no more than 10 sites per flank). Second, rabbit skin is considerably thinner than guinea pig or human skin. Therefore the investigator must be very cautious that a true intradermal (and not subcutaneous) injection is made.

Macroscopic evaluation of rabbit skin lesions was less precise than evaluation of guinea pig skin lesions. In the guinea pig there was good correlation between area of erythema (easily measured) and intensity of induration. Rabbit skin lesions demonstrated a degree of induration comparable to that seen in the guinea pig but lacked the discrete erythema that characterized the guinea pig SRF test. On microscopic examination, SRF-induced rabbit skin lesions were identical to the delayed-type hypersensitivity skin lesions classically described for this species by Gell (32) and to the SRF lesions described by Schwartz and coworkers (42).

4. PRODUCTION OF LYMPHOKINES

4.1 Problems inherent in Mitogen-induced Lymphokines

It must be appreciated that the production and assay of mitogen-induced lymphokines is fundamentally more complicated than antigen-induced lymphokine production by specifically sensitized cells. This complication arises because (unlike antigen systems) the inducing agent (Con A) affects the assay systems employed to detect lymphokines. Thus, direct injection of Con A into the skin of normal animals results in inflammation (3) whereas most antigens have no effect when they are injected into the skin of an non-immunized animal. Addition of Con A to migration chambers containing normal PEC's inhibits macrophage migration (4) whereas there is generally no effect when antigen is added to migration chambers which contain normal PEC's. Previously, most lymphokine assay systems utilized what Dumonde (1) originally termed a 'reconstituted' control. This involved addition of the inducing agent to control supernates after termination of the culture. This type of 'reconstituted' control system has obvious limitations when the inducing agent (Con A) affects the assay system. For example, normal rabbit lymph node cells were cultured with 20 μg/ml Con A. This resulted in the production of a 38% MIF activity when assayed by another (4.4) method. At the end of 24 hours in culture the supernate was removed and 0.5 ml assayed on normal guinea pig PEC. If the experiment performed above were carried on now, 10 μg of Con A would be added to the control supernate. In that case a 64% difference would be found merely between 'unreconstituted' and 'reconstituted' control supernates (due to the effect of Con A-inducing MIF production by the guinea pig lymphocytes which are always present in the normal guinea pig PEC). Therefore, any MIF actually produced by Con A-stimulated rabbit lymph node cells (38%) would be measured against a background that was already maximally responding to endogenously produced guinea pig MIF and no significant net MIF activity would be detected.

Statistical analysis of the MIF test has revealed that a major source of variability in the assay is due to variations in control migration from animal to animal (28). Small alterations in the area of control migration were found to have a disproportionately large effect on the migration ratio. This suggests that when reconstitution controls are used in mitogen-induced lymphokine systems an uncontrolled variable is introduced into the experimental protocol, increasing animal to animal variation even further.

This 'reconstitution' problem was not just restricted to mitogen-induced lymphokine systems. Many of the antisera to cells that are useful in probing the *in vitro* immune response have an effect similar to Con A on assay systems (29,30). This paragraph will describe three methods that can be used to minimize these difficulties and will supply data from specific experiments to illustrate these techniques.

A second and less troublesome problem with mitogen-induced lymphokines was the relatively narrow concentration range over which mitogens exerted a stimulatory effect. This appeared to be most pronounced for lymph node cells and least pronounced for peripheral blood lymphocytes. When Con A was injected into the skin or added directly to migration chambers this was not a significant problem (3,4). This problem was particularly severe when Con A-induced lymphokine production was studied in medium free of serum. In serum-free medium the optimal Con A dose appeared to be 5—10-fold lower. In general, optimal lymphokine production occurred over a three- to fourfold range (guinea pig lymph node: 6—24 μg/1 x 10^7 cells; rabbit lymph node: 5—25 μg/1 x 10^7 cells; and mouse spleen: 0.3—3 μg/1 x 10^7 cells). This was in contrast to the somewhat wider range over which antigen-stimulated MIF production by sensitized cells occurred (2). Supraoptimal doses of Con A appeared to inhibit lymphokine production. The reason for this high-dose Con A inhibition is not understood. It is important, therefore, to run a dose-response curve and insure that the appropriate dose of Con A is being used.

4.2 Production of MIF by Con A-stimulated Rabbit Lymph Node Cells: Use of ^{125}I-Con A to Monitor 'Appropriate' Reconstitution

Although the 'reconstitution' control assay is less than optimal, its inherent difficulties could be minimized when the amount of Con A added to control supernates was kept to a minimum. This was possible because not all the Con A added to the culture remained in the culture supernate. Con A, unlike antigen, was bound in large quantities to the cell surface. Con A also bound to macroglobulin glycoproteins in the culture medium which precipitated when the culture medium was thawed prior to assay. Therefore, if lymphoid cells were cultured with low concentrations of Con A, only minimal amounts of Con A were added to control supernates for reconstitution.

For example, rabbit lymph node cells were incubated for 24 hours in medium containing 6 μg/ml ^{125}I-Con A. The supernates of these cells contained 4 μg Con A (as determined by ^{125}I-counting) at the end of this period. Therefore the 0.5 ml aliquots (the amount added to the migration chambers) of the supernates contained 2 μg Con A. Two μg Con A in medium from unstimulated cultures caused a 29% ± 10% inhibition of macrophage migration. Presumably, inhibition of migration in excess of 29% is due to MIF produced by the Con A-activated rabbit lymph node cells. It was, therefore, noteworthy that migration was inhibited 57% ± 6% when 0.5 ml of supernate from Con A-stimulated rabbit lymph node cells was added to guinea pig peritoneal exudate cells. Thus, Con A caused lymph node cells to produce 28% (57%—29%) more MIF activity than was produced upon direct addition of Con A itself to the PEC.

4.3 Production of MIF by Con A-stimulated Rabbit Lymph Node Cells: Use of Gel Filtration to Detect MIF Activity

There are obvious limitations to the use of the system outlined above. For example, if the experiment considered in paragraph 4.1 was examined the 'appropriate' reconstitution system would not work. Out of 20 μg Con A added to the culture, 4.8 μg were bound to the lymph node cells leaving 15.2 μg/ml in the supernate. If a 0.5 ml aliquot of this was added to guinea pig PEC the resultant inhibition due to Con A alone in the system (7.6 μg) would be 51%. Thus there would still be significant interference with the assay — more MIF (51%) activity due to Con A in the supernate than MIF activity produced by the rabbit lymph node cells (38%).

After removal of Con A present from the supernate, MIF can be assayed without interference from MIF endogenously produced by Con A-stimulated lymphocytes in the guinea pig PEC assay system. Although the molecular weight of Con A is not very different from that of MIF, gel filtration can be used to obtain MIF activity free of contaminating Con A. This is because (i) any uncomplexed Con A (often the bulk of the Con A in supernates of cell cultures lacking serum) is bound tightly to the 'Sephadex'. (ii) In serum-containing medium there were only minute amounts of Con A present that were not complexed to serum macroglobulin glycoproteins. These complexes eluted from 'Sephadex' G-200 or G-100 in the void volume.

For example, verification that MIF was produced in the experiment cited in 4.1 was sought by gel filtration of supernate from such cultures (Figure 1). ^{125}I-Con A (20 µg/ml) was incubated with lymph node cells (2×10^7/ml) for 24 hours. Supernates of Con A- and control-incubated cells were concentrated by ultra-filtration in an Amicon cell and fractionated by gel filtration on a 2.5 x 95 cm 'Sephadex' G-200 column. As shown in Figure 1, Con A ^{125}I-activity was restricted to the fall through region. The post-albumin eluates were free of detectable radioactivity. Fractions were pooled starting in the 7S region, sterilized by passage through a .22 µ 'Millex' filter, and concentrated to 5 ml by vacuum dialysis against TCM-199. Fractions were then resterilized, and the volume adjusted to 25 ml with Complete TCM-199—10% NRS. The post-albumin region (Con A-stimulated culture supernate) contained 25% MIF activity when compared with this region of the control column (note that if the fall-through region had been assayed it too would have contained inhibitory activity because of the presence of Con A). Thus MIF activity free of contaminating mitogen was isolated from the supernates of Con A-stimulated rabbit lymph node cells.

4.4 Production of MIF by Rabbit Lymph Node Cells in Con A-poor Medium

Although the method in 4.3 gives an unequivocal demonstration of the actual presence of lymphokine activity it is too cumbersome for routine use and should only be used to verify the results of critical experiments. An alternative system has been designed to produce MIF in the absence of amounts of contaminating Con A that would affect the assay system. Con A need not be present during the second 24 hours in culture for MIF production to continue and DNA synthesis to occur. Cells were incubated in medium containing 20 µg/ml Con A for 24 hours at which time the Con A was washed out and fresh medium lacking Con A added. After the cells were incubated for another 24 hours, an assay for radioactivity indicated the medium contained only 0.8 µg ^{125}I-Con A/ml. However, 24—48 hour culture supernates of Con A-stimulated cells showed significant (38%) MIF activity although they contained amounts (0.4 µg) of Con A which did not affect the assay system (Table I, line 2).

Thus after 24 hours of stimulation of cells by Con A, these cells continued to produce MIF even when cultured in fresh medium devoid of Con A. Not enough Con A remained in these supernates (0.8 µg/ml) to affect the assay. The bio-chemical specificity of this system was demonstrated by the inhibitory effect exerted by methyl α-mannoside (Table I, line 4) (during the first 24 hours in

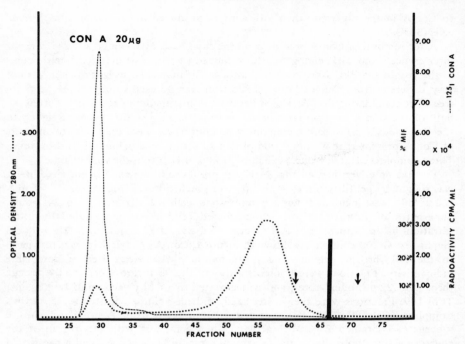

Figure 1. 'Sephadex' G-200 fractionation of supernates from 0–24 hour cultures of lymph node cells incubated with Con A (20 μg/ml). Fifty ml of Con A-stimulated or control cell supernates were concentrated by pressure dialysis in an Amicon cell to a volume of 5 ml and applied to a 2.5 × 95 cm 'Sephadex' G-200 column. The column was run with 0.14 M NaCl at 12 ml/hour and 5 ml fractions collected. Protein concentration was determined by absorbance at 280 nm (. . . . . .). Con A was determined by ^{125}I γ-radioactivity (− − − − − −). The fractions eluting after albumin (indicated ↓↓) were pooled, concentrated at least tenfold by vacuum dialysis, and dialysed for 24 hours against two changes of MEM. One-half ml aliquots of concentrate were assayed in quadruplicate for their effect upon the migration of non-immune guinea pig peritoneal exudate cells. MIF activity was calculated (Table IV) by comparing the area of migration of the 20 μg Con A fraction (the post-albumin fractions from columns run with supernates of 20 μg Con A-stimulated cells) to the area of migration of the control fractions (the post-albumin fractions of columns run with supernates of unstimulated cells). Fractions eluting with albumin showed no significant MIF activity. Results are the means of two column runs

culture) which prevented Con A from binding to the cells. Furthermore, although ATG inhibited the migration of rabbit PEC when added to migration chambers directly (data not shown), it did not alter the migration of guinea pig PEC (Table I, line 6) and it did not induce rabbit lymph node cell suspensions to produce significant amounts of MIF (Table I, line 5).

5. IDENTIFICATION OF T-CELLS AS SOURCE OF CON A-INDUCED LYMPHOKINES

The utility of Con A as a probe of cell-mediated immunity resides in the specificity of this mitogen for T-cells thus providing the closest artificial analogue

TABLE I.   Production and assay of MIF in Con A—poor medium

Test material[a] (0.5 ml aliquots of supernate from cultures of)	Area of macrophage migration[b] (mm^2)	Macrophage migration (%)
1 — Control (unstimulated cells)	1.050 ± 0.070	
2 — Control + 0.4 μg Con A (unstimulated cells)	1.104 ± 0.073	100
3 — Con A (Con A-stimulated cells)[c]	0.682 ± 0.161	61.7
4 — Con A + MAM (Con A/MAM incubated cells)	1.122 ± 0.79	101
5 — ATG[d] (ATG-incubated cells)	1.098 ± 0.081	100
6 — Control + 1500 μg ATG (unstimulated cells)	1.008 ± 0.073	90

[a]Pooled lymph node cell supernates were from cultures of 4 non-immune rabbits. During the first 24 hours in culture, cells were either unstimulated (control), stimulated with 20 μg/ml ^{125}I-Con A (Con A), treated with 3000 μg/ml ATG (ATG), or treated with both 3000 μg/ml ATG and 20 μg/ml Con A (Con A + ATG). Some Con A-stimulated cultures also contained 0.05 M methyl-α-mannoside (MAM + Con A). At the end of the first 24 hours in culture the medium was removed, the cells were washed, and cultured for an additional 24 hours in fresh medium lacking ATG or Con A. These 24—48 hour supernates lacking large quantities of Con A or ATG were assayed for MIF activity.
[b]0.5 ml aliquots of the 24—48 hour supernates (test material) were assayed for their effect upon the migration of non-immune guinea pig peritoneal exudate cells. Results are the mean ± standard error of four chambers containing duplicate capillaries. The only test material with significant migration inhibitory activity was the supernate of cells incubated 0—24 h with Con A — line 3.
[c]Con A content by γ-counting — 0.8 μg/ml.
[d]ATG content not determined but should be of the order of 4% or less of original material that is 120 μg/ml or less.

of the biologically significant event. That Con A stimulates T-cells and only T-cells to synthesize DNA is by now beyond question (6—8). The following paragraph extends the T-cell specificity of Con A to lymphokine (MIF) production. By the use of a monospecific antiserum, T-cells can be shown to be required for Con A-stimulated MIF production.

5.1 Effect of ATG Pretreatment of Lymph Node Cells upon Con A-stimulated MIF Production

Table II illustrates that incubation of lymph node cells with 6 μg/ml Con A resulted in the production of significant amounts of MIF. If cells were preincubated for 12 hours with ATG, then washed, and incubated with 6 μg Con A, MIF

TABLE II.   Effect of ATG on Con A-induced MIF production and on Con A and anti-Fab induced DNA synthesis in lymph node cells[a]

ATG (μg)	No addition (control)		Con A (6 μg/ml)		AFab (500 μg/ml)	
	DNA[b] (c.p.m.)	MIF (%)	DNA[b] (c.p.m.)	MIF[c] (%)	DNA[b] (c.p.m.)	MIF (%)
None	950[d]	<20	11,000[d]	57 ± 6	5500[d]	<20
100	750	<20	9900	40 ± 17	5500	<20
400	650	<20	3700	39 ± 7[e]	7500	<20
1000	600	<20	1200	29 ± 2[e]	8150	<20
2000	1450	<20	1450	34 ± 8[e]	5400	<20

[a]Lymph node cells from two animals were each cultured in quadruplicate. ATG was added at the beginning of culture. Cultures were washed at the end of the 12 hours incubation at 37°. The activator for T-cells (Con A) and B-cells (antibody to the Fab fragment of rabbit IgG — AFab) were then added and the cells cultured at 37° for 24 hours. Supernates were removed and assayed in quadruplicate for MIF activity. Fresh medium containing ^{3}H-thymidine was added to the cells. The cultures were incubated for a final 24 hours at 37°, and DNA synthesis was assayed.
[b]^{3}H-Thymidine incorporation between 37 and 60 hours in cultures as determined by cold TCA-precipitable radioactivity.
[c]Per cent inhibition of migration of non-immune guinea pig peritoneal exudate cells caused by 0.5 ml of 13−36 hour supernate. These supernates contain 2 μg Con A. Addition of control supernates with 2 μg added Con A to these cells results in an inhibition of migration of 29% ± 10% compared to control supernates without Con A.
[d]Average variation from tube to tube was 10% or less.
[e]Significantly different from untreated Con A control − $p \le 0.10$ for each of the two animals by Fisher's T test. These supernates do not show an inhibition of migration greater than control supernates with 2 μg added Con A.

production was inhibited (Table II). Some inhibition was observed with concentrations of ATG as low as 100 μg/ml; however, complete suppression of MIF production was achieved only at 1000−2000 μg/ml (per 2×10^7 cells) of ATG. After pretreatment with these doses of ATG, inhibition of migration beyond that expected from the residual Con A in the culture was not found — that is, only 30% migration inhibition was seen and that could be accounted for by the 2 μg Con A in the sample assayed. In these cells concomitant suppression of Con A-induced DNA synthesis followed closely the suppression of MIF production. In parallel cultures, no suppression of DNA synthesis induced by B-cell mitogen (antibody to Fab fragment of rabbit IgG) was seen. This indicated that B-cell function was not inhibited in this experiment. Neither the B-cell mitogen nor the B-cell mitogen plus ATG induced MIF production.

5.2 EFFECT OF SIMULTANEOUS ADDITION OF CON A AND ATG ON MIF PRODUCTION

At least two possible mechanisms for the action of ATG may be conceived: the binding of ATG to the T-cell membrane could 'mask' the Con A receptor sites and result in inhibition of Con A binding, or alternatively, ATG could directly destroy T-cells. To elucidate the mechanism by which ATG suppresses MIF production the following experiments were done. Con A and/or ATG were added to lymph node cells for the first 24 hours. During this time synthesis of extracellular protein (tissue culture supernate) was determined by ^{14}C-leucine incorporation into TCA-insoluble material. In parallel cultures the binding of ^{125}I-Con A to cells was assayed. At the end of 24 hours the ATG and Con A were washed out and the cells cultured for an additional 24 hours in medium free of both Con A and ATG. These supernates were assayed for MIF activity.

Table III demonstrates that whereas ATG inhibited Con A from binding to cells by only 8%, this antibody suppressed MIF production (24—48 hours) by 63%. Methyl α-mannoside (MAM) inhibited the binding of Con A by 63% and completely inhibited Con A-induced MIF production. Thus there is a distinct difference between the actions of ATG and MAM: MAM incubated cells showed a definite decrease in Con A binding. ATG incubated cells showed no decrease in Con A binding, although both ATG and MAM inhibited Con A-induced MIF production.

TABLE III.   Effect of 0—24 hour Con A and ATG upon lymph node cells[a]

Cell[a] culture	Con A bound[b] 0—24 hours (out of 20 μg)	Protein synthesis[c]	MIF production[a] (24—48 hours)
Control		12,500	
Con A	4.8γ	18,500	38.3[d]
Con A + methyl α-mannoside	1.9γ	14,400	0
Con A + ATG	4.4γ	32,200	14.2
ATG		28,200	0

[a] Same experiment as shown in Table I — refer to footnote a in Table I for explanation of conditions.

[b] Con A binding was studied using 2×10^7 cells at $37°$ without the use of glutaraldehyde or other metabolic inhibitors for 24 hours. Bound Con A was calculated as

$$\% \text{ Bound Con A} = \frac{\text{c.p.m. in pellet/Specific activity } ^{125}\text{I-Con A (c.p.m./}\mu g)}{\mu g \text{ Con A added to the culture}}$$

[c] ^{14}C-Leucine incorporation 0—24 hours in cultures as determined by TCA-precipitable radioactivity in culture supernates.

[d] Significant by Fisher's T test $p \leqslant 0.10$.

Previous studies from this laboratory showed that ATG does not lyse rabbit lymph node cells (16). Further evidence that ATG does not inhibit T-cell functions by lysing these cells or by inhibiting the binding of Con A is presented in Table III. Con A stimulated an increase in extracellular protein synthesis of 6000 c.p.m. (1.48-fold). The presence of MAM in cultures was associated with a 68% suppression of this stimulation. Thus when the binding of Con A to the cells is inhibited, the stimulation of protein synthesis is inhibited. ATG alone caused an increase of 15,700 c.p.m. (2.3-fold). Simultaneous exposure of cells to ATG and Con A caused an additive effect on protein synthesis. The expected additive increase in protein synthesis of 21,700 c.p.m. is not significantly different from the observed increase of 19,700 c.p.m. The addition of ATG to control or Con A-treated cells certainly did not cause the drop in protein synthesis expected with rapid overt cytolysis.

Thus unlike MAM, ATG-stimulated protein synthesis *per se*: it does not inhibit the stimulation of protein synthesis caused by Con A but rather contributes an additional rise of its own. In summary, these data are inconsistent with the possibility that the inhibition by ATG of Con A—induced MIF production was due to cytotoxicity of ATG for lymphocytes or due to ATG inhibiting the binding of Con A to cells.

5.3 MIF Production in the 24 Hours Following Simultaneous Addition of ATG and Con A: Use of Gel Filtration

Fractionated supernates from the first 24 hours of culture showed no significant difference in the distribution of ^{125}I-Con A whether or not the culture had been treated with ATG (compare Figure 1 with Figure 2). The recovery of Con A as a soluble, macromolecular complex eluting in the fall through peak was similar

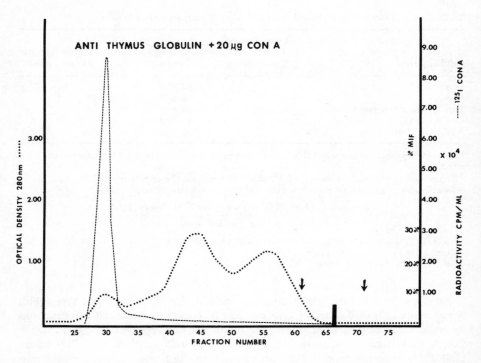

Figure 2. 'Sephadex' G-200 fractionation of supernates from 0—24 hour cultures of lymph node cells incubated with Con A (20 μg/ml) and ATG (1000 μg/ml). Supernates were concentrated, chromatographed and assayed as in Figure 1. There is no apparent change in the pattern of elution of ^{125}I-Con A from this column as compared with that of Figure 1. There is no difference between the amount of Con A eluted from this column and that eluted from the column in Figure 1 (Table IV). MIF activity was calculated and compared to similar fractions of a column run with superantes of cells incubated with ATG (1000 μg/ml). Compared with the appropriate control column (Table IV) there is no significant MIF activity in the post-albumin eluate of this column.

TABLE IV.   Effect of simultaneous addition of Con A and ATG upon Con A-induced stimulation of DNA synthesis and MIF production — use of gel filtration[a]

Addition	DNA synthesis 24—48 hours in culture	Analysis by gel filtration of 0—24 hour supernates		
		area of macrophage migration[b]	% MIF[c]	quantity of Con A in soluble complex
None (Control)	486	8.50		
20 μg Con A	6251	6.40	24.6	88.6% of the Con A in the original supernate
1000 μg ATG	759	9.35		
20 μg Con A + 1000 μg ATG	3457	8.82	5.6	92.8% of the Con A in the original supernate
500 μg AFab[d]	2017	8.45	None	
500 μg AFab + 1000 μg ATG	2520			

[a]ATG, Con A and AFab were added at the beginning of culture. Supernates were removed at the end of 24 hours of culture and analysed by gel filtration (Figures 1 and 2) for distribution of ^{125}I-Con A and for MIF activity. Fresh culture medium containing ^{3}H-thymidine was added to the cells, they were cultured for a second 24 hours, and DNA synthesis was determined.
[b]This was the area (when projected on tracing paper) of migration of quadruplicate assays from two column runs. The fractions eluting after albumin were pooled (Figures 1 and 2), concentrated tenfold, and dialysed against Complete MEM. One half ml aliquots of these fractions were then added to 1/2 ml aliquots of TCM-199—20% GPS and the combined medium added to chambers with capillaries of non-immune guinea pig peritoneal exudate cells. The migrations were projected and traced after 24 hours of culture.
[c]Fractions from columns run with supernates of unstimulated cells, ATG-treated cells, ATG-treated Con A-stimulated cells and AFab-stimulated cells all showed approximately the same effect upon the migration of the macrophage. The only fraction showing significant inhibitory activity was from the Con A column.
[d]AFab is the IgG fraction of goat antiserum to the Fab fragment of rabbit IgG and was kindly donated by Dr. M. W. Fanger, Department of Microbiology, Case Western Reserve University, Cleveland, Ohio.

(Table IV) when control and ATG-treated culture supernates were subjected to gel filtration. The 4S eluate of columns run with supernates of Con A-stimulated cells showed a significant amount of MIF activity when compared with equivalent eluates of the control column (average of four duplicate determinations, Table IV). There was no significant MIF activity in the 4S eluates of columns run with supernates of ATG or ATG-Con A-incubated cell cultures. Therefore concomitant with the reduction in Con A-induced DNA synthesis there was a suppression of MIF production after treatment of lymph node cells with ATG. This suppression occurred even within the first 24 hours after simultaneous addition of Con A and ATG. No MIF activity was observed in column fractions of cell cultures stimulated with the B-cell mitogen antibody to Fab fragment of rabbit IgG.

5.4 Evidence that Cells Other than T-cells can make Lymphokines

Although there is no conclusive *in vivo* evidence that MIF production occurs upon antigen stimulation of primed B-cells, there is mounting evidence that under certain conditions cells other than T-cells can make lymphokines. Yoshida and coworkers (39) suggested that non-immune B-cells may respond to mitogen stimulation by production of small amounts of MIF. In their work, T- and B-cell populations were separated by their different affinities for complement-coated sheep red blood cells. Guinea pig non-complement-receptor lymphocytes — presumably T-cells — produced MIF only after priming and challenge with the appropriate antigen. Stimulation of non-immune-complement-receptor lymphocytes (presumably B-cells) with E. coli endotoxin or high doses of purified protein derivative of mycobacterial culture filtrates caused production of MIF. However, these investigators did not demonstrate that MIF production by complement-receptor lymphocytes was not thymus dependent and/or due to cells bearing thymus antigens.

Wahl and coworkers (40) have studied production of the lymphokine monocyte chemotactic factor (MCF) by mitogen-stimulated B-cells. They purified T- and B-cells from guinea pig spleen cell suspensions by a variety of techniques. They stimulated purified T- and B-cells with antigen and a variety of mitogens. Their study both confirms our findings that the T-cell-specific mitogen Con A stimulates only T-cells to produce lymphokines and also confirms the studies of Yoshida and coworkers that B-cell mitogens can induce only B-cells to produce lymphokines. Several investigators have demonstrated that cells other than lymphocytes can make MIF. Continuous lines of both lymphoid and non-lymphoid cells (41) produce MIF activity that is physicochemically similar to human, antigen-induced, lymphocyte-produced MIF.

6. POSSIBLE MECHANISMS OF LYMPHOKINE PRODUCTION AND RELEASE

The activation of CMI effector cells by Con A appears to be a process involving several steps: (a) binding of inducer, (b) cell activation and (c) cellular synthesis of macromolecules with the subsequent release of these mediators from the cells. The release of mediators of immediate hypersensitivity from basophils (44) may provide a crude analogy for this production of mediators of delayed hypersensitivity (Figure 3). Both processes involve multistep mechanisms which begin with the binding of antigen to a specific receptor (IgE for basophils, unknown for T-cells) lead to activation of the cell and culminate in the release of mediators of inflammation. In both systems the specificity of the receptors for the appropriate antigen can be bypassed by the use of an inducer capable of binding to either the antigen receptor itself or to another appropriate receptor.

The initial step in Con A-induced lymphocyte stimulation is the binding of the inducer (Con A) to the lymphocyte surface. This step can be specifically inhibited by methyl α-mannoside (MAM) with resultant inhibition of all Con A-induced reactions — increased protein and DNA synthesis and MIF production (Table III). Once cellular activation is achieved the presence of Con A in the culture medium is no longer required (Table I). Activated cells do not even appear to continually require Con A on their surfaces to perform their function (47). ATG affected a process occurring after Con A binding. Simultaneous addition of Con A and ATG did not result in a decrease in Con A binding (Table III). ATG did not cause dissociation of Con A from lymphocytes. This and other data suggests that the cell surface binding site for Con A was probably not occluded by ATG. In a similar

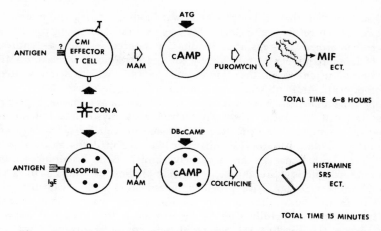

Figure 3. Parallel cellular events in immediate and delayed hypersensitivity

fashion basophils can be activated by the binding of Con A to their surfaces (43). The formation of activated basophils can also be inhibited by the presence of methyl α-mannoside.

Once an activated cell is formed the terminal process consists of the release of mediators of inflammation. In human basophils (44) this process occurs quickly (minutes), involves a microtubule-mediated contractile process (inhibitable by colchicine or vinblastine) and requires anaerobic glycolysis. These terminal events are the stage at which the greatest differences exist between Con A-induced lymphokine production and basophil degranulation. In lymphocytes the terminal steps continue for hours. They do not involve the release of preformed mediators (like histamine) but rather the synthesis (45) and release of a protein. The information for lymphokine synthesis must be only 'lightly buried' in the mammalian cell genome. MIF production (unlike antibody production) does not require prior DNA synthesis. A wide variety of cells (T-, B-, non-lymphoid) can produce lymphokines (paragraph 5.4) after appropriate stimulation. The terminal steps in Con A-stimulated lymphokine production can be inhibited by disruption of protein synthesis by puromycin (4). ATG probably does not affect these terminal steps since ATG-treated cultures continued to synthesize and export proteins (Table III). Unlike basophils, the terminal steps for lymphocytes do not appear to require microtubular function − vinblastine and colchicine have no effect on MIF production (46).

The central step − cell activation − is currently under intense investigation. In basophils (44) the central process requires Ca^{2+}, Mg^{2+}, involves the activation of an esterase, and appears to be regulated by the cellular concentration of cyclic AMP. Elevation of cyclic AMP directly by dibutryl cAMP or indirectly by catecholamines or methylxanthines inhibited histamine release by human basophils. The central process for lymphocyte activation is not well understood. Although ATG appears to affect an early step in lymphocyte activation it does not directly inhibit the initial step of lymphocyte stimulation − inducer binding. However, the data presented here do not eliminate the possibility that ATG caused a rearrangement of the cell surface such that, although Con A was bound, capping and internalization did not occur. Therefore we suspect 'ATG acts by subtly deranging one or more of the several intermediate steps after binding of inducer and before synthesis and release of lymphokines. ATG may act by altering intracellular levels of cAMP. This would be consistent with the studies of Pick (46) which demonstrated that elevation of cyclic AMP inhibits the production of antigen-induced MIF.

At one time immediate hypersensitivity was envisaged as a simple, largely passive process. Delayed hypersensitivity was viewed as a mysterious cellular activity resulting in a lump in the skin. Recent advances in cellular physiology and the biochemistry of the mediators of inflammation have made us aware that at the cellular level both processes are strikingly similar. Both histamine release and lymphokine production are initiated by the binding of antigen to a cellular receptor. Antisera to IgE have provided investigators with a valuable tool that enabled them to bypass that antigen specificity and analyse immediate hypersensitivity directly. Hopefully Con A will provide a tool that will allow us to directly probe the biochemical events in delayed hypersensitivity without the complications of antigen specificity.

7. ACKNOWLEDGEMENTS

The authors would like to thank Professor A. B. Stavitsky, Department of Microbiology, Case Western Reserve University for reagents and assistance in many of the experiments described here. He liberally gave of his time and thought to read, criticize and discuss our ideas. We also thank Ms. Holly Stinson for typing and Ms. Susan Damsel for proof-reading this manuscript

8. REFERENCES

1. Dumonde, D. C., Wolstencroft, R. A., Panayi, G. S., Matthew, M., Morley, J. and Howson, W. T. (1969) ' "Lymphokines": non-antibody mediators of cellular immunity generated by lymphocyte activation', *Nature*, 224, 38–42.
2. David, J. R. and David, R. R. (1972) 'Cellular hypersensitivity and immunity', *Prog. Allergy*, 16, 300–449.
3. Schwartz, H. J., Leon, M. A. and Pelley, R. P. (1970) 'Concanavalin A induced release of skin reactive factor from lymphoid cells', *J. Immunol.*, 104, 265–8.
4. Pelley, R. P. and Schwartz, H. J. (1972) 'The production of migration inhibitory factor by nonimmune guinea pig lymphoid cells incubated with concanavalin A', *Proc. Soc. Exp. Biol. Med.*, 141, 373–8.
5. Schwartz, H. J. and Wilson, F. (1971) 'Target cell destruction *in vitro* by concanavalin A-stimulated lymphoid cells', *Am. J. Pathol.*, 64, 295–304.
6. Stobo, J. D., Rosenthal, A. S. and Paul, W. E. (1972) 'Functional heterogeneity of murine lymphoid cells. I. Responsiveness to and surface binding of concanavalin A and phytohemagglutinin', *J. Immunol.*, 108, 1–17.
7. Owen, F. L. and Fanger, M. W. (1974) 'Studies on the human T-lymphocyte population. I. The development and characterization of a specific anti-human T-cell antibody', *J. Immunol.*, 113, 1128–44.
8. Fanger, M. W., Pelley, R. P. and Reese, A. L. (1972) '*In vitro* demonstration of two antigenically-distinct rabbit lymphocyte populations', *J. Immunol.*, 109, 294–303.
9. Daguillard, F. and Richter, M. (1969) 'Cells involved in the immune response. XII. The differing responses of normal rabbit lymphoid cells to phytohemagglutinin, goat anti-rabbit immunoglobulin antiserum and allogeneic and xenogeneic lymphocytes', *J. Exp. Med.*, 130, 1187–208.

10. Pick, E., Brastoff, J., Krejci, J. and Turk, J. L. (1970) 'Interaction between "sensitized lymphocytes" and antigen *in vitro*. II. Mitogen-induced release of skin reactive and macrophage migration inhibitory factors', *Cell Immunol.*, 1, 92—109.

11. Fanger, M. W., Hart, D. A., Wells, J. V. and Nissinoff, A. (1970) 'Requirement for cross-linkage in the stimulation of transformation of rabbit peripheral lymphocytes by antiglobulin reagents', *J. Immunol.*, 105, 1484—92.

12. Hunter, W. M. and Greenwood, F. C. (1962) 'Preparation of iodine-131 labelled human growth hormone of high specific activity', *Nature*, 194, 495—6.

13. Mittal, K. K., Mickey, M. R., Singal, D. P. and Terasaki, P. I. (1968) 'Serotyping for homotransplantation. XVIII. Refinement of microdroplet lymphocyte cytotoxicity tests', *Transplantation*, 6, 913—27.

14. McDonough, R. J. and Inman, F. P. (1970) 'Method for the preparation of normal rabbit serum immunoglobulin M (IgM)', *Anal. Biochem.*, 36, 495—504.

15. Fanger, M. W., Reese, A. C., Schoenberg, M. E., Stavitsky, A. B. and Reese, A. L. (1974) 'Evidence for T lymphocyte subpopulations in the rabbit', *J. Immunol.*, 112, 1971—80.

16. Stavitsky, A. B. and Cook, R. G. (1974) '*In vitro* anamnestic response of rabbit lymph node cells: Evidence for cell collaboration in induction and regulation', *J. Immunol.*, 112, 583—93.

17. Schlesinger, M. (1972) 'Antigens of the thymus', *Progr. Allergy*, 16, 214—99.

18. David, J. R. and David, R. (1971) 'Assay for inhibition of macrophage migration', in *In Vitro Methods in Cell-Mediated Immunity*, eds. Bloom, B. R. and Glade, P. R. (Academic Press, New York), pp. 249—58.

19. Bloom, B. R. and Bennet, B. (1971) 'The assay of inhibition of macrophage migration and the production of migration inhibitory factor (MIF) and skin reactive factor (SRF) in the guinea pig', in *In Vitro Methods in Cell-Mediated Immunity*, eds. Bloom, B. R. and Glade, P. R. (Academic Press, New York), pp. 235—48.

20. Thor, D. E. (1971) 'The capillary tube migration inhibition technique applied to human peripheral lymphocytes using the guinea pig peritoneal exudate as the indicator cell population', in *In Vitro Methods in Cell-Mediated Immunity*, eds. Bloom, B. R. and Glade, P. R. (Academic Press, New York), pp. 273—80.

21. Rocklin, R. E. (1974) 'Products of activated lymphocytes: leukocyte inhibitory factor (LIF) distinct from migration inhibitory factor (MIF)', *J. Immunol.*, 112, 1461—6.

22. Rocklin, R. E. (1975) 'Partial characterization of leukocyte inhibitory factor by concanavalin A-stimulated human lymphocytes (LIF$_{Con A}$)', *J. Immunol.*, 114, 1161—5.

23. Boros, D. L., Pelley, R. P. and Warren, K. S. (1975) 'Spontaneous modulation of granulomatous hypersensitivity in schistosomiasis mansoni', *J. Immunol.*, 114, 1437—41.

24. Boros, D. L., Warren, K. S. and Pelley, R. P. (1973) 'The secretion of migration inhibitory factor by intact schistosome egg granulomas maintained *in vitro*', *Nature*, 246, 224—6.

25. Colley, D. G. (1973) 'Eosinophils and immune mechanisms. I. Eosinophil stimulation promoter (ESP): A lymphokine induced by specific antigen or phytohemagglutinin', *J. Immunol.*, 110, 1419—23.

26. Warren, K. S., Karp, R., Pelley, R. P. and Mahmoud, A. A. F. 'The "eosinophil stimulation promoter" test in murine and human *Trichinella spiralis* infection', *J. Infect. Dis.*, Sept. 1976. In press.

27. Pelley, R. P., Karp, R., Mahmoud, A. A. F. and Warren, K. S. 'Antigen dose-response and specificity of production of the lymphokine, eosinophil stimulation promoter', *J. Infect. Dis.*, Sept. 1976. In press.

28. Bergstrand, H. and Källen, B. (1973) 'On the statistical evaluation of the macrophage migration inhibition assay', *Scand. J. Immunol.*, 2, 173—87.

29. Lamelin, J. P. and Vassalli, P. (1971) 'Inhibition of macrophage migration by a soluble factor from lymphocytes stimulated with PHA or ALS', *Nature*, 229, 426—8.

30. Pekárek, J., Svejcar, J. and Johanovský, J. (1971) 'The inhibition of formation of migration inhibiting factor (MIF) by various antilymphocyte sera', *Immunology*, 20, 895—900.

31. Gell, P. G. H. and Benacerraf, B. (1961) 'Studies on hypersensitivity. IV. The relationship between contact and delayed sensitivity: A study on the specificity of cellular immune reactions', *J. Exp. Med.*, 113, 571—85.

32. Gell, P. G. H. (1959) 'Cytologic events in hypersensitivity reactions', in *Cellular and*

Humoral Aspects of the Hypersensitive States, ed. Sherwood Lawrence, H. (Hoeber-Harper, New York), Chapter 2, pp. 43—66.

33. Kolb, W. P. and Granger, G. A. (1968) 'Lymphocyte *in vitro* cytotoxicity: Characterization of human lymphotoxin', *Proc. Nat. Acad. Sci. U.S.A.*, **61**, 1250—5.

34. Granger, G. A. (1972) 'Lymphokines — the mediators of cellular immunity', *Ser. Haematol.*, **5**, 8—40.

35. Pick, E., Krejci, J. and Turk, J. L. (1970) 'Release of skin reactive factor from guinea-pig lymphocytes by mitogens', *Nature*, **225**, 236—8.

36. Dutton, R. W. (1972) 'Inhibitory and stimulatory effects of concanavalin A on the response of mouse spleen cell suspensions to antigen. I. Characterization of the inhibitory cell activity', *J. Exp. Med.*, **136**, 1445—60.

37. Rich, R. R. and Pierce, C. W. (1974) 'Biological expressions of lymphocyte activation. III. Suppression of plaque-forming cell responses *in vitro* by supernatant fluids from concanavalin A-activated spleen cell cultures', *J. Immunol.*, **112**, 1360—8.

38. Mackler, B. F., Wolstencroft, R. A. and Dumonde, D. C. (1972) 'Concanavalin A as an inducer of human lymphocyte mitogenic factor', *Nature (New Biol.)*, **239**, 139—42.

39. Yoshida, T., Sonozaki, H. and Cohen, S. (1973) 'The production of migration inhibitory factor by B and T cells of the guinea pig', *J. Exp. Med.*, **138**, 784—97.

40. Wahl, S. M. Iverson, G. M. and Oppenheim, J. J. (1974) 'Induction of guinea pig B-cell lymphokine synthesis by mitogenic and nonmitogenic signals to Fc, Ig, and C3 receptors', *J. Exp. Med.*, **140**, 1631—45.

41. Papageorgiou, P. S., Sorokin, C. F. and Glade, P. R. (1974) 'Similarity of migration inhibitory factor(s) produced by human lymphoid cell line and phytohemagglutinin and tuberculin-stimulated human peripheral lymphocytes', *J. Immunol.*, **112**, 675—82.

42. Schwartz, H. J., Catanzaro, P. J. and Leon, M. A. (1971) 'An analysis of the effects of skin reactive factor released from lymphoid cells by concanavalin A *in vivo*', *Am. J. Pathol.*, **63**, 443—62.

43. Keller, R. (1973) 'Concanavalin A, a model "antigen" for the *in vitro* detection of cell-bound reaginic antibody in the rat', *Clin. Exp. Immunol.*, **13**, 139—47.

44. Becker, E. L. and Henson, P. M. (1973) '*In vitro* studies of immunologically induced secretion of mediators from cells and related phenomena', *Advan. Immunol.*, **17**, 93—193.

45. Sorg, C. and Bloom, B. R. (1973) 'Products of activated lymphocytes. I. The use of radiolabeling techniques in the characterization and partial purification of the migration inhibitory factor of the guinea pig', *J. Exp. Med.*, **137**, 148—70.

46. Pick, E. (1974) 'Soluble lymphocytic mediators. I. Inhibition of macrophage migration inhibitory factor production by drugs', *Immunology*, **26**, 649—58.

47. Powell, A. E. and Leon, M. A. (1970) 'Reversible interaction of human lymphocytes with the mitogen concanavalin A', *Exp. Cell Res.*, **62**, 315—25.

CHAPTER 54

Use of Concanavalin A to Enhance Immunogenicity of Tumour Antigens

W. JOHN MARTIN

1. PRINCIPLE

Many experimental tumours express cell surface components, termed tumour-associated surface antigens (TASA) which are not readily detectable on normal non-malignant cells. These TASA could presumably serve as targets for specific immunological eradication of the malignant cells. Immunotherapy is concerned with manipulating the immune system of the tumour-bearing host so that it responds more effectively against the TASA. As recorded in this volume, Con A is known to have selective effects on tumour cell surfaces and also to be capable of specifically stimulating the proliferation of T-lymphocytes. It was of interest, therefore, to explore the possible use of Con A as a means of rendering TASA capable of evoking an effective anti-tumour immune response (1—3).

2. METHODS EMPLOYED

2.1 Murine Leukaemia Cell Lines

Long-term transplantable murine leukaemias were used in these studies for the following reasons: (i) Large numbers of viable tumour cells can be readily obtained from ascitic tumours; (ii) these tumours provide excellent target cells for use in the ^{51}Cr release assay for cytotoxic lymphoid cells; and (iii) considerable information is available concerning the nature and specificity of the the various TASA expressed by these tumours. The EL4, LSTRA and L1210 tumours are leukaemias of C57BL/6, BALB/c, and DBA/2 mice respectively. Tumours were maintained in syngeneic mice by intraperitoneal passage of approximately 5×10^6 cells at 2 weekly intervals. Between 5×10^7 and 2×10^8 ascites cells can generally be harvested 10 days after tumour inoculation. The ascites cells were washed once in bicarbonate-free balanced salt solution (BSS)*.

2.2 x-Irradiation of Tumour Cells

The leukaemias used in these studies are highly virulent for syngeneic animals. In order to use tumour cells as immunogen, it is necessary to inhibit the cell's ability to proliferate in recipient mice. This can be achieved by either, (i) incubating the cells with mitomycin C (generally $50-100$ μg/ml/10^7 cells for 30 min at 37°) or (ii) exposing the cells to x-irradiation. The latter approach was used in the studies reported in this paper. EL4 cells suspended in BSS received 2000 rad at a rate of 640 rad/min using a Westinghouse Quandrocondex x-ray unit at 200 kV, 15 mA, and a half-value layer of 1 mm copper. After irradiation, the cells were washed once in BSS.

2.3 Reaction of Tumour Cells with Con A

Irradiated EL4 cells were resuspended in BSS to a concentration of 10^7 cells/ml. The cell suspension was divided into 2 aliquots. Con A (Nutritional Biochemical Corp. Cleveland, Ohio) was added to an aliquot to a final concentration of 25 μg/ml. Both aliquots were incubated at 37° for 30 min. The cells were washed once with BSS and used for immunization.

2.4 Immunization of Mice with Con A-reacted Tumour Cells

Three month old C57BL/6 mice were inoculated intraperitoneally with an initial injection of either 5×10^5 irradiated Con A-reacted EL4 cells (Con A:EL4) or 5×10^5 irradiated, but otherwise untreated, EL4 cells. Three weeks later the mice received a second intraperitoneal injection of 5×10^7 irradiated EL4 cells not reacted with Con A. Ten days later the tumour-inoculated mice were examined for

---

*BSS contains in 1 litre the following ingredients: dextrose, 1 g; KH_2PO_4, 60 mg; $Na_2HPO_4.7H_2O$, 356 mg; $CaCl_2.2H_2O$, 186 mg; KCl, 400 mg; NaCl, 8 g; $MgCl_2.6H_2O$, 200 mg; $MgSO_4.7H_2O$, 200 mg. It has a pH of 7.4.

splenic cytotoxic lymphoid cells reactive with EL4 cells, for the ability to destroy intravenously inoculated ^{125}I-iododeoxyuridine-labelled EL4 cells, and for survival after an injection of viable (non-irradiated) EL4 cells.

2.5 *In vitro* Assay for Anti-tumour Cytotoxic Lymphoid Cells

EL4 cells were resuspended to a concentration of 3×10^7 cells/ml in BSS containing 10% foetal calf serum (BSS-10). To 1 ml of cell suspension was added 100 μCi ^{51}Cr (Na$_2$^{51}CrO$_4$, Amersham/Searle, Arlington Heights, Ill.) and the mixture was incubated at 37° for 30 min. The labelled cells were washed twice with 50 ml volumes of chilled BSS-10 and resuspended to a concentration of 10^6 cells/ml.

Spleens were removed from either normal or tumour-inoculated mice and gently teased in BSS-10. The resulting cell suspensions were filtered through 100 mesh nylon net and washed once in BSS-10. Enumeration of viable lymphoid cells was performed by diluting an aliquot of the spleen cell suspension in ammonium chloride solution * to lyse erythrocytes, mixing with an equal volume of 0.25% trypan blue and transferring to a haemocytometer for phase-contrast microscopic examination of non-stained (viable) cells. Spleen cell suspensions were adjusted to 5×10^6 cells/ml in Eagles minimum essential medium containing 10% foetal calf serum. One ml aliquots of these suspensions were mixed with 0.05 ml of the suspension of ^{51}Cr-labelled EL4 cells in 35 mm dishes (Falcon plastics, no 1008). The dishes were placed on a rocking platform (Bellco Glass Co., Vineland, New Jersey) in a 37° humidified incubator in a 10% CO$_2$: 90% air atmosphere. After 4 h incubation the contents of each dish were transferred to 15×75 mm glass tubes and 1 ml of chilled BSS-10 added. The tubes were centrifuged at 2000 r.p.m. for 10 min and a 1 ml aliquot of supernatant fluid removed from each tube for determination of radioactivity. The total releaseable radioactivity was determined by 4 cycles of freeze thawing of an aliquot of the stock solution of radioactive cells. The percentage of the maximum possible ^{51}Cr release (max.) from both EL4 cells exposed to normal spleen cells (n.s.) and to immune spleen cells (i.s.) was measured in replicate dishes. The specific release of ^{51}Cr in each of the samples of i.s. was obtained by subtracting the mean percentage released from target cells exposed to n.s. from that released from target cells exposed to i.s. The standard error (SE) of the values obtained was approximated by the formula:

$$SE = \frac{1}{max.} \left[\left(\frac{Variance}{n} \right)_{i.s} + \left(\frac{Variance}{n} \right)_{n.s.} \right]^{1/2}$$

where n = no. of replicate dishes (generally n = 4–6). The SE is normally $< 2\%$ lysis.

---

*Ammonium chloride solution contains in 1 litre the following ingredients: NH$_4$Cl, 8.29 g; KHCO$_3$, 1.0 g; EDTA, 37.2 mg. It has a pH of 7.4

2.6 In *vivo* Assay for Anti-tumour Cytotoxic Lymphoid Cells

The ^{51}Cr-release assay cannot be adapted to study *in vivo* survival of intravenously inoculated tumour cells because ^{51}Cr released from dead cells is not readily excreted from the animal. In contrast the ^{125}I label in ^{125}I-iododeoxyuridine-labelled tumour cells is rapidly excreted after *in vivo* death of intravenously inoculated tumour cells. EL4 cells can be labelled with ^{125}I-iododeoxyuridine by injecting 2 μCi isotope intraperitoneally into tumour-bearing mice, 12 and 6 h prior to harvesting ascites tumour cells. Alternatively, EL4 cells maintained *in vitro* (RPMI 1640 medium $10^5 - 10^6$ cells/ml) can be labelled by the addition of 0.05 μCi/ml 12 h prior to using the cells. To test for *in vivo* immunity, 10^6 ^{125}I-labelled tumour cells are inoculated intravenously into either normal or tumour-immunized mice. After 6 h, liver and lungs are removed from each mouse and counted for radioactivity. Reduction in liver and lung radioactivity is evidence for tumour destructive *in vivo* immunity.

In addition to this study, normal and tumour immunized mice were challenged intraperitoneally with 10^5 or 10^7 viable (non-irradiated) EL4 cells and the subsequent survival of the mice recorded.

3. GENERATION OF ANTI-EL4 CYTOTOXIC LYMPHOID CELLS

Mice inoculated with either 1 or 2 injections of irradiated but otherwise untreated EL4 cells do not develop splenic cytotoxic lymphoid cells active against EL4 cells (Table I) and show no evidence of increased *in vivo* destruction of intravenously inoculated ^{125}I-iododeoxyuridine-labelled EL4 cells (Table 2). Mice initially injected with Con A:EL4 cells and boosted with irradiated EL4 cells develop high levels of anti-EL4 cytotoxic lymphoid cell activity and show significant elimination of intravenously inoculated ^{125}I-labelled EL4 cells (Tables I and II).

Mice which receive 2 injections of irradiated EL4 cells show no evidence of protection when challenged with viable EL4 cells. In contrast mice primed with

TABLE I. Generation of anti-EL4 cytotoxic lymphoid cells in mice primed with Con A-reacted EL4 cells

Immunization		Splenic cytotoxic lymphoid cell activity (% lysis)	
primary	secondary	mean	range
5×10^5 EL4	5×10^7 EL4	0.34	$-1.6-1.9$
5×10^5 Con A:EL4	5×10^7 EL4	34.2	$11.0-62.3$

The results shown are the mean and range of values obtained in 10 experiments. EL4 cells used for immunization had received 2000 rad x-irradiation. Lymph node cells and peritoneal cells of mice injected twice with EL4 cells were also tested for anti-EL4 cytotoxic lymphoid cell activity. No activity was demonstrated.

TABLE II. Reduced survival of ^{125}I-iododeoxyuridine-labelled EL4 cells in mice immunized with Con A-reacted EL4 cells

Immunization		Recovered radioactivity (% injected ±SE)	
primary	secondary	lung	liver
Nil	Nil	15.4 ± 2.2	7.5 ± 0.8
5 x 10⁵ EL4	5 x 10⁷ EL4	15.1 ± 2.9	7.9 ± 0.8
5 x 10⁵ Con A:EL4	5 x 10⁷ EL4	2.3 ± 0.5	3.6 ± 0.5

The results are expressed as percentage of the injected radioactivity recovered in the lung and in the liver of either normal mice or mice immunized as indicated. The values shown are the mean and SE of the results obtained in groups of 4 mice.

Con A:EL4 and boosted with irradiated EL4 showed significant prolongation of survival when injected with viable EL4 (Figure 1).

4. FACTORS WHICH AFFECT INDUCTION OF CYTOTOXIC LYMPHOID CELLS

4.1 Use of Con A-reacted EL4 Cells for Secondary Immunization

It was observed that if Con A:EL4 cells, rather than untreated irradiated EL4 cells, were used for the secondary immunization either no, or occasionally, a low-level, of anti-EL4 cytotoxicity developed. Additional studies indicated that mice receiving this regimen were immunized since cytotoxicity could be demonstrated against Con A:EL4 and Con A: LSTRA target cells (Table III). Mice primed with Con A:EL4 and challenged with untreated EL4 cells also develop cytotoxic reactivity against both Con A:EL4 and Con A:LSTRA cells but this reactivity is in addition to that directed against untreated EL4 cells (Table III).

4.2 Number of Con A-reacted EL4 Cells used for Primary Immunization

Mice received $10^3 - 10^7$ Con A:EL4 cells as a primary immunization and were boosted with either 10^7 or 5×10^7 irradiated EL4 cells. The levels of cytotoxic lymphoid cell activity generated in mice of the various groups are depicted in Figure 2. Regardless of the dose of Con A:EL4 cells used for primary immunization, a secondary injection of 5×10^7 EL4 was clearly more effective than 10^7 cells. The optimal dose of Con A:EL4 cells for the primary injection was $10^6 - 10^7$. Included in the above experiments were mice inoculated with $10^3 - 10^7$ irradiated EL4 cells not reacted with Con A. These mice developed no detectable anti-EL4 cytotoxicity when subsequently challenged with either 10^7 or 5×10^7 EL4 cells.

4.3 Concentration of Con A Used to React with EL4 Cells

Irradiated EL4 cells were exposed to varying concentrations of Con A prior to inoculation into mice. Three weeks later the mice received 5×10^7 irradiated EL4 cells. The results of two such experiments are shown in Table IV. The enhanced

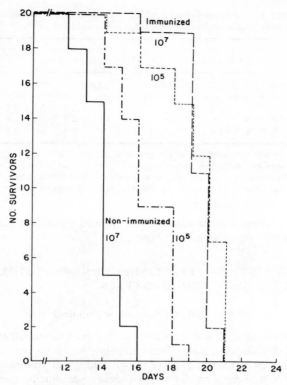

Figure 1. Survival of mice after an intraperitoneal injection of either 10^5 or 10^7 viable (non-irradiated) EL4 cells. The group of immunized mice had received an initial injection of 5×10^5 irradiated Con A:EL4 cells followed 3 weeks later with 5×10^7 irradiated EL4 cells. These mice and a group of 20 normal mice received viable tumour cells 10 days after the second immunizing injection. In repeated experiments mice injected twice with irradiated EL4 cells show no significant protection from a challenge of either 10^5 or 10^7 viable EL4 cells

TABLE III.   Specificity of cytotoxic lymphoid cells generated in mice primed with Con A-reacted EL4 cells and challenged with either EL4 cells or Con A-reacted EL4 cells

Immunization		Splenic cytotoxic lymphoid cell activity (% lysis)			
primary	secondary	EL4	Con A:EL4	LSTRA	Con A:LSTRA
5×10^5 Con A:EL4	5×10^7 EL4	30.24	21.64	2.01	15.62
5×10^5 Con A:EL4	5×10^7 Con A:EL4	−1.82	13.82	0.90	9.79

The Con A:EL4 and Con A:LSTRA target cells used in this experiment were exposed to 25 μg/ml Con A during the incubation period with ^{51}Cr.

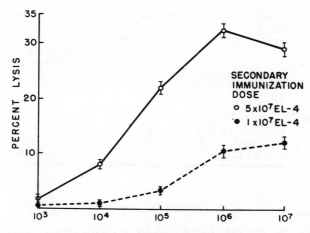

Figure 2. Splenic cytotoxic lymphoid cell activity in mice initially injected with varying numbers of irradiated Con A:EL4 cells and three weeks later with either 10^7 or 5×10^7 irradiated EL4 cells

TABLE IV. Effect of concentration of Con A used to react with EL4 cells

Experiment	Concentration of Con A (μg/ml/10^7 cells)	Splenic cytotoxic lymphoid cell activity after challenge with EL4 cells
A	0.5	1.01
	5.0	11.14
	12.5	15.41
	25.0	13.56
B	0	1.18
	25	51.80
	100	47.16
	500	40.67

immunogenicity of EL4 cells by Con A appears to be relatively little affected by concentrations of Con A between 5–500 μg/ml.

4.4 Need for Con A to be Reacted with EL4 Cells

Mice were inoculated with varying doses of soluble Con A 1 h prior to receiving irradiated EL4 cells. When these mice were subsequently challenged with irradiated EL4 cells, anti-EL4 cytotoxic lymphoid cells did not develop (Table V). Another group of mice received Con A-reacted L1210 cells rather than Con A:EL4 cells as primary immunogen. These mice were not primed to respond effectively to EL4 (Table V).

TABLE V.   Effect of separate administration of EL4 and
Con A

Primary immunization	Splenic cytotoxic lymphoid cell activity after challenge with EL4 cells
EL4	−0.2
Con A:EL4	11.9
100 μg Con A + EL4	1.2
10 μg Con A + EL4	0.8
1 μg Con A + EL4	2.0
Con A:L1210	1.4

5. GENERAL COMMENT

The results indicate that Con A can be used for the generation of anti-EL4 cytotoxic lymphoid cells in C57BL/6 mice. Significant levels of activity are consistently observed in mice given an initial injection of 5×10^5 irradiated EL4 cells reacted with 25 μg/ml Con A and a subsequent injection of 5×10^7 irradiated, but otherwise untreated, EL4 cells. Mice immunized in this way demonstrate partial resistance to an intraperitoneal challenge with viable (non-irradiated) EL4 cells. Several additional studies have been reported which suggest a possible role for Con A in enhancing the immunogenicity of TASA. Brugarolas and coworkers (4) injected Con A-reacted myeloma cells (MOPC 315) 3 days after mice had received an inoculum of viable myeloma cells. The growth rate of the primary tumour was slowed significantly in 14 of 20 mice tested. Enker and coworkers (5) observed retardation of the growth of a hepatoma in tumour-bearing rats inoculated with Con A-reacted hepatoma cells. Simmons and Rios (6) extended these observations to include a methylcholanthrene-induced fibrosarcoma, a mammary adenocarcinoma and a melanoma of mice. In none of these studies, including the study described in this paper, has the mechanism of action of Con A in enhancing the immunogenicity of TASA been determined. The nature of the interaction between the immune system and a growing tumour is poorly understood. Studies using experimental models in which it is possible to achieve even low levels of immunity are important, not only in terms of future clinical application, but also in terms of better defining the immunology of the tumour—host relationship. Con A would appear to be a most useful tool with which to investigate and to manipulate the immune response against TASA.

6. REFERENCES

1. Martin, W. J., Wunderlich, J. R., Fletcher, F. and Inman, J. K. (1971) 'Enhanced immunogenicity of chemically-coated syngeneic tumor cells', *Proc. Nat. Acad. Sci. U.S.A.,* 68, 469—72.
2. Martin, W. J. and Wunderlich, J. R. (1972) 'Immune response of mice to concanavalin A coated EL-4 leukemia' *Nat. Cancer Inst. Monogr.,* 35, 295—9.
3. Martin, W. J., Esber, E. and Wunderlich, J.R. (1973). 'Evidence for the suppression of the development of cytotoxic lymphoid cells in tumour immunized mice', *Fed. Proc.,* 32, 173—9.

4. Brugarolas, A., Takita, H. and Moore, G. E. (1972) 'Effect of syngeneic tumour cells bound to concanavalin A on tumour growth', *J. Surg. Oncol.*, 4, 123–30.
5. Enker, W. E., Craft, K. and Wissler, R. W. (1974) 'Augmentation of tumour-specific immunogenicity by concanavalin A in the Morris Hepatoma 5123', *J. Surg. Res.*, 16, 66–8.
6. Simmons, R. L. and Rios, A. (1974) 'Cell surface modification in the treatment of experimental cancer, neuraminidase or concanavalin A', *Cancer*, 34 1541–7.

CHAPTER 55

Concanavalin A as an Inflammogen

W. Thomas Shier

The inflammatory response plays a fundamental role in health as one of the body's normal defense mechanisms against injury and invasion by foreign agents whether they are simple chemicals or complex tissues, bacteria or viruses. At the same time the inflammatory response may also have pathologic consequences since it plays in important role in a number of autoimmune disease states, of which rheumatoid arthritis is a classic example.

The complicated series of events of the inflammatory response follow a similar course regardless of the cause or intensity of an injury (1). After an initial transient phase of arteriolar constriction following injury, there is an increased blood flow as chemical mediators such as histamine and certain prostaglandins, which are released as a result of the injury, cause a local dilatation of capillaries and venules. With dilatation there is an increase in permeability so that fluid and proteins leak into the interstitial spaces to produce edema (swelling). This transsudation phase is usually measured in minutes and is followed by a stasis (decreased blood flow) phase of variable but longer duration. This is followed by a complex cellular phase, the first event of which is the adherence of leukocytes, primarily polymorphonuclear (PMN) neutrophilic leukocytes, to the endothetial surface of postcapillary venules. These cells then migrate out of the venules and accumulate locally where they actively phagocytize and digest particles of dead tissue or bacteria.

Mononuclear phagocytes also pass from the bloodstream into areas of inflammation. The proportion of this type of cell in the exudate gradually increases during the course of inflammation until they predominate in the last stages of acute inflammation, when their chief function seems to be the disposal of debris. Continuing high levels of PMN leukocytes, and infiltration by conspicuous levels of lymphocytes are both indicators of chronic inflammation.

1. USES OF EXPERIMENTALLY INDUCED INFLAMMATION

1.1 Experimentally Induced Arthritis

The cause of rheumatoid arthritis remains unknown (2). Studies on the cause and pathogenesis of the disease would be greatly facilitated if a suitable and convenient animal model were available for laboratory study (3). Efforts to produce such a model were first reviewed in 1850 by Redfern (4), and since then virtually every irritant known to man has been injected into the knee joints of rabbits. The most disconcerting part about this long history of effort is that virtually every substance tested caused a pathological condition with some similarities to rheumatoid arthritis. The list includes even saline and the needle injury that occurs when nothing has been injected. The most widely used model is Freund's adjuvant-induced arthritis in rats (5), but even this model is characterized by a cellular inflammatory reaction that differs significantly from that of rheumatoid arthritis (6).

1.2 Assay of Antiinflammatory Drugs

The most important practical use of experimentally induced inflammation is in the assay of antiinflammatory drugs, since development of more effective therapy in this area is a subject of intense interest. The most widely used assay employes carrageenin (7), a partially purified, poorly characterized galactan, that induces a reproducible edema when injected in substantial quantities into the footpads of rats. The effectiveness of experimental antiinflammatory drugs can be assayed by determining the dose that causes a 50% reduction in the intensity of the inflammatory response.

1.3 Studies of the Mechanism of the Inflammatory Response

Because of its complexity the biochemistry and cell biology of the inflammatory response are still not well understood. Currently the problem of identifying the chemical causes or mediators of the physiological events provoked by various inflammatory stimuli is an active area of research. Histamine, serotonin, some prostaglandins, kinins, immunoglobulins, lysosomal enzymes, serum complement, kallikreins, plasmin and lymphokines have all been implicated in some aspect of the inflammatory process (1). The list of *in vitro* and *in vivo* models that have been established to study various aspects of the process is too long to be presented in this chapter (8).

2. INFLAMMATORY RESPONSE TO CONCANAVALIN A AND OTHER LECTINS

As observed by More and coworkers (9), as little as 10 μg of Con A induces an intense and reproducible inflammatory response. The most easily measured aspect of the inflammatory response to Con A is the intense edema which is accompanied by erythema and induration at the site of injection in the footpads of mice. This response has been determined at 2, 4, 8, 12, 24, 48 and 72 hours after injection of each of the affinity-purified lectins listed in Table I. Each of these lectins induced an edema markedly greater than that induced by the corresponding amount of bovine serum albumin. The character of the response fits conveniently into two classes. The first class is typified by Con A (10), which induces a biphasic response with a maximum edema at 2—4 hours and a second, more intense maximum occurring at 24—48 hours. The second class is typified by wheat germ agglutinin, which induces a single maximum edema at about 4 hours with an intensity comparable to the early response observed with the first class. In both cases the edema declines gradually over a period of days. Injection of large quantities of Con A (e.g. 100 μg or greater) into the footpad of a mouse induces a necrosis which may ulcerate releasing fluid and cells.

TABLE I.   Comparison of the time of appearance of maximum edema with various lectins

Lectin	Time(s) after injection to maximum edema (h)
Con A	4, 24
Ulex europaeus agglutinin[a]	2, 48
Soy bean agglutinin	2, 24
Wheat germ agglutinin	4
Pisum sativum agglutinin[b]	4
Lens culinaris agglutinin[a]	4
Limulus polyphemus agglutinin	4

[a] Obtained from G. L. Nicolson.
[b] Obtained from I. S. Trowbridge.

There is no obvious explanation for the appearance of the second phase of the response with some of these lectins. Presumably it reflects an increased involvement of some part of the cellular phase of the inflammatory response, perhaps as a result of a particular cell recruiting activity characteristic of this class of lectins. The appearance of the late response does not correlate with mitogenicity for lymphocytes, since soy bean agglutinin induces this response even though it is not a mitogen for lymphocytes (11). The observation that lectins persist for unusually long periods of time at the site of injection in tissues (10) may provide an explanation for edema persisting several days after injection, but it does not provide an explanation for the second phase of edema. For example, soy bean agglutinin, which induces the second-phase response, is the most rapidly eliminated of all lectins tested to date.

3. APPLICATIONS

3.1 Assay of Antiinflammatory Drugs in Mice

Groups of 10 Balb/c mice of the same age (4 to 6 weeks old, approximately 20 g) and sex are used to provide a sample large enough for statistical analysis. Antiinflammatory drugs, such as aspirin, indomethacin and hydrocortisone (Sigma Chemical Co.) are administered as an intraperitoneal injection of 0.25 ml of a solution (indomethacin) or finely divided suspension (aspirin and hydrocortisone) in phosphate-buffered saline (PBS, 0.15 M sodium chloride buffered at pH 7.2 by 10 mM potassium phosphate) one hour before treatment with Con A. A range of concentrations is used, and a control group of mice is included that receives an intraperitoneal injection of saline without any drug. Con A (twice crystallized) as obtained from Miles Laboratories, gives satisfactory results as the inflammogen without further purification. The commercial material is generally obtained at a concentration of approximately 40 mg/ml in saturated sodium chloride. It is prepared for use by diluting a small sample to approximately 2 mg/ml with water and dialysing 2–3 hours at 4° against PBS. This solution is diluted to 1 mg/ml, filter sterilized with a 'milipore' filter, and used immediately. Aliquots of 10 μl of the Con A solution (10 μg Con A) are injected subcutaneously into the left hind footpad of each mouse in a group without anaesthetic and aliquots of 10 μl of PBS are injected in the same manner into the right hind footpad. Six hours later the thickness of the left and right hind feet is measured at the metatarsus in three independent measurements with dial gauge calipers (National Camera Supply Co.). The edema is calculated as the average difference in thickness between the left (Con A) and right (control) feet. The dose that gives a 50% inhibition of edema (ID_{50}) can be estimated graphically (see Figure 1).

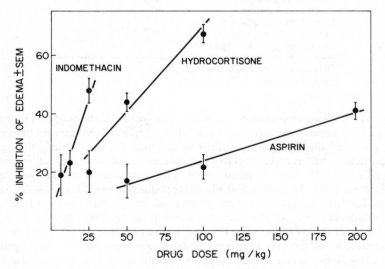

Figure 1. Inhibition of inflammation by some antiinflammatory drugs

3.2 Arthritis Model Disease in Rabbit Knee Joints

The knee joints of New Zealand white rabbits were used because of technical difficulties involved in performing intraarticular injections in smaller animals. Young females are anaesthetized with pentothal (45 mg/kg, Abbott Laboratories) administered by intraperitoneal injection. Con A solutions at a concentration of 20 mg/ml in saline are prepared by a method analogous to that described above for the testing of antiinflammatory drugs, and used immediately. Intraarticular injection is accomplished using a 27-gauge needle by bending the knee, entering just beneath the patella, and inserting the needle directly into the joint space. One knee is injected with 100 μl of a Con A solution and the other is injected with 100 μl of saline to provide a control for the response to needle-induced injury. One rabbit is injected with saline in both knees to provide a control for a sympathetic response which might occur in the control knee of an animal receiving Con A.

Tissue samples are taken at intervals for histological examination. The rabbit is sacrificed and synovial fluid withdrawn and measured using a syringe equipped with a 27-gauge needle; smears are made with the aspirated synovial fluid. Samples of meniscus, articular cartilage, patella, synovial membrane and adipose tissue are removed from the knee joints of all experimental animals and fixed in phosphate-buffered 3.7% formaldehyde. Calcified samples are treated with Jenkin's decalcifying solution. All tissue samples are infiltrated and embedded in polyester wax (12) to avoid hardening and tissue shrinkage. Serial sections are cut at 4 μ, stained with haematoxylin and eosin, and examined for cellular infiltration (10). Cartilaginous tissue is also examined for destruction of acid mucopolysaccharides by staining with toluidine blue and with safaronin O and fast green.

4. PITFALLS

The Con A preparations used to induce inflammation are supersaturated with Con A. If they are not used immediately the protein may crystallize. If crystallization occurs during the brief dialysis step described in the procedure, it is easiest to discard the preparation and begin again. There is considerable variability in the measurements of foot edema. Some of this variability can be eliminated by rejecting the animals that develop a prompt haematoma in either foot as a result of poor injection technique. In addition, measurement of the foot volume, if suitable equipment is available, should give more reproducible results than measurements of foot thickness.

If Con A is to be used as the inflammogen in studies of chronic inflammation, high doses of the lectin cannot be used because it induces necrosis in an irreproducible manner.

5. COMPARISON WITH OTHER INFLAMMOGENS

5.1 Advantages

Unlike carrageenan, turpentine, croton oil and most other commonly used inflammogen preparations, which are complex, ill-defined mixtures, Con A is a single substance with a defined structure and it is readily available in a relatively pure form.

Con A produces a more highly reproducible inflammatory response than some of the other common inflammogens.

Con A produces a much more intense inflammatory response than equal amounts of most other inflammogens, in particular carrageenan. Consequently it can conveniently be used in smaller, less expensive animals, and measurements can be made with dial gauge calipers, although they are less accurate than devices that measure the volume of the foot. The use of simpler, less expensive and more readily available equipment is advantageous in laboratories that do not routinely study inflammation.

Since Con A is retained at the site of injection for a prolonged period, it should be particularly useful for the study of chronic inflammation. This property of Con A provides the major advantage for the use of Con A in animal models of arthritis that employ intraarticular injection of an inflammogen. Since only a single injection of Con A is required to achieve chronic inflammation, the complicating effects of needle injury superimposed on the specific damage by the inflammogen are minimized.

Con A possesses several relatively well-defined biological activities that could provide useful correlations with the course of events in several phases of the inflammatory response, in particular with events in the poorly studied cellular phases of the inflammatory response.

5.2 Disadvantages

Con A is more expensive than other common inflammogens. However, very little material is used for each assay and the cost of the inflammogen is a small percentage of the total cost of an assay.

Since Con A is retained at the site of injection for a prolonged period, it is not useful for studying the terminal phases of an acute inflammatory response.

The large number of biological activities that have been reported for Con A may also constitute a disadvantage in some basic studies of the inflammatory response, since it may become difficult to sort out which activity of Con A is responsible for a given effect.

6. REFERENCES

1. Zvaifler, N. J. (1973) 'The immunopathology of joint inflammation in rheumatoid arthritis', *Advan. Immunol.*, **16**, 265—336.
2. Duthie, J. J. R. (1969) 'Rheumatoid arthritis', in *Textbook of the Rheumatic Diseases*, 4th edn., ed. Copeman, W. S. (E. and S. Livingstone, Ltd., Edinburgh), pp. 259—322.
3. Gardner, D. L. (1960) 'The experimental production of arthritis', *Ann. Rheum. Dis.*, **19**, 297—317.
4. Cited in reference 3.
5. Pearson, C. M. and Wood, F. D. (1959) 'Studies of polyarthritis and other lesions induced in rats by injection of mycobacterial adjuvant. I. General clinical and pathological characteristics and some modifying factors', *Arthritis Rheum.*, **2**, 440—59.
6. Jasin, H. E., Cooke, T. D., Hurd, E. R., Smiley, J. O. and Ziff, M. (1973) 'Immunologic models used for the study of rheumatoid arthritis', *Fed. Proc.*, **32**, 147—52.
7. Winter, C. A., Risley, E. A. and Nuss, G. W. (1962) 'Carrageenin-induced edema in hind paw of the rat as an assay for antiinflammatory drugs', *Proc. Soc. Exp. Biol. Med.*, **111**, 544—7.

8. Marx, J. L. (1972) 'Prostaglandins: mediators of inflammation?', *Science*, **177**, 780—1.
9. More, D. G., Penrose, J. M., Kearney, R. and Nelson, D. S. (1973) 'Immunological induction of DNA synthesis in mouse peritoneal macrophages', *Int. Arch. Allergy*, **44**, 611—30.
10. Shier, W. T., Trotter, J. T. III and Reading, C. L. (1974) 'Inflammation induced by concanavalin A and other lectins', *Proc. Soc. Exp. Biol. Med.*, **146**, 590—3.
11. Sharon, N. and Lis, H. (1972) 'Lectins: cell agglutinating and sugar-specific proteins', *Science*, **177**, 949—59.
12. Steedman, H. F. (1957) 'Polyester wax. A new ribboning embedding medium for histology', *Nature*, **179**, 1345.

CHAPTER 56

Concanavalin A as a Probe for the in vitro *Detection of IgE-type Antibody on the Surface of Blood Basophils and Tissue Mast Cells*

ROBERT KELLER

Antigen-induced release of histamine '*in vitro*' from the basophil blood leukocytes of allergic individuals and sensitized rabbits, or from tissue mast cells of sensitized animals, provides an excellent model system for the study of allergic reactions. The release of histamine initiated by the specific allergen is an active secretory process which requires calcium ions, is dependent on the presence of IgE-type antibody fixed to the basophil or mast cell surface, can be modulated by the levels of intracellular cyclic AMP and GMP and does not kill the cell (1–5). Recent studies suggest that Con A is capable of releasing histamine from normal or appropriately sensitized cells and that the sequence of events is similar to that elicited by the specific antigen in the allergic reaction (6–8). Although the precise molecular mechanisms involved in the release of histamine are still under discussion, Con A offers a useful, simple and reproducible tool for the detection and measurement of allergic sensitization in both man and the experimental animal.

1. .PRINCIPLE

IgE-type anaphylactic antibodies are strongly cytophilic, their Fc fragment binding selectively to plasma membranes of tissue mast cells and blood basophils. These are highly specialized cells whose numerous granules contain a variety of biologically active agents of which histamine is an important example which is regularly present (e.g. 20—30 pg/rat mast cell; 9,10). Binding of IgE-type antibody molecules to specific receptors can persist for astonishingly long periods of time. Provided that the IgE molecules attached to the plasma membrane are sufficiently close together, combination with the allergen results in bridging between the Fab portions of two neighbouring molecules (5,11,12), causing alterations in their configuration which are thought to trigger the cellular-release mechanism.

In the model system, cell suspensions containing an appropriate number of mast cells or basophils sensitized with IgE-type antibody are incubated for a limited period of time with the specific allergen or Con A. The percentage of histamine released by the two agents is then assessed.

2. METHODS

2.1 Histamine Release from Blood Leukocytes

Aliquots of blood leukocytes from individuals or from rabbits are incubated with various concentrations of Con A (1—50 μg/ml for 20—60 min at $37°$); subsequently, histamine release is assayed. This method originally described by Lichtenstein and Osler (13) has been widely adopted and has proved reliable. Details of the practical procedure and discussion of the characteristics of the reaction are to be found in refs. 8 and 14.

2.2 Histamine Release from Tissue Mast Cells

It is generally accepted that tissue mast cells are the principal source of histamine and that in most tissues, the histamine content can be related to the number of mast cells present (9,15). Usually, histamine release is studied in whole-cell suspensions obtained by washing out the peritoneal or pleural cavities or the bronchial tree. In such cell preparations, the mast cells represent only a small proportion of the total, but as other cell types do not materially interfere with the assay, this simple procedure can yield satisfactory results. If necessary, the mast cells may be greatly concentrated by differential centrifugation. Using such methods, histamine release from rat and mouse mast cells can be reliably assessed (10,16—18).

2.3 Assay of Histamine

Histamine can be assayed at very low concentration (usually below 5 ng/ml) either biologically on the isolated guinea pig ileum or by spectrofluorimetry. Details of these methods and a discussion of their advantages and disadvantages appear in ref. 19. More recently, a very precise method of histamine assay using a specific enzyme-based reaction has been introduced (20).

3. MECHANISMS INVOLVED IN CON A–INDUCED HISTAMINE RELEASE

Although data on Con A-induced histamine release from mast cells and basophils are still very limited (6,8), the available evidence strongly suggests that in many respects, the processes induced by the specific allergen and by Con A at least closely resemble one another. Thus, the extent to which in both cases histamine is released *in vitro* seems to be largely dependent upon the degree of sensitization, i.e. the density of IgE-type antibody molecules on the cell surface, the concentration of allergen or Con A and the temperature. Moreover, histamine release induced both by the specific antigen and by Con A requires calcium ions, can be modulated by agents which affect intracellular cyclic AMP levels and can be inhibited by diethylcarbamazine and chlorphenesin. Trypsinization and succinylation of Con A destroy its activity (6,21).

In view of these similarities and the evidence suggesting that the release mechanism triggered by the allergen is probably a consequence of bridging two neighbouring IgE-type antibody molecules, it is reasonable to suppose that Con A-induced histamine release is likewise triggered by bridging of adjacent IgE molecules. If it is postulated that Con A binds to every available IgE-type antibody molecule, whereas binding of the allergen remains restricted to antibody molecules specifically directed against it, it seems logical to expect that Con A would induce a pronounced release reaction than any single specific allergen.

4. PITFALLS

The following points merit special attention:

Handling of cells. Tissue mast cells and blood basophils are easily damaged and must thus be handled with special care. The cells must be processed with a minimum of delay and must not be exposed to abrupt changes in temperature. They must be suspended in appropriate isotonic solutions (8,18), must be centrifuged at low speed (400–1000 r.p.m.) and be resuspended by careful pipetting. Appropriately handled cells retain their histamine so that during 60 min incubation spontaneous release usually remains below 5% of the total histamine.

As the degree of *sensitization* and hence the sensitivity to allergen may differ considerably from the responsiveness to Con A, it is recommended that both agents are tested over a wide range of concentrations.

Possible sources of errors encountered in the biologic and spectrofluorimetric methods of *histamine assay* have been discussed by Brocklehurst (19).

5. ACKNOWLEDGEMENTS

This work was supported by the Swiss National Science Foundation (grant 3.234.74). I would like to thank Drs. W. E. Brocklehurst and James F. Riley for valuable comments.

6. REFERENCES

1. Osler, A. G., Lichtenstein, L. M. and Levy, D. A. (1968) '*In vitro* studies of human reaginic allergy', *Advan. Immunol.*, 8, 183–231.

2. Lichtenstein, L. M. (1972) 'Allergy', *Clin. Immunobiol.*, **1**, 243—69.
3. Sullivan, A. L., Grimley, P.M. and Metzger, H. (1971) 'Electron microscopic localization of immunoglobulin E on the surface membrane of human basophils', *J. Exp. Med.*, **134**, 1403—16.
4. Hastie, R. (1971) 'The antigen-induced degranulation of basophil leucocytes from atopic subjects, studied by phase-contrast microscopy', *Clin. Exp. Immunol.*, **8**, 45—61.
5. Stanworth, D. R. (1971) 'Immunoglobulin E (reagin) and allergy', *Nature*, **233**, 310—6.
6. Keller, R. (1973) 'Concanavalin A, a model 'antigen' for the *in vitro* detection of cell-bound reaginic antibody in the rat', *Clin. Exp. Immunol.*, **13**, 139—47.
7. Hook, W. A., Dougherty, S. F. and Oppenheim, J. J. (1974) 'Release of histamine from hamster mast cells by concanavalin A and phytohemagglutinin.' *Infect. & Immunity*, **9**, 903—8.
8. Siraganian, P. A. and Siraganian, R. P. (1974) 'Basophil activation by concanavalin A: Characteristics of the reaction', *J. Immunol.*, **112**, 2117—25.
9. Riley, J. F. (1959) 'The mast cells' (Livingstone, Edinburgh).
10. Keller, R. (1966) 'Tissue mast cells in immune reactions', *Monographs in Allergy*, Vol. 2, eds. Kallós, P., Goodman, H. C. and Inderbitzin, T. (S. Karger, Basel).
11. Ishizaka, K. and Ishizaka, T. (1970) 'Biological function of E antibodies and mechanisms of reaginic hypersensitivity', *Clin. Exp. Immunol.*, **5**, 25—42.
12. Ishizaka, T., Ishizaka, K., Orange, R. P. and Austen, K. F. (1971) 'Pharmacologic inhibition of the antigen-induced release of histamine and slow-reacting substance of anaphylaxis (SRS-A) from monkey lung tissues mediated by human IgE', *J. Immunol.*, **106**, 1267—73.
13. Lichtenstein, L. M. and Osler, A. G. (1964) 'Studies on the mechanism of hypersensitivity phenomena. IX. Histamine release from human leukocytes by ragweed pollen antigen', *J. Exp. Med.*, **120**, 507—30.
14. Augustin, R. (1973) 'Techniques for the study and assay of reagins in allergic subjects' in *Handbook of Experimental Immunology*, ed. Weir, D. M., 2nd ed. (Blackwell, Oxford), p. 4231.
15. Riley, J. F. and West, G. B. (1953) 'The presence of histamine in tissue mast cells', *J. Physiol.*, **120**, 528—37.
16. Archer, G. T. (1959) 'The release of histamine from the mast cells of the rat', *Austral. J. Exp. Biol. Med. Sci.*, **37**, 383—90.
17. Uvnäs, B. and Thon, I.-L. (1959) 'Isolation of 'biologically intact' mast cells', *Exp. Cell Res.*, **18**, 512—20.
18. Keller, R. and Beeger, I. (1963) 'Anaphylaxis in isolated rat mast cells. I. Effects of peptidase substrates and inhibitors', *Int. Arch. Allergy*, **22**, 31—44.
19. Brocklehurst, W. E. (1973) 'The assays of mediators in hypersensitivity reactions', in *Handbook of Experimental Immunology*, ed. Weir, D. M., 2nd ed. (Blackwell, Oxford), p. 43. 1 ff.
20. Kobayashi, Y. and Maudsley, D. V. (1972) 'A single-isotope enzyme assay for histamine', *Anal. Biochem.*, **46**, 85—90.
21. Keller, R. and Burger, M. M., unpublished.

CHAPTER 57

Chemical and Biological Properties of Dimeric Concanavalin A Derivatives

JOHN L. WANG
GARY R. GUNTHER
GERALD M. EDELMAN

The binding of the mitogenic lectin concanavalin A (Con A) to the cell surface of a lymphocyte leads to a variety of membrane events including the modulation of receptor distribution (1,2) and, under appropriate conditions, the transformation of the cell to undergo DNA synthesis and mitosis (3,4). Inasmuch as the binding events can be inhibited by the addition of simple saccharides such as glucose and mannose (5), it is assumed that the various effects of Con A are mediated by cell surface carbohydrates of the requisite specificity. Studies on the subunit and three-dimensional structure of Con A (6,7) indicate that at pH 7 it is a tetramer with four saccharide binding sites per molecule. The structural and biological data on Con A suggest, therefore, that its multimeric and multivalent nature may play an important role in its biological action on cells.

In order to explore the relation of multiple valence to the various activities of Con A, a number of chemically derivatized forms of the lectin have been explored to search for molecules with altered subunit interactions and correspondingly modified biological activities. We have found that treatment of Con A with succinic anhydride or acetic anhydride results in the formation of dimeric derivatives with unaltered carbohydrate-binding specificity (8). Although chemical modifications of the Con A molecule using a variety of group-specific reagents including acetic anhydride have been reported previously (9–13), none of these studies have suggested that the tetrameric Con A molecule can be dissociated to test the role of multiple binding and receptor cross-linkage on the cell surface. The detailed studies on dimeric succinyl-Con A (8) represent, therefore, the first use of this approach. Subsequent reports on dimeric derivatives of Con A (14,15) have confirmed many of the results.

1. PREPARATION AND CHARACTERIZATION OF SUCCINYL-CON A AND ACETYL-CON A

1.1 Succinyl-Con A

Succinyl-Con A was prepared (8) by dissolving 100 mg of Con A in 25 ml of saturated sodium acetate at room temperature. After centrifugation, the supernatant was transferred to a 50 ml flask containing 30 mg of succinic anhydride. The solution was stirred in an ice bath for 1 hour, dialysed overnight against water, and lyophilized. The lyophilized protein was subjected to a second derivatization by dissolving it in 20 ml of saturated sodium acetate. After removal of the precipitate, the protein solution was added to a flask containing 30 mg of succinic anhydride and stirred at room temperature for 90 minutes. The solution was then dialysed exhaustively against water and lyophilized.

The number of succinyl groups per polypeptide chain of Con A was determined using ^{14}C-succinic anhydride to prepare ^{14}C-succinyl-Con A. The results showed that treatment of Con A with succinic anhydride for 60 minutes resulted in the covalent coupling of 3 succinyl groups per subunit of molecular weight 26,000 (Figure 1). More extensive succinylation could be induced by rederivatization with a fresh dose of reagent for 90 minutes. The final product contained an average of 10 succinyl groups per subunit (Figure 1). This derivative, which showed a single component on gel electrophoresis under the conditions of Ornstein and Davis (16,17) as well as under dissociating conditions (18), is designated succinyl-Con A.

1.2 Acetyl-Con A

Acetyl-Con A was prepared by dissolving 25 mg of Con A in 7 ml of saturated sodium acetate. Forty microlitres of acetic anhydride were added to the protein

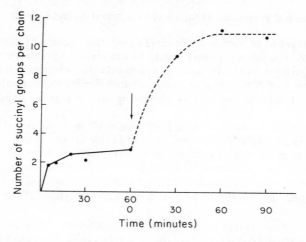

Figure 1. Average number of succinyl groups coupled per subunit as a function of time in the derivatization of Con A with succinic anhydride. At the point indicated by the arrow, the sample was dialysed against H_2O and lyophilized, and the succinylation procedure was repeated. (————) first derivatization; (— — — — — —) second derivatization. The two time coordinates in the abscissa indicate the time of the first and second derivatizations

solution, which was stirred for 1 hour in an ice bath. The protein was then dialysed against water and lyophilized. Native Con A, as well as the acetyl and succinyl derivatives, were all stable when stored in the form of lyophilized powder in the cold.

1.3 Radiolabelling of Con A using ^{14}C-succinic and ^{3}H-Acetic Anhydrides

Radiolabelled dimeric Con A derivatives permit a quantitative comparison of their cell binding properties with those of native Con A. Using ^{14}C-succinic anhydride (specific activity 0.25 mCi/mmol), we have prepared dimeric ^{14}C-succinyl-Con A with a specific activity of 0.10 μCi/mg. The labelled succinyl-Con A was used with ^{125}I-Con A to compare the number of binding sites for the lectin and its derivative on mouse spleen cells. It was found (8) that both Con A and succinyl-Con A were bound in about equal numbers.

Noonan and Burger (19) have also prepared Con A derivatized with ^{14}C-succinic anhydride, although the subunit structure of the modified lectin was not investigated. They report a specific activity of 5×10^6 c.p.m./mg using ^{14}C-succinic anhydride of specific activity 8 mCi/mmol. Using this radiolabelled Con A, they found that transformed cells bound 3.5 to 5 times more lectin molecules than normal interphase cells. In contrast, Cline and Livingston (20) found no significant difference in the number of Con A binding sites on normal and transformed fibroblasts using Con A derivatized with ^{3}H-acetic anhydride.

1.4 Physicochemical Properties of Succinyl-Con A and Acetyl-Con A

Con A is composed of identical subunits of molecular weight 26,000 (6). At $20°$ or above, Con A in solution consists largely of dimers at pH values less than 6 and of tetramers above pH 7 (21). It has also been shown that lowering the temperature to $4°$ causes the Con A tetramer to dissociate into dimers although the degree of dissociation has been reported to vary depending on the detailed conditions (22–24).

In striking contrast to native Con A, both succinyl-Con A and acetyl-Con A were found to exist as dimers at pH 7.4 and $25°$ (8) (Table I). In confirmation of these results, it has been shown by others (14,15,24) that treatment of Con A with succinic or maleic anhydride resulted in the dissociation of the tetrameric molecule into dimers. Other acetyl derivatives of Con A have been described (9,25,26) but in these latter cases, no alteration in the quaternary structure of the lectin was reported.

The affinity of succinyl-Con A for simple saccharides such as α-methyl-D-glucoside was similar to that of the native lectin (Table I). Inasmuch as native Con A and succinyl-Con A exist as tetramers and dimers, respectively, under the conditions of these measurements, the results indicate that the main effect of succinylation is an alteration of the valence of the Con A molecule with respect to saccharide binding and not a change in the specificity or affinity of the binding interaction.

2. CELL SURFACE MAPPING USING TETRAMERIC AND DIMERIC CON A

2.1 Effects of Con A and Succinyl-Con A on Surface Receptor Mobility

Native Con A has two antagonistic effects on the mobility of cell surface receptors depending on the conditions of incubation: the induction of cap formation by its own receptors and the inhibition of cap formation by its own and other receptors (1,27). Dimeric succinyl-Con A lacks both of these activities (8,27). For example, when mouse spleen cells were incubated with Con A in the cold, washed to remove unbound lectin, and then reincubated at $37°$, 62% of the cells showed cap formation by their Con A receptors (Table II, 1a). A few cells also formed caps when incubated with low concentrations of Con A at $37°$ (Table II, 1b). Similar treatments with succinyl-Con A did not show any cap formation.

At concentrations greater than $20\,\mu g/ml$ and $37°$, Con A inhibited cap formation on lymphocytes induced by a variety of receptors including the θ-antigen on thymus-derived (T) cells, the immunoglobulin (Ig) receptor on bone marrow-derived (B) cells as well as the surface carbohydrates binding to the lectin itself on both T- and B-cells (1,27). Binding of Con A at $37°$, therefore, resulted in cap formation of the Con A receptors in less than 0.2% of the cells (Table II, 1c). Similarly, the binding of Con A inhibited the capping of surface Ig molecules in 92% of the cells. In contrast, the binding of an equivalent amount of succinyl-Con A had no effect on the mobility of the surface receptors as assayed by the capping of Ig molecules (Table II, 2a).

TABLE I. Physicochemical properties of Con A, succinyl-Con A and acetyl-Con A[a]

	Conditions	M.wt.	$S_{20,w}$	K_{aff} for α-MG[b] (litre/mole)	Number of binding sites per molecule
Con A	6 M guanidine—0.1 M Tris-HCl (pH 7.0)	25,800	–	–	–
Con A[c]	0.02 M Na acetate, 0.2 M NaCl (pH 5.6)	53,000	3.9	4.1×10^3	2
Con A	PBS (pH 7.4)[d]	106,000	6.1	2×10^3	4
Succinyl-Con A	PBS (pH 7.4)	56,000	4.0	2×10^3	2
Acetyl-Con A	PBS (pH 7.4)	51,000	4.0	–	–

[a] Molecular weights (M.wt.), $S_{20,w}$ values and binding data were determined as described previously (8).

[b] α-MG represents the sugar α-methyl-D-glucopyranoside.

[c] The K_{aff} of Con A for α-MG and the number of binding sites per molecule in this case were determined by Yariv and coworkers (42). The conditions were 0.05 M Na acetate, 0.2 M NaCl, pH 5.2 at 4°.

[d] PBS is phosphate-buffered saline containing 8.00 g NaCl, 0.20 g KCl, 0.20 g KH_2PO_4 and 1.15 g Na_2HPO_4 per litre, pH 7.4.

TABLE II.   The effects of Con A and succinyl-Con A on surface receptor mobility[a]

	Conditions	% of cells forming lectin-receptor caps[b]	
		Con A	succinyl-Con A
(1a)	Lectin (170 μg/ml) preincubated in ice bath, washed and brought to 37°	62	0
(1b)	Lectin (5 μg/ml, 37°)	0.2–2	0
(1c)	Lectin (100 μg/ml, 37°)	$\leqslant$0.2	0
(1d)	Lectin (20 μg/ml) + anti-Con A (100 μg/ml)	18	82
		% inhibition of anti-Ig capping[c]	
(2a)	Lectin	92	0
(2b)	Lectin + anti-Con A	100	38
(2c)	Lectin + $(F'_{ab})_2$ anti-Con A	–	35
(2d)	Lectin + F'_{ab} anti-Con A	–	6

[a] Cap formation was assayed by fluorescence microscopy using mouse spleen cells as described (1).

[b] Capping induced by Con A was measured using fluorescein-labelled Con A (fl-Con A). Capping induced by succinyl-Con A was measured by washing the cells and incubating with 100 μg/ml fl-anti-Con A in the presence of 0.05 M NaN_3 after succinyl-Con A had been bound (1, a–c) or by adding fl-anti-Con A directly to the incubation mixture (1d). The incubation medium for cap formation was Hank's balanced salt solution containing 5% foetal bovine serum.

[c] Anti-Ig capping was measured using 100 μg/ml fluorescein-labelled rabbit anti-mouse immunoglobulin (fl-anti-Ig). The data are normalized to a value of 85% cap-forming cells with fl-anti-Ig alone. Con A and succinyl-Con A, when present, were at a concentration of 50 μg/ml. The divalent $(F'_{ab})_2$ fragment of anti-Con A was prepared (43) by cleavage of the intact immunoglobulin with pepsin. The $(F'_{ab})_2$ fragment was purified by gel filtration on 'Sephadex' G-150, and the monovalent F'_{ab} fragment was prepared by reduction and alkylation of the $(F'_{ab})_2$ fragment (43). Antibody preparations, when present, were at a concentration of 100 μg/ml. The incubation medium for cap formation was phosphate-buffered saline containing 1% bovine serum albumin.

2.2   Effect of Anti-Con A on Cell-bound Succinyl-Con A

Addition of antibodies against Con A to cells that have bound succinyl-Con A, however, restores both the induction of capping and inhibition of receptor mobility activities to the dimeric derivative (Table II, 1d, 2b). The effect of anti-Con A can also be observed using the divalent $(F'_{ab})_2$ fragment of the antibody molecule, but the monovalent F'_{ab} fragment had only a small effect (Table II, 2c, 2d). The antibody preparations alone had no effect in these experiments. These results suggest that the critical factors in the modulation of receptor mobility and distribution are the valence of the bound lectin and the formation of receptor clusters (Figure 2) (2,8).

If the inability of succinyl-Con A to modulate receptor mobility is due to a lack of sufficient cross-linkage of the Con A receptors, then the *effective* valence of succinyl-Con A on the cell surface must be reduced as compared to Con A. It is important to distinguish between this effective valence (i.e. the number of carbohydrate moieties bound by one lectin molecule on the cell surface) and the solution valence (i.e. the number of molecules of a simple sugar bound per molecule

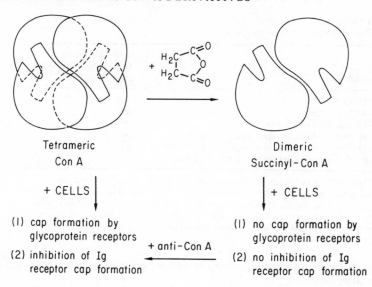

Figure 2. Schematic comparison of the effects of native tetrameric Con A and dimeric succinyl-Con A on the mobility and distribution of lympho-cyte cell surface receptors

of lectin in solution). The effective valence can be reduced in two ways: 1) by a reduction in the solution valence; and 2) by a reduction in the number of cell surface carbodydrate chains that can bind to one lectin molecule, for example, by an increase in the charge repulsion between the lectin and charged groups on the surface carbohydrates. Succinyl-Con A has reduced solution valence (Table I), and therefore its effective valence is also reduced. The net negative charge on the Con A molecule is increased by succinylation, so that it is also possible that its effective valence is decreased via increased charge repulsion. Whether or not this second mechanism is operating, it can be asserted that the modulation of receptor mobility requires the high effective valence of the Con A molecule, since these properties are restored to succinyl-Con A by anti-Con A or the divalent fragment but not by the monovalent F'_{ab} fragment (Table II).

2.3 Comparison of Pea Lectin with Succinyl-pea Lectin

Recently, we have investigated the effect of green pea lectin (28) and a succinyl derivative having the same molecular weight (14) on the mobility of cell surface receptors. It was found that the native lectin inhibited cap formation of surface Ig molecules while the succinyl derivative did not (Table III). Inasmuch as the solution valence of pea lectin and succinyl-pea lectin are assumed to be the same, the situation would appear to be different from that of Con A and succinyl-Con A. However, it was also found that antibodies to pea lectin and their divalent fragments restored the inhibitory effect when used with succinyl-pea lectin, whereas the monovalent F'_{ab} fragment did not (Table III). This suggests that a

TABLE III.   The effect of pea lectin and succinyl-pea lectin on cap
formation by anti-immunoglobulin[a]

Conditions	% inhibition of fl-anti-immunoglobulin caps
(1) Control	0
(2) Pea lectin	89
(3) Succinyl-pea lectin	0
(4) Succinyl-pea lectin + anti-pea lectin	46
(5) Succinyl-pea lectin + $(F'_{ab})_2$ anti-pea lectin	37
(6) Succinyl-pea lectin + F'_{ab} anti-pea lectin	2

[a]Pea lectin was prepared from a saline extract of green pea flour by
affinity chromatography on 'Sephadex' G-75 with elution by 0.1 M
glucose. Succinylation was performed by the method described in the text
for Con A. Other conditions and the preparation of antibody fragments
were as described in Table II, footnotes a and c.

reduction in the effective valence may be occurring exclusively via increased charge
repulsion as discussed above.

   All of the above studies are consistent with the hypothesis that extensive
cross-linkage of one class of surface molecules (e.g. the receptors for the lectins
Con A and green pea) is necessary to inhibit the mobility of other classes of surface
molecules. Extensive studies on the modulation of receptor mobility by cross-linking of a particular set of receptors induced by Con A have led to the hypothesis (2)
that surface molecules may be anchored to a common network of cytoplasmic
structures. Recent experiments (29,30) strongly suggest that this system is
composed of both microfilaments and microtubules interacting reversibly with the
membrane receptors. This cell surface modulating assembly (SMA) may also serve
as a link between cytoplasmic function and cell surface alteration, as in mitogenic
stimulation.

3. EFFECT OF CON A AND SUCCINYL-CON A ON THE
PROLIFERATION OF CELLS IN TISSUE CULTURE

3.1 Mitogenic Stimulation of Lymphocytes by Con A and Succinyl-Con A

   The dose—response curve of stimulation of mouse splenic lymphocytes by native
Con A shows a rising and a falling limb with a maximum at a concentration of
about 5 µg/ml (Figure 3A) (8). In contrast, the dose—response curve for dimeric
succinyl-Con A, which is just as mitogenic as the native lectin, showed no falling
limb over a 10-fold concentration range beyond the optimal dose. Similar results
were obtained with human peripheral lymphocytes (Figure 3B) except that, in this
case, the response to the succinyl derivative as assayed by ^{3}H-thymidine incorporation was lower than that observed with native Con A (31). The stimulation of
human lymphocytes by acetyl-Con A was also reported to be lower than that
observed with the native lectin (32). Inasmuch as it has been shown that the
amount of ^{3}H-thymidine incorporated is proportional to the number of cells

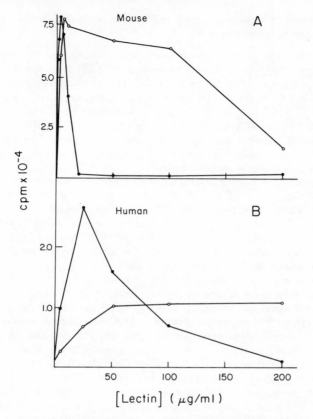

Figure 3. Dose—response curves showing the incorporation of ^{3}H-thymidine after stimulation of mouse spleen cells (A) and human peripheral lymphocytes (B) by Con A (●————●) and succinyl—Con A (○————○)

responding (4), our results suggest that dimeric succinyl-Con A was not as potent a mitogen as the native lectin for a subpopulation of human lymphocytes.

Correlation of the effects of Con A and succinyl-Con A on lymphocyte stimulation with their effects on cell surface receptor distribution indicates the following two conclusions: (a) neither capping nor inhibition of receptor mobility are strictly required for the mitogenic stimulation of lymphocytes; (b) if receptor cross-linkage is required for stimulation, the amount achieved by dimeric succinyl-Con A is sufficient.

Finally, the data obtained for both mouse and human lymphocytes (Figure 3A and B) suggest that the rising and falling portions of the dose—response curve reflect two independent events, the one representing stimulation and the other inhibition of lymphocyte proliferation. This conclusion is strongly supported by experiments using Con A and succinyl-Con A with another type of mitogen called phorbol myristate acetate (TPA).

3.2 Synergistic and Inhibitory Effects of TPA on the Stimulation of Lymphocytes by Con A and Succinyl-Con A

Phorbol esters such as TPA are potent cocarcinogens and inducers of DNA synthesis when applied to the skin of animals (33). It has been shown that TPA is independently mitogenic for bovine (34) and human (31) lymphocytes. In the experiment shown in Figure 4, 100 nM TPA was mitogenic when used alone. The addition of TPA to cultures containing suboptimal doses of Con A (Figure 4A) or succinyl-Con A (Figure 4B) greatly enhanced the response of the cells. For example, TPA and 5μg/ml succinyl-Con A led to a response higher than that observed when either reagent was used alone or than that which would be expected if the effects of two reagents were additive.

The phorbol ester led to decreased cellular response, however, when added to cultures containing doses of Con A that alone would be optimally mitogenic (Figure 4A). It is important to contrast this finding with the observations that both TPA (100 nM) and Con A (25 μg/ml), when used alone, were strongly mitogenic. In striking contrast to these findings, the addition of TPA to cultures containing succinyl-Con A only enhanced the response. No inhibition was seen even at very high concentrations of the lectin derivative (Figure 4B).

All of these observations strengthen the conclusion that the typical unimodal dose—response curve seen in the mitogenic stimulation of lymphocytes can be dissected into two portions that can be manipulated independently. It remains to be shown whether the synergism between TPA and lectins results from the interactions of several subpopulations of cells or whether the two classes of 'comitogens' act at particular points in a final common pathway of stimulation.

The fact that high doses of Con A inhibited mitogenesis as well as cell surface receptor mobility whereas succinyl-Con A had neither effect (1,8,31) suggests the possibility that the high-dose unresponsiveness of Con A-treated cultures may be correlated with the modulation of receptor mobility. It is of particular significance that the effect of low doses of TPA in reducing the cellular response to mitogenic stimulation was lectin dependent and was observed if, and only if, concentrations of the lectin higher than the optimally mitogenic dose could by themselves deliver the inhibitory signal. The recent experiments of Oliver and coworkers (35) demonstrating a role for cyclic GMP in the modulation of receptor mobility is in accord with this hypothesis, particularly because the action of TPA is thought to be mediated by the cyclic nucleotide (36). Therefore, components of the same cell surface modulating assembly (SMA) responsible for the inhibition of receptor mobility may also be involved in the regulation of the inhibitory signal for cell proliferation.

3.3 Effects of Con A and Succinyl-Con A on the Proliferation of Continuously Dividing Cells in Culture

In view of the dissociable properties of stimulatory and inhibitory events for normal lymphocytes, it was of interest to examine the effects of Con A and succinyl-Con A on a continuously dividing mouse lymphocyte tumour line, P-388 (Figure 5) (31). When these cells were cultured for 24 hours with various concentrations of Con A, DNA synthesis was slightly enhanced at low doses of the

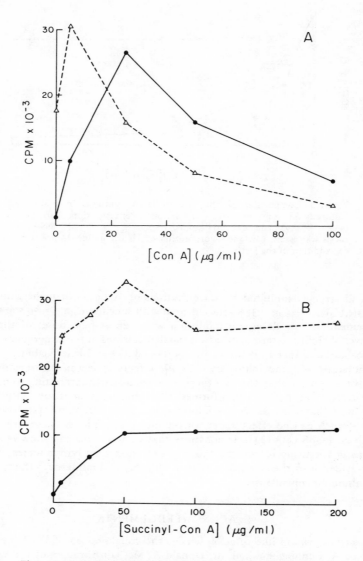

Figure 4. Comparison of the effects of the phorbol ester, TPA, on the stimulation of human lymphocytes by Con A (A) and succinyl-Con A (B). (●————●), cultures containing no TPA; (△ — — —△), cultures containing 100 nM TPA. Mitogenic stimulation was measured by the incorporation of ^{3}H-thymidine. Dimethylsulphoxide, the solvent used for TPA, was present in all cultures at a final concentration of 0.5%

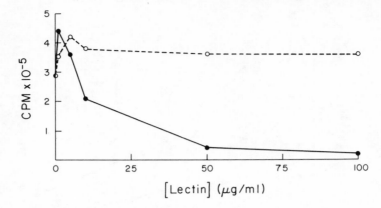

Figure 5. Comparison of the rate of DNA synthesis in P-388 cells cultured in the presence of various concentrations of Con A (●————————●) and succinyl-Con A (o — — — — — — — o). DNA synthesis was assayed by the incorporation of ^{3}H-thymidine 24 hours after the addition of the lectin

lectin and strongly inhibited at concentrations of 10 μg/ml or higher. In contrast, while DNA synthesis in P-388 cells cultured with succinyl-Con A also showed slight enhancement at low doses, no inhibition was seen at high doses of the dimeric derivative. Viability determinations on parallel cultures in the experiment shown in Figure 5 indicated that all P-388 cultures retained 87% or higher viability.

Stimulatory (37) and inhibitory (37–40) effects of lectins on transformed cell lines have also been reported by others. It is particularly interesting to note that a comparison of the data on transformed 3T3 fibroblasts in tissue culture (38,41) also showed that whereas Con A inhibited the growth of these transformed cells, succinyl-Con A had no effect under corresponding conditions. These results, as well as those on P-388 cells (31), demonstrate that tumour cells can show responses to mitogens qualitatively resembling those of untransformed lymphocytes. They lend further evidence to the conclusion that stimulatory and inhibitory effects of lectins can be altered independently.

4. ACKNOWLEDGEMENTS

The authors would like to acknowledge the collaboration of Dr. Ichiro Yahara, Dr. Bruce A. Cunningham and Mr. Donald A. McClain throughout various aspects of these studies. This work was supported by USPHS grants from the National Institutes of Health and by grants from the National Science Foundation. J. L. Wang is a Fellow of the Damon Runyon Memorial Fund for Cancer Research.

5. REFERENCES

1. Yahara, I. and Edelman, G. M. (1972) 'Restriction of the mobility of lymphocyte immunoglobulin receptors by concanavalin A', *Proc. Nat. Acad. Sci. U.S.A.*, **69**, 608–12.

2. Edelman, G. M., Yahara, I. and Wang, J. L. (1973) 'Receptor mobility and receptor—cytoplasmic interactions in lymphocytes', *Proc. Nat. Acad. Sci. U.S.A.*, 70, 1442—6.
3. Powell, A. E. and Leon, M. A., (1970) 'Reversible interaction of human lymphocytes with the mitogen concanavalin A', *Exp. Cell Res.*, 62, 315—25.
4. Gunther, G. R., Wang, J. L. and Edelman, G. M. (1974) 'The kinetics of cellular commitment during stimulation of lymphocytes by lectins', *J. Cell Biol.*, 62, 366—77.
5. Goldstein, I. J., Hollerman, C. E. and Smith, E. E. (1965) 'Protein—carbohydrate interaction. II. Inhibition studies on the interaction of concanavalin A with polysaccharides', *Biochemistry*, 4, 876—83.
6. Wang, J. L., Cunningham, B. A. and Edelman, G. M. (1971) 'Unusual fragments in the subunit structure of concanavalin A', *Proc. Nat. Acad. Sci. U.S.A.*, 68, 1130—4.
7. Edelman, G. M., Cunningham, B. A., Reeke, G. N., Jr., Becker, J. W., Waxdal, M. J. and Wang, J. L. (1972) 'The covalent and three-dimensional structure of concanavalin A', *Proc. Nat. Acad. Sci. U.S.A.*, 69, 2580—4.
8. Gunther, G. R., Wang, J. L., Yahara, I., Cunningham, B. A. and Edelman, G. M. (1973) 'Concanavalin A derivatives with altered biological activities', *Proc. Nat. Acad. Sci. U.S.A.*, 70, 1012—6.
9. Agrawal, B. B. L., Goldstein, I. J., Hassing, G. S. and So, L. L. (1968) 'Protein—carbohydrate interaction. XVIII. The preparation and properties of acetylated concanavalin A, the hemagglutinin of the jack bean', *Biochemistry*, 7, 4211—8.
10. Doyle, R. J. and Roholt, O. A. (1968) 'Tyrosyl involvement in the concanavalin A—polysaccharide precipitin reaction', *Life Sci.*, 7, part II, 841—6.
11. Hassing, G. S., Goldstein, I. J. and Marini, M. (1971) 'The role of protein carboxyl groups in carbohydrate—concanavalin A interaction', *Biochim. Biophys. Acta*, 243, 90—7.
12. McCubbin, W. D., Oikawa, K. and Kay, C. M. (1972) 'Circular dichroism studies on chemically modified derivatives of concanavalin A', *FEBS Lett.*, 23, 100—4.
13. Hassing, G. S., and Goldstein, I. J. (1972) 'Further chemical modification studies on concanavalin A, the carbohydrate binding protein of the jack bean', *Biochim. Biophys. Acta*, 271, 388—99.
14. Trowbridge, I. S. (1973) 'Mitogenic properties of pea lectin and its chemical derivatives', *Proc. Nat. Acad. Sci. U.S.A.*, 70, 3650—4.
15. Young, N. M. (1974) 'The effects of maleylation on the properties of concanavalin A', *Biochim. Biophys. Acta*, 336, 46—52.
16. Ornstein, L. (1964) 'Disc electrophoresis. I. Background and theory', *Ann. N.Y. Acad. Sci.*, 121, 321—49.
17. Davis, B. J. (1964) 'Disc electrophoresis. II. Method and application to human serum proteins', *Ann. N.Y. Acad. Sci.*, 121, 404—27.
18. Weber, K. and Osborn, M. (1969) 'The reliability of molecular weight determinations by dodecyl sulfate—polyacrylamide gel electrophoresis', *J. Biol. Chem.*, 244, 4406—12.
19. Noonan, K. D. and Burger, M. M. (1973) 'Binding of [^{3}H] concanavalin A to normal and transformed cells', *J. Biol. Chem.*, 248, 4286—92.
20. Cline, M. J. and Livingston, D. C. (1971) 'Binding of ^{3}H-concanavalin A by normal and transformed cells', *Nature (New Biol.)*, 232, 155—6.
21. Kalb, J. A. and Lustig, A. (1968) 'The molecular weight of concanavalin A', *Biochim. Biophys. Acta*, 168, 366—7.
22. McKenzie, G. H., Sawyer, W. H. and Nichol, L. W. (1972) 'The molecular weight and stability of concanavalin A', *Biochim. Biophys. Acta*, 263, 283—93.
23. Gordon, J. A. and Marquardt (1974) 'Factors affecting hemagglutination by concanavalin A and soybean agglutinin', *Biochim. Biophys. Acta*, 332, 136—44.
24. Huet, C., Lonchampt, M., Huet, M. and Bernadac, A. (1974) 'Temperature effects on the concanavalin A molecule and on concanavalin A binding', *Biochem. Biophys. Acta*, 365, 28—39.
25. Kind, L. S., Wang, H. and Lee, S. H. S. (1973) 'Activation of mouse spleen cells with acetylated concanavalin A coupled to red blood cell stroma', *Proc. Soc. Exp. Biol. Med.*, 142, 680—2.
26. Miller, I. R. and Great, H. (1972) 'Protein labelling by acetylation', *Biopolymers*, 11, 2533—6.
27. Yahara, I. and Edelman, G. M. (1973) 'The effects of concanavalin A on the mobility of lymphocyte surface receptors', *Exp. Cell Res.*, 81, 143—5.

28. Entlicher, G., Koštiř J. V. and Kocourek, J. (1970) 'Studies on phytohemagglutinins. III. Isolation and characterization of hemagglutinins from the pea (*Pisum sativum* L.)', *Biochim. Biophys. Acta*, **221**, 272—81.
29. Yahara, I. and Edelman, G. M. (1973) 'Modulation of lymphocyte receptor redistribution by concanavalin A, anti-mitotic agents and alterations in pH', *Nature*, **246**, 152—5.
30. Yahara, I. and Edelman, G. M. (1975) 'Electron microscopic analysis of the modulation of lymphocyte receptor mobility', *Exp. Cell Res.*, **91**, 125—42.
31. Wang, J. L., McClain, D. A. and Edelman, G. M. (1975) 'Modulation of lymphocyte mitogenesis', *Proc. Nat. Acad. Sci. U.S.A.*, in press.
32. Reichert, C. F., Pan, P. M., Mathews, K. P. and Goldstein, I. J. (1973) 'Lectin-induced blast transformation of human lymphocytes', *Nature (New Biol.)*, **242**, 146—8.
33. Baird, W. M., Sedgwick, J. A. and Boutwell, R. K. (1971) 'Effects of phorbol and four diesters of phorbol on the incorporation of tritiated precursors into DNA, RNA, and protein in mouse epidermis', *Cancer Res.*, **31**, 1434—9.
34. Mastro, A. M. and Mueller, G. C. (1974) 'Synergistic action of phorbol esters in mitogen-activated bovine lymphocytes', *Exp. Cell Res.*, **88**, 40—6.
35. Oliver, J. M., Zurier, R. B. and Berlin, R. D. (1975) 'Concanavalin A cap formation on polymorphonuclear leucocytes of normal and beige (Chediak-Higashi) mice', *Nature*, **253**, 471—3.
36. Goldberg, N. D., Haddox, M. K., Estensen, R., White, J. G., Lopez, C. and Hadden, J. W. (1974) 'Evidence of a dualism between cyclic GMP and cyclic AMP in the regulation of cell proliferation and other processes', in *Cyclic AMP, Cell Growth and the Immune Response*, eds. Braun, W., Lichtenstein, L. and Parker, C. W. (Academic Press, New York), pp. 247—62.
37. Mackler, B. R. (1972) 'Effects of concanavalin A on human lymphoid cell lines and normal peripheral lymphocytes', *J. Nat. Cancer Inst.*, **49**, 935—41.
38. Burger, M. M. and Noonan, K. D. (1970) 'Restoration of normal growth by covering agglutinin sites on tumor cell surface', *Nature*, **228**, 512—5.
39. Dent, P. B. and Hillcoat, B. L. (1972) 'Interaction of phytohemagglutinin with transplantable mouse lymphomas of differing malignant potential', *J. Nat. Cancer Inst.*, **49**, 373—7.
40. Ralph, P. and Nakoinz, I. (1973) 'Inhibitory effects of lectins and lymphocyte mitogens on murine lymphomas and myelomas', *J. Nat. Cancer Inst.*, **51**, 883—90.
41. Trowbridge, I. S. and Hilborn, D. A. (1974) 'Effects of succinyl-Con A on the growth of normal and transformed cells', *Nature*, **50**, 304—7.
42. Yariv, J., Kalb, A. J. and Levitzki, A. (1968) 'The interaction of concanavalin A with methyl-α-D-glucopyranoside', *Biochim. Biophys. Acta*, **165**, 303—5.
43. Nisonoff, A., Wissler, F. C., Lipman, L. N. and Woernley, D. L. (1960) 'Separation of univalent fragments from the bivalent rabbit antibody molecule by reduction of disulfide bonds', *Arch. Biochem. Biophys.*, **89**, 230—44.

CHAPTER 58

Insulin-like Effects of Concanavalin A

KWEN-JEN CHANG
PEDRO CUATRECASAS

1.1 Effects on Glucose Transport

Concanavalin A and wheat germ agglutinin are very effective in enchancing the rate of ^{14}C-glucose oxidation in isolated fat cells (Figure 1). The maximal effects are similar to those that can be achieved with insulin. The concentration required for the half-maximal effect is about 20 nM for concanavalin A (molecular weight: 100,000) and about 4 nM for wheat germ agglutinin (molecular weight; 25,000). These effects result from interactions of the lectins with high-affinity binding sites that represent only a fraction of the total number of lectin-binding sites on the fat cells, since direct binding studies with iodinated lectins demonstrate that saturation requires concentrations greater than 0.1 mg/ml ($\sim$10 μM) (1,2).

The increased rates of glucose oxidation induced by concanavalin A and by wheat germ agglutinin can be completely and selectively abolished by addition of the specific sugars α-methyl-D-mannopyranoside or N-acetyl-D-glucosamine (Table I). Since wheat germ agglutinin at low concentrations increases the binding of iodoinsulin to fat cells (3—5), the possibility exists that the lectins could be exerting their effect by facilitating effects of endogenous insulin possibly retained on the surface of the fat cells. The inability of a large excess of insulin antiserum to alter the metabolic effects of the cells in the presence or absence of the lectins (Table I) makes this possibility unlikely.

The effects on glucose oxidation of suboptimal concentrations of the lectins can be further increased by addition of insulin (4). However, the combined effect of insulin and the plant lectins does not surpass the maximal effect that can be achieved by either agent alone. Furthermore, simultaneous addition of both plant lectins in the absence or presence of insulin does not result in greater effects than can be achieved by any one of these three compounds. Trypsin treatment of fat cells, a process that decreases the apparent affinity of insulin for its biological

599

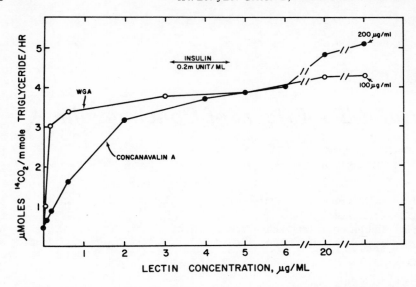

Figure 1. Effect of concanavalin A and of wheat germ agglutinin (WGA) on the conversion of ^{14}C-glucose to $^{14}CO_2$ by isolated fat cells. Fat cells were incubated for 2 h at 37° in 1.25 ml of Krebs—Ringer bicarbonate buffer containing 1% (w/v) albumin, and 0.2 mM ^{14}C-glucose (6.4 Ci/mol). The maximal insulin (0.2 munit/ml) response is indicated by the arrow. Data from ref. 4

TABLE I.   Reversal by specific sugars of plant lectin stimulation of glucose oxidation by fat cells[a]

Additions	Production of $^{14}CO_2$[b]
None	4400 ± 200
Insulin (0.2 munits/ml)	34,200 ± 1900
α-Methyl-D-mannopyranoside (25 mM)	4500 ± 400
N-Acetyl-D-glucosamine (25 mM)	4800 ± 300
Insulin + α-methyl-D-mannopyranoside + N-acetyl-D-glucosamine	32,200 ± 1100
Concanavalin A (5 μg/ml)	38,700 ± 2100
Concanavalin A + α-methyl-D-mannopyranoside	4600 ± 500
Concanavalin A + N-acetyl-D-glucosamine	36,800 ± 800
Wheat germ agglutinin (2 μg/ml)	33,000 ± 1800
Wheat germ agglutinin + N-acetyl-D-glucosamine	5400 ± 200
Wheat germ agglutinin + α-methyl-D-mannopyranoside	33,300 ± 900
Insulin antiserum (1:250 dilution)	4300 ± 300
Insulin + insulin antiserum	4200 ± 300
Concanavalin A + insulin antiserum	38,500 ± 2200
Wheat germ agglutinin + insulin antiserum	34,100 ± 2000

[a] The conditions are as described in Figure 1. The fat cell concentration is about 2 x 10⁴ cells per ml.
[b] c.p.m. per 2 h; average value ± SEM of three replications. Data from reference 4.

TABLE II.   Effect of ATP on wheat germ agglutinin and concanavalin A-stimulated rates of glucose oxidation by fat cells[a]

Lectin (μg/ml)	$^{14}CO_2$ production with addition of [b]:		
	nothing	10^{-5} M ATP	10^{-4} M ATP
Wheat germ agglutinin			
None	13,570 ± 200	12,990 ± 400	13,000 ± 500
0.08	24,200 ± 200	19,180 ± 400	15,060 ± 200
0.4	112,600 ± 700	76,800 ± 2100	61,800 ± 2500
4	139,800 ± 4000	130,000 ± 2000	121,700 ± 1000
Concanavalin A			
None	13,000 ± 400	13,100 ± 200	13,000 ± 300
4	45,900 ± 600	31,200 ± 200	22,600 ± 800
40	107,400 ± 1600	91,100 ± 5000	76,750 ± 4000
80	110,000 ± 5000	112,700 ± 9000	81,250 ± 1000

[a] Conditions are as described in Figure 1.
[b] Counts per min of $^{14}CO_2$ produced in 2 hours; average value ± S.E. Data from ref. 7.

receptor (6), also results in an apparent fall in the binding affinity of wheat germ agglutinin for the sites responsible for enhanced glucose transport (3).

It has been shown that ATP specifically inhibits the insulin-enhanced glucose oxidation and transport (7). Table II shows that ATP also inhibits the stimulatory effect of WGA and concanavalin A in a fashion similar to the inhibition of insulin-stimulated glucose oxidation (7). The data suggest that the two plant lectins may be acting by a similar mechanism, and that this mechanism may resemble that by which insulin normally exerts its effects. An insulin-like effect on glucose oxidation by concanavalin A has also been demonstrated by Czech and Lynn (8). Stimulatory effects on 3-O-methyl-D-glucose uptake in brown fat cells have been demonstrated recently (9).

1.2 Effects on Lipolysis

Concanavalin A and wheat germ agglutinin are also as effective as insulin in reversing epinephrine-induced lipolysis in fat cells (Table III). This antilipolytic property, which does not depend on the presence of glucose in the medium, is not mediated by effects on membrane transport. The plant lectins are effective at concentrations as low as those that demonstrably activate glucose transport. The antilipolytic properties are also completely reversed by the specific simple sugars.

1.3 Inhibition of Adenylate Cyclase Activity

Physiological concentrations of insulin can inhibit adenylate cyclase activity in isolated liver and fat cell membranes (10) and in fat cell ghosts (11). This effect of insulin is also mimicked by low concentrations of the plant lectins. Concanavalin A, in the concentration range of 5–50 μg/ml, effectively inhibits adenylate cylase activity in the absence or presence of epinephrine (Figure 2). Concentrations of concanavalin A greater than 50 μg/ml markedly reverse the inhibition of basal enzyme activity and cause stimulation of this enzyme.

TABLE III.   Suppression by wheat germ agglutinin and
concanavalin A of epinephrine-stimulated lipolysis in fat
cells[a]

Additions	Glycerol released[b]
None	6.8 ± 0.5
Epinephrine, 0.16 μg/ml	24.0 ± 1.1
Epinephrine, 0.16 μg/ml +	
wheat germ agglutinin, 0.11 μg/ml	20.5 ± 0.8
wheat germ agglutinin, 0.33 μg/ml	14.0 ± 0.6
wheat germ agglutinin, 1.0 μg/ml	10.3 ± 0.7
wheat germ agglutinin, 5.0 μg/ml	6.4 ± 0.5
concanavalin A, 2 μg/ml	25.7 ± 1.4
concanavalin A, 6 μg/ml	20.1 ± 0.8
concanavalin A, 30 μg/ml	10.9 ± 0.7
concanavalin A, 100 μg/ml	8.8 ± 0.9
insulin, 20 μunits/ml	6.3 ± 0.5
insulin, 5 μunits/ml	12.9 ± 0.6

[a] Fat cells (3×10^4 cells/ml) were incubated for 2 h at 37°
in Krebs—Ringer bicarbonate buffer containing 3% (w/v)
albumin.
[b] μmole of glycerol released per mmole of triglyceride;
average value ± SEM of three replications. Data from
ref. 4.

Wheat germ agglutinin (2—50 μg/ml) is also quite effective in decreasing the
basal and the epinephrine-stimulated activities of adenylate cyclase (4). The
paradoxical effects on basal enzyme activity observed with concanavalin A are not
apparent with this lectin. The paradoxical effect of concanvalin A on lymphocytes
has also been reported (12).

1.4 Stimulation of ATPase

Insulin was found to stimulate the Mg^{2+} ATPase of human lymphocytes (13) at
physiological concentrations. Jarrett and Smith (14) found recently that insulin
also stimulates slightly but significantly the Mg^{2+} ATPase of the adipocyte plasma
membrane at concentrations as low as 5 microunits per ml; maximum stimulation
occurs between 50 to 100 microunits per ml. Concanavalin A mimicks the effect of
insulin on adipocyte (14) and rat lymphocyte (15) plasma membrane Mg^{2+} ATPase.
In rat lymphocytes, the effect of concanavalin A on ATPase is inhibited by the
specific sugar, α-methyl-D-mannopyranoside. However, it is not known whether the
stimulatory effect of concanavalin A on adipocyte plasma membrane ATPase can
be inhibited by α-methyl-D-mannopyranoside.

2. CONCLUSIONS

Demonstration of a) the insulin-like effects of plant lectins on glucose transport,
lipolysis, adenylate cyclase and ATPase, b) the perturbation of the insulin-receptor
interaction by plant lectins (3) and c) the adsorption of solubilized insulin receptors

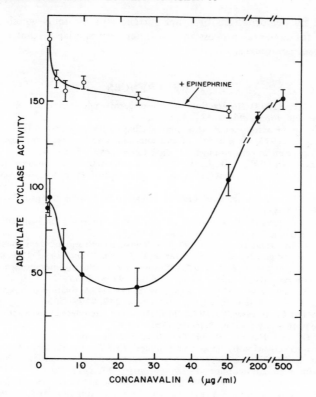

Figure 2. Effect of concanavalin A on the basal and the epinephrine (1 μM)-stimuated adenylate cyclase activities of fat cell membranes. The incubation mixture (0.1 ml) contained 50 mM Tris-HCl (pH 7.6), 7 mM MgCl$_2$, 1 mM EDTA, 2.5 mM Aminophylline, 0.1% (w/v) albumin, 3 mM α-^{32}P-ATP (1.5 μCi) and about 100 μg of membrane protein; 5 mM phosphoenolpyruvate and 60 μg/ml of pyruvate kinase were used to regenerate ATP. After 10 min at 30°, the tubes were placed in boiling water for 3 min. A recovery mixture (0.5 ml) containing cyclic ^{3}H-AMP was added to each sample. Cyclic AMP was isolated on a column containing 1 g of alumina that was eluted with 2.5 ml of 50 mM Tris-HCl (pH 7.6). Activity is expressed as pmole of cyclic AMP produced per min per mg of protein (average of triplicates ± SEM). Data from ref. 4

to concanavalin A or WGA covalently coupled to agarose, lend support to the view that the insulin receptor of fat cell and liver membranes is a glycoprotein which contains several chemically distinct sites capable of binding plant lectins in a manner which perturbs the insulin-receptor interaction. As described elsewhere (3, 5), however, the insulin-like biological activities of concanavalin A may not be modulated by its direct interaction with insulin receptors. These effects may instead be mediated by perturbations of other surface glycoproteins which in turn may affect some fundamental process in a manner very similar to that occurring with insulin-receptor complexes.

3. REFERENCES

1. Cuatrecasas, P. (1973) 'Interaction of wheat germ agglutinin and concanavalin A with isolated fat cell', *Biochemistry*, **12**, 1312—23.
2. Chang, K.-J. and Cuatrecasas, P., this volume, Chapter 23, page 201—11.
3. Cuatrecasas, P. (1973) 'Interaction of concanavalin A and wheat germ agglutinin with the insulin receptor of fat cells and liver', *J. Biol. Chem.*, **248**, 3528—34.
4. Cuatrecasas, P. and Tell, G. P. E. (1973) 'Insulin-like activity of concanavalin A and wheat germ agglutinin: Direct interaction with insulin receptor', *Proc. Nat. Acad. Sci. U.S.A.*, **70**, 485—9.
5. Cuatrecasas, P. and Chang, K.-J. (1975) 'Isolation of insulin receptors', in this volume, Chapter 45, page 421—27.
6. Cuatrecasas, P. (1971) 'Perturbation of the insulin receptor of isolated fat cells with proteolytic enzymes', *J. Biol. Chem.*, **246**, 6522—31.
7. Chang, K.-J. and Cuatrecasas, P. (1974) 'Adenosine triphosphate-dependent inhibition of insulin-stimulated glucose transport in fat cells', *J. Biol. Chem.*, **249**, 3170—80.
8. Czech, M. P. and Lynn, W. S. (1973) 'Stimulation of glucose metabolism with lectins in isolated white fat cells', *Biochim. Biophys. Acta*, **297**, 368—77.
9. Czech, M. P., Lawrence, J. C. and Lynn, W. S. (1974) 'Activation of hexose transport by concanavalin A in isolated fat cells', *J. Biol. Chem;* **249**, 7499—505.
10. Illiano, G. and Cuatrecasas, P. (1972) 'Modulation of adenylate cyclase activity in liver and fat cell membranes by insulin', *Science*, **175**, 906—8.
11. Hepp, K. D. and Renner, R. (1972) 'Insulin action on the adenyl cyclase system: Antagonism to activation by lipolytic hormones', *FEBS Lett.*, **20**, 191—4.
12. Lyle, L. R. and Parker, C. W. (1974) 'Cyclic adenosine 3′, 5′-monophosphate response to concanavalin in human lymphocytes', *Biochemistry*, **13**, 5415—20.
13. Hadden, J. W., Hadden, E. M., Wilson, E. E. and Good, R. A. (1972) 'Direct action of insulin on plasma membrane ATPase activity in human lymphocytes', *Nature (New Biol.)*, **235**, 174—6.
14. Jarrett, L. and Smith, R. M. (1974) 'The stimulation of adipocyte plasma membrane magnesium ion-stimulated adenosine triphosphatase by insulin and concanavalin A', *J. Biol. Chem.*, **249**, 5195—9.
15. Novogrodsky, A. (1972) 'Concanavalin A stimulation of rat lymphocyte ATPase', *Biochim. Biophys. Acta*, **266**, 343—9.

CHAPTER 59

Concanavalin A-Enzyme Conjugates: Site-specific Enzyme Reagents

W. THOMAS SHIER

1. RATIONALE FOR THE USE OF LECTIN–ENZYME CONJUGATES

Lectins, of which Con A is the best defined and most readily available, possess many very useful properties that have been described in other chapters of this book. Most of these uses of lectins derive ultimately from their property of being localized and concentrated on the insoluble saccharide structures to which they bind. This localizing and concentrating property is shared with antibodies, but as experimental tools lectins can be used in different situations than antibodies because (a) they are usually available in greater quantity, and (b) they have a specificity range for saccharide structures with a generally lower affinity than antibodies, and without the heterogeneity of specificity typical of most antibody preparations. However, this localizing and concentrating property is not usually shared with enzymes, although there are numerous therapeutic and experimental situations in which it would be very desirable if this property could be combined with the enzyme activity. Recently it has been observed that lectins can be coupled to enzymes through stable covalent linkages to give novel macromolecular conjugates that, as far as has been determined, retain many of the activities of both precursors.

Of the numerous enzymes with actual or potential therapeutic usefulness (1), the majority display undesirable side effects that would be minimized if the enzyme could be effectively localized at its intended site of action. The general principle of attaching enzymes to lectins in order to combine the properties of both proteins into one molecule can be illustrated by a Con A—dextranase conjugate that has been prepared by Barker and his associates (2) in an attempt to enhance the effectiveness of dextranase as an inhibitor of dental caries (3). The rationale for the use of this conjugate is that, when provided as a dietary supplement, it should be retained in contact with the dextran of dental plaque for a longer period of time and at a higher concentration than would be achieved with the free enzyme, which is rapidly removed by the normal flow of saliva. Barker and his associates anticipate that much less Con A—dextranase conjugate will be required for a significant reduction in dental caries, and that the lower level of the conjugate may minimize the adverse effect of the enzyme on the white blood cell count (2). Streptokinase and L-asparaginase are two additional examples of enzymes for which the therapeutic effectiveness could benefit from localization at the appropriate site in the body (1).

Con A has been shown to be retained for an unusually long period of time near the site of injection in every tissue that has been examined (4,5). For example, Con A requires 134 times longer for 90% elimination from mouse feet than does egg white lysozyme, 92 times longer than does E. coli L-asparaginase and 158 times longer than does bovine trypsin. Prior immunization of mice against Con A does not significantly alter the rate of clearance of the lectin from near the site of injection, suggesting that the lectin in the tissue avoids the immune elimination system by some as yet undetermined mechanism. Furthermore, it has been shown by sodium dodecylsulphate—polyacrylamide gel electrophoresis of extracts of ^{125}I-Con A-treated tissue that the lectin is retained in the tissue in an undegraded form for at least 21 days (5). In principle, conjugation of an enzyme to Con A should provide a method of localizing the enzyme in a given tissue.

Regardless of the potential therapeutic value of lectin—enzyme conjugates there are numerous experimental situations in which it is desirable to concentrate an enzyme at a particular site, and then reverse the process under mild, non-toxic conditions. Among the potential experimental uses of lectin—enzyme conjugates are site-specific reagents for studies in cell biology, reversibly immobilized enzymes and dissociable enzyme—enzyme conjugates.

2. PREPARATION OF CONJUGATES

In theory any bifunctional reagent capable of reacting with one of the functional groups present on proteins would be suitable for forming soluble protein—protein conjugates (6). In practice the ϵ-amino group of lysine, the phenol group of tyrosine and the carboxyl groups of aspartate and glutamate are almost exclusively involved in the formation of protein—protein linkages by the standard methods. The most commonly used coupling reagents are glutaraldehyde, diazobenzidine or a water-soluble carbodiimide. When it is important to retain enzyme activity in one of the components of a protein—protein conjugate, glutaraldehyde is most commonly used, since it usually causes the least enzyme inactivation under conditions that give effective cross-linking. Because the exact chemistry of the reaction of glutaraldehyde with proteins is not known (6) no explanation for this observation can be offered.

Two additional generalizations about protein—protein conjugation can be made.

First, protein—protein conjugation is favoured by the use of high protein concentrations. Secondly, the yields of the conjugation reaction are inevitably low, since the reaction product will contain each of the starting proteins in aggregated and unaggregated form as well as the desired conjugate. While some inactivation of Con A binding sites may occur as a result of glutaraldehyde reacting with functional amino acids in the binding site, this is probably not a major cause of the low yields, since addition of α-methyl-D-mannoside to the conjugation mixture does not significantly improve the yield of conjugate. A major source of the low yield is probably the steric blocking of a binding site on either component of a conjugate by the other component, thereby preventing interaction with other macromolecular substrates including the adsorbent in any affinity columns used in the purification. In cases in which one of the proteins is expensive or difficult to obtain, the yield of conjugate can be substantially increased by using an excess of the more readily available starting material.

In the paragraphs that follow, the preparation of two types of Con A—enzyme conjugates is described. The first example is the preparation of a Con A-trypsin conjugate. In this case affinity purification methods are available for both precursor proteins making possible facile isolation of the conjugate. In the second example, a Con A—L-asparaginase conjugate, affinity purification is available only for concanavalin A, so that either contamination with aggregated concanavalin A must be tolerated or additional purification procedures are required.

2.1 Preparation of a Con A—Trypsin Conjugate

All operations except assays are carried out at $0°$. Commercially available Con A (100 mg, Miles Laboratories, twice crystallized) in approximately 2 ml of saturated sodium chloride is cooled and diluted to 10 ml with cold 0.5 M sodium chloride buffered with 0.2 M Tris acetate, pH 5.0, (TAS) containing 0.01 M benzamidine hydrochloride (Aldrich) and crystalline bovine trypsin (100 mg, Sigma). Cold 25% aqueous glutaraldehyde (400 μl, Sigma) is slowly added with efficient mixing, and the mixture allowed to stand for 90 minutes. The mixture is applied directly to Con A affinity column consisting of 100 ml of 'Sephadex' G-200 in TAS. The column is washed with 10 ml of TAS, then 100 ml of TAS saturated with tyrosine, 200 ml of TAS and finally eluted with 150 ml of 0.2 M α-methyl-D-mannoside in TAS. The eluate was applied directly to a coupled trypsin affinity column consisting of 2 ml of p-aminobenzamidine—'Sepharose' conjugate in TAS. This affinity adsorbant can be prepared by coupling p-(ϵ-amino-caproylamide)benzamidine to cyanogen bromide-activated 'Sepharose' 4B (7) or by coupling p-aminobenzamide to 'Sepharose' 4B beads containing ϵ-aminocaproate spacer molecules (8). After the sample has been applied to the trypsin affinity column, the column is washed with 20 ml of TAS and eluted with 5 ml of 1 M benzamidine hydrochloride in TAS followed by 2 ml of TAS. This solution can be stored frozen in 1 ml aliquots that are individually freed of benzamidine immediately prior to analysis and use: A one ml aliquot is applied to a column of 'Bio-Gel' P-2 resin (120 × 0.9 cm) in 0.15 M sodium chloride containing 0.02 M Tris acetate, pH 5.0 and eluted with the same buffer. The fractions in the excluded volume are combined and assayed for protein by the method of Lowry (9) and for trypsin activity using *p*-tosyl-L-arginine methyl ester as a substrate (10) and crystalline trypsin in the same buffer aş standard.

This procedure typically yields a preparation containing the activity of 2 to 4 mg of trypsin and 5 to 9 mg of total protein. The conjugate has been shown to retain several activities of trypsin including esterase activity toward *p*-tosyl-L-arginine methyl ester, benzamidine binding and proteolysis toward azocasein. It has also been shown to retain several properties of Con A including saccharide binding, haemagglutination and prolonged retention at the site of injection in tissues.

Electrophoresis in the presence of 0.1% sodium dodecylsulphate on 10% to 15% gradient polyacrylamide gels (11) indicates that these preparations contain a species with the expected molecular weight for a trypsin—Con A subunit (49,000 daltons).

2.2 Preparation of a Con A—L-Asparaginase Conjugate

All operations except assays are carried out at $0°$. Commercially available Con A (5 mg in 135 μl of saturated sodium chloride, Miles Laboratories) is mixed with *E. coli* L-asparaginase (10 mg of 120 units per mg in 2 ml of 50% glycerol, pH 6.5, Sigma Chemical Co.) containing a small amount of the same enzyme labelled with ^{125}I by the method of McConahey and Dixon (12) to serve as a tracer. The mixture is dialysed against 300 ml of 0.5 M sodium chloride containing 0.05 M sodium phosphate, pH 6.5, and 0.2 M mannose, and then concentrated in the dialysis bag with dry 'Sephadex' to a volume of 2 ml. A 2% aqueous solution of glutaraldehyde (60 μl) is slowly added with efficient mixing, and the mixture allowed to stand at $0°$ for 8 h, or to the first indication of turbidity in the reaction mixture. The reaction is terminated by addition of 20 μl of a saturated aqueous solution of sodium metabisulphite, and the mixture dialysed against 3 changes of 100 ml of 0.5 M sodium chloride containing 0.05 M sodium phosphate, pH 6.5 (PBS). The mixture is applied to a Con A affinity column consisting of 10 ml of 'Sephadex' G-200 in PBS. The column is washed with 100 ml of PBS and eluted with 20 ml of 0.2 M α-methyl-D-mannoside in PBS. The eluate, which contains Con A—L-asparaginase conjugate contaminated with both aggregated and unaggregated Con A, is concentrated by vacuum dialysis and applied to a column (40 x 1.2 cm) of 'Bio-Gel' P-150 in PBS, that was previously calibrated with blue dextran and unaggregated Con A. The column is eluted with PBS and the fractions containing Con A—^{125}I-L-asparaginase determined in a gamma counter. The fractions containing most of the radioactivity are combined and assayed for L-asparaginase activity. Several assay methods for L-asparaginase are available (13). The method of Frohwein and coworkers (14) using the synthetic substrate L-aspartyl-β-hydroxamic acid (Sigma) is sufficiently sensitive for this work. On the basis of recovered radioactivity 10% to 15% of the ^{125}I-L-asparaginase in the starting mixture is coupled to Con A under these conditions and the L-asparaginase enzyme titres correspond to 55% to 80% retention of initial enzyme activity.

Treatment of radioiodinated Con A with glutaraldehyde under the same conditions results in about 50% conversion of Con A to an aggregated form that is not separated from Con A—L-asparaginase on a 'Bio-Gel' P-150 column. Assay of Con A—L-asparaginase preparations for total protein by the method of Lowry (9) indicates that they generally contain 65% to 75% aggregated Con A.

3. PITFALLS

The coupling activity of commercially available glutaraldehyde preparations varies from lot to lot and with time for a given lot. For best results a narrow range of glutaraldehyde concentrations or treatment times should be tested when either of the procedures given above is first attempted, and at intervals thereafter. Similarly, neither procedure can be applied directly to the preparation of a conjugate containing a different enzyme and/or lectin. For any extension to another system it is necessary to test a range of glutaraldehyde concentrations or treatment times in order to obtain the optimum coupling yield. In general the coupling efficiency increases when higher molecular weight proteins are used.

Sodium metabisulphite, which stops the conjugation reaction by forming bisulphite adducts with glutaraldehyde, inhibits some enzymes. It should be tested with the starting enzyme before use. If it irreversibly inhibits the starting enzyme its use can be avoided by applying the reaction mixture directly to an affinity column. This approach has the disadvantage that precise control of the time of exposure to glutaraldehyde is no longer possible, but in general this is not a serious disadvantage.

Trypsin affinity columns vary from preparation to preparation with a resulting variation in the minimum concentration of benzamidine hydrochloride required to elute trypsin. Preliminary conjugation studies with ^{125}I-trypsin indicated that when long glutaraldehyde exposure times were used, significant amounts (up to 35%) of the radioactivity were not eluted from the affinity column, while very little was lost with short glutaraldehyde exposure times. One explanation for these results is that conjugated species containing more than one functional trypsin molecule (e.g. Con A(trypsin)$_2$ species) are not eluted using the standard conditions.

4. AN ATTEMPT TO TREAT A L-ASPARAGINASE-SENSITIVE TUMOUR IN MICE WITH CON A—L-ASPARAGINASE

L-Asparaginase is used predominantly for the treatment of leukaemias and other disseminated neoplasms. However, there are several types of solid tumours that are sensitive to the enzyme (1). Treatment of the entire body with large amounts of the enzyme in order to treat a localized tumour is not only wasteful, but it stimulates the immune elimination system and it enhances the immunosuppressive properties of L-asparaginase (15). Since Con A has been shown to be retained near the site of injection in numerous tissues about 100 times longer than several diverse control proteins, and to effectively evade the immune elimination system while it is in tissues (4), a conjugate of Con A and L-asparaginase should localize the enzyme at a high concentration near the site of injection in a solid tumour and the enzyme activity should not be affected by the immune elimination system.

The Con A—[125]I-L-asparaginase conjugate prepared as described above was shown to be retained near the site of injection in the footpads of mice 95 times longer than free ^{125}I-L-asparaginase alone or mixed with free Con A (time for 90% elimination from the foot). The conjugate was tested for therapeutic effectiveness against established tumours of the L-asparaginase-sensitive lymphosarcoma 6C3HED in C3H/HeJ mice (Jackson Laboratories) using three intranodular

TABLE I.   Treatment[a] of the L-asparaginase-sensitive lymphosarcoma 6C3HED with Con A—L-asparaginase conjugate

Group number	Treatment[a] agent	Total amount of L-asparaginase activity per mouse (international units)	Fraction of mice dead after 18 days
1	Saline	0	4/6
2	L-Asparaginase + Con A	0.7	0/6
3	Con A—L-asparaginase conjugate	0.7	4/6
4	L-Asparaginase + Con A	1.4	0/6
5	Con A—L-asparaginase conjugate	1.4	4/6
6	L-Asparaginase + Con A	2.4	0/6
7	Con A—L-asparaginase conjugate	2.4	4/5

[a]C3H/HeJ mice were given subcutaneous injections of 6×10^5 viable 6C3HED cells. When the tumours grew to approximately 12 mm in diameter, on three consecutive days groups of the mice were given intranodular injections of 100 μl of saline (0.15 M sodium chloride containing 0.01 M sodium phosphate, pH 6.5) containing a total of the indicated number of units of L-asparaginase activity (a) as part of a Con A—L-asparaginase conjugate preparation containing 32.5 μg of Con A/unit L-asparaginase activity (groups 3, 5, 7) or (b) as free L-asparaginase mixed with an equivalent amount of free Con A (groups 2, 4, 6).

injections on consecutive days. Under these conditions free L-asparaginase mixed with free Con A afforded a highly significant degree of protection against the tumour (P < 0.005 by the χ^2 test with Yates' correction) (see Table I). In contrast, treatment with the same number of international units of L-asparaginase activity and the same amount of Con A in the form of a conjugate resulted in survival times that were not different from those observed in mice treated in the same manner with saline.

Several possible explanations can be given for the total lack of effect of the enzyme when conjugated to Con A. It is possible that the enzyme is very rapidly destroyed after injection, but this is unlikely since Con A injected into normal tissues is not degraded over a longer time span (5). It is possible that it is rapidly sequestered (16) into a location where neither extracellular nor intracellular asparagine can reach it. This process would have to be very rapid since tumour cells in culture show marked cytotoxic changes within an hour of exposure to L-asparaginase (17). There have been suggestions that the oncolytic effects of L-asparaginase are not solely due to the deamination of asparagine leading to depletion of intracellular asparagine in tumour cells that lack asparagine synthetase (17). Rather, the possibility has been suggested that some effects of L-asparaginase, in particular inhibition of lectin-induced blast transformation and immunosuppressive effects, may be due to a direct action of the enzyme on the cell surface (18). It is possible that the explanation for failure of Con A—L-asparaginase to kill L-asparaginase-sensitive cells *in vivo* involves the prevention of a presently unknown killing mechanism.

5. ACKNOWLEDGEMENT

The work described in this chapter was supported by U.S. Public Health Service Grants CA16123 and CA14195 from the National Cancer Institute.

6. REFERENCES

1. Cooney, D. A. and Rosenbluth, R. J. (1975) 'Enzymes as therapeutic agents', in *Advances in Pharmacology and Chemotherapy*, Vol. 12, eds. Garattlini, S., Goldin, A., Hawking, F. and Kopin, I. J. (Academic Press, New York), pp. 185–289.
2. Barker, S. A., Giblin, A. G., Gray, C. J. and Bowen, W. H. (1974) 'Preparation and properties of a conjugate containing dextranase and concanavalin A', *Carbohyd. Res.*, 36, 23–33.
3. Grenby, T. H. (1975) 'The control of dental decay. A review of protective chemicals for use as food additives', *Chem. Ind.*, 166–71.
4. Shier, W. T., Trotter, J. T. III and Reading, C. L. (1974) 'Inflammation induced by concanavalin A and other lectins', *Proc. Soc. Exp. Biol. Med.*, 146, 590–3.
5. Shier, W. T. and Trotter, J. T. III unpublished results.
6. Zaborsky, O. (1973) *Immobilized Enzymes*, Chapter 4 (CRC Press, Cleveland), pp. 61–74.
7. Sampaio, C., Wong, S.-C. and Shaw, E. (1974) 'Human plasma kallikrein. Purification and preliminary characterization', *Arch. Biochem. Biophys.*, 165, 133–9.
8. Hixson, H. J., Jr. and Nishikawa, A. H. (1974) 'Bovine trypsin and thrombin', *Methods Enzymol.*, 34B, 440–8.
9. Lowry, O. H., Rosebrough, N. J., Farr, A. L. and Randall, R. J. (1951) 'Protein measurement with the Folin phenol reagent', *J. Biol. Chem.*, 193, 265–75.
10. Worthington Enzyme Manual (1972), *Trypsin* (Worthington Biochemical Corporation, Freehold, N.J.), pp. 125–7.
11. Maizel, J. V., Jr. (1971) 'Polyacrylamide gel electrophoresis of viral proteins', *Methods Virol.*, 5, 179–246.
12. McConahey, P. J. and Dixon, F. J. (1966) 'A method of trace iodination of proteins for immunologic studies', *Int. Arch. Allergy*, 29, 185–9.
13. Jayaram, H. N., Cooney, D. A., Jayaram, S. and Rosenblum, L. (1974) 'A simple and rapid method for the estimation of L-asparaginase in chromatographic and electrophoretic effluents: comparison with other methods', *Anal. Biochem.*, 59, 327–46.
14. Frohwein, Y. Z., Friedeman, M., Reizer, J. and Grossowicz, N. (1971) 'Sensitive and rapid assay for L-asparaginase', *Nature*, 230, 158–9.
15. Hersh, E. M. (1971) 'Immunosuppression by L-asparaginase and related enzymes', *Transplantation*, 12, 368–74.
16. Goldman, R. (1974) 'Induction of vacuolation in the mouse peritoneal macrophage by concanavalin A', *FEBS Lett.*, 46, 203–8.
17. Broome J. D. (1968) 'Studies on the mechanism of tumour inhibition by L-asparaginase', *J. Exp. Med.*, 127, 1055–72.
18. Fidler, I. J. and Montgomery, P. G. (1972) 'Effects of L-asparaginase on lymphocyte surface and blastogenesis', *Cancer Res.*, 32, 2400–6.

CHAPTER 60

Differential Toxicity of Concanavalin A and PHA on Lymphoid and Haematopoietic Cell Lines

PETER RALPH

1. PRINCIPLE OF THE CYTOTOXIC ASSAY

The effect of lectins on cell lines growing exponentially can be studied in several ways. *Inhibition* can be assayed by cessation of cell growth (1), or by interference with cell metabolism such as DNA synthesis measured by incorporation of radioactive thymidine (1—4). *Toxicity* is judged by an absolute decrease in viable cell numbers, measured by trypan blue exclusion (1,3) or using size discrimination in an automatic cell counter (5). Definitions of toxicity include reduction in colony forming (cloning) ability (6), or other assays of irreversible cellular damage. We have chosen the trypan blue exclusion test to assess toxicity and in some cases the reduction in growth rate as a measure for inhibition.

2. ASSAY PROCEDURE

Cultures are set up at a concentration of 10^5 cells per ml, which is a convenient density for haemocytometer counting, and at the same time allows several doublings of control cells during the time course of the experiment. The toxic agent is added at various concentrations, and viable cells determined at time intervals thereafter. One tenth volume of 1% trypan blue in saline is added to a cell sample, and dye excluding cells counted in a haemocytometer 5 min later. In practice, viable cells can be distinguished in many cell lines by microscopic inspection without the use of a dye.

3. EFFECTS OF CON A AND PHA ON LYMPHOID AND HAEMATOPOIETIC CELLS

The specificity of Con A and PHA binding to carbohydrates has been extensively studied, and the mitogenic and haemagglutinating activities of these lectins can be inhibited by the appropriate monosaccharides. Similarly, the agglutination and killing of transformed fibroblasts (6), the inhibition of RNA synthesis in a murine myeloma line (4) and the transient stimulation and dose-dependent inhibition of DNA synthesis in human lymphoblast cell lines (3) by Con A are blocked by methyl-D-mannose. Also the inhibition of DNA synthesis in a murine T-lymphoma by PHA is reversed by N-acetyl-D-glucosamine (4). The inhibitory and toxic effects of these lectins are therefore believed to result from binding to specific receptor sites on cell surface molecules.

Our results (1, 4) demonstrate the unique sensitivity of a subset of murine and human cell lines derived from malignant T-lymphocytes to PHA and Con A, at concentrations which trigger normal T-cells into DNA synthesis. B-Lymphocyte lines are generally less sensitive, and myeloma cells are most resistant to these two plant lectins.

3.1 Cytotoxicity towards Murine Cell Lines

Table I shows the toxic effects of Con A and PHA on a variety of murine tumour cell lines derived from T- and B-lymphocytes, plasmacytes (myelomas), mast cells, myelomonocytes and erythrocyte precursors.

The growth of all haematopoietic tumour cell types, except a negative myeloma, was inhibited 50% by Con A at 2 to 30 μg/ml. The toxic concentration of Con A for each cell line was about twice the inhibitory amount. For examples of growth and killing curves, see refs. (1) and (16). Myelomas were less senstitive than T-lymphomas, seen especially when cultures grown in horse serum were compared (4). A negative myeloma XC1 (17), lacking immunoglobulin, plasmacyte antigen PC.1 and histocompatibility determinants (less than 0.002 of the H-2^k of the parent C1 myeloma), grew normally in the presence of 100 μg/ml Con A.

Sensitivity of other cell types to Con A has been described, e.g. preferential killing of SV40-transformed fibroblasts as compared to normal embryo fibroblasts (6), but this requires higher concentrations of Con A than used here for haematopoietic tumour cells.

PHA shows a more specific effect among haematopoietic cell lines. Myelomas were resistant to 100 μg/ml PHA, while T-lymphomas were very sensitive, being growth inhibited at concentrations of 3–15 μg/ml. An erythroleukaemia was also

TABLE I.   Inhibition of murine tumour cell growth by Con A and PHA[a]

Type	Strain	Induction	Sensitivity to		Reference
			Con A	PHA	
T-Lymphomas					
S49.1	BALB/c	Oil	10	3	1
S1A.12	BALB/c	Oil	20	7	1
WEHI22.1	BALB/c	Radiation	4	3	12
BW5147	AKR	Spontaneous	10	15	1
R1	C58	Spontaneous	10	6	1
T1M1.4	C57BL/6	Radiation LV	5	12	1
EL4	C57BL/6	Carcinogen	5	15	1
MBL-2	C57BL/6	Moloney LV	4	10	1
C1498	C57BL/6	Spontaneous	2	60	4
B-Lymphomas					
RAW8	BALB/c	Abelson LV	15	20	12
R7	CBF$_1$	Abelson LV	15	20	13
Myelomas					
C1.18 (X5563)	C3H	Spontaneous	15	>100	1
RPC-5.4	BALB/c	Oil	25	>100	4
S117.1BU.1	BALB/c	Oil	10	>100	1
S194/2.3.3.3	BALB/c	Oil	15	>100	1
J558.2BU.1	BALB/c	Oil	7	>100	1
MOPC315/P	BALB/c	Oil	15	>100	1
Negative myeloma					
XC1	C3H	From C1	>100	>100	4
Mastocytoma					
P815	DBA/2	Spontaneous	NT	40	12
Myelomonocytic leukaemia					
WEHI-3	BALB/c	Oil	30	40	14
Erythroleukaemia					
GM86	DBA/2	Friend LV	10	10	12

[a]Cells were grown in medium containing 10% foetal calf serum. The concentration of Con A (Calbiochem and Miles-Yeda) and PHA (PHA-P, Difco) is shown which *inhibits* growth by 50% over 3–4 days incubation, during which control cultures increase in cell numbers 10- to 50-fold. *Toxic* concentrations which will kill tumour cells within one day of incubation are 1.5 to 3 times higher than the values shown. The optimal mitogenic concentration of these preparations of Con A and PHA for mouse spleen cells is 5 μg/ml (15). C1498, described initially as a myeloid leukaemia, is classified as a T-cell here because of the presence of low amounts of θ antigen (4), shown by absorption to be identical to S49 θ antigen (R. Hyman, personal communication), and by its sensitivity to growth inhibition by unlabelled thymidine, a marker for thymocytes (16). NT = not tested.

very sensitive. B-lymphoma, mastocytoma and myelomonocytic leukaemia cell lines were intermediate. One T-lymphoma (C1498) was relatively resistant. This line resembles the normal peripheral, immunocompetent T-cell, in that it is corticoid resistant and exhibits low levels of theta antigen.

As discussed below, I believe the T- and B-lymphoid tumours retain their unique interaction with T- and B-lymphocyte mitogens, respectively, which leads to death or inhibition of tumours as contrasted with stimulation of normal lymphocytes. I would therefore assign C1498 cell line to the subset of peripheral T-cells described by Stobo, Paul and Henney as being low θ, PHA unresponsive, Con A responsive, relatively sessile and radiation resistant, and found predominately in the spleen (18). After allogeneic immunization, killer T-lymphocytes appear to have some of these properties (18). As seen in Table I, C1498 is the line most sensitive to Con A inhibition. It should be pointed out that C1498 is the only line in the table known to be contaminated with mycoplasma, and these results should be taken with caution until they can be repeated with cultures free of mycoplasma.

3.2 Cytotoxicity towards Human Cell Lines

The extreme sensitivity of murine T-lymphomas to growth inhibition by PHA, and resistance of myelomas, has parallels among human haematopoietic cell lines (Table II). Human lymphoblastoid lines, probably derived from non-malignant B-lymphocytes (21), Burkitt lymphoma lines, and the K562 line believed to be related to malignant myeloid cells (27), were all sensitive to growth inhibition by PHA at 5—10 μg/ml. A myeloma line 8226 required 4 to 8 times as much PHA for inhibition. Three human T-cell lines derived from acute lymphoblastic leukaemia patients were studied. These cells have the T-lymphocyte markers of rosette formation with sheep erythrocytes (20,22), inhibition by thymidine (unpublished) and T-cell surface antigen. These lines resembled B-lymphoid lines in sensitivity to PHA.

3.3 Independence of Con A and PHA Toxicity: Selection of a PHA-resistant Con A-sensitive Cell Line

Con A and PHA have different binding specificities and are believed to act on cells through separate receptors in some cases. Biological evidence comes, e.g. from a class of lymphocytes stimulated by Con A but not PHA (18) and from the existence of the T-tumour C1498 and myelomas (Table 1) in which sensitivity to the two lectins is not correlated. From the T-tumour S49, we have selected a line resistant to PHA toxicity by growing out surviving cells in increasing amounts of PHA (16). The resulting cell, which has lost most of its cell surface receptors for PHA, was as sensitive as the parent to killing by Con A.

3.4 Relationship of Binding of Con A and PHA to Sensitivity of the Cells

Murine T-leukaemia cells bind four times more Con A than lymph node cells on a per cell basis (30); surface areas were not compared. Cells of human lympho-

TABLE II.   Inhibition of human lymphoid cell lines by PHA[a]

Type	Origin	EBV	E rosettes	T antigen	PHA	Reference
T-Cells						
MOLT 4	ALL	−	+	+	10	20−22
HSB-2	ALL	−	+	+	10	25, 26
CEM	ALL	−	+	+	5	25
B-Cells						
Lymphoblastoid						
8866	AML	+	−	−	5	12, 21
T-CLL	T-type CLL	+	−	−	5	P.R., G. Klein
SB	ALL	+	−	−	6	25, 26
Burkitt lymphomas						
RAJI	BL	+	−	−	10	22, 23
BJA-B-1	BL	−	−	−	10	23
Myeloma						
8226.2	M	−	−	−	40	21, 22, 24
Other						
K562	CML	−	−	−	5	27

[a]Cells were grown in RPMI 1640 medium (Microbiological Associates) containing 10% foetal calf serum. Cell lines are classified according to B- or T-lymphocyte markers (20, 22−26) and the criteria of Nilsson and Pontén (21) for normal lymphoblastoid lines vs. malignant B- or T- lines. Origin: patients with ALL, acute lymphoblastic leukaemia; T-type CLL, chronic lymphocytic leukaemia; BL, Burkitt's lymphoma; M, myeloma; CML, chronic myeloid leukaemia in blast crisis. T-antigen: complement-dependent cytotoxic rabbit anti-human hypothalamus serum (gift of Dr. John Johnson, Scripps Clinic and Research Foundation). PHA: concentration in μg/ml to inhibit cell growth 50% as in Table I.

blastoid lines bind 2 to 3 times more Con A per cell than normal lymphocytes, or approximately equal amounts of Con A per unit surface area since the culture lines are larger (31). This increased binding per cell may account for the inhibition of lymphoid cell lines at Con A concentrations mitogenic for normal cells. On the other hand, comparing different malignant cell lines, lymphomas and myelomas have similar numbers of receptor sites for Con A (32), but lymphocytes and lymphomas bind much more PHA than myelomas (2,16,32). When a murine T-lymphoma cell line was selected for resistance to killing by PHA at 10 times higher concentrations than is lethal for the parent line, the variant cells were found to bind less than 1/5 of the amount of labelled PHA as the sensitive cells (16). Thus, binding data can be viewed as correlating with toxicity. However, binding is not always sufficient for a physiological event since T- and B-lymphocytes bind similar numbers of PHA and Con A molecules (32,33), and yet only T-lymphocytes are stimulated by these mitogens. The two types of lymphocytes must differ fundamentally in their triggering mechanism or signaling circuit from the surface binding site to the interior of the cells.

3.5 Relationship of Con A and PHA Cytotoxicity to Sensitivity towards Non-lectin Mitogens

We have recently shown that murine B-lymphomas are 100—1000 times more sensitive to growth inhibition by B-cell mitogens—dextran sulphate, lipopolysaccharide and PPD—than T-lymphomas, myelomas, mastocytoma or erythroleukaemia (12,13). The B-tumour lines have been classified according to their derivation from normal B-lymphocytes at various stages of differentiation (13). Binding of labelled lipopolysaccharide is detected on many cell types and is not correlated with growth inhibition. Leukaemias and lymphomas may retain the membrane mechanisms which are stimulated by mitogens. In support of this is the finding of cyclic GMP elevation in a murine B-lymphoma within one minute of addition of lipopolysaccharide (P. R. and James Watson, unpublished), similar to that found with B-lymphocytes (34). The malignant cells, already dividing rapidly, differ from lymphocytes in unknown subsequent steps, being inhibited and killed rather than stimulated. The absence of good B-cell mitogens for human lymphocytes has precluded testing this hypothesis on the many human B-cell lines.

3.6 Typing of Subclasses of Lymphocytes

Subgroups of cell lines can be distinguished within a class, such as the human T-lines shown in Table II, by the use of lectin toxicity. The subgroups may represent stages of maturation of a single lineage, or different functional subtypes, such as the helper T-cell for antibody synthesis, the antigen-specific cytotoxic T-cell and a non-specific T-suppressor cell, which seem to be three independently regulated subgroups (35). Four or five subtypes of acute lymphoid leukaemia have been described, based on morphology and clinical data (36). The ability to subtype lymphoid neoplasms by the use of lectins may further the diagnosis and treatment of these malignancies.

4. PITFALLS

4.1 Serum Effects

The toxic effects of drugs on different cell lines must be studied under the same conditions. In a routine screening of Con A inhibition of cellular DNA synthesis, it was discovered that cells growing in 10% foetal calf serum were more sensitive than when in 10% horse serum (4). Presumably foetal calf serum contains less glycoproteins that compete with cells for Con A binding sites than adult horse serum. Other examples of serum effects on the interaction of Con A and PHA (phytohaemagglutinin) with lymphocytes have been reported. The suppression of DNA synthesis in mouse T-lymphomas by Con A (7) and the binding of labelled Con A by mouse thymus cells (8) were reduced by the presence of serum, which suggests that a serum component competes with cellular receptors for Con A. A serum fraction has been isolated which blocks both PHA and Con A mitogenesis (9), and serum reduces the binding of labelled PHA to mouse T-lymphomas and myelomas (16). The toxic effects of PHA on lymphoid cell lines are the same in foetal calf serum and horse serum (4).

4.2 Secretion of Glycoproteins by Cell Lines

Another obvious complication may result from cells actively secreting glyco-proteins, such as immunoglobulins by myelona cells, some classes of which will bind Con A in the medium. The amount of immunoglobulin secreted by myeloma cells in 24 hours (up to 10 to 20 μg from 10^6 cells, refs. 10, 24) is insignificant compared to the immunoglobulin contributed by the foetal calf serum present in the medium. Furthermore, addition of 10 μg/ml murine myeloma proteins to murine T-lymphomas sensitive to killing by Con A and PHA does not alter the cytotoxicity (unpublished). However, the concentration of immunoglobulin close to a secreting myeloma cell may be very high, reducing the effective free concentration of Con A near the cell surface. Thus measurement of Con A binding sites on a cell surface performed at $0°$ may not be correlated with degree of toxicity measured over longer times at $37°$.

4.3 Comparison of Cell Lines with Different Growth Rates

The growth rate of cell cultures must also be considered, since some cells may be sensitive to a given drug only at certain phases of the cell cycle. For example, SV40-transformed fibroblasts are agglutinated by low levels of Con A while non-transformed fibroblasts generally are not (6). In mitosis however, normal fibroblasts are also agglutinated by Con A (11). Thus if Con A were toxic to normal fibroblasts only during mitosis, the lethal effects would only be seen when the cells are in growth phase during the time of the experiment. Classic examples of cell cycle-dependent toxicity are treatment of cultures with bromodeoxyuridine (BUDR) and light, high specfic activity ^{3}H-thymidine, and x-rays, all of which generally only kill cells undergoing DNA synthesis. Lymphocytes are an exception with the latter method, being very sensitive to intermitotic (non-cycling) killing by γ-irradiation (28,29).

5. ACKNOWLEDGEMENT

This work was supported by National Science Foundation Grant GB-37869 and a Yamagiwa-Yoshida Memorial International Cancer Study Grant. I thank Dr. K. Nilsson and the Dept. of Pathology, University of Uppsala, for hospitality and helpful discussion.

6. REFERENCES

1. Ralph, P. (1973) 'Retention of lymphocyte characteristics by myelomas and θ^+ lymphomas: Sensitivity to cortisone and phytohemagglutinin', *J. Immunol.*, 110, 1470–5.
2. Dent, P. (1971) 'Inhibition by phytohemagglutinin of DNA synthesis in cultured mouse lymphomas', *J. Nat. Cancer Inst.*, 46, 763–73.
3. Mackler, B. (1972) 'Effect of concanavalin A on human lymphoid cell lines and normal peripheral lymphocytes', *J. Nat. Cancer Inst.*, 49, 935–41.
4. Ralph, P. and Nakoinz, I. (1973) 'Inhibitory effects of lectins and lymphocyte mitogens on murine lymphomas and myelomas', *J. Nat. Cancer Inst.*, 51, 883–90.
5. Harris, A. W. (1970) 'Differentiated functions expressed by cultured mouse lymphoma cells', *Exp. Cell Res.*, 60, 341–53.
6. Schoham, J., Inbar, M. and Sachs, L. (1970) 'Differential toxicity on normal and

transformed cells *in vitro* and inhibition of tumor development *in vivo* by concanavalin A', *Nature*, **227**, 1244—6.

7. Dent, P. and Hillcoat, B. (1972) 'Interaction of phytohemagglutinin and concanavalin A with transplantable mouse lymphomas of differing malignant potential', *J. Nat. Cancer Inst.*, **49**, 373—7.

8. Andersson, J., Sjöberg, O. and Möller, G. (1972) 'Mitogens as probes for immunocyte activation and cellular cooperation', *Transpl. Rev.*, **11**, 131—77.

9. Chase, P. (1972) 'The effects of human serum fractions on phytohemagglutinin- and concanavalin A-stimulated human lymphocyte cultures', *Cell. Immunol.*, **5**, 544—54.

10. Nilsson, K. (1971) 'Characteristics of established myeloma and lymphoblastoid cell lines derived from an E myeloma patient: A comparative study', *Int. J. Cancer*, **7**, 380—96.

11. Fox, T. O., Sheppard, J. R. and Burger, M. M. (1971) 'Cyclic membrane changes in animal cells: Transformed cells permanently display a surface architecture detected in normal cells only during mitosis', *Proc. Nat. Acad. Sci. U.S.A.*, **68**, 244—7.

12. Ralph, P. (1974) 'Lipopolysaccharides inhibit lymphosarcoma cells of bone marrow origin', *Nature*, **249**, 49—51.

13. Ralph, P., Nakoinz, I. and Raschke, W. C. (1974) 'Lymphosarcoma cell growth is selectively inhibited by B lymphocyte mitogens: LPS, dextran sulfate and PPD', *Biochem. Biophys. Res. Commun.*, **61**, 1268—75.

14. Warner, N. L., Moore, M. A. S. and Metcalf, D. (1969) 'A transplantable myelomonocytic leukemia in BALB/c mice: Cytology, karyology, and muramidase content', *J. Nat. Cancer Inst.*, **43**, 963—82.

15. Watson, J., Epstein, R., Nakoinz, I. and Ralph, P. (1973) 'The role of humoral factors in the initiation of *in vitro*, primary immune responses. II. Effects of lymphocyte mitogens', *J. Immunol.*, **110**, 43—52.

16. Ralph, P., Hyman, R., Epstein, R., Nakoinz, I. and Cohn, M. (1973) 'Independence of θ and TL surface antigens and killing by thymidine, phytohemagglutinin, and cyclic AMP in a murine lymphoma', *Biochem. Biophys. Res. Commun.*, **55**, 1085—91.

17. Hyman, R., Ralph, P. and Sarkar, S. (1972) 'Cell-specific antigens and immunoglobulin synthesis of murine myeloma cells and their variants', *J. Nat. Cancer Inst.*, **48**, 173—84.

18. Stobo, J. D., Paul, W. E. and Henney, C. S. (1973) 'Functional heterogeneity of murine lymphoid cells. IV. Allogeneic mixed lymphocyte reactivity and cytolytic activity as functions of distinct T cell subsets', *J. Immunol.*, **110**, 652—60.

19. Gail, M. and Boone, C. (1972) 'Cell-substrate adhesivity', *Exp. Cell Res.*, **70**, 33—40.

20. Minowada, J., Ohnuma, T. and Moore, G. E. (1972) 'Rosette-forming human lymphoid cell lines. I. Establishment and evidence for origin of thymus-derived lymphocytes', *J. Nat. Cancer Inst.*, **49**, 891—5.

21. Nilsson, K. and Pontén, J. (1975) 'Classification and biological nature of established human hematopoietic cell lines', *Int. J. Cancer*, **15**, 321—41.

22. Jondal, M. and Klein, G. (1973) 'Surface markers on human B and T lymphocytes. II. Presence of Epstein—Barr virus (EBV) receptors on B lymphocytes', *J. Exp. Med.*, **138**, 1365—78.

23. Klein, G., Lindahl, T., Jondal, M., Leibold, W., Menezes, J., Nilsson, K. and Sundström, C. (1974) 'Continuous lymphoid cell lines with B-cell characteristics that lack the Epstein—Barr genome, derived from three human lymphomas', *Proc. Nat. Acad. Sci. U.S.A.*, **71**, 3283—6.

24. Matsuoka, Y., Yagi, Y., Moore, G. E. and Pressman, D. (1969) 'Isolation and characterization of free λ chain of immunoglobulin produced by an established cell line of human myeloma cell origin', *J. Immunol.*, **102**, 1126—43.

25. Kaplan, J., Shope, T. C. and Peterson, Jr., W. D. (1974) 'Epstein—Barr virus-negative human malignant T-cell lines', *J. Exp. Med.*, **139**, 1070—6.

26. Royston, I., Smith, R. W., Buell, D. N., Huang, E.-S. and Pagano, J. S. (1974) 'Autologous human B and T lymphoblastoid cell lines', *Nature*, **251**, 745—6.

27. Lozzio, C. B. and Lozzio, B. B. (1975) 'Human chronic myelogenous leukemia cell line with positive Philadelphia chromosome', *Blood*, **45**, 321—34.

28. Jacobson, L. O., Marks, E. K., Simmons, E., Hagen, L. W. and Zirkle, R. E. (1954) 'Sensitivity of hematopoietic cells to irradiation', in *Biological Effects of External X and Gamma Radiation*, ed. Zirkle, R. E. (McGraw-Hill, New York), pp. 265—76.

29. Kennison, J. and Ralph, P. (1976) 'Antibody-dependent cytotoxic effector cells have the radiation sensitivity and recovery of monocytes', submitted for publication.
30. Martin, W., Wunderlich, J., Fletcher, F. and Inman, K. (1971) 'Enhanced immunogenicity of chemically-coated syngeneic tumor cells', *Proc. Nat. Acad. Sci. U.S.A.*, **68**, 469−72.
31. De Salle, L., Munakata, N., Pauli, R. M. and Strauss, B. S. (1972) 'Receptor sites for concanavalin A on human peripheral lymphocytes and on lymphoblasts grown in long-term culture', *Cancer Res.*, **32**, 2463−8.
32. Greaves, M., Bauminger, S. and Janossy, G. (1972) 'Binding sites for phytomitogens on lymphocyte subpopulations', *Clin. Exp. Immunol.*, **10**, 537−54.
33. Stobo, J., Rosenthal, A. and Paul, W. (1972) 'Responsiveness to and surface binding of concanavalin A and phytohemagglutinin', *J. Immunol.*, **108**, 1−17.
34. Watson, J. (1974) 'The nature of the signals required for the induction of antibody synthesis', in *The Immune System: Genes, Receptors, Signals*, eds. Sercarz, E. E., Williamson, A. R. and Fox, C. F. (Academic Press, New York), pp. 511−32.
35. Janeway, Jr. C. A., Sharrow, S. O. and Simpson, E. (1975) 'T-cell populations with different functions', *Nature*, **253**, 544−6.
36. Mathé, G., Belpomme, D., Dantchev, D. (1974) 'Immunoblastic acute lymphoid leukaemia', *Biomedicine*, **20**, 333−40.

CHAPTER 61

Selection of Concanavalin A-resistant Variants of Virus-transformed Cells

LLOYD A. CULP

1. RATIONALE AND APPROACH

A variety of plant lectins, and particularly concanavalin A, have proven extremely useful in identifying differences in the architecture of the surface membrane of normal and malignant cells (e.g. see the recent review by Nicolson [1]). Cells transformed with any one of a variety of oncogenic viruses were shown to become agglutinable with Con A under conditions where the normal cell counterparts were resistant to agglutinability (1,2). The correlation of increased Con A-mediated agglutinability with loss of growth control via oncogenic virus transformation was further supported by the finding that revertant variants of virus-transformed cells, selected by FUdR treatment (3,4), were resistant to agglutinability with Con A in a manner similar to the normal growth-controlled parent cell (5). These revertant variants reacquired the flat morphology and the density-dependent regulation of growth characteristic of the untransformed parent cell, while retaining the partially expressed SV40 genome (3,4). SV40 virus rescued from revertant cells was shown to be 'wild type' in terms of its efficiency in transforming fresh populations of 3T3 cells (4); presumably revertant cells are variant in some aspect of *host* cell genome expression, rather than *viral* genome expression. More recently, temperature-sensitive revertant cells (sensitive in terms

623

of morphology and density-dependent growth control) were shown to be similarly temperature sensitive to Con A agglutinability (6).

These correlations suggested that revertant variants could be isolated using concanavalin A as a selective tool to agglutinate and possibly kill wild-type transformed cells, leaving revertant variants unaffected by the treatment. This selection technique was then used for isolating revertants in two laboratories (7,8).

Concanavalin A selection of revertants was based on a number of fortuitous discoveries. In the first place, SV40-transformed cells adherent to the tissue culture substrate were agglutinable *and* detachable from the substrate using concentrations of lectin similar to those used to agglutinate EDTA-suspended cells (2), whereas 3T3 cells were unaffected by these treatments in terms of their resistance to substrate detachment or agglutinability (8). Second, the high levels of glycoprotein in serum of cultures did not prove to be a serious obstacle to treatment of cells by binding massive amounts of lectin competitively, resulting in precipitable complexes which make the cultures difficult to handle. Finally, the agglutinated transformed cells were killed by the lectin treatment soon after formation of aggregates, as determined by their trypan stainability and their inability to be replated in new dishes even in the presence of α-methyl mannoside (Culp, unpublished data). 3T3 cells, as well as FUdR-selected revertant cells, were resistant to these toxic effects of the lectin at concentrations which were toxic to transformed cells.

Unfortunately, the mechanism of concanavalin A-mediated cytotoxicity of transformed cells has not been studied in any systematic fashion. There is good evidence now that agglutination is mediated by focal concentration of multivalent lectin molecules on the fluid surface membrane and lectin-mediated cross-linkage with a large number of sites on adjacent cells (discussed in ref. 1), but whether this phenomenon *per se* is also toxic to the viability of the cell remains unexplored. In one system, agglutinability of malignant cells with wheat germ agglutinin or soy bean agglutinin did not result in a cell killing effect, whereas Con A treatment did (11). Consideration must also be given to possible cytotoxic effects resulting from endocytosis of lectin into the cell and irreversible damage to intracellular compartments (1,9). One laboratory (10) has demonstrated marked differences in the toxicity of chemical- or virus-transformed mouse or hamster cells *in vitro* upon lectin treatment. The differential sensitivity of malignant cells to the toxic effects of lectins, such as Con A, remains a potentially fruitful area of study.

2. SELECTION PROCEDURES

There are two different approaches which can be used to select for Con A-resistant variant cells. In all selection procedures using concanavalin A, it is *imperative* to use highly purified lectin to reduce the possibility of selecting cells resistant to the effects of a contaminating component in the lectin preparation.

2.1 Selection in Complete Medium

We have shown (8) that selection can be performed with transformed cells attached to substrates in complete medium containing serum as outlined in Figure 1A. Dose—response curves, measuring reduced plating efficiency of treated

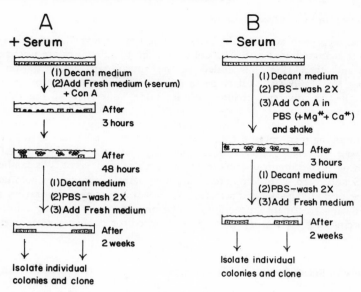

Figure 1. Protocols for selection of Con A variant cells. The approaches for selecting Con A-resistant cells are outlined using serum-containing medium (A) or serum-free medium (B)

cells as a result of the cytotoxic effects on the cells by Con A treatment, must be performed for each batch of serum used. Sera appear to be very different in their lectin-binding ability — some batches of sera bind sufficient amounts of Con A to generate insoluble precipitin complexes in the medium which interfere with analysis of cells. Other batches of serum do not generate these precipitates at Con A concentrations sufficiently high to generate a cytotoxic effect and are adequate for the selection. Dose—response relationships must also be performed with each cell type being studied, since cell lines demonstrate highly variable sensitivities to Con A-induced agglutinability and cytotoxicity.

In general, lectin-containing medium produces cell rounding, detachment and agglutination of transformed cells with Con A concentrations of 200—300 μg/ml over a period of several hours (Figure 1A). After 24—48 hours exposure to the lectin-containing medium, most cells are present in the medium as agglutinated clumps and are discarded during washing of the substrate several times with phosphate-buffered saline (PBS) solution. Fresh medium can be added to these dishes to permit cell regrowth of the very small proportion of substrate-attached cells as colonies over a 1—2 week period, or the few substrate-adherent cells can be trypsinized into fresh dishes for regrowth as colonies. Regrowth of resistant cells over a period of many generations dilutes out any lectin remaining bound to these cells (cells may also be able to digest the protein following endocytosis). Revertant variants of transformed cells have routinely been isolated with a frequency of approximately 10^{-5} from populations of virus-transformed cells (3,4,8).

2.2 Selection in Serum-free Medium

Another approach avoiding the pitfalls of serum binding of Con A in complete medium is depicted in Figure 1B. Cells attached to substrates are washed free of medium and then Con A is added at the appropriate concentration in PBS containing Mg^{2+} and Ca^{2+} (after performing a dose—response curve under these conditions) — the divalent cations permit stable adherence of the cells to the substrate over a longer period of time (12). Cell rounding, detachment and agglutination occur over a shorter span of time compared with treatment in serum-containing medium and these cytotoxic effects can be accelerated even more by agitating the cells gently and continuously during the treatment. After 1—3 hours of Con A treatment, the few remaining substrate-adherent cells can be washed with PBS and refed with complete medium to permit regrowth as colonies over a 1—2 week period. This second selection approach argues against a serum selection process during treatment in medium containing serum by binding critical serum glycoproteins with Con A, which may be required for stable cell attachment to the substrate.

2.3 Failure to Select in Suspension Cultures

The selection procedures described above apply to substrate-grown cells and result from the ability to separate agglutinated-detached cells from the resistant substrate-adherent cells by washing the substrate and regrowth of well-isolated cells as colonies. These procedures are most effective at cell densities where cells attached to the substrate are in contact with neighbouring cells where agglutination sites probably form. However, this selection approach could not be applied to suspension-grown malignant cells from which resistant variants are being sought. Treatment of suspension-grown cells would be critically dependent on the true *killing* effect of the lectin. No reports have appeared using suspension-grown cells, although resistant cells may perhaps be selected by inoculating treated populations into agar-containing (or methylcellulose-containing) medium to isolate colonies after regrowth* or by cloning after regrowth of resistant cells into mass populations. The large number of dead cells in treated suspension cultures may offer an insurmountable toxic problem to cell regrowth by the resistant cells; perhaps the large agglutinated masses of dead cells could be separated from single live cells by differential centrifugation or gravitational settling.

3. PROPERTIES OF CON A-RESISTANT CELLS

3.1 Growth Properties

The resistant cells isolated by Con A selection resembled revertant cells isolated by other techniques in several aspects as described in Table I. They are flat cells (Figure 2) which display density-dependent growth control as determined by several approaches (8). Their saturation density is much lower than that of the

---

*On the other hand, Con A-resistant cells appear to regain density-dependent growth control and may not divide as cellular aggregates in a solid matrix.

TABLE I. Properties of revertant SV40-transformed cells

Property	Swiss FUdR revertant[a]	BALB/c Con A revertant[b]
Morphology	Large, flat (similar to 3T3)	Large, flat (similar to 3T3)
Rescue of infectious SV40 virus[c]	+	+
Sialic acid composition	High, similar to Swiss 3T3 cells	High, similar to BALB/c 3T3 cells
Collagen content in confluent cultures	Low, similar to Swiss SV3T3 cells	High, similar to 'super-producer' BALB/c SVT2 cells
Cytogenetic composition	Hypertetraploid (SV3T3-hypertriploid)	Hypertetraploid (SVT2-hyperdiploid)
Quantities of substrate-attached protein and polysaccharide deposited[e]	ND[d]	High proportion, similar to BALB/c 3T3 cells
Effects of SAM on attachment kinetics[f]	ND[d]	Unaffected (SVT2 cell attachment positively affected), similar to 3T3 cells
Microfilament arrays	Large quantities of highly-ordered arrays at points of cell contact with cells and substrate, similar to 3T3 cells	Large quantities of highly-ordered arrays at points of cell contact with cells and substrate, similar to 3T3 cells

[a] Isolated and described in references 4 and 17.
[b] Isolated and described in references 8 and 18.
[c] By inactivated Sendai virus-mediated fusion with permissive monkey kidney cells.
[d] Not done.
[e] Measured by determining the ratio of radioactive protein or polysaccharide left tightly bound to the tissue culture substrate, after EGTA-mediated removal of cells, to the amounts of radioactive cell-associated protein or polysaccharide in the EGTA extract (12, 20).
[f] As compared to the kinetics of cell attachment to untreated substrates.

parental transformed cells. At their maximal density (2.5×10^5 cells/cm^2), they do not display overlapping nuclei as demonstrated by autoradiography of uniformly DNA-radiolabelled cells (Figure 3C). At the same density of cells on the dish, transformed cells exhibit marked nuclear overlap (Figure 3B). BALB/c 3T3 cells at their maximum density (1×10^5 cells/cm^2 in Figure 3A) do not display overlapping nuclei and are not as densely packed on the substrate as the revertant cells. Autoradiographic evaluation of ^{3}H-thymidine pulses to determine cellular DNA synthesis during culture growth demonstrated an approximately fourfold decrease in the proportion of revertant cells synthesizing DNA at maximal saturation density when compared to exponentially growing cells; a tenfold decrease in the proportion of 3T3 cells making DNA at their saturation density; and no decrease in the proportion of transformed cells at high density (Culp, unpublished data). Con A revertant cells therefore do not continue to divide at the normal rate at their saturation density and slough into the medium, resulting in a stable saturation density via balanced cell sloughing and division.

Con A treatment apparently selects for growth-controlled cells, and the

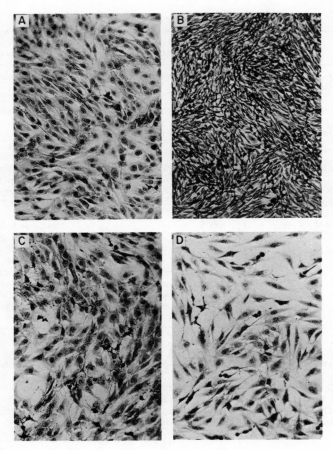

Figure 2. Morphology of BALB/c 3T3, SV40-transformed, and Con A revertant cells. (A) Confluent and growth-inhibited BALB/c 3T3 cells; (B) confluent and growing SVT2 cells; (C) confluent and growth-inhibited Con A revertant cells; (D) subconfluent and growing Con A revertant cells. Cells were stained with Giemsa after methanol fixation. (×120) [Reproduced from reference 8 with the kind permission of the editors of the *Journal of Virology*.]

frequency of Con A revertant cells in populations of transformed cells (approximately 1 in 10^5) is very similar to the frequencies observed for FUdR selection of revertants (8). This is an indication that the lectin is not generating these variant cells but is truly *selecting* for them as they arise spontaneously in populations of transformed cells by some unknown mechanism. It would be interesting to determine if other lectins can be used to select revertant cells with similar frequencies.

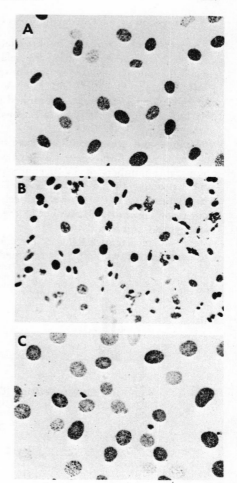

Figure 3. Packing arrangement of DNA-radiolabelled cells. Cells (approximately 2×10^5) were inoculated into 60 mm tissue culture dishes containing culture medium with 0.01 μCi/ml of ^{3}H-thymidine (40 Ci/mmole). Medium was changed every other day with ^{3}H-thymidine being present to uniformly radiolabel the nuclei of cells and until cells had reached the following densities: (A) BALB/c 3T3 cells (1×10^5 cells/cm^2) had become growth inhibited; (B) SVT2 cells (3×10^5 cells/cm^2) had confluently covered the substrate and continued to divide with piling; (C) Con A revertant cells (2.5×10^5 cells/cm^2) had become growth inhibited. At these densities, medium was decanted and the cell layer was autoradiographed using Kodak AR10 stripping film to determine the orientation of radioactive nuclei. (x 250)

3.2 Sensitivity to Con A

Con A-resistant variants have been shown to be stably resistant to agglutination and killing during a second round of treatment (7). These cells have apparently not lost their lectin binding sites since (a) they bind amounts of radioactive Con A at higher temperatures comparable to those bound to the normal and transformed cells* and (b) they become readily agglutinable with fiftyfold lower Con A concentrations after gentle trypsinization (13). The topographical distribution of Con A binding sites in these revertant cells has yet to be determined, particularly in light of the dispersed distribution of these sites on 'normal' cell surfaces and the readily focalized distribution on transformed cell surfaces(1).

*It will be important to repeat these binding experiments at 4° under conditions where endocytosis of lectin is minimized and true surface binding to high-affinity sites is being measured (9).

3.3 Expression of the Oncogenic Virus Genome

Revertant cells isolated by a variety of selection techniques continue to be valuable tools for determining the relationship of the expression of the viral and host cell genomes in terms of the control of the growth-regulated state of the cell. SV40-transformed cells yield Con A revertant cells which retain the partially expressed viral genome as determined by the presence of SV40-specific T-antigen and by rescue of wild-type infectious SV40 virus after fusion with permissive monkey kidney cells (8,13).

Ozanne and coworkers (14) have examined the proportion of SV40 DNA to cellular DNA in Con A revertant and wild-type SV40-transformed cells and found comparable numbers of SV40 copies per diploid quantity of DNA for both cell types (although the amount of SV40 DNA *per cell* in revertants was twice as high). The elevated viral DNA level per cell for revertants reflects the nearly twofold increase in chromosome number in these cells, selected by a variety of techniques (see Figure 4 and refs. 8 and 15).

Con A revertant cells also contain the same transcribed mRNA sequences observed in the wild-type transformed cells (14). Reversion is therefore not the

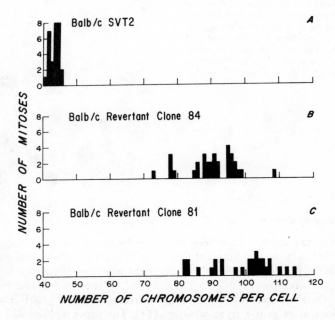

Figure 4. Cytogenetic analysis of BALB/c SVT2 and Con A revertant cells. The number of chromosomes/cell in meta-phase-arrested cells was determined by standard procedures during exponential growth of SVT2 (A), Con A revertant clone 84 (B) and Con A revertant clone 81 (C) cells. Approximately thirty mitoses were examined. [Reproduced from reference 8 with the kind permission of the editors of the *Journal of Virology.*]

result of a disproportionate amount of viral DNA in these cells or disparate transcription of that DNA. The consistently high chromosome levels demonstrated in Figure 4 suggest that revertants arise by abnormal cytokinesis during mitosis of a single transformed cell or by fusion of two transformed cells to liberate a cell with twice the number of chromosomes, some of which may be lost to yield a stable karyotype of 90—100 chromosomes. Proof for such a reversion mechanism is still lacking.

Two other cogent questions remain with regard to the state of the viral genome in revertant cells. Is the DNA of the virus still integrated into chromosomal DNA, or has it assumed an episomal relationship perhaps similar to the association of Epstein—Barr viral DNA non-covalently with chromosomal DNA in Burkitt's lymphoma cells (25)? If the viral genome remains integrated in the reverted state, integration of viral DNA may be a necessary step for transformation to the non-growth-controlled state, but not necessarily sufficient. Second, if viral DNA is integrated in revertant cell chromosomal DNA, is it located at the same chromosal site as wild-type transformed cells or has rearrangement of chromosomal material altered its effects on host functions? Experimental approaches for answering these questions may soon be developed.

3.4 Surface Membrane Properties and Substrate Adhesion

Revertant cells have presumably reacquired the minimum surface membrane properties necessary for density-dependent inhibition of growth. The fact that these cells can be isolated based on their resistance to Con A agglutinability (7,8) argues very strongly for the importance of immobility of lectin receptor sites (1) on the surface of these growth-controlled cells and/or reduced concentrations of these sites per area of membrane (9) as being very relevant to growth control. The molecular basis of immobility of these sites has not yet been established but may be due to altered composition of the fluid lipid bilayer (16) or to binding of these outer surface lectin-binding components to highly immobile inner membrane components, such as the highly organized array of subsurface microfilaments which are very prominent at cell or substrate contact sites of 3T3 and revertant cells (17—19). Isolation and characterization of Con A binding components from surface membranes of normal, virus-transformed and revertant cells remain fruitful areas of study, particularly with regard to their possible non-covalent interaction with other surface components and possibly subsurface components.

Revertant cells are being used in our laboratory to identify possible changes in the mechanism of adhesion of cells to the tissue culture substrate as a result of virus transformation. 3T3 and Con A revertant cells were shown to leave 3—5 times more protein and polysaccharide material (so-called SAM or substrate-attached material) bound to the substrate after EGTA-mediated removal of cells as compared to transformed cells (12,20). The quantities of this material and its tenacity of binding to the substrate suggested its importance in the greater adhesiveness of normal and revertant cells to the substrate. This functional role for SAM proteins and polysaccharides is being supported by a variety of experimental approaches (21—24).

In order to study the very early events in substrate adhesion of subcultured cells, an assay was developed by measuring binding of DNA- radiolabelled cells to glass

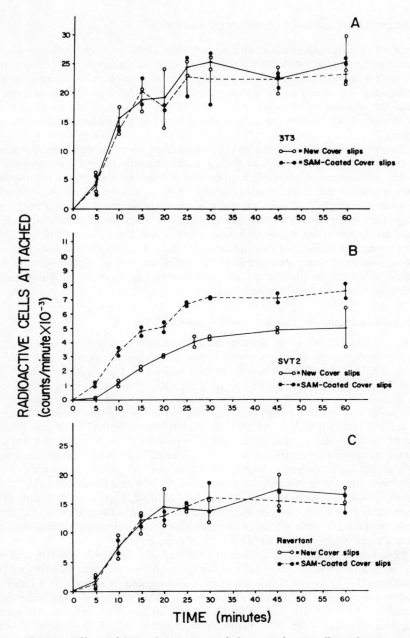

Figure 5. Effects of 3T3 substrate-attacted glycoproteins on cell-to-substrate attachment. The kinetics of attachment of DNA-radiolabelled 3T3 (A), SVT2 (B) or Con A revertant (C) cells to untreated (o———o) or 3T3 SAM-coated (•————•) coverslips at 22° has been described in reference 22. Cells for attachment experiments or for preparation of SAM-coated substrates were removed by EGTA treatment. [Reproduced from reference 22 with the kind permission of the editors of the *Journal of Cell Biology*.]

coverslips (22). As shown in Figure 5, coatings of 3T3 SAM on the coverslips stimulate the attachment of SVT2 cells by reducing the initial time lag of stable attachment and by permitting 30—35% more cells to attach at the maximal level. Attachment of revertant cells (Figure 5C) was unaffected by the SAM coating and therefore resembled 3T3 cells in this regard. Other experimental approaches have also indicated that revertant cells selected by the Con A treatment have reverted in terms of their interaction with the tissue culture substrate via quantitative and qualitative differences in the SAM being deposited. Methods have now been found for characterizing the limited number of SAM proteins synthesized by the cell and distinguishing them from a limited number of substrate-bound serum proteins which appear to be critically important for 'proper' and stable adhesion of cells to the substrate (24).

Variant cells, resistant to the agglutinability and the cytotoxic killing with the lectin concanavalin A, continue to be extremely valuable tools for identifying the surface membrane components which are critically important in regulating the growth and social behaviour of cells. Revertant variants of oncogenic virus-transformed cells may also prove to be highly important in determining the sensitive balance between expression of the viral and host cell genomes in terms of control over formation of growth-regulatory surface membrane components.

4. ACKNOWLEDGEMENTS

Some of these experiments were initiated during my stay in the laboratory of Dr. Paul Black at the Massachusetts General Hospital, to whom I am deeply indebted. I would also like to acknowledge financial support as a Harry H. Pinney Fellow in Cancer Research at Case Western Reserve University, research grant 5-R01-CA 13513 from the National Cancer Institute, training grant 5-T01-GM-00171 from the U.S. Public Health Service to the Department of Microbiology, and career development award 1-K04-CA 70709 from the National Cancer Institute.

5. REFERENCES

1. Nicolson, G. L. (1974) 'The interactions of lectins with animal cell surfaces', *Int. Rev. Cytol.*, 39, 89—190.
2. Burger, M. M. (1969) 'A difference in the architecture of the surface membrane of normal and virally transformed cells', *Proc. Nat. Acad. Sci. U.S.A.*, 62, 994—1001.
3. Pollack, R. E., Green, H. and Todaro, G. J. (1968) 'Growth control in cultured cells: selection of sublines with increased sensitivity to contact inhibition and decreased tumor-producing ability', *Proc. Nat. Acad. Sci. U.S.A.*, 60, 126—33.
4. Culp, L. A., Grimes, W. J. and Black, P. H. (1971) 'Contact-inhibited revertant cell lines isolated from SV40-transformed cells. I. Biologic, virologic and chemical properties', *J. Cell Biol.*, 50, 682—90.
5. Pollack, R. E. and Burger, M. M. (1969) 'Surface-specific characteristics of a contact-inhibited cell line containing the SV40 viral genome', *Proc. Nat. Acad. Sci. U.S.A.*, 62, 1074—6.
6. Noonan, K. D., Renger, H. C., Basilico, C. and Burger, M. M. 'Surface changes in temperature-sensitive simian virus 40-transformed cells', *Proc. Nat. Acad. Sci. U.S.A.*, 70, 347—9.
7. Ozanne, B. and Sambrook, J. (1971) 'Isolation of lines of cells resistant to agglutination by

concanavalin A from 3T3 cells transformed by SV40', *Lepetit Colloq. Biol. Med.*, **2**, 248—53.

8. Culp, L. A. and Black, P. H. (1972) 'Contact-inhibited revertant cell lines isolated from simian virus 40-transformed cells. III. Concanavalin A-selected revertant cells', *J. Virol.*, **9**, 611—20.

9. Noonan, K. D. and Burger, M. M. (1973) 'The relationship of concanavalin A binding to lectin-initiated cell agglutination', *J. Cell Biol.*, **59**, 134—42.

10. Shoham, J., Inbar, M. and Sachs, L. (1970) 'Differential toxicity on normal and transformed cells *in vitro* and inhibition of tumor development *in vivo* by concanavalin A', *Nature*, **227**, 1244—6.

11. Inbar, M., Ben-Bassat, H. and Sachs, L. (1972) 'Inhibition of ascites tumor development by concanavalin A', *Int. J. Cancer*, **9**, 143—9.

12. Culp, L. A. and Black, P. H. (1972) 'Release of macromolecules from BALB/c mouse cell lines treated with chelating agents', *Biochemistry*, **11**, 2161—72.

13. Ozanne, B. (1973) 'Variants of simian virus 40-transformed 3T3 cells that are resistant to concanavalin A', *J. Virol.*, **12**, 79—89.

14. Ozanne, B., Sharp, P. A. and Sambrook, J. (1973) 'Transcription of simian virus 40. II. Hybridization of RNA extracted from different lines of transformed cells to the separated strands of simian virus 40 DNA', *J. Virol.*, **12**, 90—8.

15. Pollack, R. E., Wolman, S. and Vogel, A. (1970) 'Reversion of virus-transformed cell lines: Hyperploidy accompanies retention of viral genes', *Nature*, **228**, 967—70.

16. Wisnieski, B. J., Parker, J. G., Huang, Y. O. and Fox, C. F. (1974) 'Physical and physiological evidence for two-phase transitions in cytoplasmic membranes of animal cells', *Proc. Nat. Acad. Sci. U.S.*, **71**, 4381—5.

17. McNutt, N. S., Culp, L. A. and Black, P. H. (1971) 'Contact-inhibited revertant cell lines isolated from SV40-transformed cells. II. Ultrastructural study', *J. Cell Biol.*, **50**, 691—708.

18. McNutt, N. S., Culp, L. A. and Black, P. H. (1973) 'Contact-inhibited revertant cell lines isolated from SV40-transformed cells. IV. Microfilament distribution and cell shape in untransformed, transformed and revertant BALB/c 3T3 cells', *J. Cell Biol.*, **56**, 412—28.

19. Perdue, J. F. (1973) 'The distribution, ultrastructure, and chemistry of microfilaments in cultured chick embryo fibroblasts', *J. Cell Biol.*, **58**, 265—83.

20. Terry, A. H. and Culp, L. A. (1974) 'Substrate-attached glycoproteins from normal and virus-transformed cells', *Biochemistry*, **13**, 414—25.

21. Culp, L. A., Terry, A. H. and Buniel, J. F. (1975) 'Metabolic properties of substrate-attached glycoproteins from normal and virus-transformed cells', *Biochemistry*, **14**, 406—12.

22. Culp, L. A. (1974) 'Substrate-attached glycoproteins mediating adhesion of normal and virus-transformed mouse fibroblasts', *J. Cell Biol.*, **63**, 71—83.

23. Culp, L. A. (1975) 'Topography of substrate-attached glycoproteins from normal and virus-transformed mouse fibroblasts', *Exp. Cell Res.*, **92**, 467—77.

24. Culp, L. A. and Buniel, J. F. (1976) 'Substrate-attached serum and cell proteins in adhesion of mouse fibroblasts', *J. Cell Physiol.*, in press.

25. Nonoyama, M. and Pagano, J. S. (1972) 'Separation of Epstein—Barr virus DNA from large chromosomal DNA in non-virus-producing cells', *Nature (New Biol.)*, **238**, 169—71.

APPENDIX:

Commercially Available Reagents and Concanavalin A Derivatives

1. *Concanavalin A*
   Concanavalin A, freeze dried

   Calbiochem
   Miles
   Pharmacia
   SERVA
   SIGMA

   Concanavalin A in saturated NaCl solution

   Boehringer
   Miles

   Concanavalin A in sterile 5% glucose — Calbiochem

   Concanavalin A in ammonium sulphate suspension — Miles

2. *Concanavalin A – agarose*
   'Glycosylex'-A, Con A content 10 mg/ml swollen gel — Miles

   Concanavalin A–'Sepharose', Con A content about
   8 mg/ml swollen gel — Pharmacia

   Concanavalin A–agarose, Con A content 300 mg/g
   agarose in suspension — SIGMA

3. *Fluorescein-labelled concanavalin A*
   Fluorescein isothiocyanate–Con A 14–18 mg
   protein/ml — Miles

4. *Concanavalin A–ferritin conjugate*
   Protein concentration 0.3–0.5%, molar ratio
   Con A/ferritin 15.0 — Miles

5. *Immunochemicals*
   Rabbit antibody to concanavalin A — Calbiochem
   Rabbit antibody to horse ferritin — Calbiochem
   Ferritin-labelled anti-IgG — Biogenzia Lemania SA
   Fluorescein-labelled anti-IgG — Behringwerke

*We have listed the sources of biochemicals that we have found useful. This is, however, *not* a complete list of commercially available preparations nor of suppliers.

6. *Concanavalin A—metal derivatives*
   Con A—Mn^{2+}, Con A—Zn^{2+}, Con A—Co^{2+} Miles

7. *Radiolabelled concanavalin A*
   Con A—^{63}Ni 0.03 μCi/mg   Miles
   Concanavalin A (acetyl ^{3}H) 75—125 Ci/mmol   New England Nuclear

Addresses of Companies

Behringwerke AG	Marburg, Germany
Biogenzia Lemania SA	27 Avenue Morges, Lausanne, Switzerland
Boehringer & Söhne GmbH	Mannheim, Germany
Calbiochem	P.O. Box 12087, San Diego, CA 92112, USA
Miles Laboratories Inc.	Elkart, Indiana 46514, USA
Miles Laboratories Ltd.	Stoke Court, Stoke Poges, Slough, SL2 4LY, England
Miles-Yeda Ltd.	Kyriat Weizmann, Rehovot, Israel
New England Nuclear	575 Albany Street, Boston, Mass. 02118, USA
Pharmacia Fine Chemicals AB	Box 175, S-751 04, Uppsala, Sweden
SERVA Feinbiochemica GmbH	D69 Heidelberg, Postfach 1505, Germany
SIGMA London, Chemical Company Ltd.	12 Lettice St., London, S.W.6, England
SIGMA Chemical Company	P.O. Box, Saint Louis, USA

INDEX

Acetyl Con A, 191, 585
Acyl azide method, 325
Adenylate cyclase, 601
Adhesion, 235, *236*, 307, 316
Affinity chromatography, 18, 99, *323, 333,*
    339, 349, 355, 429, 435, 467
  in the presence of detergent, 326, *389,*
    399, *409, 421*
Agglutination
  agitation, effect of, 253
  alignment model, 234
  cell layer assay, *307,* 309
  composite reaction model, 235
  erythrocytes, *271,* 284, 302
  facilitated adhesion, 237, *239*
  fibroblasts, 286, 302, *307*
  heterologous, 316
  kinetics, 251, *279,* 293, 298, *307*
  lymphocytes, 82, *267*
  measurement with particle counter, *285,*
    293
  mechanism, *231*
  multiple lectins, 315
  proteases, effect of — on agglutinability,
    251
  rapid reactions, *281*
  rates, 251, 255
  reversibility, 243
  temperature, effect of, 252, 257, 317
  time dependence, 252
  titration curve, 249, 273, 282
  sea urchin embryo cells, 298
  virus, *479,* 485
Alkylation, 382, 390
Allergic reaction, 581
Anti Con A sera, 71, 110, 143, *144,* 153
  169, 590
Anti ferritin antibodies, 110
Anti IgG antibodies, 143, 145, 146
Anti T-cell serum, *544*
Antigens, viral, *355*
Antiinflammatory, 574, 576
Arthritis model disease, 574, 577
L-asparaginase, 606, 608

B-cells, 507, 515, 614, 616
Blood group glycoproteins, 333
Brain membranes, 379, *389,* 400, 435

Capping, 70, 104, 588, 591
Carcinoembrionic antigen, 333
Cell fractionation, *447, 467*
Cell mediated immunity, *539, 565*
Cell surface modulating assembly (SMA),
  592
Cell surface replica, *123,* 130
Chloramin T method, 187
Chorionic gonadotropin, 339
Chromaffin granules, *350*
Chromium (^{51}Cr) labelling, *525*
Clustering of Con A receptors, 103
Concanavalin A
  affinity chromatography on Sephadex,
    18
  aminoacid analysis, 36
  binding, 55, 105, 112, 162, 165, *173,*
    196, 198, 201, 203, 213, 214,
    218, 273, 274
    erythrocytes, 198, 273, 274
    fat cells, 203
    fibroblasts, 105, 198
    filtration assay, 201, *214*
    glycolipids, 61
    glycoproteins, 59, 162, 165, 333
    lymphocytes, 105
    membranes (isolated), 112, *213,* 214
    monosaccharides, 56
    polysaccharides, 59, 162, 165
    specificity, *55*
    Triton X-100, effect of, 218
  chemical structure, *33*
  coated fibres, *447*
  crystallization, *17*
  cytotoxicity, *613*
  denaturation, 25
  diffusion constant, 21
  dimeric derivative, *585*
  enzyme conjugates, *605*
  extinction coefficient, 21
  ferritin conjugate, *95, 103,* 105, *117*
  fluorescein conjugate, *69,* 74, *79,* 80, 81
  gel electrophoresis, 22
  inflammation, induced by, *273*
  insolubilized, *323, 447,* 506, 510, 516
  insulin like effects, *599*
  ion content, 22, 36, 40, 41
  iron dextran technique, *137*

637